WEATHER
and Climate

WEATHER
and Climate

an introduction

Sheila Loudon Ross

OXFORD

UNIVERSITY PRESS

OXFORD
UNIVERSITY PRESS

Oxford University Press is a department of the University of Oxford.
It furthers the University's objective of excellence in research, scholarship,
and education by publishing worldwide. Oxford is a registered trade mark of
Oxford University Press in the UK and in certain other countries.

Published in Canada by
Oxford University Press
8 Sampson Mews, Suite 204,
Don Mills, Ontario M3C 0H5 Canada

www.oupcanada.com

Library and Archives Canada Cataloguing in Publication
Ross, Sheila Loudon, 1958–
Weather and climate : an introduction / Sheila Ross.

Includes bibliographical references.
ISBN 978–0–19–544587–9

1. Meteorology—Textbooks. 2. Climatology—Textbooks.
I. Title.

QC861.3.R68 2013   551.5   C2012-906444-0

Cover image: Design Pics/Richard Wear/Getty Images

Oxford University Press is committed to our environment.
This book is printed on Forest Stewardship Council™ certified paper
and comes from responsible sources.

Printed and bound in the United States of America.

1 2 3 4 — 16 15 14 13

Brief Contents

Contents

5 | Radiation   86

6 | Energy Balance   122

7 | Water Vapour   150

12 | General Circulation of the Atmosphere   292

13 | Air Masses and Fronts   318

14 | Storms   338

15 | Weather Forecasting   376

16 | Global Climate   406

17 | The Changing Atmosphere   442

From the Publisher

When we think about how weather and climate affect our lives on a daily basis, we might think of having to carry around an umbrella when the forecast calls for rain or put on extra sunscreen when UV ratings are high. But the effects of weather and climate can also cause great upheaval. Even if we haven't experienced an extreme weather event, we've seen many covered on the news—a freak ice storm that has caused millions of dollars' worth of damage, a sudden outbreak of tornadoes that has destroyed crops, or even a violent hurricane that has left thousands of people displaced, wounded, or deceased. We're usually prepared for "extreme" aspects of the climate in which we live—most Canadians, for example, own winter coats and live in well-insulated homes. However, with the impacts of climate change becoming more and more apparent, it seems that we might not be able to depend on the regularity of climate patterns for much longer. Thus, our need to understand how and why the processes that drive these patterns operate is becoming increasingly urgent.

With this growing urgency in mind, Oxford University Press is proud to be publishing this in-depth, scientifically grounded, thoroughly Canadian introduction to the study of weather and climate. Sheila Ross's rigorous yet accessible approach to the discipline illuminates the often-hidden processes that create and influence weather and climate across the globe. Moreover, the focus on contemporary Canadian issues and research, which is balanced by examples from around the world, makes this text particularly relevant and relatable to students in Canada today.

Features

Designed with students new to the discipline in mind, this thorough introduction includes a wide array of features that will help students make the most of their learning experience.

Learning goals give students an at-a-glance overview of what they will encounter in each chapter, while chapter summaries remind students of the most significant concepts that have been covered.

during the day (Figure 6.16a). The division of energy between the two convective heat fluxes will depend to a large extent on the amount of moisture available. The more energy that goes into evaporation, the less that is available for raising the temperature. For moist surfaces, the latent heat flux will often be higher than the sensible heat flux.

DEW
Atmospheric water vapour that has condensed onto a cool surface during the night.

FROST
Atmospheric water vapour that has collected on a surface as ice crystals.

At night, when the net radiation, Q^*, is negative, the surface begins to cool because the loss of radiation is greater than the gain. The surface becomes cooler than the air and the ground, the temperature gradients are reversed, and sensible heat is transferred toward the surface (Figure 6.16b). Now the conductive sensible heat flux is upward, and the convective sensible heat flux is downward. The latter will operate only under windy conditions, as it requires *mechanical* convection to *force* warm air from above toward the surface. Thus, the surface will get much cooler under

calm conditions than it would under windy conditions. In some cases, a water vapour gradient will be directed toward the surface at night, so that the latent heat flux will flow from the air to the surface. When this happens, water vapour can condense on the cool surface to form **dew**. If temperatures are below freezing, **frost** will form instead.

The typical daily variation of net radiation, Q^*, is shown in Figure 6.15. This curve becomes the starting point for the energy balance diagram shown in Figure 6.19, which also includes curves representing the three heat fluxes, Q_H, Q_E, and Q_G. Figure 6.19 shows that the three heat fluxes are of roughly equal magnitude in the morning. However, as the surface warms, the atmospheric temperature and water vapour gradients strengthen, and the winds pick up. The result is that the convective heat fluxes become the dominant terms in the afternoon. Of these two, the latent heat flux is largest because, in this case, the surface is fairly moist. In mid-afternoon, the conductive heat flux becomes negative, indicating that heat is flowing toward the surface instead of away, as it was earlier. This flow of heat adds to the amount of energy available at the surface, so that evaporation continues into the evening. Notice that, in the evening, the latent heat flux exceeds the net radiation. At night, both convective heat fluxes are almost zero. With a temperature gradient directed toward the surface, and with light winds, heat transfer by the air is negligible. As a

Remember This

As a result of a daytime radiative surplus, the surface will *warm*, and

- sensible heat will flow from the surface to the ground by conduction,
- sensible heat will flow from the surface to the atmosphere by convection, and
- latent heat will flow to the atmosphere by convection as water evaporates at the surface and condenses in the air.

As a result of a nighttime radiative deficit, the surface will *cool*, and

- sensible heat will flow from the ground to the surface by conduction,
- sensible heat will flow from the atmosphere to the surface by convection, and
- latent heat will flow from the atmosphere by convection as water vapour condenses on the surface as dew.

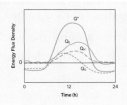

FIGURE 6.19 | Idealized energy balance for a moist vegetated surface on a clear day in summer.
Source: Bailey, Oke, & Rouse, 1997, p. 29.

Marginal definitions of key terms ensure that students have a full understanding of the more discipline-specific terminology used in each chapter. Expanded definitions also appear in an end-of-text glossary, for quick reference.

"Remember This" boxes draw students' attention to key points and provide valuable tools for review.

"Question for Thought" boxes ask students to take a moment to make their own connections and consider how key topics relate to real-world processes.

...eat loss from the surface is bal... a flow of sensible heat from the ...applies to a fairly typical moist ...t as surface types and atmos... vary, so too can the energy bal... are endless.

In fact, an added complication is the possibility that heat, moisture, or both could be carried *horizontally* over a surface by advection. Equation 6.12 assumes that there is no advection, but this is not to say that advection is not important to the surface energy balance. In actuality, advection can influence the temperature and moisture gradients by bringing in warmer or colder, or drier or wetter, air. For example, if cold air blows over a warm surface, the temperature gradient will increase and drive a greater convective sensible heat flux. Similarly, if dry air blows over a moist surface, the water vapour gradient will increase, increasing the flow of latent heat. Advection can also change the *direction* of the heat flows. For example, if warm, dry air flows over a cool, moist surface, sensible heat might begin to flow *toward* the surface. In addition, this flow of heat will greatly enhance evaporation from the moist surface.

Question for Thought

Which surface characteristics influence the surface energy balance? How does each characteristic affect the radiative, convective, and/or conductive heat fluxes?

6.6.3 | Examples of Two Surface Climates

To illustrate the way in which the energy balance influences surface climates, we will consider the energy balance of a desert surface and that of an ocean surface, each located in the subtropics (Figure 6.20). These two surfaces are both relatively simple: they are smooth, extensive, and non-vegetated. In addition, they represent extreme opposites in terms of climate.

Being located in the subtropics, both surfaces will receive high amounts of solar radiation, $K\downarrow$. Because deserts have high albedos of 30 to 40 per cent and

FIGURE 6.20 | *Top:* A sea surface in the Bahamas and *bottom:* a desert surface in Mauritania. What accounts for the difference in daily temperature ranges between the two surfaces?

oceans have low albedos of less than 10 per cent (Table 5.2), the ocean will absorb considerably more of the incident solar radiation than will the desert. This means that K^* will be higher for the ocean. Further, net longwave radiation, L^*, will be at a greater loss for the desert because the hotter desert surface will emit more longwave radiation, $L\uparrow$, than will the cooler ocean surface, and the drier desert air will result in less incoming longwave radiation, $L\downarrow$. With a lower absorption of shortwave radiation and a higher loss of longwave radiation, the net radiation, Q^*, will be less for the desert surface than it will be for the ocean surface (Figure 6.21).

Despite absorbing less radiation, the desert surface can reach much higher daytime temperatures

In the Field

Using Global Climate Model Output for Future Scenarios of Climate Change
Adam Fenech and Neil Comer, University of Prince Edward Island, Climate Collaborating Unit

Our present understanding of the climate system and how it is likely to respond to increasing concentrations of greenhouse gases in the atmosphere would be impossible without the use of global climate models (GCMs). GCMs are powerful computer programs that incorporate physical processes to simulate, as accurately as possible, the functioning of the global climate system in three spatial dimensions and in time.

The results from 24 GCMs were used in the deliberations of the Intergovernmental Panel on Climate Change's (IPCC) Fourth Assessment Report (4AR), which was released in 2007. Each of the modelling centres provided future projections for at least two, sometimes three, emission scenarios (scenarios that describe how greenhouse gas emissions could evolve over the next 100 years). Thus, the report was able to take into account about 72 possible future outcomes for a location's climate (figures 1 and 2).

While the models are all in agreement on the direction of temperature change, results between models can vary widely, and models each contain their own inherent biases. The differences in results exist because of the differences between each GCM's model resolution, model formulations, and model parameterization. Differences also arise depending on which emission scenario of future greenhouse gases is selected.

There are several caveats to using GCMs for future projections. The resolution of the models varies and is completely determined by the modelling centre; that is, there is no "standard" grid size or projection method. The output from the model represents an average of the entire grid cell area—a grid cell with a large resolution of about 250 by 250 km. This approximation means that even the distribution of land/water grid cells differs between models. For many locations, this can have important implications because many climates are influenced significantly by the moderating effect of proximity to large water bodies. Regardless of these caveats, models remain the best option available for creating future climate projections—versus, for example, a simple extrapolation of historical trends.

Three approaches have been developed to provide some direction for determining which of the many future projections of climate should be used in planning: the extremes (max./min.) approach; the ensemble approach; and the validation approach. The *extremes (max./min.) approach* suggests that it is best to plan within the full range of possibilities that the GCMs present. It takes the projection for the maximum change as well as the projection for the minimum change and uses both as the range of considerations when planning. The *ensemble approach* suggests that it is best to plan for the average change of all the models. It uses a mean or median of all the models (or many models) to reduce the uncertainty associated with any individual model. The *validation approach* suggests that only those

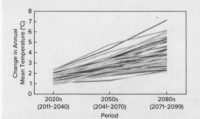

FIGURE 1 | Future projections of annual mean temperature changes at the City of Toronto grid cell in the 2020s, 2050s, and 2080s using all 24 models contributing to the IPCC 4AR (change in degrees Celsius from 1961–90 baseline period).

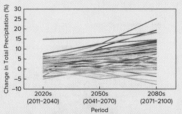

FIGURE 2 | Future projections of annual total precipitation changes at the City of Toronto grid cell in the 2020s, 2050s, and 2080s using all 24 models contributing to the IPCC 4AR (change in percentage from 1961–90 baseline period).

"In the Field" boxes—written by top researchers from across Canada—offer students insight into the types of research meteorologists and climatologists do every day. Topics covered include measuring levels of aerosols in the atmosphere, studying precipitation patterns on regional and global scales, assessing the impact of forest fires on air quality, and understanding and monitoring atmospheric characteristics of Arctic environments.

Equations and formulas make complex theories and processes accessible by breaking them down into their component parts, while "Example" boxes show students how to apply key equations to actual problems. Equations that students will find particularly useful for solving end-of-chapter problems are highlighted in shaded boxes throughout the text and printed inside the front and back covers of this text.

Example 7.15

A psychrometer reading gives a dry-bulb temperature of 16°C and a wet-bulb temperature of 14°C. Determine the vapour pressure, relative humidity, and dew-point temperature for this air.

Use Table 7.1 to obtain the saturation vapour pressure at the wet-bulb temperature, then use Equation 7.16 to calculate vapour pressure.

$$e = 1500 \text{ Pa} - 65 \text{ Pa/K} (16°C - 14°C)$$
$$= 1370 \text{ Pa}$$
$$= 1.4 \text{ kPa}$$

Use Table 7.1 to obtain the saturation vapour pressure for 16°C, then use Equation 7.11 to calculate relative humidity.

$$RH = \left(\frac{1.4 \text{ kPa}}{1.8 \text{ kPa}}\right) \times 100\%$$
$$= 78\%$$

Note that you can also obtain relative humidity from Table 7.3 or Figure 7.17.

Use Table 7.1, and the vapour pressure of 1.3 kPa obtained above, to determine that the dew-point temperature is about 11°C.

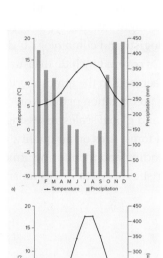

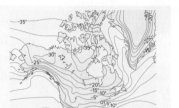

FIGURE 12.12 | Map of January temperatures in North America. Notice how the isotherms bend southward over the continent, especially toward the east.
Source: Adapted from Strahler and Archibold, 2011, p. 84.

FIGURE 12.11 | a) Estevan Point, British Columbia, is located on the west coast of Canada. b) Sydney, Nova Scotia, is located on the east coast. The annual temperature range in Estevan Point is smaller than that in Sydney. Why do oceans tend to have a moderating influencing on temperature? Why does Estevan Point receive more precipitation than Sydney receives?

the southern hemisphere. In fact, without land masses to slow the flow, the westerly winds can reach high speeds as they blow over the southern oceans and circle the continent of Antarctica. Like the doldrums and the horse latitudes, these winds were named by sailors, who referred to them as the roaring forties, the furious fifties, and the screaming sixties, after the latitudes at which they are found.

Just as warm air flows toward the poles from the *subtropical* highs, cold air flows toward the equator from the *polar* highs. This flow is deflected by the Coriolis force, producing the polar easterlies. Because these winds cover only a small portion of Earth's surface, they are a relatively insignificant part of the general circulation. However, they do play an important role in forming front... polar easterlies me... ally somewhere betw... fronts are sometimes... ous front called the... they are not continu... over cold air, creati... can develop into low... These systems are k... as *mid-latitude cyclon*... simply *frontal systems*... they are responsibl... most of the precipit...

Notice that Figure 12.3 shows a much more regular westerly flow in the southern hemisphere than in the northern hemisphere. This regularity is due to there being almost no land masses in the mid-latitudes of

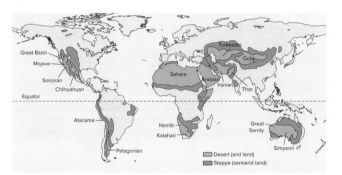

FIGURE 12.8 | The deserts of the world. The deserts of the subtropics form due to the subtropical highs. Most of the mid-latitude deserts are rain-shadow deserts.

12.2.4 | The Extratropical Circulation

While surface air diverging from the subtropical highs flows toward the ITCZ, forming the trade winds, it also flows poleward, creating the mid-latitude westerlies. These winds are not nearly as steady as the trade winds; in fact, they are westerly only *on average*. This is because the flow is disturbed by the travelling cyclones and anticyclones responsible for the variable weather of the mid-latitudes. The predominantly westerly flow means that climates of

west-coast locations in the mid-latitudes are more strongly influenced by the moderating effects of the ocean than are climates of east-coast locations. The annual temperature range on west coasts, therefore, is usually much smaller than it is on east coasts. In particular, east-coast winters are colder than west-coast winters (Figure 12.11). For example, for the same latitude, winter temperatures are much higher on Canada's west coast than they are on Canada's east coast (Figure 12.12).

FIGURE 12.9 | The Sahara Desert. What accounts for the extreme diurnal temperature ranges of subtropical deserts? (See Section 6.6.3.)

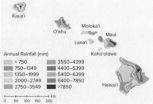

FIGURE 12.10 | A map of precipitation for the Hawaiian Islands.
Source: Based on the Online Rainfall Atlas of Hawai'i at http://rainfall.geography.hawaii.edu/

Figures, photos, and maps bring the discussion to life by illustrating how key processes operate and relate to the world beyond the page.

End-of-chapter review questions and problems help students synthesize what they have learned, while suggestions for further research encourage students to develop their research skills by investigating key topics related to each chapter.

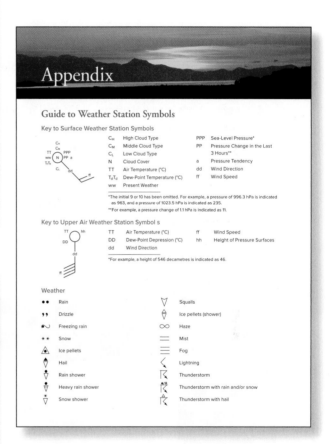

A list of weather station symbols, included in an appendix, gives students the information they need to perform in-depth analyses of weather maps.

A cloud chart, located at the end of the text, provides a concise comparison of the most common types of clouds, to help students classify clouds when they make their own observations of the atmosphere.

---

290 | Weather and Climate

Review Questions

1. What determines the magnitude and direction of the pressure gradient force?

2. What is the relationship between the Coriolis force and wind speed? What is the relationship between the Coriolis force and latitude? In what direction does the Coriolis force act?

3. In what direction does the centripetal force act? What determines the magnitude of this force?

4. What determines the magnitude of the friction force?

5. Why does the effect of surface friction extend upward into the atmosphere? What factors determine the depth of the atmosphere that will be influenced by surface friction?

6. How could you apply Newton's first law to explain why geostrophic winds parallel isobars?

7. Why are gradient winds faster around high-pressure areas than they are around low-pressure areas?

8. Why do surface winds cross isobars at an angle?

9. How do convergence and divergence lead to vertical motions? Describe the patterns of convergence, divergence, and vertical motions in high- and low-pressure systems.

10. Why are low-pressure systems associated with cloudy weather, while high-pressure systems are associated with clear weather?

11. How do temperature gradients produce upper-level pressure gradients? Why do such pressure gradients increase with height?

12. What is the relationship between changes in wind direction with height and thermal advection?

13. How do land–sea breeze circulation systems develop? How do valley and mountain winds develop?

Problems

1. Calculate the Coriolis parameter, f_c, for
   a) 10° latitude and
   b) 80° latitude.

2. For each latitude in Problem 1, calculate the Coriolis force if the wind speed is 15 m/s. If the wind is blowing from west to east in the northern hemisphere, what is the direction of the Coriolis force?

3. a) Calculate the geostrophic wind for the height contours shown in the diagram, below. The heights are given in decametres, the distance between the contours is 500 km, and the Coriolis parameter is 1×10^{-4} s^{-1}.
   b) At which point—1, 2, or 3—does the wind speed calculated in a) best apply?

4. Calculate the geostrophic wind velocity in each case.
   a) The latitude is 50°, the air density is 0.8 kg/m³, and the pressure gradient is 1.7 kPa/100 km.
   b) The Coriolis parameter is 1×10^{-4} s^{-1} and the height gradient on the 500 hPa surface is 0.8 m/km.

5. For a geostrophic wind speed of 12 m/s, a radius of curvature of 1000 km, and f_c of 7×10^{-5} s^{-1}, calculate
   a) the gradient wind speed for a cyclone and
   b) the gradient wind speed for an anticyclone.

6. Use a geostrophic wind velocity of 35 m/s to do the following.
   a) Calculate the speed of this wind around a cyclone at a radius of curvature of 500 km.
   b) Calculate the speed of this wind around an anticyclone at a radius of curvature of 500 km.
   c) Account for the difference in your answers to a) and b).

7. Calculate the magnitude of the thermal wind for a change in thickness of 380 m over 1000 km. The Coriolis parameter is 1×10^{-4} s^{-1}.

Appendix

Guide to Weather Station Symbols

Key to Surface Weather Station Symbols

C_H	High Cloud Type	PPP	Sea-Level Pressure*
C_M	Middle Cloud Type	PP	Pressure Change in the Last 3 Hours**
C_L	Low Cloud Type		
N	Cloud Cover	a	Pressure Tendency
TT	Air Temperature (°C)	dd	Wind Direction
T_dT_d	Dew-Point Temperature (°C)	ff	Wind Speed
ww	Present Weather		

*The initial 9 or 10 has been omitted. For example, a pressure of 996.3 hPa is indicated as 963, and a pressure of 1023.5 hPa is indicated as 235.
**For example, a pressure change of 1.1 hPa is indicated as 11.

Key to Upper Air Weather Station Symbols

TT	Air Temperature (°C)	ff	Wind Speed
DD	Dew-Point Depression (°C)	hh	Height of Pressure Surfaces
dd	Wind Direction		

*For example, a height of 546 decametres is indicated as 46.

Weather

● ●	Rain		Squalls
,,	Drizzle		Ice pellets (shower)
	Freezing rain	∞	Haze
✳ ✳	Snow		Mist
	Ice pellets		Fog
	Hail	<	Lightning
	Rain shower		Thunderstorm
	Heavy rain shower		Thunderstorm with rain and/or snow
	Snow shower		Thunderstorm with hail

Cloud Chart

Altitude (m)

12,000
10,000
8,000
6,000
4,000
2,000
0

Cirrus
Cirrostratus
Cirrocumulus
Altocumulus
Altostratus
Cumulonimbus
Nimbostratus
Cumulus Congestus
Stratocumulus
Stratus
Cumulus Humilis

High Clouds
Middle Clouds
Low Clouds

CIRRUS CLOUDS appear as fibrous streaks scattered across the sky. Because they form high in the troposphere, where it is both dry and cold, cirrus are relatively thin clouds, composed of ice crystals.

CIRROCUMULUS CLOUDS are high-altitude clouds made of ice crystals. They are layer clouds that have been destabilized to produce tiny individual clumps of cloud that are usually arranged in a regular pattern, between which blue sky can be seen.

Robust Online Ancillary Suite

Weather and Climate: An Introduction is supported by a wide range of supplementary items for students and instructors. These resources are guaranteed to enrich and complete the learning and teaching experience.

For Students

A comprehensive study guide with supplemental material related to each chapter offers students additional content to help them understand the key concepts presented in the text and review for tests and exams.

For Instructors

The following instructor's resources are available to qualifying adopters. Please contact your OUP Canada sales representative for more information.

* **An instructor's manual** offers additional resources—including sample syllabi, solutions for in-text questions and problems, and additional suggestions for research—to help instructors make learning material as relevant and engaging as possible for students.
* **PowerPoint slides** for use in classroom lectures provide concise overviews of material covered in the text.
* **A test generator** offers a variety of options for sorting, editing, importing, and distributing questions.
* **An image bank** includes a wide selection of stunning photos that will enhance the vibrancy of slides and handouts.

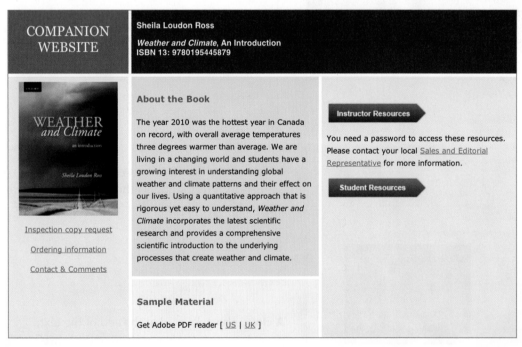

 www.oupcanada.com/Ross

Preface

This book is a guide to understanding the atmosphere and how it produces weather and its longer-term counterpart climate. Largely responsible for weather and climate are the transfers and transformations of energy, mass, and motion that are continually taking place within the atmosphere. But the atmosphere does not produce these effects alone. Instead, it is part of a system—referred to in this book as the *Earth-atmosphere system*—in which it constantly interacts with the water, life, and rocks of this planet. For example, clouds form as air ascends and cools *within the atmosphere*, but clouds cannot form without the water and particles that are supplied by processes operating *at Earth's surface*. Although we have quite a good understanding of many of the processes associated with cloud formation that operate within the atmosphere itself, much research still needs to be done to understand the relationships between cloud formation and such factors as soil moisture or the type and abundance of particles released into the atmosphere. Thus, in order to fully understand clouds, we must fully understand *all* the processes that produce them, for if any of these processes change, so will clouds. It follows, by extension, that viewing Earth as a system allows us to better understand global change.

Although such a holistic approach to understanding Earth has ancient roots, many twentieth-century scientific thinkers and researchers tended toward specialization. This specialization likely contributed to the period's great scientific and technological advancements, but it neglected to take in the wider picture. Therefore, as we are becoming increasingly aware, specialization does not provide enough understanding of how Earth operates to adequately predict or deal with the widespread environmental degradation that we now face as a result of our actions. Thus, the 1970s saw the emergence of the discipline of Earth system science and the formulation of the Gaia hypothesis. Both of these approaches to Earth science recognize that Earth behaves as a system characterized by complex interactions and feedbacks.

In recent decades, undoubtedly as a result of this shift in perspective, the word *interdisciplinary* has increasingly appeared in university and college course descriptions and textbooks. Further, research is now often interdisciplinary in nature. This trend is illustrated in some of the "In the Field" boxes scattered throughout this book; in these boxes, a variety of Canadian atmospheric scientists describe their research, which often involves investigating linkages between the atmosphere and processes operating *outside* the atmosphere. As the discussions in these boxes reveal, this sort of broader-based research, which falls outside traditional disciplinary boundaries, can offer many exciting challenges and rewards.

As you read through this book, then, you would do well to take note of the many linkages within the Earth-atmosphere system, but you should also keep in mind that we are far from having a complete understanding of most of them. As we take a systems-based approach, we will undoubtedly come to understand Earth with greater clarity. Hopefully, such clarity will make it possible to more accurately predict the consequences of our actions, to reverse—or at least lessen—the impact of the environmental degradation that we have already caused, and, ultimately, to prevent further damage.

Acknowledgements

My academic career in physical geography, in particular the atmospheric sciences, began thanks to Dr Timothy Oke, professor emeritus of the University of British Columbia. After completing my graduate studies under his supervision, I was hired as a geography instructor at Capilano University, in North Vancouver. Over the years, my colleague, Karen Ewing, has provided valuable mentoring and has always been willing to share ideas.

I was able to do a large portion of the work on this book thanks to a paid educational leave from Capilano University. Robert Campbell, dean of arts and sciences, continued to encourage me to see the project through to completion. Karen provided valuable feedback on the first drafts of some of the earlier chapters, while another colleague, Christopher Gratham, provided assistance with the chapter on weather forecasting.

I would also like to thank the many students I have worked with over my 25 years of teaching at Capilano University. It is through their feedback that I have come to learn what works and what doesn't, and that enthusiasm is key. In particular, my thanks go to my Geography 214 class of the 2011 spring term. During that term, as I was completing chapters for review, these students showed an interest in my project and made use of my draft chapter summaries as study guides.

It is the people at Oxford University Press that have ultimately made this book possible. Katherine Skene accepted my book for publication and welcomed me. Peter Chambers saw the book through its developmental stages. Janice Evans's attention to detail, advice on diagrams, and excellent editorial skills greatly improved my work. An expert team of designers, illustrators, and formatters transformed my rough sketches into the diagrams you will encounter throughout the book.

For the last few years, I have been writing this book while teaching full time. I thank my family, including my parents, for supporting me in this. My daughter was company late at night while she studied and I wrote. My son provided technical help, his knowledge of computers being far greater than mine. My husband accepted that I had to work most evenings.

Finally, I, along with the publisher, would like to thank the following reviewers, as well as those who wish to remain anonymous, whose thoughtful comments and suggestions helped shape this book:

Jacqueline Binyamin, University of Winnipeg
Jim Bowers, Langara College
Danny Harvey, University of Toronto
Mark Moscicki, University of Windsor
Ian Strachan, McGill University
Stanton E. Tuller, University of Victoria
John Yackel, University of Calgary

Sheila Ross,
Capilano University

WEATHER
and Climate

1

The Study of the Atmosphere

Learning Goals

After studying this chapter, you should be able to

1. *appreciate* Earth's atmosphere;

2. *distinguish* between weather and climate;

3. *compare* scientific laws to scientific theories;

4. *appreciate* how scientific models can be used;

5. *outline* the steps of the scientific method;

6. *appreciate* the importance of math in science;

7. *explain* how the units of the derived dimensions—force, energy or work, power, and pressure—are obtained; and

8. *describe* the changes in pressure, density, and temperature with height in the atmosphere.

Viewed from space at sunset, Earth's atmosphere appears to be a fairly simple, uniform layer that is quite separate from the surface of the planet. In actuality, the atmosphere is part of a complex system in which it continually interacts with the surface, effectively shaping climate and, ultimately, the planet's ability to sustain life.

In the Earth sciences, we use the term **atmosphere** to refer to the layer of gases that surrounds a planet or other celestial body. Interestingly, we also use the word *atmosphere* more generally when referring to the *feeling* we get from our surroundings. Planetary atmospheres certainly shape the *feeling* at the surface of a planet, and Earth's atmosphere is no exception.

ATMOSPHERE
The layer of gases surrounding a planet or celestial body.

Earth would be an entirely different place without its atmosphere. The "sky" above would be a black emptiness, and the planet itself would appear far less bright and beautiful from space (Figure 1.1). Without an atmosphere, there would be no air to breathe, and no ozone layer to prevent the sun's dangerous ultraviolet radiation from reaching Earth's surface. Temperatures would reach extreme highs and lows, as there would be nothing to reflect away solar radiation by day, and nothing to keep in warmth by night. Because sound won't travel in a vacuum, it would be very quiet. There would be no wind, and there would certainly be no clouds, nor would there be rain or, perhaps, any liquid water at all. In short, there would be no weather. All of these characteristics of the atmosphere, and more, result from the operation of processes that are mostly invisible.

FIGURE 1.1 | Planet Earth from space. The blue haze around the edge of Earth is its atmosphere. Although Earth's atmosphere is thin in comparison to the size of the planet, this layer of gases has a dramatic impact on the conditions at Earth's surface.

1.1 | Weather and Climate

Atmospheric science is our quest to understand these processes. This science is often divided into two subdisciplines: **meteorology**, which is the study of **weather**, and **climatology**, which is the study of **climate**. *Weather* is the state of the atmosphere at a specific time and place. We describe the state of the atmosphere in terms of the *elements* of weather, which include atmospheric pressure, clouds, precipitation, wind, temperature, and humidity. An important characteristic of the weather elements is their ability to quickly change. In fact, the atmosphere is in a constant state of change, and this is what makes weather so changeable.

METEOROLOGY
The study of the atmospheric processes responsible for weather.

WEATHER
The state of the atmosphere at a given place and time.

CLIMATOLOGY
The study of climate.

CLIMATE
The average conditions of the atmosphere.

Weather is an integral part of life on this planet. It affects how we feel and what we wear. It might cause us to cancel our plans, and it gives us something to talk about with strangers. It creates an ever-changing skyscape to fill us with wonder and delight when we take the time to pay attention. However, the effects of weather can also be far from trivial. Weather can ruin crops, flatten buildings, sink ships, bring floods or droughts, and much more (Figure 1.2). Most importantly, weather provides an essential service: as clouds form and rain falls—in other words, as water changes phase—weather filters and replenishes our water supplies. A practical motivation to study the atmosphere is to understand what causes weather to change and, thus, to be able to make forecasts (Chapter 15).

While weather is characterized by change, climate is less about change and more about what stays the same. As such, climate is often defined as "average

FIGURE 1.2 | An aerial view of the flooding around Morris, Manitoba, in April of 2009. About 1800 km² of land were flooded in the province, with 2500 people registering as evacuees.

weather." The difference between weather and climate is nicely illustrated by comparing a weather map to a climate graph (Figure 1.3). While the weather map depicts the conditions at a *moment* in time, the climate graph is based on *decades'* worth of data. (Climate graphs are included throughout this book to illustrate factors that influence climate.)

Climate is a direct result of the *location* of a place. The most important control on climate is latitude: low latitude climates are warmer than high latitude climates. In addition, most coastal climates experience milder temperatures and more rainfall than do most continental climates. It can even make a difference to climate if a place is located on a west coast or an east coast. For example, in Canada, prevailing winds from the west mean that west coast climates are more strongly influenced by the ocean than are east coast climates. Mountain ranges also affect climate. Some of the wettest places on Earth are on the windward sides of mountains, while some of the driest places are on their leeward sides (Figure 1.4). There is even the familiar correlation between increasing altitude and decreasing temperature.

Climate is studied at a variety of scales. The smallest scale climates occur over areas of a few square metres to several square kilometres. Such **microclimates** include anything from a cool, shady spot under a tree to the climate created by a city. In Section 6.6 we will see how the characteristics of a surface influence energy flows, thus creating microclimates. The largest scale we consider in climatology is the *planetary* scale; the average temperature of our planet is currently just below 15°C. This temperature is largely a result of the composition of the atmosphere and the reflectivity of the planet. In Section 6.4 we will look at the flows of energy at the planetary scale. Today we are extremely motivated to understand the complexities of climate at the planetary scale because human activities are causing changes in both the composition of the atmosphere *and* the reflectivity of the surface. The net result of these changes is an observed increase in temperature, known as **global warming**. Such **anthropogenic** climate change is addressed in Section 17.3.2.

MICROCLIMATE
The climate of a small area at Earth's surface.

GLOBAL WARMING
The increase in Earth's temperature caused by increasing concentrations of greenhouse gases associated with human activities.

ANTHROPOGENIC
Related to human activities.

Question for Thought

Which human actions might have the greatest impact on the atmosphere?

Remember This

We have two good reasons to understand the atmosphere:

- forecasting the weather and
- predicting how human actions might influence climate at a variety of scales.

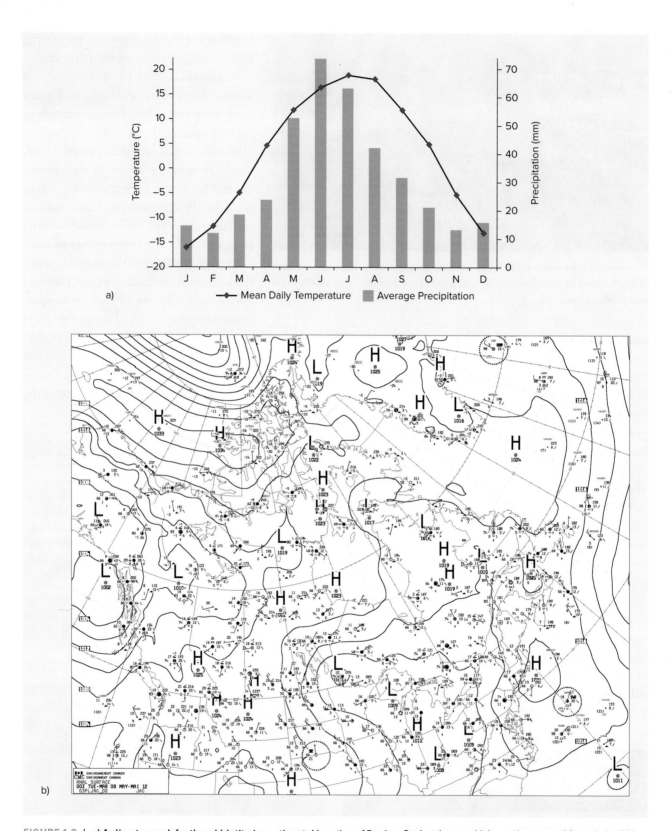

FIGURE 1.3 | a) A climate graph for the mid-latitude continental location of Regina, Saskatchewan. **b)** A weather map of Canada for 8 May 2012 at 0000 UTC.

Sources: a) Data from Environment Canada, National Climate Data and Information Archive, "Canadian Climate Normals 1971–2000" for Regina Int'l A, at www.climate .weatheroffice.gc.ca/climate_normals, accessed 9 May 2012; b) Environment Canada, Weather Office, www.weatheroffice.gc.ca/analysis/index_e.html, 8 May 2012.

FIGURE 1.4 | Mountains affect climate. *Top:* Situated on the windward side of the Coast Mountains, Vancouver receives high levels of precipitation, with an average of **167 mm** of precipitation in November. *Bottom:* Osoyoos, in the Southern Okanagan, is located in the rain shadow of these mountains and receives an average of only **26 mm** of precipitation in the same month.

1.2 | The Earth-Atmosphere System

In creating weather and climate, the atmosphere does not act independently. Instead, it is part of a **system** that we call the *Earth-atmosphere system*. The four major parts of the Earth-atmosphere system are the **lithosphere**, the **biosphere**, the **hydrosphere**, and, of course, the atmosphere itself. The lithosphere is made up of the rocks of the planet, the biosphere comprises all life on the planet, and the hydrosphere includes all the water of the planet. (Here the ice of the planet has been included with the hydrosphere; sometimes it is considered separately as the **cryosphere**.) As in any system, the components continually interact, in this case through the cycling of both energy and matter (Figure 1.5). As rocks weather, certain gases are added to and removed from the atmosphere. Volcanic eruptions eject gases and particles into the atmosphere. Water is continually cycled between the atmosphere and the hydrosphere through the phase changes of **evaporation** and **condensation**. Life exchanges gases with the atmosphere through the processes of **respiration** and **photosynthesis**. In fact, throughout Earth's long history, life has significantly impacted the composition of the atmosphere, and thus influenced the climate. These interactions are considered in detail in Chapter 2.

Simply due to the sheer number of interactions involved, the study of Earth systems can be a daunting task. This task is further complicated by the operation of **feedback effects**. A *positive* feedback effect is a mechanism that operates within a system to *amplify* the effects of an initial change. A *negative* feedback effect is a mechanism that operates within a system to *lessen* the effects of an initial change. While negative feedback effects can be stabilizing, positive feedback effects can be destabilizing. As mentioned above, human actions are causing Earth's temperature to increase; this change is likely triggering several feedback effects. One is the *ice-albedo feedback effect*, which occurs as warming causes

SYSTEM
An interrelated set of parts.

LITHOSPHERE
The rocks of Earth.

BIOSPHERE
Life on Earth.

HYDROSPHERE
The water of Earth.

CRYOSPHERE
The ice of Earth.

EVAPORATION
The process by which a substance, usually water, changes phase from a liquid to a gas.

CONDENSATION
The process by which a substance, usually water, changes phase from a gas to a liquid.

RESPIRATION
The life process in which oxygen is removed from the atmosphere and carbon dioxide is returned.

PHOTOSYNTHESIS
The life process in which energy from the sun is used to convert carbon dioxide and water to oxygen and carbohydrates.

FEEDBACK EFFECT
A mechanism that operates within a system to either amplify or lessen an initial change.

ice to melt, in turn decreasing the reflectivity of the planet and increasing the absorption of the sun's radiation. Because this process will accelerate the warming, it is a *positive* feedback effect (Figure 1.6). We will explore feedback effects in our discussion of climate change in Chapter 17.

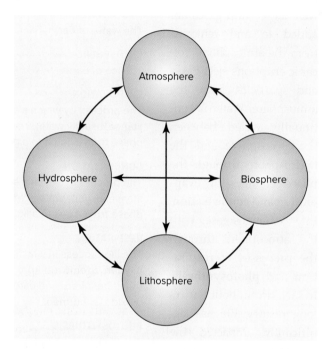

FIGURE 1.5 | **A schematic depiction of the interactions between the four spheres of the Earth-atmosphere system.**

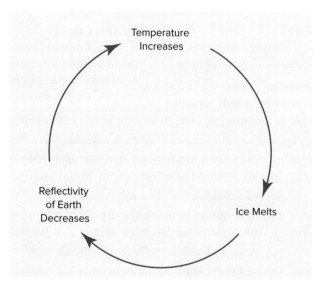

FIGURE 1.6 | **The loop associated with the ice-albedo feedback effect.**

1.3 | The Role of Science

The purpose of science is to understand the complexities in the world around us by trying to find order, or patterns. In fact, some scientists have marvelled at the fact that there *is* order in nature. In practising science, we begin by making observations that provoke us to ask questions and investigate further. These investigations can ultimately lead us to the development of **scientific laws** and **theories** that can be used to describe and explain how the world works. Laws and theories are similar in several ways: they are widely accepted by the scientific community, they can be used to make predictions, and they can be rejected if shown to be untrue. Yet there are important differences between laws and theories as well.

Laws provide a *description* of how nature works, and they can be expressed in words, mathematical equations, or both. An example of a scientific law used in atmospheric science is the radiation law known as the *Stefan–Boltzmann law*. In *words*, this law tells us that the amount of radiation emitted by an object increases with the fourth power of the temperature of that object. The *mathematical* expression of this law is given in Equation 5.4. In either form, this law *describes* the relationship between temperature and the amount of radiation emitted. It was determined by experiment, and it can be used in its mathematical form to make predictions.

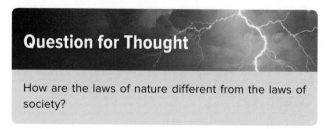

Question for Thought

How are the laws of nature different from the laws of society?

Theories provide an *explanation* of how nature works. Theories are generally far more complex than laws, as they tend to encompass a large body of knowledge. Unlike laws, they cannot be expressed as a single equation, although they can include equations. An example of a theory described in this book is the *polar front theory*, which provides a detailed description of the structure and development of the storms known as mid-latitude cyclones. (Section 14.2.1). This theory came about as a result of the work carried out by several Norwegian scientists interested in being able to forecast the weather. Over the years, the polar front theory has been modified as further investigation has increased our understanding of the processes that create the weather of the mid-latitudes.

Scientific investigation follows the **scientific method** and begins with observation. Observation can be as simple as perceiving some aspect of the world around us and wondering how it operates, but there are many possible sources of observations, such as the results of an experiment or unexplained patterns in a set of measurements. If an observation arouses our curiosity, we will ask a question. After further observation, we might form a **hypothesis**—an initial attempt to answer the question posed. Here is an example of how this method might operate: first, we *observe* that the sky is blue; next, we *ask why* the sky is blue; then, we *speculate on a reason* for the sky being blue. Once we make a hypothesis, we must test it. If testing shows the hypothesis to be true, we will accept it, although we have not yet *proven* it. If testing shows the hypothesis to be untrue, we will reject it.

We can test a hypothesis by performing an experiment, making further observations, or utilizing a **model**. Models are representations of reality that scientists use to describe, explain, and predict. They are particularly useful in atmospheric science, where traditional experiments are often difficult—if not impossible—to perform. The essential purpose of a scientific model is to simplify reality so that we can understand it. There are three types of models: conceptual, physical, and mathematical.

Conceptual models—also known as *mental models*—represent our attempts to present our ideas about how something works. They are often used in teaching to show relationships or to explain abstract concepts, and they are generally presented as diagrams. A very simple example is the loop diagram used to illustrate the feedback mechanism described in Section 1.2 (Figure 1.6). Another example is the diagram, created by the Norwegian meteorologists as part of their polar front theory, that shows the stages in the life cycle of a mid-latitude cyclone (Figure 14.6).

An example of a *physical model* in atmospheric science is a rotating annulus, which has been used to perform "dishpan" experiments (Figure 12.23). In these experiments, the annulus is filled with water to represent the fluid character of the atmosphere. It is then heated at the outer edge to represent the equator and cooled in the middle to represent the poles, and it is rotated to represent Earth's rotation. Dishpan

SCIENTIFIC LAW
A precise statement that describes the behaviour of nature and is believed to always hold true.

SCIENTIFIC THEORY
A body of knowledge that provides a detailed explanation for a set of observations.

SCIENTIFIC METHOD
A series of steps followed in scientific investigation.

HYPOTHESIS
A tentative explanation for an observation.

MODEL
A representation of reality used to help in understanding complex or abstract natural phenomena.

Question for Thought

What observations could you make about clouds? What questions could you derive from your observations? How might you follow the scientific method to investigate those questions?

Remember This

There are five steps in the scientific method:

1. making an observation;
2. asking a question;
3. formulating a hypothesis;
4. making further observations, conducting an experiment, and/or utilizing a model; and
5. reaching a conclusion, by which science moves forward.

experiments allow us to investigate the *causes* of the large-scale wind patterns over Earth by varying certain conditions. For example, by varying the speed of rotation of the annulus we can investigate the effect of Earth's rotation on global wind patterns (Section 12.3). Another example of a physical model is a cloud chamber, which—as its name suggests—allows us to simulate, and thus explain, processes at work in clouds. Cloud chambers have helped us to understand the processes that turn cloud drops into raindrops (Section 10.3).

Mathematical models allow us to investigate the effects of changing conditions by varying the numerical inputs to the model. These models can be as simple as one equation with one variable, or as complex as a computer program that makes use of vast sets of equations and takes hours to run. Many of the equations introduced in this book could be thought of as models. For example, in Chapter 6 we will derive a simple equation for calculating the temperature of a planet based on the output of the sun, the planet's distance from the sun, and the planet's reflectivity (Equation 6.3). Each of these variables can be changed to explore its effects on the planet's temperature. Working on a larger scale, computer models known as **general circulation models**, or GCMs, are attempts to simulate the workings of Earth's entire climate system. These models are currently being used to make predictions about how climate might change if human activities lead to a doubling of atmospheric carbon dioxide concentrations (Section 17.3).

Our investigations will certainly not always result in the formulation of laws and theories—in fact,

GENERAL CIRCULATION MODEL (GCM)
A computer program that represents the physics of the atmosphere through a set of equations.

most of the time they don't—but they do often lead to important advances in science or technology. They can also result in significant warnings to humankind, as the following story illustrates. F. Sherwood Rowland and Mario J. Molina's now-famous work on the depletion of the ozone layer was triggered by an *observation* made by another scientist, James Lovelock. In the early 1970s, Lovelock was making measurements of atmospheric composition with a new device he had invented, which could detect minute concentrations of gases. His measurements showed that a group of synthetic gases known as chlorofluorocarbons (CFCs) was accumulating in the atmosphere. Although he reported his findings, he was not concerned about them because, at the time, CFCs were regarded as harmless. Rowland and Molina, however, were curious about what might ultimately end up happening to these gases. By conducting *experiments*, they determined that once CFCs reached the layer of the atmosphere known as the stratosphere, they would likely be broken down by ultraviolet radiation and release chlorine.

Curious about what would happen to this chlorine, Rowland and Molina made use of a *mathematical model* that had been developed by previous researchers. The results showed that one chlorine atom is capable of destroying hundreds of thousands of ozone molecules and that a product of these reactions is chlorine monoxide. This work led Rowland and Molina to put forward the *hypothesis* that CFCs could destroy Earth's protective ozone layer. Although this hypothesis caused alarm and led some countries to ban the use of CFCs in aerosol sprays, there was no proof that the ozone layer was indeed being destroyed until two separate research groups discovered the ozone hole over Antarctica in 1985. Further *observations* showed that the ozone hole was not only an area of low ozone concentrations, it was also an area containing high amounts of chlorine monoxide. From this, researchers drew the conclusion that ozone was indeed being destroyed by the chlorine from CFCs. Shortly thereafter, action was taken to ban CFCs on a global level and, for their discovery, Rowland and Molina won the 1995 Nobel Prize in Chemistry.

Question for Thought

What are the limitations of models?

1.4 | The Role of Math— A Language and a Tool

As shown above, most scientific laws can be written as mathematical equations, and many scientific models are mathematical. In science, math provides a way of communicating the *form* of a relationship. For this reason, math has often been described as the *language* of science. The famous physicist Richard Feynman was referring to math when he eloquently wrote "If you want to learn about nature . . . it is necessary to learn the language that she speaks in. She offers her information only in one form" (1967, p. 52). In order to understand science, then, it is essential to understand math.

In addition to being the *language* of science, math is also an important *tool* of science. First, math provides us with a set of numbers that we can use to quantify our observations. For example, most of the elements of weather can be expressed quantitatively. We can say that the temperature is *11°C* or that *3 mm* of rain fell in the last *6 hours*. Numbers allow us to keep records and look for patterns. Second, as mentioned above, math provides us with a means to make predictions. For example, as we will see in Chapter 8, parcels of air cool at a regular rate as they rise in the atmosphere; thus, we can use math to predict the temperature an air parcel will have once it has risen a certain distance (Example 1.1).

Example 1.1

We know that an unsaturated air parcel cools at a rate of 10°C for every kilometre it rises in the atmosphere. If the air temperature at Earth's surface is 26°C, what will be the temperature of this air if it rises 2.3 km?

$$26°C - 2.3\ km\ (10°C/km) = 3°C$$

Using math, we predict that if the air rises 2.3 km, it will cool to 3°C.

Mathematical equations are introduced throughout this book; those that are particularly important are highlighted in shaded boxes. In most cases, a new equation is followed by an example to illustrate its use. For each of these examples, consider that the equation is being used as a language to help you understand a new concept, as a tool to make a prediction, or both.

1.5 | Dimensions and Units

In order to make measurements and perform calculations, we need a consistent, generally accepted set of units with which to work. In this book, we will use the system of units known as the MKS system, which essentially corresponds to the internationally adopted International System of Units (SI). This system is based on seven basic dimensions; we will use four of them: length, mass, time, and temperature (Table 1.1). Notice that the name "MKS" comes from the first letters of the base units corresponding to the first three dimensions: metres, kilograms, and seconds. The base units for temperature are degrees kelvin. A kelvin degree is the same size as a Celsius degree, but while 0 K corresponds to absolute zero, 0°C corresponds to the freezing point of water. Both units will be used in this book. All other dimensions are derived from the basic ones. An example of a derived dimension is **density**, ρ, which is defined as the amount of mass, m, in a unit volume, V ($\rho = m/V$). Thus, using the basic units, we derive the units of

DENSITY
The amount of mass in a unit volume.

TABLE 1.1 | Four of the basic dimensions of the MKS system, along with their units.

Dimension	Unit
Length	metre (m)
Mass	kilogram (kg)
Time	second (s)
Temperature	kelvin (K)

FORCE
An action capable of accelerating an object.

ENERGY
The capacity to do work.

WORK
The transfer of energy by mechanical means.

POWER
The rate at which energy is transferred, or work is done.

PRESSURE
The force per unit area.

ENERGY FLUX DENSITY
The rate of the flow of energy per unit area of surface.

density to be kilograms per cubic metre (kg/m^3). Four derived dimensions that are commonly used in atmospheric science are **force**, **energy** or **work**, **power**, and **pressure** (Table 1.2).

The units used for *force* can be explained using Sir Isaac Newton's first two laws of motion. His first law states that if an object is stationary, it will remain stationary unless an unbalanced force is applied to make it move, and if an object is moving at a constant speed and direction, it will continue to do so unless an unbalanced force is applied to change its speed, direction, or both. In the context of physics, a change in speed or direction is an *acceleration*. Therefore, we can use Newton's first law to define *force* as an action capable of accelerating an object. Newton's second law then tells us that the force, F, required to accelerate a given mass is simply the product of the mass, m, and the acceleration, a.

$$\text{Force} = F = m\,a = \text{kg} \cdot \text{m} \cdot \text{s}^{-2} = 1 \text{ newton} \quad (1.1)$$

TABLE 1.2 | Four derived dimensions of the MKS system.

Derived Dimension	Unit	Basic Unit
Force	newton (N)	kg·m·s^{-2}
Energy (Work)	joule (J)	kg·m^2·s^{-2}
Power	watt (W)	kg·m^2·s^{-3}
Pressure	pascal (Pa)	kg·m^{-1}·s^{-2}

Equation 1.1 shows that the basic units of force are kg·m·s^{-2}, or the units for mass (kg) multiplied by the units for acceleration (m·s^{-2}). The name given to these units is, appropriately, newtons (N).

The definition of *energy* as the capacity to do *work* shows that energy and work are related. Through this definition, energy and work both have the same units. To determine these units, we use the definition of work, which is that work, W, is done when a net force, F, moves an object through a distance, d.

$$\begin{aligned}\text{Work} = W = F\,d &= \text{kg} \cdot \text{m} \cdot \text{s}^{-2} \cdot \text{m} \\ &= \text{kg} \cdot \text{m}^2 \cdot \text{s}^{-2} = 1 \text{ joule}\end{aligned} \quad (1.2)$$

Equation 1.2 gives the basic units of energy and work as kg·m^2·s^{-2}. The name given to these units is joules (J). Just as the units for force recognize Newton's work on motion, the units for energy recognize the work of James Prescott Joule on energy.

Power is the rate at which energy flows, or work is done. Using this definition, power is energy, E, or work, W, divided by time, t.

$$\begin{aligned}\text{Power} = \frac{E}{t} \text{ or } \frac{W}{t} &= \text{kg} \cdot \text{m}^2 \cdot \text{s}^{-2} \cdot \text{s}^{-1} \\ &= \text{kg} \cdot \text{m}^2 \cdot \text{s}^{-3} = 1 \text{ watt}\end{aligned} \quad (1.3)$$

Equation 1.3 shows that the basic units of power are kg·m^2·s^{-3}. The name given to these units is watts (W) after James Watt, an engineer known for his work in improving the efficiency of steam engines. Watt also developed the units of *horsepower*: one horsepower is equivalent to 746 watts. Related to power is **energy flux density**, the rate of energy flow per unit area of surface. The units used for energy flux density are watts per square metre (W/m^2). In climatology, energy flux density is used to quantify energy flows through a surface.

Finally, *pressure*, P, is defined as the force, F, per unit area, A.

$$\begin{aligned}\text{Pressure} = P = \frac{F}{A} &= \text{kg} \cdot \text{m} \cdot \text{s}^{-2} \cdot \text{m}^{-2} \\ &= \text{kg} \cdot \text{m}^{-1} \cdot \text{s}^{-2} = 1 \text{ pascal}\end{aligned} \quad (1.4)$$

Equation 1.4 shows that the basic units of pressure are kg·m^{-1}·s^{-2}, or pascals (Pa), after Blaise Pascal, who clarified the meaning of pressure. **Atmospheric** pressure is the force exerted by the atmosphere on Earth's surface. This force results from the *weight* of the atmosphere. Using Newton's second law, weight, W, can be defined as the force on an object due to gravity.

$$W = m\,g \tag{1.5}$$

In Equation 1.5, *g* is 9.8 m/s², which is the acceleration due to gravity. Atmospheric pressure is regularly measured as part of routine weather observations. In Canada, it is usually reported in *kilo*pascals (kPa). In Canadian meteorological offices, however, both millibars (mb) and hectopascals (hPa) are used. There are 100 pascals in a hectopascal, and one hectopascal is equivalent to one millibar. The average atmospheric pressure at sea level is 101.325 kPa, or 1013.25 mb.

Example 1.2

Given that the average pressure at sea level is 101.325 kPa, and that Earth's radius is 6.37 × 10⁶ m, approximate the mass of the atmosphere.

First, convert the pressure to pascals: 101.325 kPa = 101,325 Pa. Then use Equation 1.5 in Equation 1.4, and solve for *m*.

$$P = \frac{m\,g}{A}, \quad m = \frac{P\,A}{g} = \frac{[(101{,}325 \text{ Pa})\, 4\pi\, (6.37 \times 10^6 \text{ m})^2]}{9.8 \text{ m/s}^2}$$
$$= 5.27 \times 10^{18} \text{ kg}$$

Question for Thought

If Earth was larger, but the mass of the atmosphere was the same, how would the atmospheric pressure at Earth's surface change?

1.6 | The Structure of the Atmosphere

The **standard atmosphere** is a set of values representing the average vertical distribution of pressure, temperature, and density in the atmosphere

(Table 1.3). These values are calculated, but they correspond well with averages of observations made by balloons and aircraft. Figure 1.7, created using the data in Table 1.3, provides a visual representation of the atmosphere's *average* structure. Notice that pressure and density decrease almost exponentially with height, while the change in temperature with height defines four distinct layers—two in which temperature

ATMOSPHERIC PRESSURE
The force exerted by the atmosphere on Earth's surface.

STANDARD ATMOSPHERE
A set of values that represents the average vertical distribution of pressure, temperature, and density in the atmosphere.

TABLE 1.3 | The standard atmosphere.

Height (km)	Temperature (°C)	Pressure (kPa)	Density (kg/m³)
0	15.0	101.325	1.23
1	8.5	89.874	1.11
2	2.0	79.501	1.01
3	−4.5	70.121	0.91
4	−11.0	61.660	0.82
5	−17.5	54.048	0.74
6	−23.9	47.217	0.66
7	−30.5	41.105	0.59
8	−36.9	35.651	0.53
9	−43.4	30.800	0.47
10	−49.9	26.499	0.41
11	−56.5	22.699	0.37
12	−56.5	19.400	0.31
14	−56.5	14.170	0.23
16	−56.5	10.352	0.17
18	−56.5	7.565	0.12
20	−56.5	5.529	0.09
25	−51.6	2.549	0.04
30	−46.6	1.197	0.02
35	−36.1	0.559	0.01
40	−22.1	0.288	0.004
45	−8.1	0.152	0.002
50	−2.5	0.078	0.001
60	−26.2	0.021	0.0003
70	−53.6	0.005	0.00008
80	−74.6	0.001	0.00002

decreases with height, and two in which temperature *increases* with height.

1.6.1 | Pressure and Density

Pressure decreases with height in the atmosphere because, as shown in Example 1.2, pressure results from the weight of the overlying atmosphere. Pressure is greatest at Earth's surface because the weight of the entire atmosphere lies above. Higher up, there is less overlying atmosphere, so the pressure is less. In a similar way, water pressure is greatest on the ocean floor because of the entire weight of the water lying above. In fact, divers and submarines must adjust for the increase in pressure as they go deeper, or they will be crushed.

Density decreases with height in the atmosphere because air is compressible (Section 3.3). Because air near the surface is compressed by the weight of the air above, density is greatest at the surface. It is because the atmosphere rapidly thins with height that high-altitude hikers must carry oxygen (Figure 1.8). An important result of the atmosphere's compressibility is that half the mass of the atmosphere is squeezed into its lowest 5.5 km, while the other half is spread over many tens of kilometres (Figure 1.7). Table 1.3 shows that the pressure at 5.5 km is about 50 kPa, which is about half of the pressure exerted by the *entire* mass of the atmosphere.

The fact that *density* decreases with height explains why the decrease in *pressure* with height is almost exponential, rather than linear. In the higher-density air near the surface, pressure will drop very quickly; higher up, where air molecules are farther apart, pressure will drop more slowly.

Question for Thought

Warm air rises because it is less dense. Notice that the standard atmosphere shows that surface air is warmer than air above. What keeps this surface air from continually rising?

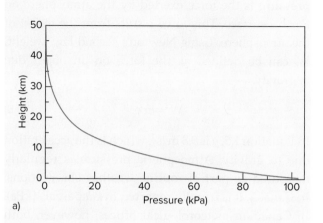
a)

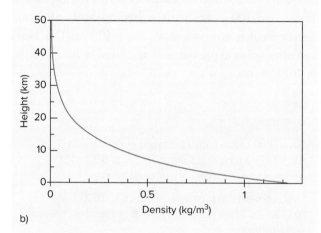
b)

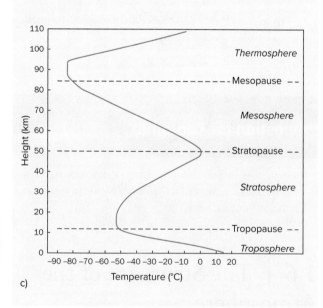
c)

FIGURE 1.7 | a) The change in pressure with height in the atmosphere. b) The change in density with height in the atmosphere. c) The change in temperature with height in the atmosphere.

1.6.2 | Temperature

The change in temperature with height shows alternating layers of decreasing and increasing temperature (Figure 1.7). The lowest layer, the **troposphere**, extends from Earth's surface to the **tropopause**, located at an average height of 11 km. As shown by the standard atmosphere, temperature decreases through this layer at an average rate of 6.5°C/km (Table 1.3). The troposphere is warmest on the bottom because it is heated by Earth's surface, in much the same way as a stove element warms the air above. Because tropical latitudes are warmer than polar latitudes, the effects of Earth's surface will extend higher into the atmosphere in tropical latitudes. Thus the tropopause can be as high as 16 km in the tropics, while it reaches only about 8 km above the poles. Additionally, because it is warmed from the bottom, the troposphere is a turbulent layer throughout which heat, water vapour, and particles originating at Earth's surface are thoroughly mixed. Both turbulence and water vapour are needed for weather and, as a result, the troposphere is the layer in which all our weather occurs and almost all clouds form.

The next layer of the atmosphere, the **stratosphere**, stretches from the tropopause to the **stratopause**, located at an altitude of about 50 km. This layer contains very little moisture because almost all of the water vapour from Earth's surface is removed by the weather processes operating in the troposphere. From the tropopause to an altitude of about 20 km, temperature remains constant with height. Above this *isothermal* layer, temperature begins to increase with height, a condition called an **inversion**. Because of this inversion, the stratosphere is not turbulent—with the warmest air at the top, there is little vertical movement and, therefore, little turbulence. If particles from Earth's surface make it to the stratosphere, they can remain there for several years. For example, volcanoes can have a longer lasting impact on climate if the gases and dust they eject reach the stratosphere (Figure 1.9). The stratosphere contains most of the atmosphere's ozone, in what we know as the *ozone layer*. Because the ozone layer absorbs ultraviolet radiation from the sun, it not only protects life on Earth from the harmful effects of this radiation, it also warms the stratosphere. Since the stratosphere is heated by the sun, it is warmest on top.

Above the stratosphere is the **mesosphere**, a layer in which temperature decreases with height from the relatively warm stratopause up to

FIGURE 1.8 | Climbers on the Khumbu Icefall, Mount Everest, heading for their base camp, which lies at an altitude of just over 5000 m. High-altitude hikers such as these carry oxygen to compensate for the thinning atmosphere on the ascent.

TROPOSPHERE
The layer of the atmosphere extending from Earth's surface to an average height of 11 km.

TROPOPAUSE
The top of the troposphere.

STRATOSPHERE
The layer of the atmosphere extending from, on average, 11 to 50 km above Earth's surface.

STRATOPAUSE
The top of the stratosphere.

INVERSION
An increase in temperature with altitude.

MESOSPHERE
The layer of the atmosphere that extends from about 50 km above Earth's surface to about 85 km above the surface.

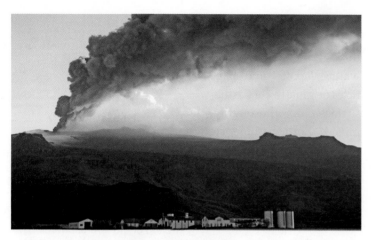

FIGURE 1.9 | When the Eyjafjallajokull Volcano in Iceland erupted in April 2010, it sent an explosion of ash high into the atmosphere.

MESOPAUSE
The top of the mesosphere.

THERMOSPHERE
The top layer of the atmosphere. Its base is located at an altitude of about 85 km; it has no well-defined top.

the **mesopause** at about 85 km. This height marks the bottom of the **thermosphere**, a layer in which temperature begins to increase again due to strong absorption of the shortest wavelengths of the sun's radiation. At the bottom of the thermosphere, most of this very short wavelength radiation has been absorbed, leaving little to warm the mesosphere. There is, of course, no clearly defined boundary between the top of the atmosphere and space. Due to the nature of gases, this transition is very gradual. The "top" of the atmosphere is often said to be 500 km, and at this height temperatures can reach upward of about 1000°C.

Despite the very high temperatures, the thermosphere would not *feel* warm, because the air is so thin.

High temperatures mean molecules are moving fast, but for heat to be transferred, molecules must collide. At Earth's surface, a person is hit by billions of molecules every second, and heat is rapidly transferred. In the thermosphere, such collisions are relatively rare, so there is little heat transfer. Thus, although the molecules at the upper limits of the atmosphere are moving at the speed of 1000°C molecules, there are not enough of them to transfer this heat. In fact, because of air's compressibility, 80 per cent of the atmosphere's mass lies in the troposphere, and 99.9 per cent of its mass lies below the stratopause. Of the remaining 0.1 per cent, 99 per cent is in the mesosphere, meaning that there is very little air left at the altitude of the thermosphere.

Until very recently, people were confined to Earth's surface and knew next to nothing about conditions throughout most of the atmosphere. Undoubtedly, some observers would have intuitively realized that pressure should decrease with height above Earth's surface, but the changes in temperature with height would have been far less easy to anticipate. By climbing mountains, people realized that temperature decreased with altitude, and many assumed it would continue to do so. Others argued, however, that since the upper atmosphere is closer to the sun, temperatures must start to increase again at some point. With our advanced technology, we have not only reached the top of the atmosphere but beyond to space, and we now know that both suppositions were correct.

As our frontiers have expanded, so has our knowledge, but the more observations we make, the more there is to explain.

1.7 | Chapter Summary

1. Processes at work in the atmosphere create weather and climate. Weather describes the state of the atmosphere at a given time and place. The elements of weather include temperature, pressure, humidity, winds, clouds, and precipitation. Climate describes the average conditions of the atmosphere on a variety of scales, from the micro scale to the planetary scale.

2. The atmosphere is part of a complex system that also includes the rocks, the water, and the life of the planet. This system is made even more complicated by

feedback effects, which are mechanisms that develop within systems and reinforce or lessen the effects of an initial change.

3. Science is about trying to understand and explain the observations we make about the world. Scientific investigation follows the scientific method. It has led to our discovery of laws that *describe* the way nature works, most of which can be expressed as mathematical equations. Scientific investigation has also led to our development of theories—large bodies of knowledge use to *explain* the way nature works.

4. Math acts as both a language and a tool in science. Measurements and calculations make use of the internationally adopted MKS system of units. Four derived dimensions are important in atmospheric science: force, energy or work, power, and pressure.

5. Pressure and density decrease with height in the atmosphere. Pressure decreases with height because the mass of the atmosphere lying above decreases with height. Density decreases with height because air is compressible. The decrease in density with height causes pressure to decrease exponentially with height.

6. The change in temperature with height defines four atmospheric layers: the troposphere, the stratosphere, the mesosphere, and the thermosphere. The troposphere is the layer closest to the surface. It contains more moisture and is more turbulent than the stratosphere, the layer directly above. In comparison, the stratosphere contains more ozone.

Review Questions

1. What are some reasons for studying the atmosphere?

2. How are weather and climate different from one another?

3. What characteristics of a place might influence its climate?

4. In what ways are scientific laws similar to scientific theories? In what ways do they differ?

5. What are the three types of models that researchers use? How do they use these models to advance science?

6. How can we account for the almost exponential decrease in both pressure and density with height in the atmosphere?

7. How are the troposphere and the stratosphere different from one another? What accounts for these differences?

Suggestions for Research

1. Investigate the sorts of studies that atmospheric researchers do in the field. To find this information, you could consult journal articles and abstracts, visit websites maintained by Environment Canada or other organizations involved in conducting atmospheric research, or even talk to some of the professors at your university who have participated in atmosphere-related research projects.

2. Find specific examples of conceptual models, physical models, and mathematical models that have been used in atmospheric research. For each model, describe how it represents reality, and explain how researchers have used it to make predictions and/or test hypotheses.

References

1. Feynman, R. (1967). *The character of physical law.* Cambridge, MA: MIT Press.

Volcanic eruptions—such as the eruption of Chile's Villarrica volcano, shown here—are one of the many natural process that release gases and aerosols into the atmosphere. In addition to geologic processes, life processes have also helped to shape the composition of the atmosphere throughout Earth's long history.

2

The Composition of the Atmosphere

Learning Goals

After studying this chapter, you should be able to

1. *distinguish* between constant and variable gases, and provide examples of each;

2. *list* the sources and sinks for the most common atmospheric gases;

3. *appreciate* the important roles of carbon dioxide, water vapour, and ozone in Earth's atmosphere;

4. *describe* the various types and sources of atmospheric aerosols;

5. *provide an account* of the evolution of Earth's atmosphere; and

6. *explain* why Earth's atmosphere is different from the atmospheres of Venus and Mars.

In Chapter 1, you learned about the *structure* of the atmosphere; in the present chapter, you will learn about the *composition* of the atmosphere. With the exception of Mercury, all of the planets in our solar system have atmospheres, but Earth's atmosphere is a unique mix that makes life possible: it directly supports most life processes, it protects life from harmful radiation from the sun, and its composition helps produce the ideal climate for life. In fact, as you will see in Chapter 5, reflection and absorption of radiation in the atmosphere helps keep Earth from becoming too warm or too cold.

You will also find in this chapter the story of how our atmosphere has evolved. As you will see, life has played an important role in creating the atmosphere that supports it. Ironically, human activities are now influencing the composition of the atmosphere in ways that could be detrimental to life. These changes are the topic of Chapter 17.

2.1 | The Chemical Composition of Air

Our atmosphere is a mixture of gases, commonly referred to as *air*. The major gases in Earth's atmosphere are listed in Table 2.1. Together, nitrogen (N_2) and oxygen (O_2) make up 99 per cent of the atmosphere, while argon (Ar) makes up most of the remaining 1 per cent. Next in abundance are water vapour (H_2O) and carbon dioxide (CO_2). The concentrations of all other atmospheric gases are very small; for this reason they are usually expressed in parts per million (ppm) rather than percentages (1 per cent equals 0.0001 ppm). Only one group of gases listed in Table 2.1, the chlorofluorocarbons (CFCs), does not occur naturally. These gases have been gradually increasing in the atmosphere since they were first synthesized in the 1920s. There are also hundreds of other gases, not listed, that occur in even smaller quantities; these can be detected only by very sensitive instruments.

CONSTANT GASES
Gases that have consistent concentrations across the atmosphere, up to a height of about 80 km.

VARIABLE GASES
Gases that have different concentrations in different areas of the atmosphere and at different times.

TABLE 2.1 | The composition of Earth's atmosphere.

Constant Gases	Per Cent of Dry Air by Volume
Nitrogen (N_2)	78.08
Oxygen (O_2)	20.95
Argon (Ar)	0.93
Neon (Ne)	0.00182
Helium (He)	0.00052
Hydrogen (H_2)	0.00006
Variable Gases	
Water Vapour (H_2O)	0.25
Carbon Dioxide (CO_2)	0.039
Methane (CH_4)	0.0002
Nitrous Oxide (N_2O)	0.00003
Carbon Monoxide (CO)	0.000009
Ozone (O_3)	0.000004
Chlorofluorocarbons (cfcs)	0.00000002

Question for Thought

Express the concentration of atmospheric carbon dioxide given in Table 2.1 in ppm. Why do you think that ppm measurements are more useful when working with small concentrations?

Table 2.1 is divided between **constant gases** and **variable gases**. Constant gases are well-mixed throughout the atmosphere up to a height of about 80 km. On time scales of hundreds of years, they exist in approximately the same concentrations. No matter when or where a sample of air is taken, the concentration of a constant gas shouldn't change. For example, in a sample of air containing 100 molecules, 78 should be nitrogen. Nitrogen, oxygen, and the inert gases are all constant gases. Variable gases are those that have concentrations that *vary* in time and space. The concentration of a variable gas will likely

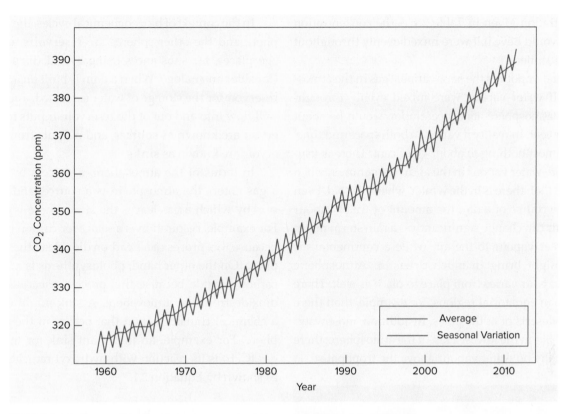

FIGURE 2.1 | Mean carbon dioxide measurements at Mauna Loa Observatory in Hawaii.

Source: Data from NOAA, Earth System Research Laboratory, www.esrl.noaa.gov/gmd/ccgg/trends, 5 May 2012.

be different depending on when, or where, a sample of air is taken.

Carbon dioxide is a variable gas. Because plants use carbon dioxide for photosynthesis, the concentration of this gas is lower in summer and higher in winter. In addition to this seasonal variation, the amount of carbon dioxide in the atmosphere has increased roughly 35 per cent since pre-industrial times, due to both fossil fuel combustion and deforestation. Over the last decade, the average *rate* of increase of carbon dioxide concentration has been close to 2 ppm per year (NOAA, 2012). Figure 2.1 shows the increase in carbon dioxide concentration since measurement began in 1958. Superimposed upon this general upward trend is the seasonal variation of carbon dioxide. Surprisingly, although this graph is representative of carbon dioxide concentrations for the entire planet, the seasonal variation is still evident. These variations are greatest in the northern hemisphere, because the southern hemisphere is mostly ocean.

Ozone (O_3) varies with height in the atmosphere (Figure 2.2). It is most abundant in the stratosphere, where it forms the *ozone layer*. The concentration of ozone in the stratosphere is about 10 ppm. The ozone

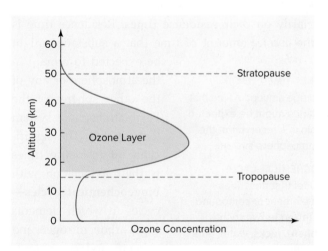

FIGURE 2.2 | The variation of ozone with height in the atmosphere.

Source: Adapted from R.P. Turco. (2002). *Earth under siege* (2nd ed.). New York, NY: Oxford University Press, p. 411.

concentration given in Table 2.1 is the concentration ozone would have if it were mixed evenly throughout the atmosphere.

Water vapour is the *most* variable gas in the atmosphere. If water vapour were mixed evenly throughout the atmosphere, its concentration would be about 0.25 per cent. In reality it varies, in both space and time, from almost nothing to about 4 per cent. There is usually more water vapour in the air in the summer, when it is hot, than there is in the winter, when it is cold. Even over the course of a day, the amount of water vapour in the air can change significantly—a rain storm might add water vapour to the air, while a continental air mass might bring in much drier air. Atmospheric water vapour varies from place to place as well. There is more in equatorial regions, for example, than there is over deserts or at the poles. In addition, most water vapour lies in the lowest 2 km of the atmosphere; there is virtually no water vapour above the tropopause.

Question for Thought

Why are polar air and desert air so dry? (Note that they are dry for different reasons.)

2.2 | Residence Time

Whether gases are constant or variable depends primarily on their **residence times**. Residence time is the *average* amount of time that a substance might be expected to remain in the atmosphere or any of the other spheres of the Earth-atmosphere system (Section 1.2). The significance of residence times is their association with **biogeochemical cycles**—cycles in which elements (e.g., carbon, nitrogen, and oxygen) and compounds (e.g., water) are continuously transferred between the atmosphere and the rocks, water, and life of our planet.

RESIDENCE TIME
The average amount of time that a substance might be expected to remain in a reservoir of the Earth-atmosphere system.

BIOGEOCHEMICAL CYCLE
The model that describes how an element or compound is transferred between the atmosphere, rocks, water, and life of Earth.

In the context of biogeochemical cycles, the atmosphere and the other spheres are **reservoirs**, or storage places, for substances being cycled (Figure 2.3). Consider an analogy. When a dam is built on a river, a reservoir for the storage of water is formed, and water will flow into and out of the reservoir. Inputs to a reservoir are known as **sources**, and outputs from a reservoir are known as **sinks**.

In terms of the atmosphere, a process by which a gas enters the atmosphere is a source, and a process by which a gas leaves the atmosphere is a sink. For example, respiration is a *source* for carbon dioxide because this process *adds* carbon dioxide to the atmosphere. On the other hand, photosynthesis is a *sink* for carbon dioxide because this process *removes* carbon dioxide from the atmosphere. A sink might also be a chemical transformation that occurs in the atmosphere. For example, an important sink for methane gas (CH_4) is its reaction with hydroxyl radicals (OH), as shown by Equation 2.1.

$$CH_4 + OH \rightarrow CH_3 + H_2O \qquad (2.1)$$

A condition known as **steady state** exists when the inflows to a reservoir are equal to the outflows from that reservoir. Over the short term, and under conditions free from such artificial interventions as industrial activity, most atmospheric gases are close to being in steady state. Under steady state conditions, we can calculate residence time for a particular gas, as shown in Equation 2.2 and Example 2.1.

$$\text{Residence Time} = \frac{\text{Reservoir Size}}{\text{Inflow/Outflow Rate}} \qquad (2.2)$$

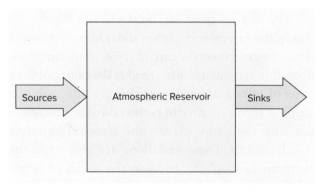

FIGURE 2.3 | The atmosphere as a reservoir, with sources and sinks.

In this equation, the reservoir size is the *amount* of the gas in the atmosphere. As the equation demonstrates, residence time will be longest for gases that exist in large quantities and/or pass through the system at low rates. The latter is generally the case for gases with low reactivity.

Example 2.1

The methane sink shown in Equation 2.1 occurs at a rate of about 400 megatonnes (Mt) per year. Considering that the total amount of methane in the atmosphere is about 4000 Mt, what is the residence time of methane?

Use Equation 2.2.

$$\text{Residence Time} = \frac{4000 \text{ Mt}}{400 \text{ Mt/yr}} = 10 \text{ yrs}$$

This is a relatively short residence time.

(This example is based on Turco, 2002, p. 376.)

Table 2.2 gives residence times for some atmospheric gases. As you can see, the residence times for the constant gases—nitrogen and oxygen—are much longer than those for the variable gases. Nitrogen has the longest residence time because it is abundant and not very reactive. Oxygen has a shorter residence time because there is less of it and it is considerably more reactive. As we will see in Chapter 17, residence time is an important concept in assessing the impact of anthropogenic changes to atmospheric composition.

TABLE 2.2 | Residence times for select atmospheric gases.

Atmospheric Gas	Residence Time
Nitrogen (N_2)	1,600,000 years
Oxygen (O_2)	3000–4000 years
Water Vapour (H_2O)	10 days
Carbon Dioxide (CO_2)	100 years[a]
Methane (CH_4)	12 years
Nitrous Oxide (N_2O)	114 years
Ozone (O_3)	days to weeks
Chlorofluorocarbons (CFCs)	100 years[b]

[a] Because of the complexity of the carbon cycle, this value is difficult to estimate. It is most often cited as roughly 100 years.
[b] This residence time refers to CFC-12 specifically.

2.3 | Nitrogen (N_2)

The *sink* for atmospheric nitrogen is a process known as **nitrogen fixation**. This process is essential to life because it converts nitrogen gas (N_2), which is inaccessible to most organisms, into a form of nitrogen that organisms can use to make protein. Most often, this process is carried out by certain soil bacteria that produce *fixed* forms of nitrogen, such as nitrate ions. Nitrogen can also be fixed in the atmosphere itself by lightning. The nitrate ions thus formed are rained out of the atmosphere and into the soil. Because they are soluble in water, these ions can be taken up by plants. Nitrogen fixation is a part of the nitrogen cycle whereby nitrogen is transferred between the atmosphere and the biosphere. Because nitrogen exists more or less in steady state, this sink must be balanced by a source.

The *source* for atmospheric nitrogen is a process called **denitrification**, in which certain other soil bacteria convert the fixed forms of nitrogen back into nitrogen gas or nitrous oxide gas (N_2O). The only sink for nitrous oxide is **photodissociation** in the stratosphere, which produces either nitrogen gas or nitric oxide gas (NO). Because it needs to reach the stratosphere before it will be broken apart, nitrous oxide has a relatively long residence time—about 114 years (Table 2.2). Since we began using nitrogen fertilizers in the early half of the twentieth century, we have been adding increasing amounts of *artificially* fixed nitrogen to our soils. With more fixed nitrogen, denitrification has increased, and more nitrous oxide is being returned to

RESERVOIR
A storage place.

SOURCE
A process by which a substance enters a reservoir.

SINK
A process by which a substance leaves a reservoir.

STEADY STATE
A condition that exists when the inflows to a reservoir are equal to the outflows from the reservoir.

NITROGEN FIXATION
The process by which nitrogen gas is removed from the atmosphere and converted to a soluble form of nitrogen that can be taken up by plants.

DENITRIFICATION
The process by which bacteria convert nitrogen in the soil to nitrogen gas or nitrous oxide gas.

PHOTODISSOCIATION
A process in which a molecule is split apart by the absorption of radiation.

the atmosphere. The resulting increase of nitrous oxide in our atmosphere is of concern for two reasons. First, nitrous oxide is a greenhouse gas (Section 17.3) and second, it can be transformed to nitric oxide, which destroys stratospheric ozone (Section 17.2).

2.4 | Oxygen (O_2)

The major *source* of oxygen for the atmosphere is photosynthesis. Photosynthesis is the process through which plants use energy from the sun to convert carbon dioxide and water into organic matter, or carbohydrate (CH_2O), and oxygen (Figure 2.4). This process is shown in Equation 2.3.

$$CO_2 + H_2O \leftrightarrow CH_2O + O_2 \qquad (2.3)$$

FIGURE 2.4 | **A deciduous forest canopy. Through the process of photosynthesis, plants remove carbon dioxide from the atmosphere and release oxygen.**

Through photosynthesis, the *inorganic* carbon in carbon dioxide is converted into the *organic* carbon of carbohydrate, and carbon is transferred from the atmosphere to the biosphere. The double arrow in the equation shows that this reaction can go either way.

In fact, three of the atmospheric *sinks* for oxygen are represented by the reverse of Equation 2.3. One of these sinks is respiration. During respiration, the carbohydrates in animals combine with oxygen to produce carbon dioxide and water. A second important sink for oxygen is decomposition. As organic material, or carbohydrate, is decomposed, the bacteria responsible consume oxygen and produce carbon dioxide. When oxygen is not available, as in a wetland, **anaerobic decomposition** occurs, and the bacteria produce methane *and* carbon dioxide. Equation 2.4 is a simplified chemical equation for this process.

$$2CH_2O \rightarrow CO_2 + CH_4 \qquad (2.4)$$

ANAEROBIC DECOMPOSITION
A process of decay that occurs when oxygen is unavailable.

OXIDATION
The addition of oxygen to a compound, which is accompanied by a loss of electrons.

GREENHOUSE GAS
A gas that allows the shorter wavelength radiation from the sun to pass through the atmosphere, while it absorbs the longer wavelength radiation leaving Earth's surface.

GREENHOUSE EFFECT
An increase in the temperature of a planet due to the presence of greenhouse gases in its atmosphere.

A third sink for oxygen is combustion. As organic matter is burned, oxygen is used and carbon dioxide is produced.

Oxidation of minerals is also a *sink* for oxygen, although it is a fairly minor sink compared to the other three. A familiar example of oxidation is rusting, which is the oxidation of iron. Minerals in Earth's crust that contain iron can be oxidized, producing either ferric iron (Fe^{3+}) or ferrous iron (Fe^{2+}). As can be seen from the above processes, oxygen is quite reactive; were it not for photosynthesis, the atmosphere might be quickly depleted of this gas.

2.5 | Carbon Dioxide (CO_2)

Like nitrogen and oxygen, carbon dioxide is essential to life. In addition, because it is a **greenhouse gas,** carbon dioxide influences climate. The presence of greenhouse gases in the atmosphere makes Earth warmer than it would be without them, thus the term **greenhouse effect** (Section 5.6). Without greenhouse gases in its atmosphere, Earth would have an average surface temperature of about −18°C instead of about 15°C; this gives Earth a greenhouse effect of 33°C. Without the greenhouse effect, Earth would be less suited to supporting life. Yet there is concern that the ongoing increase in atmospheric levels of carbon dioxide is strengthening the greenhouse effect and causing changes that may be detrimental to life (Section 17.3.2).

The **carbon cycle** is an extremely complex biogeochemical cycle. It is linked to the *organic* processes of photosynthesis, respiration, and decomposition, *and* to the *inorganic* processes of rock weathering and plate tectonics (Figure 2.5). A further complication is that some of these are short-term processes, while others are very long term. This complexity makes it difficult to pinpoint the residence time of carbon dioxide, although it is usually given as about 100 years (Table 2.2).

In the organic carbon cycle, the sources for carbon dioxide are sinks for oxygen, and the sources for oxygen are sinks for carbon dioxide (Equation 2.3). When carbon dioxide is removed from the atmosphere through photosynthesis, it can be quickly returned by decomposition or respiration. In these cases, carbon is stored for the short term as organic matter of the biosphere. However, under circumstances in which decomposition cannot occur, carbon will be stored for the long term—several hundred thousand years or more—in the rocks of the lithosphere.

Remember This

- Photosynthesis is a *source* for oxygen and a *sink* for carbon dioxide.
- Respiration is a *source* for carbon dioxide and a *sink* for oxygen.
- Decomposition is a *source* for carbon dioxide and a *sink* for oxygen.
- Combustion is a *source* for carbon dioxide and a *sink* for oxygen.

Such long-term storage of carbon can occur if organic matter is buried in such a way that it is not exposed to oxygen. Under these conditions, the organic matter will not decompose, and carbon dioxide will not return to the atmosphere—at least, not right away. The organic

CARBON CYCLE
The model that describes the processes by which carbon is transferred between the various reservoirs of the Earth-atmosphere system.

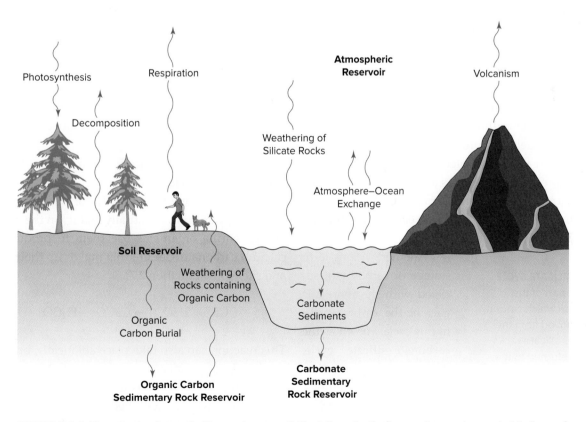

FIGURE 2.5 | The natural carbon cycle. How are human activities influencing the flows and reservoirs seen in this diagram?

matter might be buried in mud that eventually forms the sedimentary rock, shale. When there is a lot of organic matter present, and the conditions are right, what we know as fossil fuels—coal, oil, and natural gas—might form in the shale. This burial of organic carbon represents a long-term storage of carbon and the second-largest reservoir of carbon in the Earth-atmosphere system.

Changes in the rate of organic carbon burial can affect climate. For example, during the Carboniferous Period, about 300 million years ago, large coal beds were formed. (Most oil and gas deposits were formed more recently.) The swampy conditions of the time meant that the rate of burial of organic carbon was particularly high. As a result, atmospheric carbon dioxide levels dropped and climate cooled, leading to an ice age. Another consequence was high levels of oxygen in the atmosphere.

Question for Thought

Why would high rates of organic carbon burial lead to high oxygen levels?

Organic carbon stored in rocks as fossil fuels is eventually returned to the atmosphere through chemical weathering. This is a natural process that takes place over millions of years. When we mine and burn fossil fuels, the process is essentially the same, but it is greatly accelerated. In fact, burning fossil fuels at present rates returns as much carbon dioxide to the atmosphere in one year as would be returned by hundreds of thousands of years of weathering (Wallace & Hobbs, 2006, p. 45).

Another long-term storage of carbon is associated with the inorganic part of the carbon cycle. In this cycle, known as the **carbonate–silicate cycle**, the *source* for carbon dioxide is volcanic eruptions, and the *sink* for carbon dioxide is the chemical

CARBONATE–SILICATE CYCLE
The inorganic part of the carbon cycle, in which carbon dioxide is removed from the atmosphere as silicate rocks weather and returned to the atmosphere hundreds of thousands to millions of years later by volcanic eruptions.

weathering of rocks containing silica (Figure 2.6). It is important to make a clear distinction here. The chemical weathering of rocks containing organic carbon, which is part of the organic carbon cycle, is a *source* for carbon dioxide. In direct contrast, the chemical weathering of silicate rocks is a *sink* for carbon dioxide.

Remember This

- The weathering of rocks containing organic carbon is a *source* for carbon dioxide.
- The weathering of silicate rocks is a *sink* for carbon dioxide.

The first step in the weathering of many rocks occurs when carbon dioxide dissolves in rain water to form carbonic acid (H_2CO_3), as shown in Equation 2.5.

$$CO_2 + H_2O \leftrightarrow H_2CO_3 \qquad (2.5)$$

The next step is the dissociation of carbonic acid into hydrogen ions (H^+) and bicarbonate ions (HCO_3^-), as shown in Equation 2.6.

$$H_2CO_3 \leftrightarrow H^+ + HCO_3^- \qquad (2.6)$$

The resulting acidic rain water then reacts with rocks at Earth's surface, forming solid products as well as ions in solution.

The ions of importance in the context of the carbonate–silicate cycle are calcium ions (Ca^{2+}) from calcium silicate ($CaSiO_3$) in rocks. These ions, along with the bicarbonate ions, eventually end up in the oceans. There, marine organisms use them to produce the calcium carbonate ($CaCO_3$) that makes up their shells and skeletons (Equation 2.7).

$$Ca^{2+} + 2HCO_3^- \rightarrow CaCO_3 + H_2CO_3 \qquad (2.7)$$

This reaction can also take place abiotically, but most of the time organisms are involved. When these organisms die, their shells and skeletons sink to the ocean floor where they eventually form carbonate rock, the most common form of which is limestone.

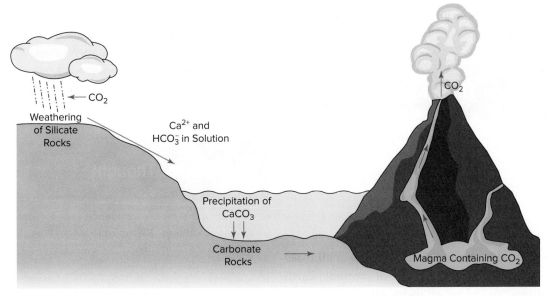

FIGURE 2.6 | The carbonate–silicate cycle. Carbon dioxide is removed from the atmosphere as it dissolves in precipitation and returned to the atmosphere as it erupts from volcanoes. Note that this cycle is part of the carbon cycle depicted in Figure 2.5.

This rock is the largest reservoir for carbon in the Earth-atmosphere system.

Remember This

Carbon is stored for the long term through the burial of organic carbon and the weathering of silicate rock.

As part of oceanic crust, carbonate rock is eventually subducted at convergent plate boundaries. Subduction exposes the rock to increasingly higher temperatures. This leads to metamorphism or, with further heating, to melting and the formation of magma. These processes release carbon dioxide that is eventually returned to the atmosphere through volcanic eruptions. If carbon dioxide were not returned,

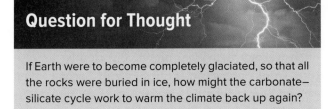

Question for Thought

If Earth were to become completely glaciated, so that all the rocks were buried in ice, how might the carbonate–silicate cycle work to warm the climate back up again?

Earth would slowly cool because atmospheric carbon dioxide would be gradually depleted by the weathering of rocks.

Interestingly, some scientists now believe that the carbonate-silicate cycle is part of a negative feedback mechanism that has helped maintain Earth's climate throughout its long history. This feedback operates because carbon dioxide levels influence climate, and climate, in turn, influences weathering. When atmospheric carbon dioxide levels increase, temperatures increase. Warm temperatures tend to increase the rate of chemical reactions and, therefore, the chemical weathering of rock. Suppose that a period of active volcanism increases the amount of carbon dioxide in the atmosphere. This will increase temperatures and, therefore, weathering rates. As weathering rates increase, the amount of carbon dioxide in the atmosphere will, in turn, gradually decline. This will cause temperatures to drop back down again, as shown in Equation 2.8.

$$CO_2 \uparrow \rightarrow \text{temperature} \uparrow \rightarrow \text{weathering} \uparrow$$
$$\rightarrow CO_2 \downarrow \rightarrow \text{temperature} \downarrow \qquad (2.8)$$

This mechanism is a *negative* feedback because the initial temperature *increase* leads to a chain of events that ends in a temperature *decrease* (Section 1.2). A *decrease* in carbon dioxide levels would, of course, trigger the same feedback effect, but in reverse.

Question for Thought

On the short time scale, how is atmospheric carbon dioxide removed and replenished? How is it removed and replenished on the longer time scale?

2.6 | Water Vapour

Water is the only substance that can exist in all three phases—solid, liquid, and gas—at normal Earth temperatures. As water changes phase, it absorbs and releases **latent heat** (Section 4.1.2), which is important in atmospheric heat transfer and as a source of energy for storms. The phase changes of water are responsible for most of what we know as weather, and weather is a part of the most familiar biogeochemical cycle of all: the **hydrologic cycle** (Figure 2.7).

Because it involves phase changes rather than chemical changes, the hydrologic cycle is much

LATENT HEAT
The energy associated with phase changes.

HYDROLOGIC CYCLE
The model that describes the processes by which water is transferred between the various reservoirs of the Earth-atmosphere system.

simpler than the cycles of nitrogen, oxygen, and carbon. The *source* of water vapour for the atmosphere is evaporation, and the *sink* is condensation, which leads to precipitation. Water vapour tends to condense fairly quickly, making the residence time for water vapour in the atmosphere only about 10 days (Table 2.2).

Question for Thought

How do you think the residence time for water in the oceans compares to the residence time for water vapour in the atmosphere?

Once water falls back to Earth as precipitation, it can run over the ground, into a river, and, ultimately, into the ocean. The only way back to the atmosphere is through evaporation. Water may be evaporated directly from the ocean, other water bodies, or the soil, or it may be pulled up by plant roots and returned to the atmosphere as vapour through the plant's leaves. Soil water that is not evaporated will continue its journey downward, becoming groundwater. Groundwater eventually returns to the surface as it seeps into rivers and lakes.

Question for Thought

How do you think the water cycle and the carbon cycle might be linked?

Again, through its phase changes, water is responsible for much of what we know as weather. In addition, water, in its various forms, strongly influences climate. To begin, water vapour is a powerful greenhouse gas that contributes about double the warming of carbon dioxide. Just like greenhouse gases, water *droplets* that make up clouds absorb longwave radiation leaving Earth. For this reason, cloudy nights are often warmer than clear nights. But clouds can also cool. Think of how cool it can suddenly become when a cloud passes over the sun on a warm summer afternoon. This cooling occurs because the water droplets

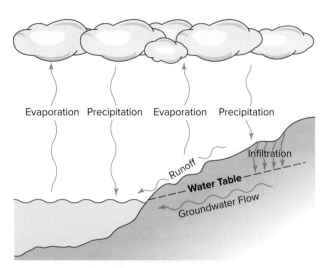

Evaporation   Precipitation   Evaporation   Precipitation

Runoff

Infiltration

Water Table

Groundwater Flow

FIGURE 2.7 | The hydrologic cycle.

in clouds *reflect* the sun's shortwave radiation. In Section 6.4, we will see that the effect of clouds on Earth's climate is not straightforward: some cloud types have a net warming effect, while others have a net cooling effect.

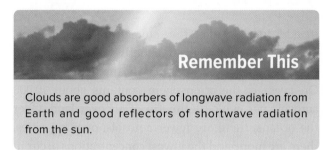

Remember This

Clouds are good absorbers of longwave radiation from Earth and good reflectors of shortwave radiation from the sun.

The reflectivity of the water on Earth's surface also has an impact on climate. As ice, water is very reflective of the sun's radiation (Section 5.4). Earth's polar ice caps, for example, reflect a significant amount of solar radiation back into space. In lakes or oceans, on the other hand, water has a very low reflectivity, making water bodies very good absorbers of solar radiation. If the amount and type of clouds, or the amount of ice, or the surface area of the oceans changes, so does Earth's reflectivity and, therefore, its climate.

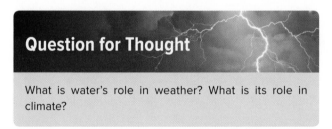

Question for Thought

What is water's role in weather? What is its role in climate?

2.7 | Ozone

Ozone has several roles in Earth's atmosphere. First, it absorbs **ultraviolet radiation** from the sun, thus protecting life on Earth from this harmful radiation. Second, it absorbs Earth's longwave radiation; thus, it is a greenhouse gas. Finally, because of its highly reactive nature, ozone is a pollutant. Even at very low concentrations, on the order of only parts per billion, it is damaging to certain materials, like rubber and plastic, and it is harmful to animals and plants.

The ozone layer of the stratosphere forms in two steps. First, oxygen molecules absorb ultraviolet radiation in the 0.1 to 0.2 µm wavelength range (1 µm (micron) = 10^{-6} metres), splitting the oxygen molecules into oxygen atoms by the photodissociation reaction shown in Equation 2.9.

$$O_2 + \text{ultraviolet radiation} \rightarrow O + O \qquad (2.9)$$

Second, the newly formed oxygen atoms collide with oxygen molecules, forming ozone (Equation 2.10).

$$O + O_2 + M \rightarrow O_3 + M \qquad (2.10)$$

M represents a molecule that is needed to carry away excess energy; this molecule can be any gas molecule in the atmosphere.

Once formed, ozone absorbs ultraviolet radiation in the wavelength range from 0.2 to 0.3 µm and, in the process, is split into an oxygen atom and an oxygen molecule (Equation 2.11).

$$O_3 + \text{ultraviolet radiation} \rightarrow O + O_2 \qquad (2.11)$$

While equations 2.9 and 2.10 together represent the *source* for ozone, Equation 2.11 represents the *sink*. Notice that the oxygen atom formed in Equation 2.11 is once again available to form ozone; thus, there is a continuing cycle of ozone production and destruction that maintains a steady level of ozone in the stratosphere. Because these reactions occur very quickly, ozone has a short residence time of days to weeks (Table 2.2).

As a result of the above sequence of events, ozone, together with oxygen, completely absorbs ultraviolet wavelengths shorter than about 0.3 µm. This process is highly effective, and very little of this radiation reaches Earth's surface. Some of the longer wavelengths of ultraviolet radiation—those between 0.3 and 0.4 µm—do reach Earth's surface, but they are less harmful.

The steps in ozone formation given by equations 2.9 and 2.10 suggest why the ozone layer is in the stratosphere, rather than higher or lower in the atmosphere. First, the process shown by Equation 2.10 requires collisions. *Above* the stratosphere, density is so low that few collisions occur, so there is negligible formation of ozone.

ULTRAVIOLET RADIATION
Radiation with wavelengths ranging from 0.1 to 0.4 µm.

HOMOSPHERE
The lower atmosphere, in which the constant gases are thoroughly mixed.

HETEROSPHERE
The upper atmosphere, in which the heaviest molecules are on the bottom and the lightest are on the top.

Question for Thought

In what ways does the atmosphere support life? What role does ozone play?

2.8 | Compositional Layers

In Section 1.6.2, you saw how atmospheric scientists traditionally divide the atmosphere into four layers, based on the way in which temperature changes with height. We can also divide the atmosphere into two layers, based on its *composition* (Figure 2.8).

There are two competing forces that influence the distribution of gases in the atmosphere. The first is *molecular diffusion*. Acting alone, molecular diffusion would cause the gases to settle out according to their molecular weights, and the atmosphere would be layered with the heaviest gas on the bottom and the lightest on the top. Such layers do not exist in the lower atmosphere because a second force—*mixing*—is also at work. When mixing is the dominant force, the molecular weight is no longer a factor, and the constant gases are evenly distributed.

The region of the atmosphere dominated by mixing reaches a height of about 80 km. This layer is known as the **homosphere** because its composition, at least in terms of the constant gases,

Second, the process shown by Equation 2.9 requires ultraviolet radiation. *Below* the stratosphere, there is very little ultraviolet radiation left to form oxygen atoms, as most of it has been absorbed by ozone in the stratosphere.

is homogeneous. Above 80 km, molecular diffusion dominates, and the molecules settle out with the heaviest on the bottom and the lightest on the top. This layer is known as the **heterosphere**.

2.9 | Atmospheric Aerosols

In addition to gases, the atmosphere also contains **aerosols**. Although we are usually unaware of them, these particles are very abundant in our atmosphere. On average, the atmosphere contains about 10,000 aerosols per cubic centimetre of air over land. Over the oceans, there are fewer aerosols; over big cities, there are more; and the lowest concentrations occur in polar regions. In addition, there are far fewer aerosols in the stratosphere than in the troposphere.

Aerosols are traditionally classified by size (Table 2.3), because their size influences their prevalence and their role in the atmosphere. The number of aerosols in the atmosphere decreases rapidly with their size; in other words, there are far more small aerosols than large ones. Particles with radii larger

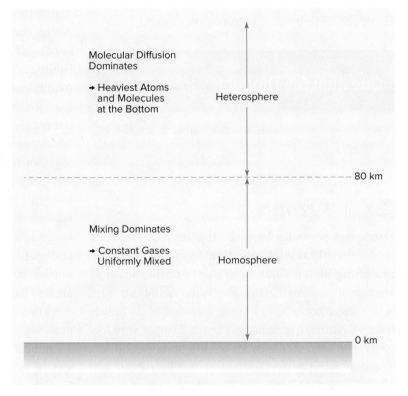

FIGURE 2.8 | Compositional layers of the atmosphere. Where do the tropopause, the stratopause, and the mesopause fit in?

TABLE 2.3 | Classification of atmospheric aerosols by size.

Classification	Radius (μm)
Aitken Aerosols	< 0.1
Large Aerosols	0.1 to 1.0
Giant Aerosols	> 1.0[a]

[a] Aerosols with a radius larger than 10 μm generally settle out of the atmosphere quickly.

than 10 μm tend to quickly settle out of the atmosphere. Aerosols are also classified as either **primary** or **secondary**, depending on whether they *are emitted* into the atmosphere or *form* in the atmosphere.

Aerosols have two particularly significant roles in the atmosphere. First, they serve as surfaces upon which atmospheric water vapour can condense to form water droplets. Without aerosols, water vapour would not condense in the atmosphere unless very high supersaturations were reached. In this role, the solubility of the aerosol is also important (Section 9.1.2). Second, aerosols both scatter and absorb radiation. Both scattering and absorption influence climate (Section 17.3), while the scattering of sunlight produces **haze** (Figure 2.9).

2.9.1 | Sources

Sources for aerosols are numerous and diverse. Some sources are natural, while others are anthropogenic. The largest *natural* source of primary aerosols is the oceans. Salt particles enter the atmosphere through the evaporation of sea spray (Figure 2.10). The largest source of primary aerosols from the land is dust blown up into the atmosphere from deserts (Figure 2.11). Such mineral dust is produced by the weathering of rock, and its dryness allows it to be easily picked up by the wind. There are

AEROSOLS
Tiny solid or liquid particles suspended in the atmosphere.

PRIMARY AEROSOLS
Aerosols that are emitted directly into the atmosphere.

SECONDARY AEROSOLS
Aerosols that form in the atmosphere.

HAZE
A reduction of visibility caused by the scattering of visible radiation in the atmosphere.

FIGURE 2.10 | These breaking waves at Long Beach, Pacific Rim National Park, on the west coast of Vancouver Island, release salt aerosols into the atmosphere through sea spray.

FIGURE 2.9 | This early morning haze in Hampton, Prince Edward Island, is the result of an accumulation of aerosols in the atmosphere.

FIGURE 2.11 | Sandstorms—such as this one in Eritrea, near the Sudanese border—release dust aerosols into the atmosphere.

also biological sources of natural primary aerosols. Biological aerosols include such things as seeds, pollen, spores, bacteria, and viruses. Finally, volcanoes are a significant, albeit sporadic, source of natural aerosols. Volcanic eruptions spew both gases and ash into the atmosphere. The gases contribute to the formation of secondary aerosols, as discussed below. The ash provides a source of mineral aerosols. Like salt, dust, and most of the other natural aerosols (aside from bacteria and viruses, which tend to be smaller), these aerosols are usually bigger than 1 μm in radius.

Anthropogenic sources of dust aerosols include agriculture, mining, transportation, and other industrial activities. Such dust appears in a wide range of sizes. Another major anthropogenic source of aerosols is combustion, which includes biomass burning—for example, in forest fires—and fossil fuel burning. On average, the particles released by combustion tend to be smaller than 0.1 μm in radius. Incomplete combustion produces *soot*, or *black carbon*, which seems to absorb, rather than scatter, sunlight. While dust and combustion products enter the atmosphere naturally, human activities have significantly increased both of these sources.

Secondary aerosols form in the atmosphere through chemical transformations known as *gas-to-particle conversions*. These aerosols can be of natural or anthropogenic origin, and they are typically smaller than most primary aerosols, ranging in size from about 0.01 to about 0.1 μm in radius.

Many secondary aerosols are produced from sulphur gases and from nitrogen oxide gases. Sulphur dioxide gas (SO_2) is released by fossil fuel burning, biomass burning, and volcanic eruptions. It is also produced by the oxidation of hydrogen sulphide (H_2S) and dimethyl sulphide (C_2H_6S), which are emitted by certain biological processes, such as the activity of phytoplankton in the oceans. In the atmosphere, sulphur dioxide gas is converted to sulphuric acid (H_2SO_4), which condenses to form droplets. The

> **VOLATILE ORGANIC COMPOUNDS (VOCs)**
> Carbon-containing compounds that easily vapourize.
>
> **HYDROCARBONS**
> Substances containing hydrogen and carbon, the most common being methane.

most common nitrogen oxides—nitric oxide (NO) and nitrogen dioxide (NO_2)—are most often produced naturally by lightning and artificially by fossil fuel burning. Through gas-to-particle conversion processes, they form droplets of nitric acid (HNO_3).

These sulphate and nitrate aerosols can contribute to acid rain because the acidic droplets provide surfaces for water vapour to condense on in the atmosphere. If these droplets become large enough, they will fall as rain. In addition, sulphate aerosols produced by sulphur dioxide gas from volcanoes can remain in the stratosphere for a few years after an eruption. These aerosols are good at scattering sunlight, so they can cause cooling. For example, after the 1991 eruption of Mount Pinatubo in the Philippines, Earth's average temperature dropped by about 0.5°C over the next two years, due to the accumulation of sulphuric acid droplets in the stratosphere.

Some carbon-based aerosols are produced by gas-to-particle conversions of **volatile organic compounds** (VOCs). Many VOCs are **hydrocarbons**; our fuels—coal, oil, natural gas, and wood—are all examples of VOCs. Complete combustion of these fuels should produce only carbon dioxide and water vapour, but combustion is rarely complete, and unburned, or partially burned, VOCs are also emitted. In addition, VOCs are emitted through the evaporation of fuels and solvents. The strong smell of gasoline at the gas pump comes from the evaporation of the gasoline. In nature, VOCs are given off by vegetation. Aerosols formed from gas-to-particle conversions of VOCs can consist of liquid or solid particles.

2.9.2 | Sinks

Like the atmospheric gases, atmospheric aerosols have sinks as well as sources. About 80 to 90 per cent of aerosols are removed from the atmosphere through precipitation. This process occurs in three steps. First, atmospheric water vapour condenses on the surface of certain aerosols—particularly large and giant aerosols—which thus form the centres of cloud droplets. Second, more water vapour moves toward these cloud droplets, pulling additional aerosols along with it. Often, tiny Aitken aerosols are incorporated into

In the Field

Assessing the Impact of Forest Fires on Air Quality

Ian McKendry, University of British Columbia

Air-pollution meteorologists are increasingly focusing on air pollution that is transported *into* urban areas, sometimes from sources that are very far away. These pollutants may include particles from desert dust storms, or they may be generated by other natural events such as volcanic eruptions or forest fires.

Wildfires are an essential, natural disturbance in a range of fire-dependent ecosystems, including many coniferous boreal or temperate forests, particularly in North America and Eurasia; tropical forests, such as those in Asia and Africa; eucalyptus forests in Australia; most vegetation types in Mediterranean climates; some oak-dominated forests; grasslands; savannas; and marshes.

Forest fires occur both naturally (e.g., as a result of lightning strikes) and as a result of human activities (e.g., the clearing of land or deliberate ignitions). In many parts of the world, the incidence and spread of wildfire is complicated by the development of fire-suppression activities. Ultimately, though, wildfires are closely linked to high temperatures and dry conditions in the presence of ample forest "fuel."

As with any other combustion source (e.g., factories or automobiles), wildfires generate emissions of aerosols, in the form of carbon-based particles, and gases, including carbon monoxide, carbon dioxide, and oxides of nitrogen and hydrocarbons. Some of these particles and gases in the atmosphere may be transformed during transport away from their source. For example, oxides of nitrogen and hydrocarbons in forest-fire smoke plumes may react photochemically to produce ozone (O_3), a pollutant known to have a significant impact on human respiratory health. Health studies during the 2003 Kelowna wildfires in British Columba showed that particles in the atmosphere lead to severely degraded air quality and an immediate and significant increase in hospital admissions and physician visits for respiratory-related illnesses.

Other recent research in the Pacific Northwest region of North America—where forest fires occur frequently in the summer months—has shown significant impacts of forest fires on air quality and visibility. For example, in the region surrounding Vancouver, forest-fire smoke is experienced over the region on up to 30 per cent of summer days. On these days, visibility is reduced considerably, daily ground-level concentrations of particles less than 2.5 µm in diameter double, and maximum daily ozone concentrations increase by about 25 per cent compared to similar days on which fire smoke is not present in the lower atmosphere. These increases likely contribute to the violation of air-quality standards and have an impact on human health.

Much of our current concern with the impacts of forest fires on air quality is closely tied to the larger problem of impending global climate change. As global temperatures increase and the incidence of extreme drought increases over many continental areas, it is likely that the incidence of forest fires will also increase. Recent work with the Canadian Global Climate Model suggests that over the next century, the length of the fire season, as well as the potential for damaging wildfires to occur, will increase significantly. Globally, such models suggest that the potential for damaging wildfires will increase from low to moderate in North America, central Asia, and southern Europe, and from moderate to high in South America, southern Africa, and Australia. Finally, regional-scale computer models (similar to those used to forecast daily weather) are now being developed and used in areas such as British Columbia to forecast the impacts of fires on air quality on a day-to-day basis. These forecasts provide the public with timely warnings to limit their outdoor activities and thereby reduce the impacts of forest-fire smoke on human health.

IAN MCKENDRY is a professor in the Department of Geography at the University of British Columbia. He is a member of the Atmospheric Science Programme and the UBC-CAR (Centre for Aerosol Research). His current research focuses on using LIDAR to study intercontinental and regional pollution transport.

clouds in this way. Third, once the cloud droplets grow into rain drops, they fall, picking up still more aerosols. Because of this removal process, the air is often much clearer after a rain. Of the remaining 10 to 20 per cent of aerosols, most tend to simply settle out of the atmosphere under the influence of gravity. Yet this can occur only with giant aerosols, as smaller aerosols are too easily kept aloft by updrafts.

These sinks are quite efficient in the troposphere, and residence times are usually less than a week. As mentioned above, typical residence times in the stratosphere can be much higher, on the order of a few years. This is because there is no rain, and very little mixing, in the stratosphere.

2.10 | The Formation and Evolution of the Atmosphere

The composition of Earth's atmosphere has not always been the same as it is today; rather, it is the result of a long evolution. Planet Earth, along with the entire solar system, was created about 4.6 billion years ago from a *nebula*, a rotating cloud of gas and dust formed by an exploding star (Figure 2.12). We are currently observing the formation of a solar system in another part of our galaxy. Such observations provide us with valuable evidence about the creation of our own solar system.

Our solar system began to form when much of the mass of the nebula was pulled into its centre by the force of gravity. In the process, the gas and dust were heated by compression and the release of gravitational energy. The resulting high temperatures enabled hydrogen atoms to fuse to form helium atoms. Such fusion reactions release great amounts of energy. From this our sun was born. Based on models from the field of astronomy, however, the sun would have been producing about 30 per cent less energy when it first formed than it is today. Over a period of about 100 million years, the dust left behind after the formation of the sun gradually collided and accreted to form planetesimals, which eventually became planets (Figure 2.13). Over the next several million years, Earth was bombarded by meteors that continued to add to

its mass. This period in Earth's history is referred to as the *heavy bombardment period*. Evidence from craters on the moon indicates that this period ended about 3.8 billion years ago.

Because of the immense amount of time involved, it is difficult to know anything with certainty about the origin and early evolution of Earth's atmosphere. The evidence that we have comes mostly from ancient rocks, meteorites, and the present-day atmospheres of

FIGURE 2.12 | This optical image of the Orion Nebula shows thousands of young stars, the largest of which illuminate the centre of the nebula.

Venus and Mars, our closest neighbours in the solar system. Scientists believe that these planets likely formed in the same way as did Earth, thus their atmospheres may have initially been the same as Earth's, although they are now very different.

There is still controversy over whether or not Venus, Earth, and Mars had atmospheres when they first formed. If they did, the gases in these primitive atmospheres would likely be the same as the gases in the nebula from which the solar system was created. Studies of the gases in the sun tell us about the inert gases that were in the original nebula. Because inert gases do not easily react, any inert gases in a possible original atmosphere would still be in our atmosphere today. They would not have dissolved in the oceans or reacted with the rocks. Since the amounts of these

gases in Earth's present atmosphere are much less than they are in the sun, we can conclude that either Earth did not have an atmosphere when it first formed or, if it did, this atmosphere was quickly lost.

The primitive atmosphere from which today's atmosphere evolved had two probable sources. The first involves **outgassing** from the Earth's interior. The planetesimals from which Earth formed likely contained *volatiles* that became trapped inside our growing planet. As Earth was forming, it could have heated up to its melting point due to the impact of planetesimals, the accretion process, and radioactivity. At this point, the volatiles would have been released as gases to form an atmosphere. These gases could also have been released into the atmosphere through volcanic eruptions; Earth is thought to have been more volcanically active when it was young than it is now. The second probable source of gases for Earth's atmosphere was the comets and asteroids that were impacting Earth during the heavy bombardment period. For example, comets contain ice, so comets could have added water vapour to Earth's developing atmosphere.

OUTGASSING
The release of gases dissolved in rock.

If we assume we are right about these sources' role in forming Earth's early atmosphere, we can approximate the composition of this atmosphere based on information we gather about Earth's interior, volcanic activity, and the meteors that collide with Earth today. For example, we can assume that the gases in the early atmosphere would be of the same types, and in the same proportions, as the gases erupting from volcanoes today. Based on this sort of evidence, we can conclude that it is likely that our primitive atmosphere contained mostly water vapour and carbon dioxide, with small amounts of nitrogen, sulphur, and other gases.

Not only was this early atmosphere quite different from today's in composition, it seems that it was also much denser. Evidence for a denser atmosphere comes from the observation that the rocks of today's Earth contain large amounts of carbon, which would have come from the carbon dioxide in the early atmosphere. In addition to this loss of carbon

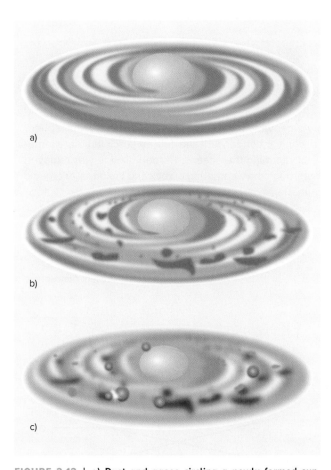

FIGURE 2.13 | a) Dust and gases circling a newly formed sun. b) Particles of dust gradually collide and accrete to form planetesimals. c) Planetesimals collide with one another and continue to accrete smaller particles, eventually forming planets.

dioxide, the evolution of our atmosphere involved the loss of water vapour and the introduction of oxygen (Figure 2.14). The *absolute* amount of nitrogen probably hasn't changed; instead, the *relative* amount of nitrogen would have increased as the atmosphere thinned.

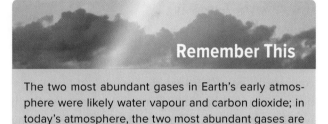

Remember This

The two most abundant gases in Earth's early atmosphere were likely water vapour and carbon dioxide; in today's atmosphere, the two most abundant gases are nitrogen and oxygen.

It is likely that the atmosphere lost most of its water vapour rather early on. As the young Earth cooled, the water vapour would have condensed and rained down enough water to form the oceans. Indeed, ripple marks preserved in sedimentary rocks indicate the presence of liquid water as far back as 3.8 billion years ago. Water that began in the atmosphere is now stored mostly in the oceans, making the oceans the largest reservoirs of water on the planet.

Carbon dioxide was probably lost more gradually than was water vapour. There are two processes through which this might have happened. The first is the chemical weathering of rocks. As water rained down, it would have dissolved atmospheric carbon dioxide, producing carbonic acid (equations 2.5 and 2.6). This acidic water would have then weathered the rocks of the crust and carried the resulting ions to the oceans, where they were deposited as carbonate sedimentary rock (Equation 2.7). The second process is the burial of organic matter. The presence of liquid water would have made life possible, and we have evidence that life, in the form of bacteria, began around the time that the heavy bombardment period ended. These early bacteria may have developed a form of photosynthesis that used atmospheric carbon dioxide to produce organic matter. If the organic matter did not decompose, it would accumulate, storing carbon in the process. As a result of both these sinks, Earth's carbon was slowly transferred from the atmosphere to the rocks. Recall that the rocks are now the largest reservoirs of carbon on the planet.

Remember This

Water vapour was removed from the early atmosphere when it condensed and fell as precipitation, eventually forming the oceans. Carbon dioxide was removed through the weathering of rock and the burial of organic matter, eventually accumulating in rocks.

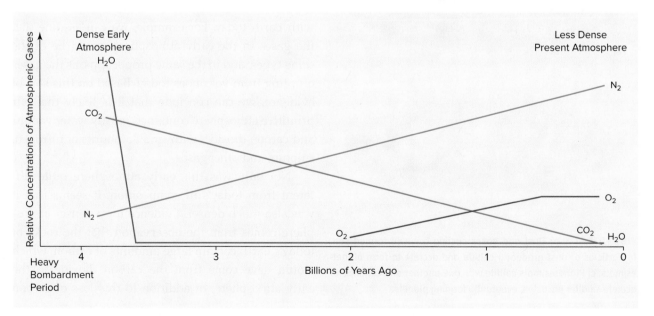

FIGURE 2.14 | The change in the *relative* proportion of major gases of the atmosphere over time.

The evolution of the atmosphere also involved the introduction and accumulation of oxygen. Photodissociation can produce oxygen from water and sunlight, but this process alone cannot account for the abundance of oxygen in today's atmosphere. Instead, the oxygen comes mostly from photosynthesis. Although the earliest photosynthesis produced sulphur rather than oxygen, it was not long before cyanobacteria evolved, and oxygen production began (Equation 2.3). Oxygen was probably first produced in this way about 3.5 billion years ago, in the oceans.

Prior to the introduction of oxygen, iron released from the weathering of rocks accumulated in the oceans as *reduced* iron, which is soluble in water. When cyanobacteria living in the oceans began to produce oxygen, the oxygen reacted with the dissolved iron, forming layered rocks known as *banded iron formations*. Once this oxygen had reacted with all the reduced iron in the oceans, it could accumulate and diffuse into the atmosphere. When oxygen entered the atmosphere, it reacted with iron released by the weathering of rocks making the iron insoluble so that it accumulated in soils, forming iron deposits known as *redbeds*. Modern researchers have found that there are no banded iron formations *younger* than about 2 billion years old (most are between 3.5 billion and 2 billion years old), and there are no redbeds *older* than about 2 billion years old. (Figure 2.15). This transition is a good indication that oxygen began accumulating in the atmosphere about 2 billion years ago.

Since about 600 million years ago, oxygen levels have fluctuated above and below 21 per cent. This fluctuation is believed to be associated with the variation in rates of organic carbon burial. When these rates are high, oxygen levels increase because buried organic matter does not decompose. On the other hand, when the rates of organic carbon burial are low, oxygen levels decrease (Section 2.5). For example, when the rate of organic carbon burial was very high about 300 million years ago, atmospheric oxygen concentrations were likely about 35 per cent, which is approximately 50 per cent higher than they are today (Kump, Kasting, & Crane, 2004, p. 225).

There is speculation that it was the accumulation of oxygen in our atmosphere that set off the dramatic rise in the diversity of life. So, just as life had produced oxygen, oxygen had, in turn, made further life forms possible. Not only did the presence of oxygen allow for the process of respiration, it also influenced life in another very important way: it led to the formation of ozone. Mathematical models suggest that by about 1.9 billion years ago, Earth had a substantial ozone layer. With an ozone layer, the possibilities for life further increased because life no longer had to protect itself from ultraviolet radiation.

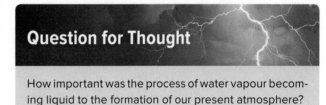

Question for Thought

How important was the process of water vapour becoming liquid to the formation of our present atmosphere?

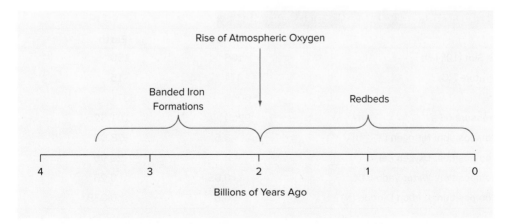

FIGURE 2.15 | The transition from the occurrence of banded iron formations to the occurrence of redbeds 2 billion years ago marks the rise of atmospheric oxygen.

2.11 | Earth as a Habitable Planet

While Earth is suitable for life, its nearest neighbours, Venus and Mars, are not. Although the primitive atmospheres of all three planets were likely the same, Earth's atmosphere is now very different (Section 2.10). Whereas Earth's atmosphere is high in nitrogen and oxygen, Venus' and Mars' atmospheres are high in carbon dioxide, and neither of these planets' atmospheres contains a significant amount of oxygen (Table 2.4). In addition, Venus' atmosphere is almost 100 times denser than Earth's, while Mars' atmosphere is about 1000 times *less* dense than Earth's.

Question for Thought

Considering that there is no oxygen in Venus' atmosphere and very little in Mars', how do you think the temperature structures of these atmospheres might differ from Earth's?

The explanation for the differences in the three planets' atmospheres begins with the difference in their relative distances from the sun. Earth's distance from the sun provides Earth with a temperature that allows water to become liquid. Venus, being closer to the sun, is too hot, and Mars, being farther from the sun, is too cold. As a result, Earth has sometimes been named the "Goldilocks Planet," as it is just the right temperature for liquid water. As we saw in Section 2.10, water vapour became liquid water early in the evolution of Earth's atmosphere; this made life possible and allowed for the removal of much of the carbon dioxide from our atmosphere.

In contrast, as Venus' atmosphere evolved, water vapour did not condense but rather remained in the atmosphere, where photodissociation gradually broke apart the water molecules into hydrogen and oxygen. The hydrogen atoms, being very small in mass, escaped to space. The oxygen probably reacted with the rocks of the Venusian crust. As a result, the present atmosphere of Venus contains practically no water vapour.

Without water, carbon dioxide is not removed from the atmosphere; thus, through volcanic eruptions powered by plate tectonics, carbon dioxide has accumulated in Venus's atmosphere. Venus and Earth have roughly the same amount of carbon: on Venus, it is in the atmosphere, in the form of carbon dioxide; on Earth, it is in the rocks. With so much carbon dioxide in its atmosphere, Venus has a large greenhouse effect, making Venus much hotter than can be accounted for by its distance from the Sun alone.

Like Venus, Mars has lost water vapour from its atmosphere through photodissociation; but, unlike Venus, Mars has retained some water, which is frozen into ice caps. In fact, Mars is so cold that some carbon dioxide freezes out of the atmosphere, causing seasonal fluctuations of the ice caps. It is possible that Mars has not always been as cold as it is today. The evidence for this is that there appear to be channels on Mars that look as though they were carved by

TABLE 2.4 | Characteristics of Venus, Earth, and Mars

	Venus	Earth	Mars
Distance to the Sun (10^6 km)	108	150	228
Surface Temperature (°C)	474	15	−53
Radius (km)	6049	6371	3390
Atmospheric Pressure (kPa)	9200	101.325	0.7
Atmospheric Composition: Nitrogen (%)	3.5	78.08	2.7
Atmospheric Composition: Oxygen (%)	0	20.95	0.15
Atmospheric Composition: Water Vapour (%)	0.05	0.25	0.10
Atmospheric Composition: Carbon Dioxide (%)	96.5	0.039	95.3

Sources: Adapted from C.N. Hewitt & A.V. Jackson, eds. (2009). *Atmospheric science for environmental scientists*. West Sussex, England: Wiley-Blackwell, p. 27, and W.H. Schlesinger. (1997). *Biogeochemistry: An analysis of global change* (2nd ed.). San Diego, CA: Elsevier Academic Press, p. 41.

running water. Mars may have had a stronger greenhouse effect before its atmosphere thinned.

The thinning of Mars' atmosphere likely came about because Mars is a small planet, about half the size of Earth. A smaller planet has a smaller gravitational field. This makes it easier for gases to escape from Mars' atmosphere and may be the reason that Mars has lost most of its nitrogen. Atmospheric nitrogen can undergo photodissociation, producing nitrogen atoms, and Mars' gravitational field is not strong enough to prevent these atoms from escaping to space.

With the loss of both water vapour and nitrogen from Mars' atmosphere, only carbon dioxide is left, but there is very little even of this gas. Mars does have a greenhouse effect, but it is very small. Some carbon dioxide may have been lost to space, and some may have formed carbonate rocks when, and if, Mars had liquid water. Another consequence of being a small planet is that Mars' interior would have cooled more quickly than Earth's or Venus', so that it could no longer support active plate tectonics. Without plate tectonics, carbon dioxide cannot be added to Mars' atmosphere. Whereas on Venus there is virtually no *sink* for carbon dioxide, on Mars there is no *source*. Earth has both.

On Earth, it has been liquid water and the life it supports that have made our atmosphere what it is.

Without both liquid water and life, the carbon dioxide in our atmosphere would not have been drawn down as the sun's output increased (Section 2.10), and Earth might have eventually become too hot to support life. In addition, the same photosynthetic bacteria that consumed carbon dioxide also produced the oxygen that allowed for the evolution of other life forms and the development of a protective ozone layer. It seems that life and the atmosphere have co-evolved.

It was partly this realization that led James Lovelock to propose the Gaia hypothesis in the 1970s. Lovelock based this hypothesis on the observation that Earth's atmosphere would be different if there was no life on the planet. For example, methane and oxygen could not exist together in our atmosphere without a continuous supply of each, and both are supplied by life. The Gaia hypothesis proposes that the various parts of the Earth-atmosphere system—the lithosphere, the atmosphere, the biosphere, and the hydrosphere—operate together as a self-regulating system. In other words, life did not evolve as an adaptation to an already prepared environment, as had been previously supposed. Instead, life evolved *with* its environment and, being self-regulating, the Earth-atmosphere system has been able to maintain a temperature that is ideal for life.

2.12 | Chapter Summary

1. Atmospheric gases are either constant or variable. Those that are constant are found in the same concentrations throughout the atmosphere up to a height of about 80 km; this lower compositional layer is known as the homosphere. Above 80 km, in the layer known as the heterosphere, the gases settle out based on their molecular weights. The variable gases are those that occur in different amounts in different times and places.

2. Elements and compounds are transferred through the four spheres of the Earth-atmosphere system in biogeochemical cycles. The average amount of time that a gas spends in the atmosphere, or other reservoir, is known as its residence time. Constant gases have long residence times, while variable gases have shorter residence times.

3. Nitrogen entered the early atmosphere through outgassing. Today, the source of atmospheric nitrogen is denitrification, while the sink is nitrogen fixation.

4. The major source of oxygen for the atmosphere is photosynthesis. The sinks for oxygen are respiration, decomposition, combustion, and oxidation.

5. In the short-term carbon cycle, the sources are respiration, decomposition, and combustion, while the sink is photosynthesis. In the long-term carbon cycle, the sources are volcanism and weathering of organic carbon, while the sink is silicate weathering. Carbon dioxide is a greenhouse gas. The carbonate–silicate cycle controls the amount of carbon dioxide in the atmosphere. Therefore, it is thought that this cycle may have been an important regulator of the climate throughout most of Earth's history.

6. Water vapour is the most variable of the atmospheric gases; it is also a greenhouse gas. Water exists in all three phases at normal Earth temperatures, and it absorbs and releases latent heat as it changes phase.

7. The residence time for ozone in the atmosphere is very short because it is continually created and destroyed by photochemical reactions in the stratosphere. In the process, most of the sun's ultraviolet radiation is absorbed, preventing it from reaching Earth's surface. Ozone is also a greenhouse gas and a pollutant.

8. Atmospheric aerosols are solid or liquid particles that can remain suspended in the atmosphere. The sources of aerosols are numerous and diverse; they come from the land and the ocean, they result from life processes, and they are produced by human activities. Primary aerosols are emitted directly into the atmosphere, while secondary aerosols form in the atmosphere.

9. Earth's primitive atmosphere was high in water vapour and carbon dioxide, and low in nitrogen. There was no oxygen. As water vapour condensed, it left the atmosphere in the form of precipitation and formed the oceans. Some of the carbon dioxide dissolved in rain water and was removed by rock weathering. Much of the rest of the carbon dioxide was removed by photosynthesis and subsequent burial of organic carbon. As a result of both processes, most of Earth's carbon is now stored in rocks. As the originally dense atmosphere thinned, nitrogen became the most abundant gas. Even though oxygen was being produced by photosynthesis as far back as 3.5 billion years ago, this gas did not begin to accumulate in our atmosphere until about 2 billion years ago. Some of the oxygen in the atmosphere reacted to form the ozone layer.

10. Comparison of Earth's atmosphere to those of Venus and Mars suggests why Earth is habitable while the other two planets are not. Earth's distance from the sun, combined with its size, probably caused its atmosphere to evolve quite differently from those of the other two planets. Once life established itself on Earth, it seems to have played an important role in further shaping the atmosphere.

Review Questions

1. What is the difference between constant gases and variable gases? Give an example of each. How is residence time an important determinant of whether or not a gas will be constant or variable?

2. What sinks for oxygen are sources for carbon dioxide? What sinks for carbon dioxide are sources for oxygen?

3. How can photosynthesis lead to a) short-term carbon storage and b) long-term carbon storage?

4. How does the carbonate–silicate cycle influence climate? How does climate influence the carbonate–silicate cycle?

5. How does the ozone layer form?

6. What is the importance of each of the following variable gases in our atmosphere: water vapour, carbon dioxide, and ozone?

7. What are the differences between primary aerosols and secondary aerosols? Give examples of each.

8. What two important roles do aerosols play in the atmosphere?

9. How is the composition of the troposphere similar to and different from that of the stratosphere?

10. What are the two most abundant gases in today's atmosphere? What were the two most abundant gases in Earth's primitive atmosphere? How do you account for this change?

11. What evidence do we have that oxygen did not begin to accumulate in the atmosphere until about 2 billion years ago?

12. What is the largest reservoir in the Earth-atmosphere system for a) carbon, b) water, and c) nitrogen?

13. What is the significance of each of the following to the evolution of Earth's atmosphere: a) distance from the sun, b) life, c) plate tectonics, d) Earth's size, and e) the solubility of carbon dioxide in water?

Suggestions for Research

1. Explore how either the carbon cycle or the nitrogen cycle is being altered by human activities. Note the short- and long-term consequences of these changes, and find out what, if anything, is being done to slow or reverse the changes.

2. Research different theories—both past and current—that account for the evolution of Earth's atmosphere. Evaluate the evidence behind each theory, and comment on why such theories are continually being revised.

References

Hewitt, C.N., & A.V. Jackson, eds. (2009). *Atmospheric science for environmental scientists*. West Sussex, England: Wiley-Blackwell.

Kump, L.R., J.F. Kasting, & R.G. Crane. (2004). *The earth system* (2nd ed.). Upper Saddle River, NJ: Pearson Prentice Hall.

NOAA. (April 2012). Trends in atmospheric carbon dioxide. Retrieved from www.esrl.noaa.gov/gmd/ccgg/trends

Schlesinger, W.H. (1997). *Biogeochemistry: An analysis of global change* (2nd ed.). San Diego, CA: Academic Press.

Turco, R.P. (2002). *Earth under siege* (2nd ed.). New York, NY: Oxford University Press.

Wallace, J.M., & P.V. Hobbs. (2006). *Atmospheric science: An introductory survey* (2nd ed.). Boston, MA: Elsevier Academic Press.

The behaviour of the atmosphere varies as temperature, pressure, and density change. Understanding the relationships between these variables can help us explain many atmospheric processes, such as the formation of clouds around these mountain peaks in Alberta.

3

The Behaviour of the Atmosphere

Learning Goals

After studying this chapter, you should be able to

1. *apply* the ideal gas law to explain how changes in temperature and pressure influence density;

2. *discuss* the implications of the hydrostatic equation to the change in pressure with height in the atmosphere;

3. *apply* the various forms of the hypsometric equation;

4. *explain* how temperature affects the distribution of pressure in both the horizontal and vertical directions; and

5. *recognize* the features shown on weather maps.

Chapter 2 was about the *chemical* state of the mixture of gases that makes up our atmosphere, while this chapter is about the *physical* state of this mixture of gases. The physical state of the atmosphere is generally represented by the three *state* variables: temperature, pressure, and density. For a gas, these variables are related by the **ideal gas law**. This law is fundamental to an understanding of the *behaviour* of the atmosphere because it allows us to predict how the atmosphere will respond to a change in one, or more, of the state variables. The implication of the name *gas law* is, of course, that gases behave differently from liquids or solids. The uniqueness of their behaviour is accounted for by the **kinetic theory of matter**, which tells us that the differences between gases, liquids, and solids begin at the molecular level.

One of the state variables, pressure, is of particular importance to weather. High pressure is normally associated with clear skies, while low pressure is normally associated with cloudy skies. Because of the significance of pressure to weather, one of the main purposes of weather maps is to depict the distribution of pressure at Earth's surface. Figure 3.1 is an example of a weather map for Canada. You may have seen simplified versions of such maps on television and in newspapers. Using **isobars**, these maps show where the pressure is highest and where it is lowest. In addition, the spacing of the isobars depicts the degree of *variation* of pressure over the map area. Where the isobars are closer together, pressure changes rapidly with distance; where they are farther apart, pressure changes more slowly with distance. Weather maps also often show a variety of **fronts**, the simplest of which are **warm fronts** and **cold fronts**. The symbols used to label warm and cold fronts are shown on Figure 3.1.

A large part of a weather forecaster's job is to determine how pressure will change. But, because the atmosphere is a **fluid** and, like all fluids, subject to changes that are both rapid and difficult to predict, this is not a simple matter. As air moves, and warms and cools, the pressure we measure at the surface changes. Surface pressure is, therefore, a function of what is happening throughout the depth of the air column above the surface. Because the entire depth of the troposphere is, thus, important in forecasting, forecasters use not only maps of surface pressure but also maps showing the distribution of pressure at various levels in the atmosphere. Unlike surface maps, these upper-air maps rarely appear on television or in newspapers. To begin to understand the *vertical* structure of the atmosphere, we use the concept of **hydrostatic balance**. This concept tells us that the change in pressure with height, or depth, in a fluid is dependent on the density of the fluid. In turn, the ideal gas law tells us that the density of a gas depends on its temperature and pressure.

IDEAL GAS LAW
A scientific law that provides the relationship between the pressure, temperature, and volume (or density) of a gas.

KINETIC THEORY OF MATTER
A scientific theory that states that matter is composed of molecules and that these molecules are in constant motion.

ISOBARS
Lines of constant pressure.

FRONT
A narrow zone of transition between air of different properties.

WARM FRONT
A front at which warm air is advancing and replacing cold air.

COLD FRONT
A front at which cold air is advancing and replacing warm air.

FLUID
A substance that can flow; liquids and gases are both fluids.

HYDROSTATIC BALANCE
The state of a stationary fluid when the vertical forces on it are balanced.

3.1 | The Kinetic Theory of Matter

The *kinetic theory of matter* was formulated in 1738. It states, among other things, that matter is composed of very tiny particles called molecules and that these molecules are in constant motion. This theory provides a means of understanding matter at the molecular level. It provides a link between observable, or macroscopic, properties and the microscopic properties that provide important clues as to what it is that makes matter behave the way it does. As such, we can use this theory to account for the *differences* between the three phases of matter and, in particular, to begin to understand gases.

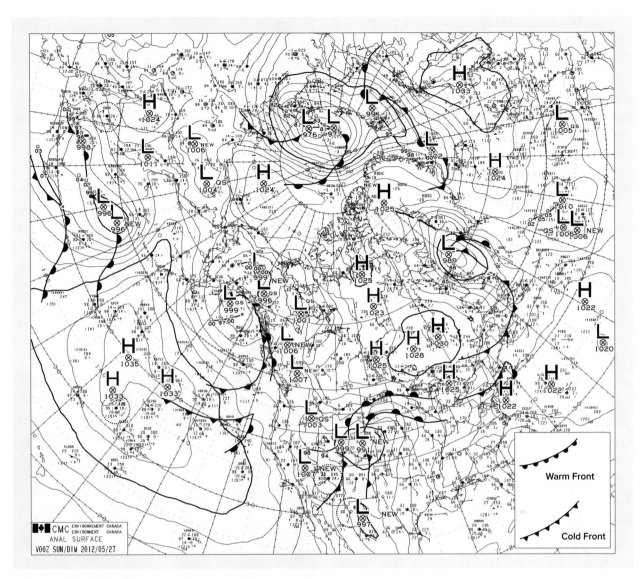

FIGURE 3.1 | Weather map of the northern hemisphere (Canada is near the centre) showing isobars, areas of high and low pressure, and fronts.

Source: Environment Canada, Weather Office, www.weatheroffice.gc.ca/analysis/index_e.html, 28 May 2012.

At the molecular level, we can differentiate between solids, liquids, and gases based on the amount of freedom of their molecules. In solids, the molecules are strongly attracted to one another and thus held tightly together; these molecules can vibrate, but they are not free to move from one place to another. In liquids, the attraction between molecules is slightly less than it is in solids, and the molecules can move more independently from each other. Molecules in gases have the most freedom to move independently, as the attraction between them is very small. On the macroscopic level, we generally discuss such differences in attraction in terms of *density*, which is an indirect measure of the spacing of molecules (Section 1.5). For example, the density of dry air is about one thousand times less than that of liquid water. Gases are mostly empty space.

We can use the kinetic theory to understand the meaning of **temperature** and pressure for a gas. Temperature is a measure of the *average* kinetic energy, or speed, of the molecules. Recall that

> **TEMPERATURE**
> A measure of the average kinetic energy of the molecules in a substance.

pressure is defined as force per unit area (Section 1.5). This force results from the continuous collisions of the gas molecules with a surface. In the atmosphere, the movement of molecules within a parcel of air exerts a pressure on the air molecules surrounding that parcel. It follows from the above that temperature depends on the speed of the molecules, while pressure depends on the number of molecules per volume *and* the speed of these molecules. The more molecules there are, and the faster they are moving, the more pressure they exert. This is a good example of how macroscopic properties, such as temperature and pressure, can be explained by processes operating at the microscopic level.

3.2 | Early Contributions to the Kinetic Theory of Gases

Like all theories, the kinetic theory of gases—a part of the kinetic theory of matter—arose as a result of observation and experiment. The early experiments involved are all examples of controlled experiments, in that one variable—temperature, pressure, or volume—was held constant while the others were varied.

3.2.1 | Constant Temperature (Boyle's Law)

One of the first of these experiments was performed by Robert Boyle (1627–1691) (Figure 3.2). Boyle's work led him to conclude that *if the temperature of a gas is held constant and its volume is increased, then the pressure it exerts will decrease.* The reverse is also true. In short, the volume and pressure of a gas are inversely proportional.

The simple relationship between the pressure, P, and volume, V, of a gas can be expressed mathematically as follows.

$$(PV)_{time1} = (PV)_{time2} \qquad (3.1)$$

Equation 3.1 states that the product of pressure and volume at time 1 will be the same as it is at time 2, as long as temperature does not change. This relationship is known as *Boyle's law*. This law, which Boyle formulated in 1662, is shown graphically in Figure 3.3.

Question for Thought

How does the kinetic theory of gases account for Boyle's law?

3.2.2 | Constant Pressure (Charles's Law)

Experimenting some years later, Jacques Charles (1746–1823) (Figure 3.4) made another discovery that contributed to the development of the kinetic theory of gases. He discovered that *if pressure is held constant, an increase in temperature will cause an increase in volume.* In other words, with pressure constant, temperature and volume are directly proportional; if temperature doubles, volume must also double. As shown in

FIGURE 3.2 | Seventeenth-century Irish chemist Robert Boyle conducted one of the first controlled experiments leading to the kinetic theory of gases.

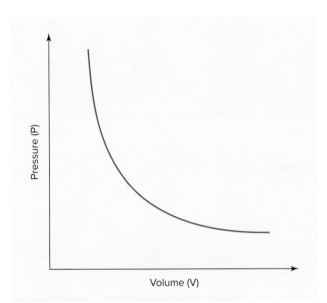

FIGURE 3.3 | A graphic representation of Boyle's law, which states that when temperature is constant, the volume and pressure of a gas are inversely proportional to one another.

Figure 3.5, the relationship between temperature and volume is linear.

$$\left(\frac{V}{T}\right)_{time\ 1} = \left(\frac{V}{T}\right)_{time\ 2} \qquad (3.2)$$

Equation 3.2 states that the relationship between volume and temperature at time 1 is the same as it is at time 2, as long as pressure is constant. This relationship became known as *Charles's law*; it is thought that Charles formulated it in 1787.

Question for Thought

How does the kinetic theory of gases account for Charles's law?

3.2.3 | Constant Volume

A third important gas law was discovered in the early 1700s, although our records do not clearly show who first formulated it. This third law states that *if volume is held constant, an increase in temperature will cause an increase in pressure.* That is, with volume

FIGURE 3.4 | In 1787, French scientist Jacques Charles formulated his law, which was a major contribution to what would eventually become the kinetic theory of gases.

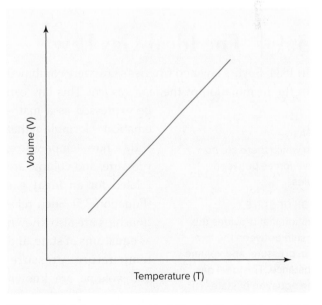

FIGURE 3.5 | A graphic representation of Charles's law, which states that when pressure is constant, temperature and volume are directly proportional to one another.

constant, temperature and pressure are directly proportional (Figure 3.6). This relationship is expressed in Equation 3.3.

$$\left(\frac{P}{T}\right)_{time\ 1} = \left(\frac{P}{T}\right)_{time\ 2} \tag{3.3}$$

Question for Thought

Compare Figure 3.3 to figures 3.5 and 3.6. What do the graphs tell you about the difference between these relationships? Try drawing these graphs using equations 3.1 through 3.3. (Hint: To do this, you will first need to use the equations to create sets of numbers.)

Remember This

The following relationships are true for gases:

- with temperature constant, volume and pressure are inversely proportional;
- with pressure constant, temperature and volume are directly proportional; and
- with volume constant, temperature and pressure are directly proportional.

3.3 | The Ideal Gas Law

In 1834, Boyle's law and Charles's law were combined in the formulation of the *ideal gas law*. This law can be expressed as a mathematical formula that shows how temperature, pressure, and volume are related for an **ideal gas** (Equation 3.5). Such relationships are also known as **equations of state**, and temperature, pressure, and volume are known as state variables because they describe the state of a substance. There is a relationship between the

IDEAL GAS
A gas in which there are *no* attractive forces between molecules.

EQUATION OF STATE
An equation that provides the relationship between the temperature, pressure, and volume of a substance. The ideal gas law is an equation of state.

IDEAL GAS CONSTANT
The constant, *R*, in the ideal gas equation.

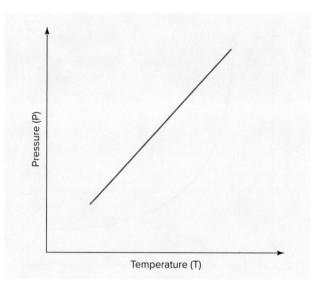

FIGURE 3.6 | A graphic representation of the law that states that when volume is constant, temperature and pressure are directly proportional to one another.

three state variables for all substances, but that for gases is the simplest.

In reality, there are no ideal gases, although gases at high temperatures and low pressures come close to the ideal because it is with these conditions that the molecules are farthest apart and moving fastest. Despite this seeming restriction, the gas law works well under most conditions, so it is generally useful for making predictions about the behaviour of gases and, therefore, the atmosphere.

To obtain an equation for the ideal gas law, we can combine equations 3.1 and 3.2 to give Equation 3.4.

$$\left(\frac{PV}{T}\right)_{time\ 1} = \left(\frac{PV}{T}\right)_{time\ 2} = constant \tag{3.4}$$

Notice that if temperature is held constant, Equation 3.4 reduces to Equation 3.1, and if pressure is held constant, Equation 3.4 reduces to Equation 3.2. In the mathematical statement of the ideal gas law (Equation 3.5), the constant in Equation 3.4 is given the symbol *R*, which is recognized as the **ideal gas constant**.

$$\frac{PV}{T} = R \tag{3.5}$$

Equation 3.5 implies that no matter how many experiments one does altering the pressure, volume, or temperature of a gas, PV/T will always equal the same number, R. This is similar to what one would find by dividing the circumference by the diameter of many different circles; the result will always equal the constant π.

Equation 3.5 is widely applied in physics and chemistry and, depending on its application and the system of units used, is expressed in slightly different forms and with different values of R. In atmospheric sciences, we cannot do controlled experiments, so density, ρ, is more convenient to use than volume (recall from Section 1.5 that $\rho = m/V$, so that $V = m/\rho$). Substituting for volume into Equation 3.5, we obtain Equation 3.6 for a unit mass.

$$\frac{P}{T\rho} = R_d \tag{3.6}$$

R_d represents the value of the gas constant for *dry* air. With pressure in pascals, temperature in degrees kelvin, and density in kilograms per cubic metre, the value of R_d is $287\ \text{J}\cdot\text{kg}^{-1}\cdot\text{K}^{-1}$.

Equation 3.6 is the form of the gas law most applicable to the atmosphere. In the atmosphere, pressure and temperature are easily measurable, while density is not. Therefore, the gas law is useful for calculating density from pressure and temperature (Example 3.1). In addition, the gas law is often used to substitute pressure and temperature for density in certain equations; this makes the equations easier to apply.

Example 3.1

What is the density of a parcel of air that has a pressure of 98 kPa and a temperature of 32°C?

Convert kilopascals to pascals and degrees Celsius to degrees kelvin, then use Equation 3.6.

$$\rho = \frac{P}{T\,R_d}$$

$$= \frac{98000\ \text{Pa}}{(305\ \text{K})\,(287\ \text{J}\cdot\text{kg}^{-1}\cdot\text{K}^{-1})}$$

$$= 1.1\ \text{kg/m}^3$$

Question for Thought

How would you confirm that the units of R_d are $\text{J}\cdot\text{kg}^{-1}\cdot\text{K}^{-1}$?

As written, Equation 3.6 applies to *dry* air but, because the atmosphere is rarely dry, we use a concept known as **virtual temperature** to account for the presence of atmospheric water vapour. We use temperature because it, like water vapour, affects the density of air. When air is warmed, its density decreases; likewise, when water vapour is added to air, the density of that air decreases. Virtual temperature is the temperature dry air would need to be to have the same density as moist air. The more water vapour, the higher the virtual temperature needs to be. This means that virtual temperature is greater than actual temperature by an amount proportional to the amount of water vapour in the air.

> **VIRTUAL TEMPERATURE**
> The temperature used in the ideal gas law to account for the fact that moist air is less dense than dry air.

$$T_V = T\,(1 + 0.61r) \tag{3.7}$$

In Equation 3.7, T and T_v are the air temperature and the virtual temperature, respectively. They are both expressed in degrees kelvin. The letter r represents the mixing ratio for water vapour, which is the ratio of the mass of water vapour to the mass of dry air; in this case, it must be expressed in grams of water vapour per gram of dry air (g/g) (Section 7.5.1.2). If air at 20°C has a mixing ratio of 0.015 g/g, its virtual temperature will be 22.7°C. This means that, because this air contains some water vapour, it has the same density as dry air at 22.7°C.

Question for Thought

How could you determine how important it is to use a concept like virtual temperature?

Two important properties of air are evident from the gas law. First, the gas law shows that, at constant temperature, an increase in pressure will result in an increase in density. This means that *air is compressible*. Second, the gas law shows that, at constant pressure, an increase in temperature will cause a decrease in density. In other words, *warm air is less dense than cold air*. We could demonstrate both these relationships using a balloon. The density of air in a balloon will decrease as the balloon rises in the atmosphere and expands because the pressure on it is decreasing. (Recall from Section 1.6.1 that density decreases with height in the atmosphere because air near the surface has a greater pressure on it than air higher up.) The density will also decrease if the air in the balloon is heated. In fact, this is how hot air-balloons work (Figure 3.7).

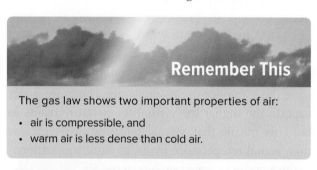

Remember This

The gas law shows two important properties of air:

- air is compressible, and
- warm air is less dense than cold air.

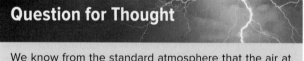

Question for Thought

We know from the standard atmosphere that the air at Earth's surface is both warmer *and* denser than the air above it. How can this be? (For reference, see Table 1.3.)

These two properties make the behaviour of gases distinct from the behaviour of liquids. First, with much more space between the molecules of a gas than between the molecules of a liquid, gases are much more easily compressed. In fact, liquids are not compressible unless they are subjected to very high pressures. For example, at the average depth of the oceans, about 4 km, pressure exerted by the water above would be about 40,000 kPa. But even at this extreme pressure, the density of the water would be only about 1.8 per cent greater that it is at the surface. (The effects of colder temperatures at this depth have not been accounted for here.) An interesting result of this difference in compressibility is that in a compressible fluid, like the atmosphere, the change in pressure with height follows an *exponential* relationship, while in an incompressible

FIGURE 3.7 | The flight of hot-air balloons, such as these ones floating above Saint-Jean-sur-Richelieu in Quebec, is possible because the warm air in the balloon is less dense than the surrounding cooler air.

fluid, like the ocean, the change in pressure with depth follows a *linear* relationship (Figure 3.8).

Second, although liquids and gases *both* expand when heated, gases expand much more than liquids. With only weak attractive forces between gas molecules, it is easy for these molecules to spread far apart. As air warms from 0°C to 20°C, it will expand about 7 per cent; as water warms over the same interval, it expands only about 0.06 per cent.

Remember This

- Density in a gas depends on both pressure and temperature.
- Density in a liquid depends mostly on temperature.

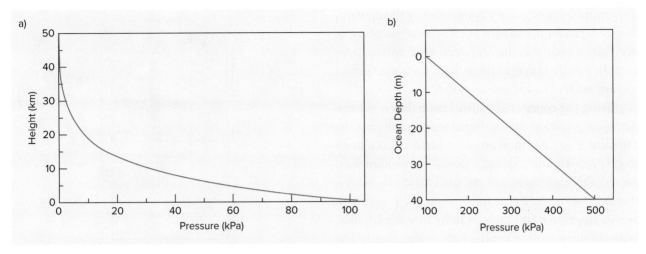

FIGURE 3.8 | a) Pressure decreases exponentially with height in the atmosphere. b) Pressure increases linearly with depth in the ocean. The pressure of 100 kPa shown at the ocean's surface is due to the pressure exerted by the atmosphere.

The discussion to this point provides a good example of the differences between laws and theories. The gas *law* has been shown, by experiment, to always hold true, for any *ideal* gases, under any conditions. Further, by using the gas law, we can make accurate *predictions*. For example, if pressure and temperature change, we can use Equation 3.6 to determine exactly how much density will change. But the gas law does not tell us *why* density will change. For an *understanding* of this change, we need the kinetic *theory* of gases, which represents gases as molecules in constant motion. In fact, this theory became accepted *because* it could explain the observations scientists made about gases. Thus, while the kinetic theory cannot be proven, and we can use it to make only theoretical predictions, it plays an essential role in the scientific process by providing an *explanation* for our observations.

3.4 | Hydrostatic Balance

Two forces act on the atmosphere in the vertical direction: gravity, which pulls downward, and the **pressure gradient force**, which pushes upward as a result of the decrease in pressure with height. A pressure gradient force occurs where pressure differences exist in a fluid; this force causes the fluid to move from high pressure to low pressure. For example, when a tire is punctured, the air will flow out of it until the pressure inside the tire is the same as the pressure outside the tire. In Section 11.1.1, we will see that this same force is what creates winds. If the pressure gradient

force acted alone, the atmosphere would rush off into space. On the other hand, if gravity acted alone, the atmosphere would collapse in a dense layer at Earth's surface. Instead, these two forces are balanced; this balance is known as *hydrostatic balance* (Figure 3.9). **Hydrostatics**, or fluid statics, is the study of *stationary* fluids and is part of a branch of physics known as *fluid mechanics*. Recall that Newton's first law tells us that for something to be stationary, the forces on it must be balanced (Section 1.5). Because the

PRESSURE GRADIENT FORCE
A force that occurs due to differences in pressure. The magnitude of this force is proportional to the pressure gradient, and its direction is from high pressure to low pressure.

HYDROSTATICS
The study of stationary fluids.

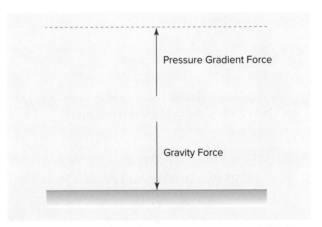

FIGURE 3.9 | The atmosphere is close to being in hydrostatic balance because the downward force of gravity is balanced by the upward-directed pressure gradient force.

downward force of gravity is balanced by the upward-directed pressure gradient force, the atmosphere is normally close to being in hydrostatic balance. An exception to this occurs where there is strong vertical motion, such as in a vigorous storm.

Using the concept of hydrostatic balance, we can derive an equation that is useful for determining the pressure at any depth in an *incompressible* fluid. In an incompressible fluid, density would be constant with depth; although there are no such fluids in reality, this is close to being true for liquids. The pressure at any depth, z, in a fluid that is in hydrostatic balance, is due to the weight of the fluid above that depth. The force exerted due to this weight is $F = m\,g$, where m is mass and g is the acceleration due to gravity (9.8 m/s^2) (Figure 3.10). The mass can be expressed as $m = V\,\rho$, where V is volume and ρ is density. In turn, the volume can be expressed as $V = z\,A$, where A is the area, so that $F = \rho\,g\,z\,A$. Because pressure is force per area, we obtain Equation 3.8 for determining the pressure at any depth in a fluid.

$$P = \frac{F}{A} = \frac{\rho\,g\,z\,A}{A} = \rho\,g\,z \qquad (3.8)$$

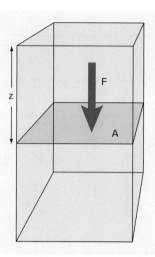

FIGURE 3.10 | A column of fluid with cross-sectional area, *A*. At depth *z*, the column of fluid exerts a force, *F*, on area *A*. This force per area is pressure. How and why would the pressure on *A* change if the column were deeper?

Example 3.2

The density of water is 1000 kg/m³. What is the water pressure at a depth of 2 m?

Use Equation 3.8.

$$P = (1000 \text{ kg/m}^3)\,(9.8 \text{ m/s}^2)\,(2 \text{ m})$$
$$= 19,600 \text{ Pa}$$
$$= 19.6 \text{ kPa}$$

Example 3.3

The average atmospheric pressure at sea level is 101.325 kPa. What depth of water is required to produce the same pressure?

Use Equation 3.8.

$$101,325 \text{ Pa} = (1000 \text{ kg/m}^3)\,(9.8 \text{ m/s}^2)\,z$$
$$z = 10.34 \text{ m}$$

Why does it require such a small depth of water to produce the same pressure as the much deeper atmosphere?

Although Equation 3.8 works for fluids in which density is constant with depth, it does not work for gases, which are compressible, because their density varies considerably with depth. We can derive a form of Equation 3.8 that is more suitable for compressible fluids, like the atmosphere, by considering a slab of fluid with a small height, Δz, and an area, A, (Figure 3.11). The assumption is that this slab is thin enough that the variation of density over its height is not significant. If the pressure on the top of the slab is P, then the pressure pushing up on the bottom of the slab is $P + \Delta P$. The forces acting *downward* on this slab are the weight of the slab, $\rho\,g\,\Delta z\,A$, and the force due to the pressure from above, $P\,A$. The force acting *upward* on the bottom of the slab is due to the pressure of the fluid below, $(P + \Delta P)\,A$. Because the fluid is stationary, the forces on the slab are equal.

$$\rho\,g\,\Delta z\,A + P\,A = (P + \Delta P)\,A$$

$$\frac{\Delta P}{\Delta z} = -\rho\,g \qquad (3.9)$$

Equation 3.9 is often referred to as the *hydrostatic equation*. It gives the rate of change of pressure with height for a small change in height. The negative sign indicates that pressure *decreases* as height *increases*. (It can be ignored in most calculations.)

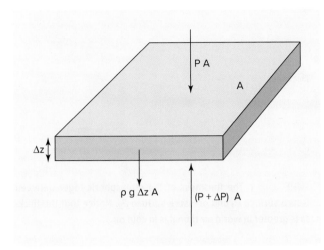

FIGURE 3.11 | The forces acting on a "slab" of the atmosphere.

The hydrostatic equation shows that *the change in pressure with height in a fluid is directly proportional to the fluid's density.* As density increases, the change in pressure with height will increase because, with greater density, air molecules will be more tightly packed than they would be with lower density. Recall that the gas law tells us that density, in turn, depends partly on temperature. Since warm air is less dense than cold air, it follows that *pressure decreases more quickly with height in cold air than it does in warm air.* We can most easily see the effect of temperature on the change in pressure with height by applying an equation known as the *hypsometric equation.*

Example 3.4

The CN Tower in Toronto is 553.3 m high. What is the approximate *difference in pressure* between the top and bottom of the tower? You can assume that the density of air at the surface is 1.23 kg/m³ and that the density of air at 500 m is 1.17 kg/m³.

Use Equation 3.9 and an average value of 1.2 kg/m³ for density.

$$\Delta P = (1.2 \text{ kg/m}^3)(9.8 \text{ m/s}^2)(553.3 \text{ m})$$

$$= 6506.8 \text{ Pa}$$

$$= 6.5 \text{ kPa}$$

There is a decrease in pressure of about 6.5 kPa from the bottom to the top of the tower. Why would the accuracy of such a calculation decrease if we were to consider greater depths of the atmosphere?

3.5 | The Hypsometric Equation

Hypsometry is the science of measuring heights. It is commonly associated with the study of the variation in elevation of the land surface. The application of hypsometry to atmospheric science is somewhat more abstract. In the atmosphere, instead of land surfaces, we envision **pressure surfaces** (Figure 3.12). Applied to the atmosphere, hypsometry is about the relationship between height and pressure.

HYPSOMETRY
The science of measuring heights.

PRESSURE SURFACE
An imaginary surface in the atmosphere upon which the pressure is the same everywhere.

We can derive the hypsometric equation by substituting the gas law into the hydrostatic equation to eliminate density. The change in pressure with height is then expressed as a function of pressure and temperature, instead of a function of density, as shown in Equation 3.10.

$$\frac{\Delta P}{\Delta z} = -\left(\frac{P}{R_d \overline{T_v}}\right) g \qquad (3.10)$$

Equation 3.10 explicitly shows that the change in pressure with height is greater in cold air than it is in warm air. This has important implications for the

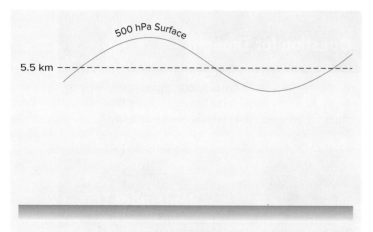

FIGURE 3.12 | The 500 hPa pressure surface, or the surface upon which pressure is 500 hPa. Recall from Section 1.5 that hPa stands for hectopascals and that 10 hPa is equal to 1 kPa. On average, this pressure surface occurs at a height of 5.5 km (Table 1.3).

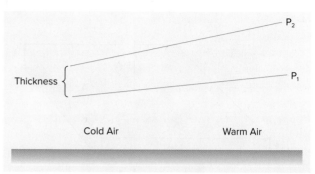

FIGURE 3.13 | The thickness of the atmospheric layer between pressure surface P_1 and pressure surface P_2. Notice that the thickness is greater in warm air than it is in cold air.

vertical structure of pressure systems, and for driving atmospheric circulations (Section 3.6).

Because temperature and pressure both vary with height, Equation 3.10 is integrated from height z_1, where the pressure is P_1, to height z_2, where the pressure is P_2. We thus obtain the following form of the hypsometric equation.

$$\Delta z = z_2 - z_1 = \left[\frac{R_d \, \overline{T_v}}{g}\right]\left[\ln\left(\frac{P_1}{P_2}\right)\right]$$

(3.11)

In Equation 3.11, Δz represents the height difference, or **thickness**, between two pressure surfaces. The thickness of an atmospheric layer is the difference in height between two pressure surfaces. At the bottom of this layer, the pressure is P_1; at the top of this layer, the pressure is P_2 (Figure 3.13). Equation 3.11 shows that the thickness of an atmospheric layer is proportional to the *average* virtual temperature, $\overline{T_v}$, of that layer. The greater the temperature, the thicker the layer because pressure decreases more slowly with height in warm air than it does in cold air. Thicknesses are often plotted on upper-air weather maps to indicate, among other

THICKNESS
The difference in height between two pressure surfaces in the atmosphere.

things, the average virtual temperature of a layer of the atmosphere. Thickness is useful because it gives an indication of temperature for a *layer* rather than for a specific height. The hypsometric equation has other practical applications in meteorology.

Example 3.5

What is the thickness of the atmospheric layer from 1000 hPa to 500 hPa if the average virtual temperature of the layer is 0°C?

Use Equation 3.11 and convert the temperature to degrees kelvin.

$$\Delta z = \left[\frac{287 \times 273}{9.8}\right]\left[\ln\left(\frac{1000}{500}\right)\right]$$

$$= 5541.7 \text{ m}$$

What would be the thickness of this layer if it cooled to –5°C?

$$\Delta z = \left[\frac{287 \times 268}{9.8}\right]\left[\ln\left(\frac{1000}{500}\right)\right]$$

$$= 5440.2 \text{ m}$$

Note that it would be more accurate to use a spreadsheet to do the above calculation for a number of thin layers, as long as you know the temperature of each thin layer.

We can apply the hypsometric equation to convert station pressure observations to sea level pressure, a practice known as "reducing to sea level pressure"

Question for Thought

How would the answers to the questions in Example 3.5 be affected if you used temperature in place of virtual temperature? All else being equal, would thickness be greater in moist air or dry air?

(Figure 3.14). Because pressure decreases so rapidly with height, pressure observations are strongly affected by the altitude of the observing station. For the purposes of weather forecasting, we need to eliminate the effect of altitude; if we don't, the effect of altitude will obliterate the effects of the weather systems we are trying to find and analyze. For example, a weather observing station at an elevation of 1000 m above sea level might record a pressure of 93 kPa. Even in a hurricane, it would be unlikely for sea level pressure to be as low as this. When roughly converted to sea level pressure, this value becomes 103 kPa, a fairly high pressure normally associated with clear, blue skies. *Reduction to sea level pressure* standardizes

pressure observations so that they are comparable, thus allowing meteorologists to determine the location and extent of high- and low-pressure systems. The pressure values plotted on surface weather maps (Figure 3.1) have been reduced to sea level pressure.

To use the hypsometric equation to reduce station pressure to sea level pressure, we rearrange Equation 3.11 and take the exponential of both sides to eliminate the logarithms. The result is given as Equation 3.12.

$$\ln P_1 - \ln P_2 = \left(\frac{g}{R_d\,\overline{T_v}}\right)(z_2 - z_1)$$

$$P_1 = P_2 \exp\left[\left(\frac{g}{R_d\,\overline{T_v}}\right)(z_2 - z_1)\right] \tag{3.12}$$

We can simplify this equation to give Equation 3.13. The variables have been redefined as shown in Figure 3.15. In this equation, g/R is a constant equal to 0.0342 K/m; it is designated by a.

$$P_0 = P\,e^{\left(\frac{a}{\overline{T_v}}\right)z} \tag{3.13}$$

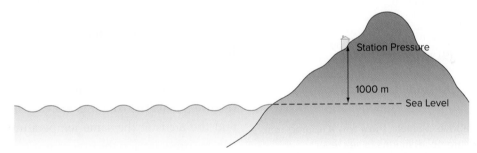

FIGURE 3.14 | The pressure observed at an observing station—in this case, at an altitude of 1000 m—is the *station pressure*. To eliminate the effects of altitude on pressure, we reduce this value to *sea level pressure*.

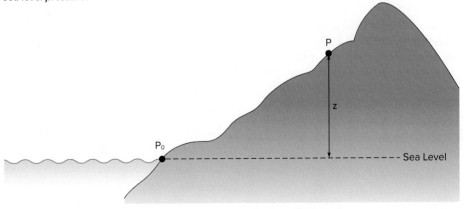

FIGURE 3.15 | The pressure at sea level is P_0. The pressure at height z (at the observing station) is P.

Alternatively, Equation 3.13 can be solved for P and used to determine the pressure at any height above sea level, given sea level pressure.

$$P = \frac{P_0}{e^{\left(\frac{a}{T_v}\right)z}}$$

(3.14)

Equations 3.13 and 3.14 both make the limiting assumption that the atmosphere is *isothermal*, or constant in temperature, between sea level and height z. Because the real atmosphere is *not* isothermal, these equations are most accurate for small height changes. Notice also that Equation 3.14 shows that for an isothermal atmosphere, the change in pressure with height would be *exactly* exponential. In the real atmosphere, the change in pressure with height is *close to* exponential.

Example 3.6

A weather observing station is located at an elevation of 950 m. The pressure at this station is observed to be 91 kPa. If the average virtual temperature of the air layer between sea level and 950 m is 10°C, what is sea level pressure for this station?

Use Equation 3.13.

$$P_0 = (91 \text{ kPa}) \, e^{\left(\frac{0.0342}{283 \text{ K}}\right)(950 \text{ m})}$$

$$= 102.1 \text{ kPa}$$

How does the above example show why it is important to reduce pressure observations to sea level?

Example 3.7

If sea level pressure is 99.7 kPa, what is the pressure at an elevation of 645 m, when the average temperature from the surface to 645 m is –1°C?

Use Equation 3.14.

$$P = \frac{99.7 \text{ kPa}}{e^{\left(\frac{0.0342}{272 \text{ K}}\right)645 \text{ m}}}$$

$$= 92.3 \text{ kPa}$$

Question for Thought

Use Equation 3.14 to construct a graph showing the decrease in pressure with height for an isothermal atmosphere with a virtual temperature of 0°C. How is the graph you have drawn different from the change in pressure with height in the real atmosphere?

3.6 | The Influence of Temperature on Pressure Distribution

As noted at the beginning of this chapter, the ideal gas law and the hydrostatic equation are fundamental to an understanding of the atmosphere. The gas law shows how temperature, pressure, and density are related, while the hydrostatic equation shows, among other things, that pressure decreases more quickly with height in cold air than it does in warm air. Both have significant implications for the horizontal and vertical distribution of atmospheric pressure.

Consider two columns of air side by side (Figure 3.16a). The columns are each 1 km in height, and the air is at the same temperature. Initially, the surface pressure and the pressure at 1 km are the same in both columns. To begin, assume that these columns are confined horizontally, but not vertically. That is, air can move up or down, but not sideways. Imagine that column A is warmed, while the temperature of column B remains constant (Figure 3.16b). As a result, column A expands upward so that pressure decreases more slowly with height than it does in column B. The surface pressures do not change because, in each case, the same weight of air still lies above. Notice, however, that as a direct result of this expansion, the pressure at any height *above* the surface is higher in column A than it is in column B. In other words, *at any height above the surface, pressure will be higher in warm air than it*

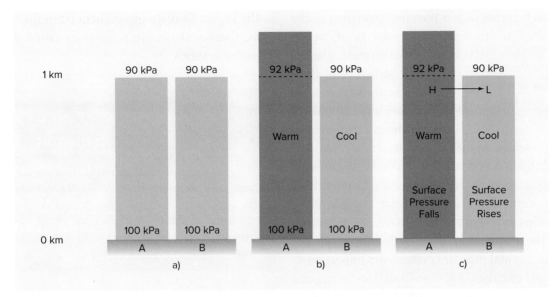

FIGURE 3.16 | a) The decrease in pressure with height is the same in both air columns. b) The air in column A is warmed, causing it to expand upward so that at a height of 1 km the pressure in column A is higher than that in column B. At this point, air is prevented from moving horizontally. c) Air can now move horizontally, and because the expansion of column A created a pressure gradient above the surface from the warm air column to the cold air column, air will flow along this pressure gradient. As a result, the surface pressure in the warm column decreases, while that in the cold column increases.

is in cold air. This creates an upper-air pressure gradient from the warm air to the cold air.

Now imagine that the air is allowed to move horizontally. As a result of the pressure gradient from the warm air column to the cold air column, air will flow out of the warm air column into the cold air column (Figure 3.16c). As air leaves the warm column, surface pressure in this column will *decrease*, and as it flows into the cold column, surface pressure in this column will *increase*. It follows that *warming can decrease surface pressure, and cooling can increase surface pressure.*

This may seem to contradict what the gas law tells us: that warming a gas will *increase* pressure and cooling a gas will *decrease* pressure. In a lab, where volume is kept constant, this is true, but in the atmosphere, where the air is unconfined, the heated air expands out of the air column, causing density—and thus surface pressure—to drop. The result of such air movement caused by temperature differences is a **thermal pressure system**. Thermal low-pressure systems are caused by warming; they are warm-cored systems. Thermal high-pressure systems are caused by cooling; they are cold-cored systems.

THERMAL PRESSURE SYSTEMS Shallow areas of high or low pressure that are created by cooling or warming, respectively.

Thermal pressure systems are *shallow*, meaning that with height above the surface they will gradually weaken until eventually the pressure pattern reverses itself. In other words, a low at the surface will gradually become a high aloft, and a high at the surface will gradually become a low aloft. This occurs because in the warm air of the thermal low, pressure is decreasing more slowly with height than it is in the surroundings (Figure 3.17). As a result, what was low pressure near

Remember This

- Pressure decreases more slowly with height in warm air than it does in cold air.
- At any height above the surface, pressure will be higher in warm air than it is in cold air.
- Warming can lower the surface pressure, and cooling can raise the surface pressure.

the surface becomes higher pressure, compared to the surroundings, with height. The reverse is, of course, true in a thermal high.

The structure of thermal pressure systems has important implications for atmospheric circulation at a variety of scales. At the small scale, for example, sea-breeze circulation systems develop due to differences in heating between the land and the sea (Section 11.4). At the opposite extreme, the planetary-scale circulation develops as a result of the differences in heating between the equator and the poles (Section 12.3).

While thermal pressure systems are important to atmospheric circulation systems and also to tropical weather, they are *not* particularly important weather producers in the mid-latitudes. Instead, weather of the mid-latitudes is mostly controlled by the development and movement of *deep* low- and high-pressure systems that form as a result of the complex interaction of many atmospheric processes (Section 14.2). These systems extend throughout the troposphere and intensify with height. In contrast to thermal pressure systems, these lows are cold-cored and the highs are warm-cored. It is this attribute that makes them deep (Figure 3.18). In the cold-cored lows, pressure is decreasing much more quickly with height than it is in the surroundings, while the opposite occurs in the highs. To distinguish them from thermal systems, these systems are sometimes called **dynamic pressure systems**.

> **DYNAMIC PRESSURE SYSTEMS**
> Deep high- and low-pressure systems that develop as a result of complex air motions.

> **SYNOPTIC WEATHER MAP**
> A weather map that gives a visual synopsis of the weather conditions that are occurring at a given time.

Remember This

- Thermal pressure systems are *shallow* because thermal lows are warm-cored, while thermal highs are cold-cored.
- Dynamic pressure systems are *deep* because dynamic lows are cold-cored, while dynamic highs are warm-cored.

3.7 | Weather Maps

We can apply some of the content of this chapter to understanding weather maps. Maps compiled from weather observations made simultaneously all over the world are known as **synoptic weather maps**. These maps, which are also known as charts, are important to weather forecasting. In essence, forecasting involves projecting the conditions shown on the maps forward in time. As simple as that may sound, weather forecasting is, in practice, a complex process requiring human expertise combined with tremendous computing power (Chapter 15).

The observations necessary for producing weather maps are coordinated by the World Meteorological

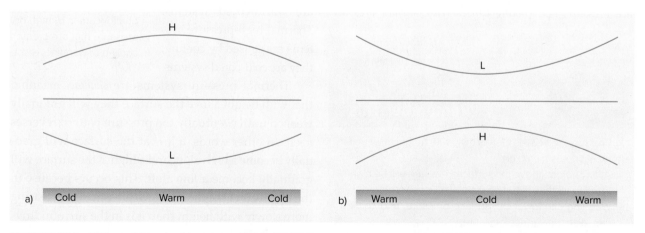

FIGURE 3.17 | a) A thermal low-pressure system. **b)** A thermal high-pressure system. In both diagrams, the lines are isobars representing pressure surfaces. The spacing between these lines represents the change in pressure with height. How do these diagrams show that thermal pressure systems are shallow?

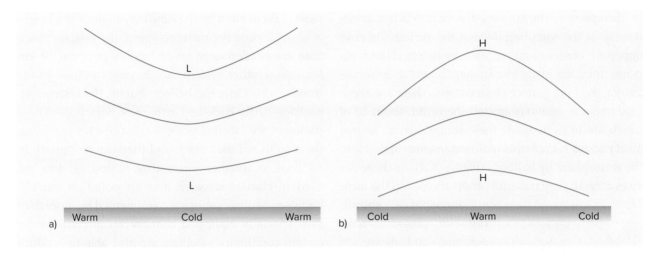

a) Warm   Cold   Warm

b) Cold   Warm   Cold

FIGURE 3.18 | a) A dynamic low-pressure system. b) A dynamic high-pressure system. How do these diagrams show that dynamic pressure systems are deep?

Organization (WMO). Weather observations typically made at the surface are listed in Table 3.1. For this, there are over 10,000 land stations worldwide, about 800 of which are in Canada. Most of these stations are now automated (Figure 3.19). Observations over the oceans come from roughly 1000 moored or drifting buoys and, on average, over 7000 commercial ships. Considering that the oceans cover 70 per cent of Earth's surface, there are often large gaps in observational data that cause problems for weather forecasters. For example, weather forecasting along Canada's west coast is often hampered by the sparsity of data over the Pacific Ocean. In such cases, satellite imagery can help forecasters fill in some of the gaps (Figure 3.20).

TABLE 3.1 | Commonly observed weather variables.

Variable	Measure
Temperature	degrees Celsius (°C)
Pressure	kilopascals (kPa)
Dew-Point Temperature	degrees Celsius (°C)
Cloud Cover	tenths (e.g., 1/10) or eighths (e.g., 1/8)
Cloud Type	labels (e.g., cirrus, cumulus, stratus)
Wind Speed	kilometres per hour (km/h)
Wind Direction	direction (e.g., N, W, SE)
Precipitation Amounts	millimetres (mm)

FIGURE 3.19 | Automated weather stations such as this one in Fernie, British Columbia, can measure such things as rainfall, humidity, barometric pressure, temperature, and wind speed and direction.

Sampling of the air *above* the surface is just as necessary as the sampling done at the surface. Because upper-air observations require more effort and expense, they are made less frequently and at fewer locations than are surface observations. There are about 1000 upper-air stations around the world, about 50 of which are in Canada. At these stations, small instrument packages, known as **radiosondes**, are carried into the atmosphere by balloons (Figure 3.21). As these devices ascend, they transmit observations—in the form of **atmospheric soundings**—of pressure, temperature, and dew-point temperature every few seconds to radio-receiving stations at the surface. In addition, radiosonde ascents can be tracked using ground-based radar to obtain information about the change in wind speed and direction with height. As the balloons rise and encounter lower and lower pressure, they expand to several times their original size. As a result,

most of them burst by the time they make it to a height of about 25 km. For meteorological analysis, however, data are needed only up to heights of about 10 km because weather systems are largely confined to the troposphere. Once the balloon bursts, the instrument package drifts back to Earth on a parachute. These packages are labelled with instructions for returning them so that if they are found they can be reused. In addition to weather forecasting, sounding data are used in climate research and air pollution models. Radiosonde observations are augmented by more than 3000 specially equipped commercial aircraft. Under certain conditions, satellites are also able to produce some useful atmospheric soundings; it is likely that one day they will replace radiosondes altogether.

To ensure that the observations are simultaneous, they are based on the Coordinated Universal Time (UTC) system. UTC is the time in Greenwich, England. The most important observations are those made at 0000, 0600, 1200, and 1800 UTC. Radiosonde observations are made only twice a day, at 0000 and 1200 UTC. It is from all these observations that both surface and upper-air charts are compiled; a chart labelled 0000 UTC shows the state of the atmosphere at that time.

The easiest map to understand, and probably the most familiar from television and newspapers, is the *surface* weather map (Figure 3.1). The main purpose of these maps is to show the variation of sea level pressure. (Remember that station pressures will have been converted to sea level pressure using Equation 3.13). The first step in the construction of weather maps is the plotting of weather observations using **weather station symbols** (Section 15.5, Appendix). These symbols provide the information needed to draw isobars, which give a picture of the variation of pressure at sea level (Figure 3.1). The information provided by the symbols is also used to draw *fronts*. While isobars are drawn by computers, fronts are drawn by meteorologists. To do this, meteorologists look primarily for changes in temperature, but they also look for areas of cloud and precipitation, shifts in wind direction, and changes in humidity (Section 13.5).

Upper-air charts also show pressure variation, but this is done differently on these charts than it

RADIOSONDE
A package of instruments that measure pressure, temperature, and moisture and send this information back to the surface through radio transmissions.

ATMOSPHERIC SOUNDING
Measurement of the change with height of certain atmospheric variables, such as temperature, pressure, humidity, and wind.

WEATHER STATION SYMBOLS
Symbols plotted on a weather map to provide information about observed weather elements.

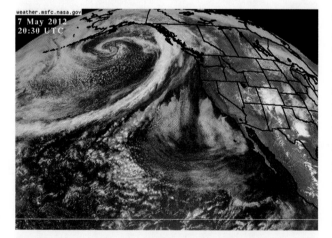

FIGURE 3.20 | A visible satellite image showing a portion of the Pacific Ocean and Canada's west coast. Such images provide forecasters with information about weather systems over the ocean that might otherwise be missed.

FIGURE 3.21 | An atmospheric researcher launches a weather balloon carrying a radiosonde near the Halley Research Station in Antarctica.

is on surface charts. Surface charts show the *variation of pressure* at sea level, while upper-air charts show the *variation in height* of a pressure surface. Constant-pressure charts have been in use since the 1950s. To illustrate a constant-pressure chart, consider the specific example of the 500 hPa pressure surface, which is located at approximately 5.5 km above Earth's surface (Figure 3.12). Recall that pressure surfaces are not flat or stationary; instead, they undulate up and down in the atmosphere. Because temperature influences the rate of decrease of pressure with height, temperature can influence the height of pressure surfaces. Where the air is warm, a pressure surface will be a little higher than normal; where the air is cool, the pressure surface will be a little lower than normal (Figure 3.22a). It is this variation in the height of pressure surfaces that is plotted on upper-air maps (Figure 3.22b).

Remember This

- On surface charts, we use *isobars* to map the variation of pressure at sea level.
- On upper-air charts, we use *contours* to map the variation in height of a pressure surface.

To show this variation on an upper-air chart, we use contours, or lines of constant altitude, instead of isobars. Despite this difference, upper-air charts can be interpreted in a similar way to surface maps because where heights are high, pressures are high, and where heights are low, pressures are low. You can see this by tracing a constant height across Figure 3.22a. Notice also that high pressure is associated with warm air, while low pressure is associated

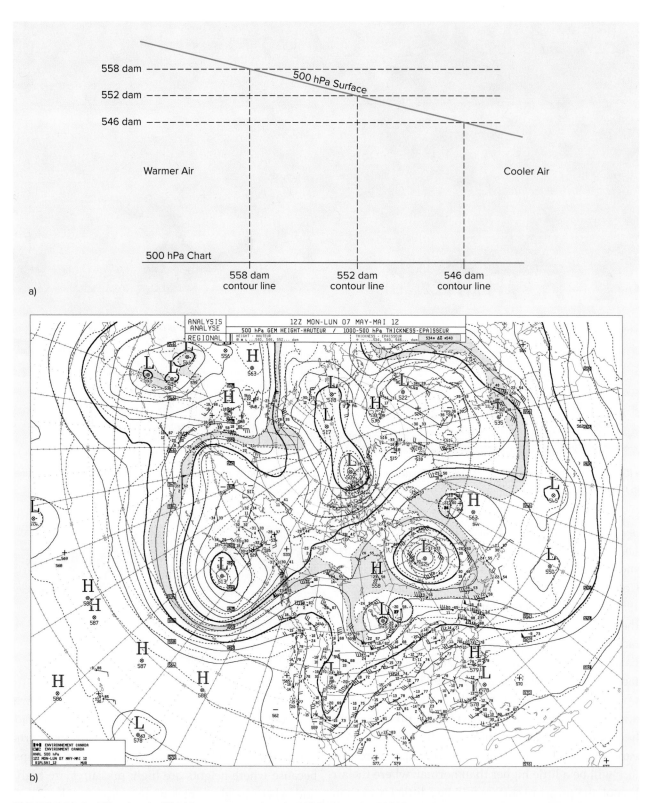

FIGURE 3.22 | a) Mapping the 500 hPa pressure surface. In this diagram, the 500 hPa surface is represented by a plane sloping downward from where the air is warm to where it is cooler. The changing height of this surface is shown on an upper-air chart using contours. In Canada, these upper-air charts have contour intervals of 6 dam (decametres) (1 dam is equal to 10 m). b) A 500 hPa chart. The solid lines are height contours, and the dashed lines are contours representing the thickness of the 1000 to 500 hPa layer. The height contours indicate that this pressure surface slopes downward toward the North Pole where the air is colder.

Source: b) Environment Canada, Weather Office, www.weatheroffice.gc.ca/analysis/index_e.html, 7 May 2012.

TABLE 3.2 | The four most used upper-air charts, their average heights, and commonly plotted secondary fields.

Pressure Surface	Approximate Height	Secondary Field
850 hPa	1500 m	temperature (°C)
700 hPa	3000 m	temperature (°C)
500 hPa	5500 m	thickness of the 1000 hPa to 500 hPa layer (dam)
250 hPa	10,000 m	wind speed (knots)

with cold air. Keep in mind, however, that this is not necessarily the case at the surface.

Combined with a surface map, upper-air charts create a package that aids in visualizing the three-dimensional nature of the atmosphere. Conventionally, four different pressure surfaces are mapped: the 850 hPa surface, the 700 hPa surface, the 500 hPa surface, and the 250 hPa surface (Table 3.2). The highest of these, the 250 hPa surface, is located near the top of the troposphere, at a height of about 10,000 m (Figure 3.23). The 250 hPa chart is the chart that most clearly shows the location of the **polar front jet stream** in the northern hemisphere (Section 12.2.5.2). Each upper-air chart also includes a secondary field, which provides additional information that is valuable in weather forecasting (Section 15.5).

POLAR FRONT JET STREAM
A narrow band of very fast westerly wind that occurs in the mid-latitudes in the upper portion of the troposphere.

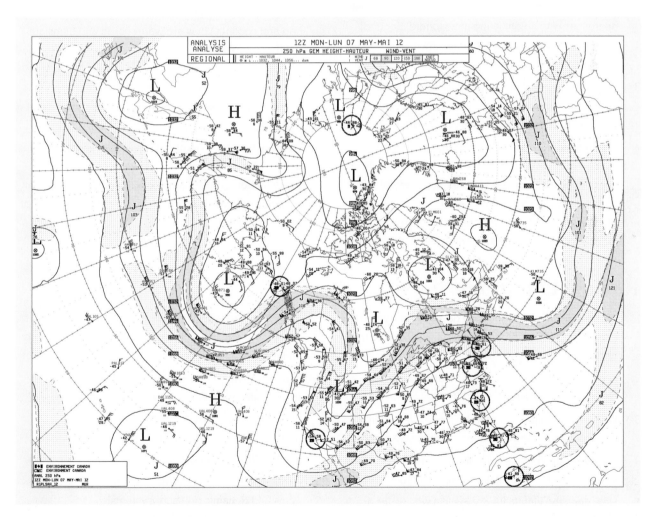

FIGURE 3.23 | A 250 hPa chart. The closely spaced contours indicate the location of the polar front jet stream of the northern hemisphere. The dashed isolines indicate wind speeds.

Source: Environment Canada, Weather Office, www.weatheroffice.gc.ca/analysis/index_e.html, 7 May 2012.

3.8 | Chapter Summary

1. The kinetic theory of matter states that all matter is made up of molecules in constant motion. The molecules in gases move more independently than do the molecules in solids and liquids. There is, therefore, a unique relationship between temperature, pressure, and density for gases as expressed by the ideal gas law.

2. Two important properties of air arise from the gas law. First, gases are compressible. This is why density decreases with height in the atmosphere. Second, gases expand much more than do liquids and solids when heated.

3. The atmosphere is normally in hydrostatic balance. The force of gravity pulling downward balances the pressure gradient force pushing upward.

4. The hydrostatic equation shows that the rate of change of pressure with height in a fluid depends on the density of the fluid. Because the density of the atmosphere decreases with height, the rate of change of pressure with height is not constant; it decreases quickly at first, then more slowly. In addition, because cold air is denser than warm air, pressure decreases more quickly with height in cold air than in warm air.

5. The hypsometric equation is derived from the hydrostatic equation. This is done using the gas law to substitute temperature and pressure for density. The hypsometric equation shows that the thickness of an atmospheric layer increases with temperature.

6. There are two types of pressure systems: thermal and dynamic. Thermal pressure systems result from heating and cooling. Lows are warm-cored and highs are cold-cored. This makes these systems shallow. In contrast, dynamic low-pressure systems are cold-cored while dynamic highs are warm-cored. This makes these systems deep.

7. Using weather observations taken simultaneously around the world, weather maps are constructed for forecasting purposes. Surface weather maps show the variation of sea level pressure and the location of fronts. Upper-air charts show the variation in height of a pressure surface.

Review Questions

1. At the molecular level, how do gases differ from solids and liquids? How do these differences influence the behaviour of gases?

2. How could you use the kinetic theory to define temperature and pressure?

3. What relationship is expressed in Boyle's law? What relationship is expressed in Charles's law? Describe both.

4. How does each of the following influence the density of air: a) temperature, b) pressure, c) moisture content?

5. How does the kinetic theory explain why gases are far more compressible than liquids?

6. What determines the rate of change of pressure with height, or depth, in a fluid? How and why is the rate of change of pressure with height in the atmosphere different from the rate of change of pressure with depth in the oceans?

7. How is the hypsometric equation different from the hydrostatic equation? Why is the hypsometric equation more applicable to the atmosphere?

8. What does the phrase "reduce to sea level pressure" mean? Why is this necessary?

9. What is the difference between thermal pressure systems and dynamic pressure systems?

10. What are the steps involved in producing a synoptic weather map? How are surface charts fundamentally different from upper-air charts?

Problems

1. a) Use the gas law and the standard atmosphere to calculate average sea level pressure.
    b) Given that the total mass of the atmosphere is 5.14×10^{18} kg and that the mean radius of Earth is 6.37×10^6 m, calculate average sea level pressure.

2. If a sealed rigid container is heated, it will explode. Do the following calculation to see why. A container is at 25°C, and the pressure of the air inside is the same as the atmospheric pressure of 102 kPa. What is the pressure of the air in the container if it is heated to 75°C?

3. What is the density of air in each case below?
    a) $P = 101.3$ kPa, $T = 15$°C
    b) $P = 90$ kPa, $T = 9$°C
    Account for the change in density between a) and b).

4. Calculate the virtual temperature in each case below.
    a) $T = 25$°C, $r = 10$ g/kg
    b) $T = 25$°C, $r = 3$ g/kg
    Explain your results.

5. a) Calculate the decrease in pressure from sea level to 100 m.
    b) Calculate the decrease in pressure from 5000 m to 5100 m.
    Account for your findings.

6. If the atmosphere were incompressible, with a constant density equal to the density at the surface, how high would the atmosphere need to be to exert the same pressure as it does now? (This height is known as the *scale height* of the atmosphere.)

7. Calculate the thickness of the 1000 hPa to 900 hPa layer when the average virtual temperature of the layer is as follows.
    a) 0°C
    b) 20°C
    Account for your findings.

8. Convert the following station pressures to sea level pressure. Assume the average virtual temperature is 0°C.
    a) 93.7 kPa at 793 m
    b) 892 hPa at 1100 m

9. Given that sea level pressure is 101.325 kPa and average virtual temperature is 0°C, what is the pressure at the following heights?
    a) 1400 m
    b) 1 km

Suggestions for Research

1. Investigate how weather elements such as rainfall, humidity, barometric pressure, temperature, and wind speed and direction are measured at weather stations. List as many instruments as you can find, and note the benefits and drawbacks to using automated weather stations.

2. Visit the website for Environment Canada's Weather Office, and find examples of surface and upper-air weather charts. Attempt to interpret these maps using what you have learned in this chapter. As part of your interpretation, consult the key for weather station symbols in this book's appendix.

4

Energy

Learning Goals

After studying this chapter, you should be able to

1. *distinguish* between heat, internal energy, and temperature;

2. *distinguish* between sensible heat and latent heat;

3. *explain* how energy is transferred by work;

4. *explain* how the first law of thermodynamics illustrates the conservation of energy;

5. *distinguish* between a constant pressure process and an adiabatic process; and

6. *explain* how energy is transferred as heat.

Energy transfers occur throughout the atmosphere, but most occur close to Earth's surface. One of the more visible impacts of energy transfer is the melting of Arctic ice, which occurs as the ice absorbs latent heat from its surroundings. The melting icicles shown here—below an icy overhang in Lancaster Sound, Nunavut—are evidence that energy is being transferred.

In addition to the gas law and the concept of hydrostatic balance, *energy* and *energy transfer* are fundamental to our understanding of the atmosphere. Although most of us have some idea of what energy is, it remains a rather abstract concept. To understand this term, it is helpful to consider its origins. The English word *energy* comes from the Greek word *energos*, which means "active" or "working." This origin suggests that energy is associated with making things happen or getting things done. Somewhat similar is the definition of energy traditionally used in physics, which states that energy is *the capacity to do work*. Another way to understand energy is to consider how we *perceive* it. We can *see* energy as light. We can *feel* energy as warmth. We can *experience* energy as movement.

LAW OF CONSERVATION OF ENERGY
The law that states that energy cannot be created or destroyed but can be transformed from one form to another and transferred as heat or work.

HEAT
Energy in the process of being transferred from a warmer object to a cooler object.

CONDUCTION
The transfer of heat between molecules in contact.

CONVECTION
Motions in a fluid that transfer the properties (e.g., heat) of that fluid.

RADIATION
The emission of energy as electromagnetic waves. This term is also used to denote the energy that travels in this way.

KINETIC ENERGY
Energy associated with motion.

The **law of conservation of energy** is a fundamental law of science. It states that *energy can never be created or destroyed, but it can be* transformed *from one form to another and* transferred *from one object to another*. Even the sun's energy is not created; it results from the *transformation* of mass. This is why the sun, and all the other stars, have finite lives; they will eventually use up their supplies of mass (Figure 4.1). Albert Einstein put the relationship between mass and energy into his famous equation: $E = mc^2$. In this equation, E is energy, m is mass, and c is the speed of light. Since the speed of light is a very large number, approximately 3×10^8 m/s, it follows that a small amount of mass can be converted to a large amount of energy.

Energy can be *transferred* in two ways. First, it is transferred from warmer to colder objects as **heat**. This transfer occurs through the processes of **conduction**,

FIGURE 4.1 | An X-ray image of the sun. Nuclear fusion reactions in the sun convert mass into the energy we sense as heat and light.

convection, and **radiation**. Second, energy is transferred mechanically as *work* (Section 1.5).

Question for Thought

How would you show that the units on either side of the equation $E = mc^2$ are equivalent?

4.1 | Heat

Heat is energy in the process of being transferred from one object to another as a result of a temperature difference between them. As heat flows from the warmer object to the cooler one, the molecules in the cooler object will begin to move faster as they gain **kinetic energy**. Recall that temperature is a measure of the average kinetic energy of the molecules (Section 3.1).* This energy, contained in the molecules of a substance, contributes to what is known as **internal energy**. That part of the internal energy of a substance that comes from the kinetic energy of the molecules is sometimes called **thermal energy** because of the association between kinetic energy and temperature. However, not all of the internal energy is associated with motion; some of it is **potential energy** resulting from the attractive forces between the molecules. Internal energy, therefore, is the *total* energy of the molecules in a substance, as it includes their kinetic *and* potential energy.

Internal energy occurs at the *microscopic* scale. At the *macroscopic* level, a change in internal energy can produce a temperature change, as mentioned above, but it can also produce a phase change. Further, it is possible for temperature changes and phase changes to occur together, the increased internal energy being divided between the two. When heat transfer results in a temperature change, the heat involved is known as **sensible heat**. When heat transfer results in a phase change, the heat involved is known as **latent heat**.

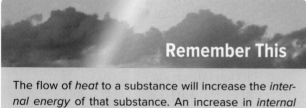

Remember This

The flow of *heat* to a substance will increase the *internal energy* of that substance. An increase in *internal energy* will result in a *temperature* increase, a phase change, or both.

4.1.1 | Sensible Heat

We can measure sensible heat with a thermometer. We can also *feel* this sort of heat. When sensible heat is transferred *to* an object, the object will feel warmer; when sensible heat is transferred *from* an object, the object will feel colder. Given the same input of heat, however, some things will warm more than others.

*There are three kinds of kinetic energy at the molecular level: translational, rotational, and vibrational. Only translational kinetic energy is measured as temperature. Like potential energy, rotational and vibrational energy do not contribute to temperature.

The relationship between heat flow and temperature change was discovered through experiment in the eighteenth century. It is expressed in the following simple equation, which applies only if no phase change occurs:

$$Q = m\,c\,\Delta T \qquad (4.1)$$

In Equation 4.1, Q is heat flow in joules, m is mass in kilograms, ΔT is temperature change in degrees kelvin, and c is the **specific heat**. Specific heat is a *property* of a substance; its value varies from one substance to another, as was first shown by Scottish scientist Joseph Black in the eighteenth century (Table 4.1). Substances with high specific heats will warm or cool *less* than substances with low specific heats, given the same input or output of heat.

Table 4.1 shows that water has a very high specific heat. The reason for this is the strong attractive forces between water molecules. In other words, the potential energy associated with water molecules is high. As a result, when water is heated, most of the increased internal energy is used to weaken the bonds between the molecules. This leaves only a small amount of energy for increasing the kinetic energy of the molecules. It follows that water will contain more internal energy than an equivalent mass of another substance at the same temperature. This means that water can store large amounts of energy. Examples 4.1 and 4.2 illustrate the significance of specific heat and, in particular, the high specific heat of water.

INTERNAL ENERGY
The total energy contained within the atoms and molecules of a substance.

THERMAL ENERGY
That part of the internal energy of a substance that is associated with the kinetic energy of the molecules.

POTENTIAL ENERGY
Energy associated with position.

SENSIBLE HEAT
Heat associated with a temperature change.

LATENT HEAT
Heat associated with a phase change.

SPECIFIC HEAT
The amount of heat, in joules, required to raise the temperature of 1 kg of a substance by 1 K.

TABLE 4.1 | Specific heats of various substances.

Substance	Specific Heat (J·kg^{-1}·K^{-1})
Asphalt	920
Concrete	880
Glass	670
Steel	500
Gold	130
Sand	800
Dry Soil	800
Wet Soil	1480
Granite	790
Basalt	840
Light Wood	1420
Dense Wood	1880
Ice (at 0°C)	2100
Water	4186
Steam	2040

Remember This

For two substances with the same mass and temperature, the one with the higher specific heat will contain more internal energy.

Example 4.1

How much heat is needed to raise the temperature of 10 kg of a) sand and b) water by 20°C?

Because we are dealing with a temperature *change*, 20°C equals 20 K.

Use Table 4.1 to look up the specific heats for sand and water, then use Equation 4.1.

a) $$Q = (10 \text{ kg}) (800 \text{ J·kg}^{-1}\cdot\text{K}^{-1}) (20 \text{ K})$$
$$= 160{,}000 \text{ J}$$

b) $$Q = (10 \text{ kg}) (4186 \text{ J·kg}^{-1}\cdot\text{K}^{-1}) (20 \text{ K})$$
$$= 837{,}200 \text{ J}$$

Because the specific heat of water is about five times higher than the specific heat of sand, it requires about five times more heat to warm the water by 20°C. It follows that in the process of warming, the internal energy of the water has increased considerably more than the internal energy of the sand.

Question for Thought

Considering water's high specific heat, why do you think water is a good substance to use in home heating systems?

Example 4.2

Imagine that 300 mL of coffee at 90°C is poured into a ceramic mug with a temperature of 20°C and a weight of 200 g. Heat will be transferred from the coffee to the mug until they reach the same temperature, or the *equilibrium temperature*. What will be the equilibrium temperature in this case? The specific heat of ceramic is about 1090 J·kg^{-1}·K^{-1}, and we can assume that the specific heat of coffee is the same as that for water.

Using the conservation of energy, and assuming that the mug is isolated from its surroundings, we can state that all the heat lost from the coffee flows into the mug.

At equilibrium:

Heat lost by coffee = Heat gained by mug

Since 1 L of water weighs 1 kg, the mass of the coffee is 0.3 kg.

Use Equation 4.1, and let *T* be the equilibrium temperature.

$$(0.3 \text{ kg}) (4186 \text{ J·kg}^{-1}\cdot\text{K}^{-1}) (90°\text{C} - T) =$$
$$(0.2 \text{ kg}) (1090 \text{ J·kg}^{-1}\cdot\text{K}^{-1}) (T - 20°\text{C})$$

$$T = 79.6°\text{C}$$

At equilibrium, both the mug and the coffee will be 79.6°C. Because of the coffee's high specific heat, it *cooled* only 10.4°C, while the mug *warmed* 59.6°C.

4.1.2 | Latent Heat

As mentioned above, heat flow can also produce phase changes. For every substance, there are specific combinations of temperature and pressure at which it will change phase. For example, at 0°C and normal atmospheric pressure, ice will *melt*, changing from solid to liquid. At normal atmospheric pressure and a temperature of 100°C, water will *boil*, changing from liquid to gas. However, even at temperatures below 100°C, liquid water can change to vapour through *evaporation*.

Evaporation occurs when some surface molecules in a liquid have enough energy to become vapour (Section 7.2). In any substance, there is quite a range in the speed of the molecules; in liquids, the fastest ones will be able to escape from the surface, to become gas. At higher temperatures, more molecules are moving fast, so more can vaporize. Evaporation rates, therefore, increase with temperature. (Other factors influence evaporation rates as well.) Because the fastest moving molecules escape, the average speed of those remaining will be lower. This means that the temperature of a liquid drops when evaporation is taking place. Thus, we say that *evaporation is a cooling process*.

Recall that heat involved in phase changes is *latent* heat (Section 2.6). In 1871, Joseph Black, who also determined that different substances have different *specific* heats, recognized the existence of latent heat. He noticed that when heat was added to ice at 0°C, the temperature of the ice did not increase; instead, the heat was used to melt the ice. Because there was no temperature change, this heat was called *latent*, meaning "present, but not visible" or "hidden". During phase changes from solid to liquid, liquid to gas, or solid to gas, latent heat is *absorbed from* the surroundings (Figure 4.2). For example, when our sweat *evaporates*, we cool down because the process absorbs latent heat from our bodies and the air next to us. During phase changes in the opposite direction, latent heat is *released to* the surroundings. For example, when *condensation* forms on a glass containing a cold drink, the drink will warm because heat is released during the condensation process (Figure 4.3).

The amount of heat involved in phase changes between liquid water and water vapour is known as the **latent heat of vaporization**, L_V; it is equal to 2.26×10^6 J/kg of water (Figure 4.2). This value applies at 100°C. The latent heat of vaporization increases slightly with decreasing temperature. At 20°C, for example, it is about 2.5×10^6 J/kg. To calculate the amount of heat, Q, involved in evaporation or condensation, we can simply multiply the mass, m, of water by the latent heat of vaporization, as shown in Equation 4.2a:

LATENT HEAT OF VAPORIZATION The amount of heat associated with the phase change of a substance between liquid and gas.

$$Q = L_v \, m \qquad (4.2a)$$

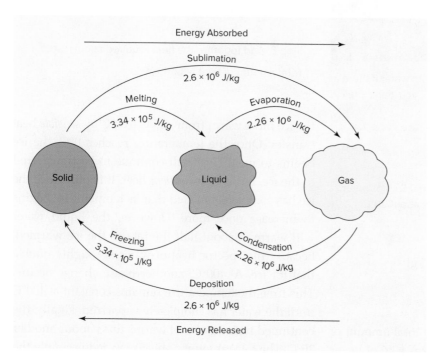

FIGURE 4.2 | The amount of latent heat involved in the phase changes of water. Why do you think there is more latent heat associated with the change between liquid and gas than there is with the change between solid and liquid?

FIGURE 4.3 | The release of latent heat through condensation on the outside of the glass will warm the water; in turn, the ice will melt as it absorbs heat from the liquid water.

The amount of heat involved in the phase change between ice and liquid water is known as the **latent heat of fusion**, L_F. It is equal to 3.34×10^5 J/kg. We can multiply this value by the mass of water to determine the amount of heat involved in melting or freezing, as shown in Equation 4.2b:

LATENT HEAT OF FUSION
The amount of heat associated with the phase change of a substance between solid and liquid.

$$Q = L_F\, m \qquad (4.2b)$$

It is also possible for water to change directly between the solid and gaseous phases (Figure 4.2). When the phase change is from ice to vapour, it is called **sublimation**; when it is in the opposite direction, it is called **deposition**. Either way, the amount of latent heat that is involved is simply the sum of the latent heat of vaporization and the latent heat of fusion.

DEPOSITION
A phase change from gas to solid.

SUBLIMATION
A phase change from solid to gas.

Question for Thought

Sometimes farmers spray their trees with water if there is a threat of overnight frost. Why do you think they do this?

Example 4.3

How much latent heat is absorbed when 0.5 kg of water evaporates? Assume the temperature is about 20°C.

Use Equation 4.2a.

$$Q = (2.5 \times 10^6 \text{ J/kg})\,(0.5 \text{ kg})$$
$$= 1{,}250{,}000 \text{ J}$$

In Example 4.4, we calculate the total amount of heat that must be transferred to warm 1 kg of water from ice at −25°C to steam at 125°C (Figure 4.4). First

Example 4.4

Calculate the total heat involved in warming 1 kg of water from an ice cube at −25°C to vapour at 125°C.

Step 1: Warm the ice 25 K. Use Equation 4.1.

$$Q = (1 \text{ kg})\,(2100 \text{ J}\cdot\text{kg}^{-1}\cdot\text{K}^{-1})\,(25 \text{ K})$$
$$= 52{,}500 \text{ J}$$

Step 2: Melt the ice. Use Equation 4.2b.

$$Q = (1 \text{ kg})\,(3.34 \times 10^5 \text{ J/kg})$$
$$= 334{,}000 \text{ J}$$

Step 3: Warm the water 100 K. Use Equation 4.1.

$$Q = (1 \text{ kg})\,(4186 \text{ J}\cdot\text{kg}^{-1}\cdot\text{K}^{-1})\,(100 \text{ K})$$
$$= 418{,}600 \text{ J}$$

Step 4: Vaporize the water. Use Equation 4.2a.

$$Q = (1 \text{ kg})\,(2.26 \times 10^6 \text{ J/kg})$$
$$= 2{,}260{,}000 \text{ J}$$

Step 5: Warm the vapour 25 K. Use Equation 4.1.

$$Q = (1 \text{ kg})\,(2040 \text{ J}\cdot\text{kg}^{-1}\cdot\text{K}^{-1})\,(25 \text{ K})$$
$$= 51{,}000 \text{ J}$$

Step 6: Add together the heat required at each step.

Total amount of heat = 3,116,100 J.

the ice must warm from −25°C to 0°C by *sensible* heat transfer. Once the temperature reaches 0°C, the ice begins to melt. Heat will continue to be transferred to the ice, but it is now *latent* heat. It is only once the ice has completely melted that its temperature begins to increase once more. However, the liquid water will warm at about half the rate as the ice warmed, because the specific heat of water is roughly double that for ice. At 100°C, another phase change occurs. This time the temperature remains constant at 100°C until the water has completely vaporized. Finally, the continued transfer of heat warms this vapour another 25°C. The total amount of heat required is simply the sum of the heat required in each individual step.

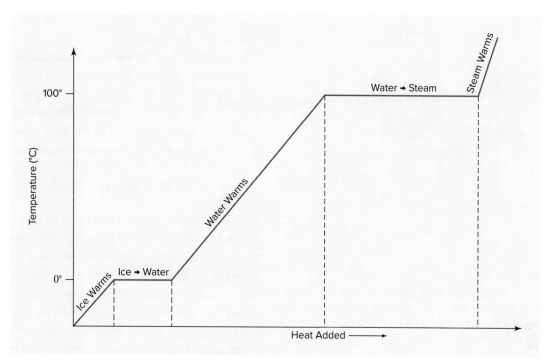

FIGURE 4.4 | Graph showing the temperature and phase changes of water as heat is added to it. Why is the horizontal line representing the phase change from ice to liquid shorter than the horizontal line representing the phase change from liquid to gas?

4.2 | Work

Whereas heat transfers energy as a result of temperature differences, work transfers energy by mechanical means. This could be something as simple as a push or a pull. When a person pushes a cart, for example, that person transfers chemical energy—which he or she has obtained from food—to the cart, at which point it becomes kinetic energy.

In the atmosphere, the concept of work is far more abstract than it is in the example given above. Gases perform work when they expand or contract. For example, an expanding gas does work because it *pushes* on its surroundings. This work transfers internal energy away from the expanding gas; consequently, the gas cools. The opposite is also true. When a gas is compressed, it warms because, in this case, the surroundings have performed work on the gas. Notice that when a gas is performing work on its surroundings, its volume increases due to expansion; conversely, when work is being done on a gas, its volume decreases due to compression. Therefore,

work performed upon, or by, a gas involves changes in volume.

To express this work mathematically, consider a gas in a container fitted with a movable piston (Figure 4.5). The gas law shows that if the gas is heated, it will do work on the piston, pushing it up. Because work is equal to force multiplied by distance (Equation 1.2), the work, W, done can be expressed as the product of the force, F, of the gas on the piston and the distance over which the piston moves, Δx. That is, $W = F \Delta x$. In turn, the force is equal to pressure, P, multiplied by the area, A, of the piston, so that $W = P A \Delta x$. The area of the piston times the change in distance, $A \Delta x$, is equal to the change in volume of the gas, ΔV. By following these steps, we obtain Equation 4.3:

$$W = P \Delta V \qquad (4.3)$$

Equation 4.3 shows that the work, in joules, performed by a gas is equal to the pressure of the gas multiplied by the change in the volume of the gas. Usually it is only gases that perform work on their surroundings in

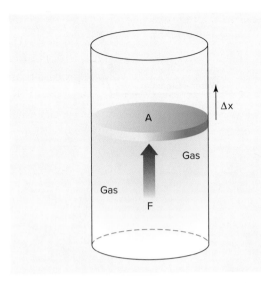

FIGURE 4.5 | When the gas in the cylinder expands, it produces a force, F, that performs work on the piston. The piston has area, A, and the distance that the gas moves the piston is Δx.

this way, because expansion in liquids and solids is so small that the amount of work performed is negligible.

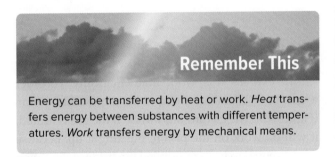

Remember This

Energy can be transferred by heat or work. *Heat* transfers energy between substances with different temperatures. *Work* transfers energy by mechanical means.

4.3 | The First Law of Thermodynamics

Because energy can be transferred as heat or by work, there are two ways to raise the temperature of a gas—by adding heat or by compressing the gas. To raise the temperature of a gas in a container, for example, you could place a heat source such as a flame below the container, or you could use a piston to compress the gas (Figure 4.6). Either way,

FIRST LAW OF THERMODYNAMICS
A law that states that a change in the internal energy of a substance is associated with the transfer of energy as heat or by work.

the internal energy and, therefore, the temperature of the gas will increase.

The **first law of thermodynamics** is about the relationship between the transfer of energy as heat or by work, and the associated changes in internal energy. This law was determined by experiment, and it was first formally stated by Rudolf Clausius in 1850. It can be expressed mathematically as Equation 4.4:

$$\Delta U = Q - W \qquad (4.4)$$

In this equation, ΔU is the change in internal energy, Q is heat transfer, and W is work. When heat is added to the system, Q is positive; when heat is removed, Q is negative. When work is done by the system, W is positive, and when work is done on the system, W is negative. For example, Equation 4.4 shows that if heat is added to a gas and, as a result, the gas performs work on its surroundings by expanding, then the change in internal energy will be the difference between the heat added and the work done.

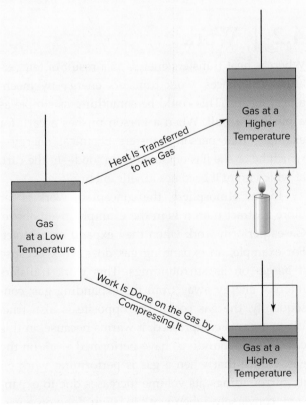

FIGURE 4.6 | To increase the temperature of a gas, heat can be transferred to the gas or work can be done on the gas.

For our purposes, it is most convenient to write Equation 4.4 as shown below:

$$Q = \Delta U + W = m\,c\,\Delta T + P\,\Delta V \qquad (4.5)$$

In this equation, $m\,c\,\Delta T$ represents the amount of heat used to change the internal energy of the air (Equation 4.1), and $P\,\Delta V$ represents the amount of heat involved in work (Equation 4.3). Equation 4.5, therefore, shows that all heat transferred to, or from, a gas must be accounted for by a temperature change, work, or both (Figure 4.7). Thus, the first law of thermodynamics is a statement of the conservation of energy.

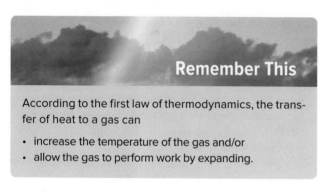

Remember This

According to the first law of thermodynamics, the transfer of heat to a gas can

- increase the temperature of the gas and/or
- allow the gas to perform work by expanding.

If a gas is confined, all added heat will increase its temperature; if a gas is unconfined, some of the added heat will cause the gas to expand, leaving less

heat to increase the temperature. The former is a *constant volume process*, while the latter is a *constant pressure process*. For the same heat transfer, temperature will increase more in a constant volume process than it will in a constant pressure process. Because of this, there are two specific heats for gases. The specific heat for a constant volume process, c_v, is 717 J·kg⁻¹·K⁻¹, and the specific heat for a constant pressure process, c_p, is 1004 J·kg⁻¹·K⁻¹. Therefore, we can write Equation 4.5 for a constant volume process (Equation 4.6) and for a constant pressure process (Equation 4.7):

$$Q = m\,c_v\,\Delta T + P\,\Delta V = m\,c_v\,\Delta T \qquad (4.6)$$

$$Q = m\,c_v\,\Delta T + P\,\Delta V = m\,c_p\,\Delta T \qquad (4.7)$$

Because the air cannot expand in a constant volume process, ΔV in Equation 4.6 is zero, and Equation 4.6 reduces to Equation 4.1. In other words, no work is done; *all* the heat transferred to the gas will be used to increase its internal energy, or temperature. In contrast, Equation 4.7 shows that during a constant pressure process, some of the heat is used in expansion of the gas, so ΔV is not equal to zero. As a result, the temperature increase for a constant pressure process will be less than that for a constant volume process.

Question for Thought

Why is c_v less than c_p?

Although constant volume processes cannot occur in the atmosphere, constant pressure processes can. In Example 4.5, heat is being transferred from Earth's surface to the air above. Some of the heat is used to raise the temperature of the air; some of the heat is used in expansion. As the air expands, the pressure it exerts on the surrounding air does not change; thus, this is a constant pressure process. (But recall from Section 3.6 that heating will reduce the *surface* pressure because as the air expands there will be less of it above a given area of surface.)

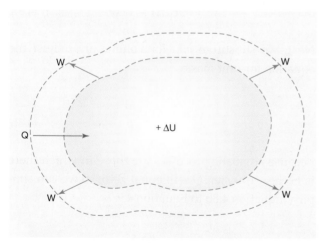

FIGURE 4.7 | The first law of thermodynamics applied to a parcel of air. *Q* is the heat added to the air parcel. This added heat increases the internal energy, *ΔU*, of the air parcel and allows it to perform work, *W*, by expanding. The inner dashed line represents the original air parcel; the outer dashed line represents the expanded air parcel. How would this diagram be different if heat were *lost* from the air parcel?

Example 4.5

Imagine a column of air with a height of 1 km and a cross-sectional area of 1 m². Assume that the density of the air in this column is 1.2 kg/m³. How much heat must be added to this air to raise its temperature by 2°C? How much of this heat is used in expansion of the air?

First, calculate the mass of the air column.

$$1000 \text{ m}^3 \times 1.2 \text{ kg/m}^3 = 1200 \text{ kg}$$

Second, find out how much heat would be needed to raise the temperature of the air by 2°C if it *didn't* expand. In other words, find out how much heat would be needed in a constant volume process. Use Equation 4.6, with $\Delta V = 0$.

$$Q = (1200 \text{ kg}) (717 \text{ J} \cdot \text{kg}^{-1} \cdot \text{K}^{-1}) (2 \text{ K})$$
$$= 1{,}720{,}800 \text{ J}$$

Third, find out how much heat would be needed to raise the temperature of the air by 2°C, allowing for the fact that in the atmosphere it *will* expand. This is a constant pressure process. Use Equation 4.7.

$$Q = (1200 \text{ kg}) (1004 \text{ J} \cdot \text{kg}^{-1} \cdot \text{K}^{-1}) (2 \text{ K})$$
$$= 2{,}409{,}600 \text{ J}$$

Finally, to find out how much heat is used in the expansion of the air, rearrange Equation 4.7 to solve for work.

$$W = P \, \Delta V = m \, c_p \, \Delta T - m \, c_v \, \Delta T$$
$$= 2{,}409{,}600 \text{ J} - 1{,}720{,}800 \text{ J}$$
$$= 688{,}800 \text{ J}$$

This example shows that more heat is needed, for the same temperature change, in a constant pressure process than in a constant volume process. This is because, in a constant pressure process, some of the added heat is doing work.

Another thermodynamic process that is common in the atmosphere is an **adiabatic process**, in which temperature changes result from pressure changes rather than heat transfer. A true adiabatic process can be achieved only when a substance is thermally insulated from its environment, but vertical motions in the atmosphere are generally rapid enough that the amount of heat the air parcel exchanges with its surroundings is very small. Because of this, we can assume that such vertical motions are adiabatic. Due to the compressibility of air, air parcels adjust to the changing pressure of their surroundings by expansion and contraction. This work causes temperature changes—a rising air parcel expands and cools, while a sinking air parcel is compressed and warms. Since pressure, temperature, and density all change in rising and sinking air parcels, it follows that in an adiabatic process all three variables will change (Figure 4.8).

ADIABATIC PROCESS
A thermodynamic process in which temperature changes without a transfer of heat.

Question for Thought

In order to keep temperature constant, Boyle's experiments (Section 3.2.1) would have required that heat be added or removed as the volume of the gas changed. Why?

To derive an equation for an adiabatic process, therefore, we start with a form of the first law of thermodynamics equation that allows pressure, temperature, and density to vary:

$$Q = m \, c_p \, \Delta T - V \, \Delta P \tag{4.8a}$$

Next, we substitute m/ρ for volume and adjust the equation for unit mass:

$$Q = c_p \, \Delta T - \frac{\Delta P}{\rho} \tag{4.8b}$$

Because adiabatic processes are those in which there is no heat transfer, Q will equal zero, so we can simplify Equation 4.8b to Equation 4.9:

$$c_p \, \Delta T = \frac{\Delta P}{\rho} \tag{4.9}$$

In this equation, $c_p \, \Delta T$ represents the change in internal energy, and $\Delta P/\rho$ represents work done due to expansion or compression of the gas. This equation

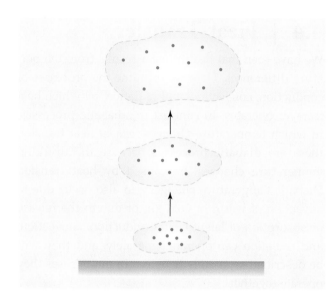

FIGURE 4.8 | The adiabatic ascent of an air parcel. As the parcel rises, it gradually expands as the pressure on it decreases. This expansion causes the air to cool and, in the process, the density of the air parcel also decreases.

shows that expansion, or a decrease in pressure, will result in a decrease in temperature, and compression, or an increase in pressure, will result in an increase in temperature.

Question for Thought

How could you use Equation 4.9 to confirm that a decrease in pressure will cause cooling and vice versa?

We can derive a general equation relating temperature and pressure for an adiabatic process from Equation 4.9 by making use of the gas law to replace density. Then, with integration, we can obtain Equation 4.10, which is known as **Poisson's equation**:

$$\frac{T_2}{T_1} = \left(\frac{P_2}{P_1}\right)^{\frac{R_d}{c_p}}$$

(4.10)

In this equation, the subscripts for temperature and pressure indicate initial and final states, and the exponent is the ratio of the gas law constant, R_d, to the specific heat of air at constant pressure, c_p. This

ratio is equal to 0.286. Temperature must be in degrees kelvin. Recall from Section 3.2.3 that if volume is constant, pressure and temperature are directly proportional. Poisson's equation also shows that pressure and temperature are directly proportional, but the exact relationship between them is different in this case because volume, or density, is *not* constant. (We will apply Equations 4.9 and 4.10 in Section 8.1, as they are fundamental to an understanding of atmospheric stability.)

POISSON'S EQUATION
An equation relating temperature and pressure for an adiabatic process.

Example 4.6

Determine the temperature of an air parcel after it rises from sea level to a height of 2000 m. At sea level, the parcel's temperature is 23°C.

Use Table 1.3 to estimate the sea level pressure as 101.325 kPa and the pressure at 2000 m as 79.501 kPa, then use Equation 4.10 (Poisson's equation).

$$\left(\frac{T_2}{296\ K}\right) = \left(\frac{79.501}{101.325}\right)^{0.286}$$

$$\left(\frac{T_2}{296\ K}\right) = (0.785)^{0.286}$$

$$\left(\frac{T_2}{296\ K}\right) = 0.933$$

$$T_2 = 276\ K$$

$$= 3°C$$

As the air parcel rises, it encounters lower pressure and expands. This expansion causes it to cool to 3°C.

Question for Thought

If a gas is heated, it will expand. If a gas expands due to decreasing pressure, it will cool. Why?

We can also apply the first law of thermodynamics in the form of Equation 4.4 to phase changes because phase changes, like temperature changes, are associated with changes in internal energy. For example, when water changes from a liquid to a gas, the molecules will have considerably more internal energy. Not only that, but when water becomes a gas, it expands to almost 1700 times its original volume. The work resulting from this expansion uses some of the energy that would otherwise have gone into internal energy. This is shown in Example 4.7.

Example 4.7

a) If 1 L of water is boiled until it has completely vaporized, how much latent heat is required? (Note that 1 L of water weighs 1 kg).

Use Equation 4.2a.

$$Q = (2.26 \times 10^6 \text{ J/kg}) (1 \text{ kg})$$

$$= 2.26 \times 10^6 \text{ J}$$

b) How much does the internal energy of the water molecules increase as a result of this phase change? (Note that when 1 L of liquid water becomes a gas, its volume increases to 1671 L. Also note that 1 L of water has a volume of 1000 cm^3 or 0.001 m^3.)

First, use Equation 4.3 to determine how much work is done as a result of the expansion that occurs when water becomes gas. Assume that this process is occurring at the standard atmospheric pressure of 101.3 kPa.

$$W = (1.013 \times 10^5 \text{ Pa}) (1.671 \text{ m}^3 - 0.001 \text{ m}^3)$$

$$= 1.69 \times 10^5 \text{ J}$$

Now use Equation 4.4 to determine the change in internal energy.

$$\Delta U = 2.26 \times 10^6 \text{ J} - 1.69 \times 10^5 \text{ J}$$

$$= 2.1 \times 10^6 \text{ J}$$

The water molecules have experienced a large increase in internal energy as a result of becoming vapour, but not quite as much as they would have experienced if vaporization were not associated with expansion.

4.4 | Heat Transfer

We have seen that heat transfer results from temperature differences. Here we examine the processes of conduction, convection, and radiation, by which heat transfer operates. In contrast to adiabatic processes, in which temperature changes *without* heat transfer, these are **diabatic processes** because the resulting temperature changes *are* caused by heat transfer. Diabatic temperature changes can also occur due to advection of warm or cold air, or due to the release or absorption of latent heat. Conduction, convection, and radiation can operate separately, and they will be described separately here, but in many cases they operate together.

4.4.1 | Conduction

Conduction transfers heat from molecule to molecule through a substance or between substances. According to kinetic theory, when molecules are warm, they move faster than when they are cold. When molecules are in contact, faster moving molecules will pass their motion on to slower moving ones, transferring heat in the process.

Because molecules are close together in a solid, **conductivity** is generally highest for solids; in contrast, gases—such as air—generally have low conductivities because their molecules are farther apart (Table 4.2). Of the solids, metals have the highest conductivities. For example, when you use a metal spoon to stir a pot of soup, heat is conducted from the end of the spoon that is in the soup to the other end very quickly. A spoon made of wood, which is a poorer conductor, conducts heat much more slowly.

Materials that are poor conductors make good insulators. Air, for example, is an excellent insulator. Things that keep us warm, like comforters and down jackets, generally contain a lot of air space. Double-glazed windows help reduce heat loss because they are made with a tiny air space between two panes of glass. Snow is also a good insulator because it contains pockets of air; once plants have a cover of snow, they retain some of their heat, and the danger of frost damage becomes less.

TABLE 4.2 | Conductivities of various substances.

Substance	Conductivity (W·m⁻¹·K⁻¹)
Still Air (10°C)	0.025
Water (10°C)	0.62
Ice (0°C)	2.24
Fresh Snow	0.08
Old Snow	0.42
Dry Sand	0.15–0.25
Moist Sand	0.25–2.0
Granite	2.9
Limestone	1.3
Light Wood	0.09
Dense Wood	0.19
Stainless Steel	16
Aluminum	200

Question for Thought

How can the kinetic theory be used to account for the differences in conductivity between solids, liquids, and gases?

Because of the low conductivity of air, conduction is not an important heat transfer process in the atmosphere as a whole. However, heat *is* conducted between the surface and the thin air layer directly in contact with it. This layer is known as the **laminar boundary layer**, and it is normally just a few millimetres thick.

Heat is also conducted between the surface and the ground below. During the day, when the surface is normally warmer than the ground, heat will be conducted downward. At night, the surface is normally colder than the ground, and heat is conducted back up. On a daily basis, this heat flow influences depths of at *most* 10 or 20 cm. Even on an annual basis, surface heating does not penetrate far into the ground. The annual heating cycle may reach depths of only a few metres, and thus, at relatively shallow depths in the ground, temperatures change very little.

As a result of heat flow between the surface and the ground, the conductivity of the ground material can have an important influence on surface temperatures. If this material is a poor conductor, the surface can get quite hot during the day because heat is not easily conducted downward. At night, this surface will rapidly cool because there is little heat stored in the ground to be conducted back up. Because the surface temperature is an important determinant of the heat flow into the atmosphere, the conductivity of the ground can influence daily temperature ranges in the air above. For example, the poor conductivity of dry sand is one reason for the large daily temperature ranges over deserts (Figure 4.9).

DIABATIC PROCESS
A process in which temperature changes as a result of heat transfer.

CONDUCTIVITY
The property of a substance that describes its ability to conduct heat.

LAMINAR BOUNDARY LAYER
The layer of air in contact with Earth's surface, through which heat is transferred by conduction.

Question for Thought

Considering what you have learned about conductivity, specific heat, and latent heat, how would you explain why dry sand can be very hot to walk on?

FIGURE 4.9 | Daily temperature ranges in the Kalahari Desert can exceed 30°C.

4.4.2 | Convection

Convection transfers heat through the motions of liquids and gases. While conduction requires contact between molecules, convection involves whole groups of molecules moving around together. Hotter groups of molecules move to areas where the molecules are colder, transferring heat in the process. Because convection requires an ability to flow, it does not commonly occur in solids. There are two types of convection: **thermal convection** and **mechanical convection**. Both types are associated with **turbulence**.

Thermal, or free, convection is driven by the density differences that result from temperature differences. Warmer, less dense fluids have a tendency to rise, while cooler, denser fluids have a tendency to sink. This movement creates convection currents that transfer heat. For example, a pot of water on the stove will quickly be heated through because of convection (Figure 4.10). Initially, the water in contact with the bottom of the pot will heat by conduction; then, as the warmed water rises and the cold water above sinks in a continual cycle, the rest of the water will heat by convection.

A similar process occurs in the atmosphere and, on a warm day, can carry heat from the surface to heights of 2 to 3 km. As we have seen when the sun heats Earth's surface during the day, this heat is conducted to air molecules in contact with the surface—those molecules that make up the laminar boundary layer. From here, *thermal convection* takes over: as the density of the warmed surface air decreases, this air rises and is replaced by cooler air that sinks from above. Thus, warming induces free convection, which carries this warmth upward.

On the other hand, winds induce *mechanical*, or forced, convection. In our homes, fans can force air to move around. On a hot summer day, a fan can blow away the warm air in contact with our skin, making us feel cooler; in winter, fans force warm air from a furnace into a room. In the same way, winds can force mixing in the atmosphere.

The turbulence generated by winds increases with **wind shear** and **surface roughness**. *Wind shear* refers to a change in wind speed, direction, or both in any direction across the flow (Figure 4.11). Wind shear generates vertical motions in otherwise horizontal flow. *Surface roughness* refers to the character of the surface. For example, while a frozen lake is a very smooth surface, urban areas and forests are very

THERMAL CONVECTION
Convection that is driven by the density differences that result from temperature differences.

MECHANICAL CONVECTION
Convection that is driven by mechanical forces.

TURBULENCE
Random, irregular motions in a fluid.

WIND SHEAR
The change in wind speed and/or direction in any direction across the flow.

SURFACE ROUGHNESS
The degree of irregularity of a surface.

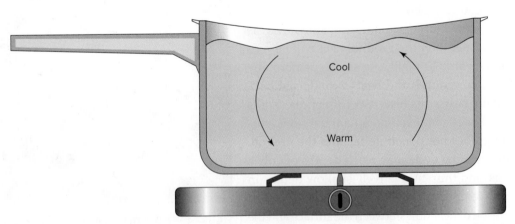

FIGURE 4.10 | Thermal convection. This diagram shows how convection currents transfer heat through a pot of water on the stove. A similar process occurs in the atmosphere.

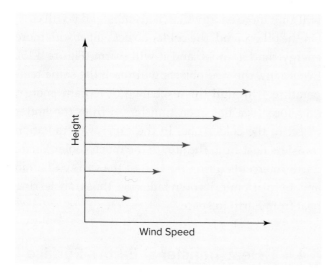

FIGURE 4.11 | The increase in wind speed with height produces wind shear.

rough surfaces. Roughness, like shear, can generate vertical motions in the flow that mix the air, *forcing* warm air down and cold air up.

Although thermal convection can quite effectively carry heat from the surface upward into the atmosphere, mechanical convection can significantly increase this transfer. As a result, on a windy day, temperature is often more uniform with height than it is on a calm day (Figure 4.12). Consider that soup sitting in a pot on the stove will burn on the bottom if it is not stirred, but it can be brought to a constant temperature throughout by stirring. The behaviour of the atmosphere, upon being stirred by wind, is similar to this, but far more complex (Section 8.1.1).

On a clear night, the surface air can become much colder than the air above. Since cold air cannot freely rise, thermal convection cannot occur. A wind, however, can *force* the cold air up, helping to distribute heat more evenly. That is why phenomena like dew, frost, and fog, which occur under conditions of strong surface cooling, are more likely to develop on calm nights than on windy nights.

Similar effects can occur in water bodies, but in reverse, because they are heated from the top rather than the bottom. On a warm day, a warm layer of water will form at the surface of an ocean or a lake. Because this warm water will be less dense than the water below, it will not sink. Swimmers know that the water deeper than half a metre or so below the surface can be quite cold. On a windy day, such temperature stratification is less likely to develop. The surface water won't feel quite as comfortable for swimming, but the water below won't feel quite as cold (Figure 4.13). Mixing tends to distribute heat more effectively.

The above focuses on the transfer of *sensible* heat by convection, but convection can also transfer *latent* heat. In fact, on a global scale, the convective transfer of latent heat is greater than is the convective transfer of sensible heat (Section 6.4). When water evaporates at Earth's surface, heat is absorbed in the form

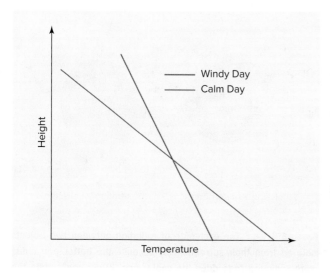

FIGURE 4.12 | The change in temperature with height in the lower atmosphere on a calm day and on a windy day.

FIGURE 4.13 | The surface warmth swimmers enjoy in an ocean or a lake on a calm day can be mixed—by mechanical convection—with the cooler water below on a windy day.

of latent heat. As air rises in the atmosphere, water vapour is carried upward. Eventually, this water vapour will condense to form clouds, and latent heat will be released. The process of cloud formation, therefore, effectively transfers heat from Earth's surface up into the atmosphere.

4.4.3 | Radiation

Radiation is a form of energy that travels as **electromagnetic waves** (Section 5.1) and, in the process, can transfer heat. Some familiar examples of electromagnetic waves are light, microwaves, radio waves, and X-rays. Because electromagnetic waves can travel in a vacuum, radiation is the only form of heat transfer that can operate in space. It is by radiation, therefore, that heat is transferred from the sun to Earth, and from Earth to space.

ELECTROMAGNETIC WAVES
Waves that are formed and propagated by oscillating electric and magnetic fields.

Radiation can transfer heat because substances both *emit* and *absorb* radiation (Figure 4.14). In fact, the radiation laws tell us that *everything* emits radiation, and that hotter substances emit more radiation than do cooler substances (Section 5.2). To understand how radiation transfers heat, imagine two objects exchanging radiant energy. If these two objects are isolated from their surroundings, the hotter object will emit more energy than it absorbs, and it will cool. On the other hand, the colder object will absorb more energy than it emits, and it will warm (Figure 4.15). Eventually, the two objects will reach the same temperature. Through this exchange of radiant energy, therefore, heat has been transferred from the hotter object to the colder one. In the same way, radiation transfers heat from the sun to the much cooler Earth. Earth intercepts a tiny portion of the sun's radiation and, in turn, emits its own radiation, thus transferring heat from Earth to space.

4.4.4 | Heat Transfer at Earth's Surface

As Earth's surface absorbs the sun's radiation during the day, it warms. (Some of the sun's radiation is absorbed as it passes through the atmosphere, but most of it is absorbed by the surface.) Heat transfer then takes place *away* from the warmer surface, up into the cooler atmosphere by convection, and down into the cooler substrate by conduction (Figure 4.16a). In addition to these non-radiative processes, radiative heat transfer operates to transfer heat away from the surface. Like convection, this process transfers heat up into the air above the surface. This occurs as radiation emitted by Earth is absorbed by gases and aerosols, including cloud droplets, in the atmosphere. If radiation was the only process that transferred heat

FIGURE 4.14 | These young campers are warmed as they absorb the radiant energy emitted by the fire.

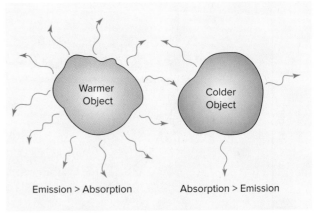

Emission > Absorption   Absorption > Emission

FIGURE 4.15 | The exchange of radiation between two objects isolated from their surroundings. Because the hotter one emits more radiation than does the cooler one, it will cool, while the cooler object will warm.

heat upward from the ground, and, if there is a wind, *forced* convection carries heat downward from the air above (Figure 4.16b).

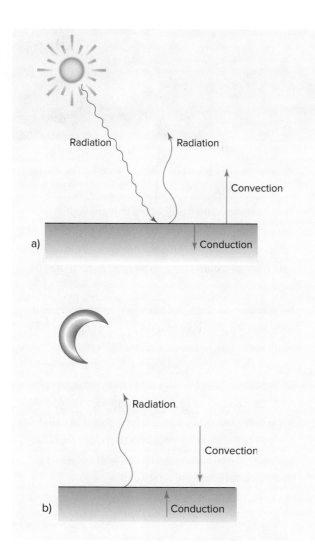

FIGURE 4.16 | Heat transfer at Earth's surface by a) day and b) night, under ideal conditions.

to the atmosphere, far less heat transfer would occur; as a result, the surface would be much hotter—over 60°C on average—and the temperature decrease with height in the troposphere would be greater.

At night, the surface continues to emit radiation to the cooler atmosphere and, since it is now emitting more radiation than it is absorbing, gradually cools. As the surface cools, the non-radiative heat transfer processes carry heat *toward* it: conduction carries

Question for Thought

How is heat transferred away from the surface of the moon (Figure 4.17)?

FIGURE 4.17 | The moon.

Question for Thought

How has the kinetic theory (Section 3.1) helped you to understand the following topics discussed in this chapter: the distinction between heat, internal energy, and temperature; the process of evaporation; the amount of energy associated with phase changes; and the non-radiative heat transfer processes?

4.5 | Chapter Summary

1. The law of conservation of energy states that energy can never be created or destroyed but can be transformed from one type to another and transferred from one object to another. Two forms of energy that are important in the atmosphere are internal energy and radiant energy.

2. Energy can be transferred as heat, work, or both. Heat transfers energy as a result of temperature differences. Work transfers energy by mechanical means. In the atmosphere, work is usually associated with the expansion or compression of air.

3. Sensible heat is associated with temperature changes, while latent heat is associated with phase changes. Specific heat is a property of a substance that determines how much its temperature will change in response to sensible heat transfer.

4. The first law of thermodynamics is a statement of the law of conservation of energy. It states that changes in internal energy are equivalent to transfers of energy as heat, work, or both.

5. Two thermodynamic processes that are important in the atmosphere are constant pressure processes and adiabatic processes. An adiabatic process is one in which temperature changes without adding or removing heat.

6. Heat is transferred by conduction, convection, and radiation. The transfer of heat by conduction is far more important in the ground than it is in the air. Convection transfers heat in fluids and is, therefore, an important heat transfer process in the atmosphere. Radiation can transfer heat through a vacuum, making it the only method by which heat transfer occurs from the sun to Earth, and from Earth to space. Together, convection and radiation transfer heat between Earth's surface and the atmosphere.

Review Questions

1. How are heat, internal energy, and temperature different from one another?

2. How are specific heat and sensible heat different from one another?

3. How does a substance's specific heat influence its response to heat transfer?

4. How are sensible heat and latent heat different from one another?

5. Why is the heat involved in phase changes called latent, or hidden, heat?

6. How do gases perform work?

7. In what ways are heat and work the same? In what ways are they different?

8. What makes the first law of thermodynamics (Equation 4.4) a statement of the law of conservation of energy?

9. Given the same input of heat, why will the temperature of an air parcel increase more in a constant volume process than it will in a constant pressure process? Which process is applicable to the atmosphere? Why?

10. In what way does an adiabatic process obey the law of conservation of energy?

11. How is heat transferred by a) conduction, b) convection, and c) radiation?

12. Does heat transferred from the surface penetrate farther into the ground or the air? Why?

Problems

1. If 10 mL of water condenses on a glass holding 300 mL of a cold drink, how much will the drink warm? Assume that all the latent heat released by condensation warms the drink and that the specific heat of the drink is the same as the specific heat for water. (Note that 1 L of water weighs 1 kg.)

2. A commercial jet is flying at an altitude where the air pressure is 20 kPa and the temperature is −60°C. If the air is brought inside the aircraft and pressurized to 80 kPa, what will be its temperature?

3. How much heat does a freezer need to remove for 500 g of water at 20°C to be made into a block of ice at −15°C?

4. If 10,000 J of heat are added to 1 m³ of air at constant pressure, how much will the air warm? How much of the 10,000 J is used in expansion of the air? (Assume the density of the air is 1.2 kg/m³.)

5. How much heat would be released into the atmosphere if 1 km³ of ocean water cooled by 1°C? (This problem shows that water stores a lot of heat!)

6. 10,000 J of heat are transferred to a) 1 kg of water and b) 1 kg of dry soil. How much will each warm?

Suggestion for Research

1. Investigate the experiments done by Joseph Black that led Black to recognize that substances have different specific heats and that phase changes are associated with latent heat.

2. Look into Rudolf Clausius's work in the field of thermodynamics, and explain why Clausius's statement of the first law of thermodynamics has remained so central to our understanding of temperature changes in the atmosphere.

We cannot see all types of radiation but the radiation that we can see often produces spectacular displays in the sky. The red appearance of the sky at sunset is one such display—it results because the atmosphere scatters some colours of visible radiation, or light, from the sun more than it scatters others.

5

Radiation

Learning Goals

After studying this chapter, you should be able to

1. *distinguish* between ultraviolet, visible, and infrared radiation;

2. *describe* the relationships given by the four radiation laws;

3. *distinguish* between the radiation emitted by the sun and the radiation emitted by Earth;

4. *define* the term *albedo*, and *explain* how it might be important to climate;

5. *describe* the effects of Rayleigh scattering and Mie scattering;

6. *understand* how atmospheric gases and aerosols interact with shortwave radiation and longwave radiation;

7. *explain* Earth's greenhouse effect;

8. *explain* why we have seasons; and

9. *account* for all the factors that influence the amount of solar radiation absorbed at Earth's surface.

In Chapter 4, you learned that radiation is a type of energy that travels as waves and that the absorption and emission of radiation can result in the transfer of energy as heat. All substances emit radiation, and scientific laws governing the emission of radiation tell us that temperature determines both the *rate* of emission and the *wavelengths* that will be emitted. These radiation laws allow us to distinguish between the radiation emitted by the sun, and that emitted by Earth.

WAVELENGTH
The distance between any two like points on a wave.

As radiation travels through the atmosphere, it is absorbed and reflected by gases and aerosols. In this chapter, and in Chapter 6, you will see that these processes are fundamentally important to climate. In particular, the selective absorption of radiation by atmospheric gases is responsible for the *greenhouse effect*, first described in Section 2.5. You will also follow the journey of the sun's radiation as it travels through space to the top of the atmosphere, and eventually to Earth's surface. In doing this, you will be able to account for the variation, in both time and space, of solar radiation at Earth's surface.

ELECTROMAGNETIC SPECTRUM
The continuous spectrum of wavelengths of electromagnetic radiation.

5.1 | The Electromagnetic Spectrum

Radiation is energy that travels in the form of electromagnetic waves. The mechanism responsible for electromagnetic radiation is molecular motion that creates an electric field. Electric fields, in turn, induce magnetic fields. As electric and magnetic fields oscillate together, they produce electromagnetic waves that travel outward from their source like the ripples that form when a rock is thrown into a pond. For example, an alternating electric current in a power line will produce electromagnetic radiation, which travels outward from the wire in all directions. Electromagnetic waves are similar in *form* to sound waves and waves on the water but, unlike sound and water waves, they do not require the presence of a medium to travel. As

a result, they are able to transfer energy through the vacuum of space.

The existence of electromagnetic radiation was predicted in the 1860s by James Clerk Maxwell from a set of equations he had derived describing the behaviour of electric and magnetic fields. The equations showed, among other things, that electromagnetic radiation travels in waves at a speed of 3×10^8 m/s. At the time it was already known that this is the speed of light, so Maxwell rightly concluded that light is a form of electromagnetic radiation. However, not all electromagnetic radiation is light.

About 20 years later, Heinrich Hertz produced and detected radio waves, another form of electromagnetic radiation. He also found that radio waves have much longer wavelengths than light. In fact, all that separates one form of electromagnetic radiation from another is wavelength, which is measured as the distance between any two like points on a wave (Figure 5.1).

The range of wavelengths across the **electromagnetic spectrum** is tremendous (Figure 5.2). It runs from wavelengths of a millionth of a micron through to wavelengths measuring in kilometres. Although the spectrum is continuous, there are major differences in the properties of electromagnetic radiation across this spectrum. These differences have led us to divide the spectrum into *groups* of wavelengths. The shortest wavelengths of radiation, which we call *gamma rays*, are produced by radioactivity and can penetrate solid materials, such as thick concrete. *X-rays* have slightly longer wavelengths and are produced by firing electrons at a target in a vacuum tube. X-rays can penetrate living tissue and, as such, have well-known

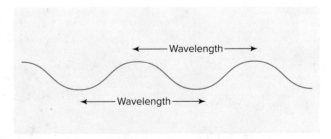

FIGURE 5.1 | A wavelength is measured as the distance between two like points on a wave.

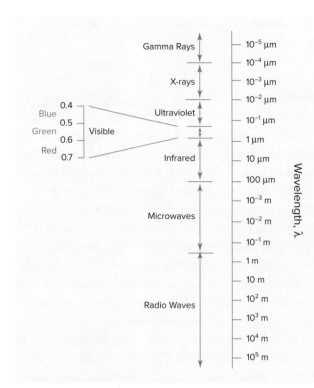

FIGURE 5.2 | The electromagnetic spectrum.

applications in medicine. Still longer in wavelength is *ultraviolet radiation*. This radiation is familiar to us as the radiation from the sun that gives us a suntan, and it can be artificially produced by black lights.

Radiation with wavelengths from roughly 0.4 to 0.7 μm is the only radiation that we can see; thus, we call this group of wavelengths *visible radiation*, or light. Visible radiation can be further subdivided into the colours of the rainbow (Figure 5.3), with the three primary colours of light being blue, green, and red, from shortest wavelength to longest. When we see all of the colours of light together, they appear white. It is likely *because* we can see it that visible radiation was the first part of the electromagnetic spectrum to be recognized by Maxwell in the 1860s.

The next group on the spectrum, *infrared radiation*, is most noticeable as warmth, because our bodies preferentially absorb this sort of radiation. (For this reason, when infrared radiation was first detected, it was called "heat rays.") Slightly longer wavelength *microwave radiation* is, of course, most commonly used

FIGURE 5.3 | This rainbow near Lac-Saint-Jean, Quebec, reveals the spectrum of colours that make up visible radiation.

to warm food in microwave ovens. *Radio waves*, the longest wavelengths of the spectrum, can be both produced and detected by antennas. Because radio waves are not strongly absorbed in the atmosphere, they are the wavelengths used for broadcasting. In atmospheric sciences, we are concerned only with the middle portion of the electromagnetic spectrum—namely, ultraviolet, visible, and infrared radiation.

Remember This

The following wavelengths of electromagnetic radiation are most important in the atmosphere:

- ultraviolet radiation: 0.1 to 0.4 µm;
- visible radiation: 0.4 to 0.7 µm; and
- infrared radiation: 0.7 to 100 µm.

PLANCK'S LAW
A radiation law stating that, for any substance, the rate of emission per wavelength increases with temperature, and the lengths of the waves emitted decrease with temperature.

STEFAN–BOLTZMANN LAW
A radiation law stating that the rate of emission of radiation by a substance will increase with the temperature of the substance.

WIEN'S LAW
A radiation law stating that the wavelength at which a substance emits the most energy will decrease as the temperature of the substance increases.

BLACK BODY
A hypothetical substance that does not reflect or transmit radiation but instead absorbs all of the radiation incident on it.

ABSORPTIVITY
A measure of a substance's ability to absorb incident radiation.

Although the various forms of electromagnetic radiation are very different, they have certain characteristics in common: they all *travel at the speed of light*, they can all *travel in a vacuum*, and they all *transfer energy*. (Recall that the way in which radiation transfers energy was described in Section 4.4.3.)

Shorter wavelength radiation can transfer more energy than longer wavelength radiation can. This is because, in addition to behaving as waves, radiation also behaves as particles. It is as though the energy of radiation is packaged in bundles; these bundles of energy are called *photons*. Equation 5.1 shows that the amount of energy, *E*,

in a photon is proportional to the frequency, *f*, of the radiation.

$$E = h\,f \qquad (5.1)$$

In Equation 5.1, *h* is Planck's constant; at roughly 6×10^{-34} joule-seconds, it is the smallest constant in physics. Frequency is a measure of the number of waves that pass a given point in a second. The units used to measure frequency are waves per second (s^{-1}), or hertz, named after Heinrich Hertz. Because frequency is inversely proportional to wavelength, Equation 5.1 shows that radiation with shorter wavelengths carries more energy than does that with longer wavelengths. X-rays, for example, carry considerably more energy than do microwaves. The more energy it carries, the more damaging the radiation is to life. As wavelength decreases, radiation begins to become dangerous at the ultraviolet wavelengths. Ultraviolet radiation coming from the sun, and reaching Earth's surface, is easily absorbed by living things, with potentially damaging effects. This is why the ozone layer (Section 2.7) is so important for the protection it provides life.

5.2 | Emission of Radiation

Everything—from the sun and Earth right down to the tiniest gas molecule in the atmosphere—emits electromagnetic radiation. It is obvious to us that the sun emits radiation because we *see* this radiation as light, and *feel* it as heat. Although Earth also emits radiation, this radiation is far less obvious, mostly because it is invisible. The differences between the radiation emitted by the sun and that emitted by Earth can be explained by three radiation laws: **Planck's law**, the **Stefan–Boltzmann law**, and **Wien's law**. These laws tell us about the relationships between the temperature of a substance and the radiation it emits. Strictly speaking, these laws apply only to **black bodies**—substances that have an **absorptivity** of 1.0, or 100 per cent. Black bodies are perfect emitters, as they emit radiation exactly as the radiation laws predict. No true black bodies exist but, in most cases, the radiation laws work well because most things are close to being black bodies.

The characteristics of the radiation emitted by an object are best shown graphically. Figure 5.4 shows

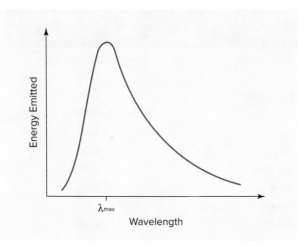

FIGURE 5.4 | Graph showing the emission of radiation as a function of wavelength. The resulting curve is known as *Planck's curve*.

that objects emit radiation over a *continuous* spectrum, or range of wavelengths, but the rate of emission of radiation is not the same at each wavelength. (This does not apply to gases; their emission spectrums are *discontinuous*.) Instead, as wavelength increases, the rate of radiation emission *rapidly* increases to a peak at the **wavelength of maximum emission** (λ_{max}); then, as longer wavelengths are approached, the rate of radiation emission *gradually* decreases.

The exact *lengths* of the waves emitted and *rates* of emission at each wavelength have not been labelled on Figure 5.4 because they both depend on temperature. For hotter objects, the wavelengths emitted are shorter and the emission rates are higher than they are for cooler objects. The curve shown in Figure 5.4 is known as **Planck's curve**. The formulation of this characteristic curve, and the equation that describes it, represents the culmination of almost 50 years of work, by five scientists in particular: Gustav Kirchhoff (1824–1887), Josef Stefan (1835–1893), Ludwig Boltzmann (1844–1906), Wilhelm Wien (1864–1928), and Max Planck (1858–1947).

5.2.1 | The Stefan–Boltzmann Law

In 1861, Gustav Kirchhoff used principles of thermodynamics to prove that the emission of radiation by a black body varies with wavelength and temperature, but he was unable to determine the exact mathematical form of the relationship. Almost 20 years later, in

1879, Josef Stefan discovered, by experiment, that the rate of emission of radiation, *E*, is proportional to the fourth power of the absolute temperature, *T*.

$$E \sim T^4 \tag{5.2}$$

To determine this, Stefan measured the emission of radiation from two substances with different temperatures. He discovered that the ratio of the temperatures of these substances, raised to the fourth power, is equal to the ratio of the energy emitted by each. Example 5.1 is a hypothetical illustration of his work.

In 1884, Ludwig Boltzmann, a student of Stefan's, used thermodynamics to derive the *mathematical* form of Stefan's relationship. The result has become known as the Stefan–Boltzmann law.

$$E = \sigma T^4 \tag{5.3}$$

> **WAVELENGTH OF MAXIMUM EMISSION**
>
> The wavelength at which the rate of emission of radiation is highest.
>
> **PLANCK'S CURVE**
>
> The graphical representation of Planck's law; it shows that the rate of emission of radiation per wavelength rises rapidly with increasing wavelength, reaches a peak (λ_{max}), and then decreases gradually with further increases in wavelength.

> **Example 5.1**
>
> During experimentation, Stefan might have noted that a substance with a temperature of 500°C (773 K) emitted 20,250 W/m^2 of radiation, while a substance with a temperature of 1000°C (1273 K) emitted 148,900 W/m^2 of radiation. (Recall from Section 1.5 that W/m^2 is the rate of flow of energy per unit area of surface, or the flux density.) How does this data indicate that the amount of radiation emitted is proportional to the fourth power of temperature?
>
> $$\left(\frac{773 \text{ K}}{1273 \text{ K}}\right)^4 = \left(\frac{20{,}250 \text{ W/m}^2}{148{,}900 \text{ W/m}^2}\right)$$
>
> $$0.14 = 0.14$$
>
> In other words, if the ratio of the temperatures is raised to the fourth power, it is equal to the ratio of the radiation emitted.

In Equation 5.3, E represents the flux density of the energy emitted by a black body with a temperature of T. Flux density is measured in watts per square metre (W/m²), and temperature is measured in degrees kelvin. The symbol σ represents the Stefan–Boltzmann constant, which is 5.67×10^{-8} W·m⁻²·K⁻⁴. (Boltzmann determined this constant mathematically, rather than experimentally.) Equation 5.3 shows that as temperature increases, so does the rate of flow of radiation (Figure 5.5).

EMISSIVITY
The ratio of radiation emitted by a real substance to the amount emitted by a black body at the same temperature.

As written, Equation 5.3 applies only to black bodies. To account for the fact that real things are not *perfect* black bodies, we use a dimensionless number known as **emissivity**, ε.

$$E = \varepsilon \, \sigma \, T^4 \qquad (5.4)$$

Emissivity is a characteristic of a substance. The emissivity of a black body is 1.0 so, for a black body, Equation 5.4 becomes Equation 5.3. Table 5.1 confirms that most things are close to being black bodies, as most have emissivities above 0.9.

TABLE 5.1 | Approximate infrared emissivities for various substances.

Substance	Infrared Emissivity
Water	0.98
Fresh Snow	0.99
Old Snow	0.82
Ice	0.97
Wet Soil	0.95
Dry Soil	0.92
Sand	0.90
Grass	0.90
Forest	0.98
White Paper	0.93
Highly Polished Silver	0.02
Aluminum	0.03
Glass	0.92
Human Skin	0.98

Question for Thought

How does the shiny coating of emergency blankets make these blankets particularly effective in keeping people warm? (Use Table 5.1 to answer this question.)

FIGURE 5.5 | This thermal image of apartment buildings was created by detecting the amount of radiation *emitted* by various areas of the scene. Because warm objects emit more radiation than do cooler objects, emitted radiation provides information about temperature. The dark blue on the thermogram indicates the coldest temperature (–41°C), while green, yellow, and orange indicate increasingly warmer temperatures, and red shows the warmest (3°C).

Example 5.2

Calculate the amount of radiation emitted by

a) a grass surface with a temperature of 15°C and an emissivity of 0.9.

b) a sandy surface with a temperature of 35°C and an emissivity of 0.9.

Use Equation 5.4.

a) $E = (0.9)(5.67 \times 10^{-8} \text{ W·m}^{-2}\text{·K}^{-1})(288 \text{ K})^4$
$= 351.1 \text{ W/m}^2$

b) $E = (0.9)(5.67 \times 10^{-8} \text{ W·m}^{-2}\text{·K}^{-1})(308 \text{ K})^4$
$= 459.2 \text{ W/m}^2$

The hot sand emits considerably more radiation than the cooler grass surface.

5.2.2 | Wien's Law

The work of Wilhelm Wien provided the next link in the development of the radiation laws. Wien's experiments in 1893 led him to discover that the wavelength of maximum emission of a substance is inversely proportional to the temperature of the substance. This law became known as *Wien's displacement law*, or simply *Wien's law*, and it is written mathematically as follows:

$$\lambda_{max} = \frac{2897}{T}$$

(5.5)

In Equation 5.5, λ_{max} is the wavelength of maximum emission, and 2897 is an experimentally derived constant with units of μm K. Wien's law shows that as temperature increases, the wavelengths emitted decrease. As a result, as temperature increases, Planck's curve (Figure 5.4) shifts to the left, so that both the wavelength of maximum emission and all emitted wavelengths are shorter.

As an object's temperature increases, it will gradually emit shorter wavelengths. At normal Earth temperatures of about 290 K, the radiation emitted by the object will fall mostly in the infrared portion of the spectrum, and there will not be enough of it to be noticeable. As the temperature rises higher, and the amount of infrared radiation being emitted by the object increases, we will begin to feel heat. When the temperature reaches about 1000 K, the object will glow red as it starts to emit significant amounts of visible light. Eventually, the object will be emitting strongly in all the colours and will appear yellow or white, like a light bulb. Incandescent light bulbs work by heating a tungsten filament to about 2800 K, so that visible light is emitted. Wien's law shows that the wavelength of maximum emission for something at this temperature is about 1 μm, which is in the infrared portion of the spectrum. Planck's curve, however, indicates that enough visible light is being emitted from objects at 2800 K that they will appear yellow.

We can apply Wien's law to determine the temperature of stars. It is difficult to determine the temperature of a star based on its brightness, or the *amount* of radiation we sense coming from it, because the star's brightness does not depend solely on the star's temperature, but also on its size and distance from us. We can, however, measure the spectrum of radiation coming from the star and use that information to determine the temperature of the star. In other words, we make use of the fact that a star's temperature determines its *colour*. Stars that appear reddish are cooler than stars that appear bluish. In Example 5.3, a star's temperature is calculated based on measurement of its wavelength of maximum emission.

Example 5.3

The wavelength of maximum emission of a star is 0.1 µm.
  a) What is the temperature of the star?
  b) What is the colour of the star?

a) Rearrange Equation 5.5 to solve for temperature.

$$T = \frac{2897 \ \mu m \ K}{\lambda_{max}}$$

$$= \frac{2897 \ \mu m \ K}{0.1 \ \mu m}$$

$$= 28{,}970 \ K$$

b) This star will appear bluish. Although it emits most of its radiation at 0.1 µm, which falls in the ultraviolet portion of the spectrum, Planck's curve shows that it also emits visible radiation, of which it emits more blue light than red or green light.

After Wien had formulated his law, he worked to discover the complete relationship between temperature, wavelength, and rate of emission. In one attempt to do this, in 1896, he combined his law and

Remember This

Together, the Stefan–Boltzmann law and Wien's law tell us that as temperature increases,

- the rate of emission of radiation increases and
- the wavelength of maximum emission, along with all the emitted wavelengths, get shorter.

the Stefan–Boltzmann law. The equation he derived did a good job of predicting emission of radiation at short wavelengths, but it did not work well for the longer wavelengths.

5.2.3 | Planck's Law

In 1900, Max Planck finally determined the relationship linking the emission of radiation to temperature *and* wavelength. He did this by making some modifications to the equation formulated by Wien, and by making what was at the time a radical hypothesis: that electromagnetic radiation travels not only in waves but also in what he called *photons*. (Remember that we now see electromagnetic radiation as being wave-like *and* particle-like.) At last an equation had been found that fit the experimental data and the originally theoretical findings of Kirchhoff; this equation has become known as *Planck's law* (Equation 5.6). It gives the rate of emission of radiation at a single wavelength, E_λ, for a black body at a given temperature.

$$E_\lambda = \frac{c_1}{\lambda^5 \left[\exp\left(\dfrac{c_2}{\lambda T}\right) - 1 \right]} \tag{5.6}$$

In this equation, c_1 is equal to 3.74×10^8 W·m^{-2}·μm^4 and c_2 is equal to 1.44×10^4 μm/K. Equation 5.6 is, of course, the equation for Planck's curve (Figure 5.4). Both the Stefan–Boltzmann law and Wien's law can be derived from Planck's law. The Stefan–Boltzmann law is the equation for the *area under the curve*, and Wien's law is the equation for the *peak of the curve*. We can now use these three laws to return to the explanation of the differences between the radiation emitted by the sun and that emitted by Earth.

5.2.4 | Shortwave Radiation vs Longwave Radiation

Although Earth and the sun are not perfect black bodies, they are close enough to being black bodies that we can apply the radiation laws to both. Example 5.4 shows that the sun emits roughly 64,000,000 W/m^2 and that its wavelength of peak emission is about 0.5 μm. Using Planck's law, we can show that the sun

emits most of its radiation in the wavelength range of 0.15 to 3.0 μm; this range includes ultraviolet, visible, and infrared radiation. Example 5.4 also shows that Earth emits about 390 W/m^2. This is an average value for Earth, as it is calculated using Earth's average temperature of 15°C (288 K). The wavelength of peak emission for Earth is about 10.0 μm and, according to Planck's law, Earth emits most of its radiation over the range of wavelengths from 3.0 μm to 100.0 μm. All of this radiation falls in the infrared portion of the electromagnetic spectrum.

Example 5.4

Determine the rate of radiation emission and the wavelength of maximum emission for the sun and for Earth. Assume that they are both black bodies, and use a temperature of 5800 K for the sun and 288 K for Earth.

Use Equation 5.3 to determine the rate of emission for the sun.

$$E = (5.67 \times 10^{-8} \text{ W·m}^{-2}\text{·K}^{-4}) (5800 \text{ K})^4$$
$$= 64,164,532 \text{ W/m}^2$$

Use Equation 5.5 to determine the wavelength of maximum emission for the sun.

$$\lambda_{max} = \frac{2897 \text{ μm·K}}{5800 \text{ K}}$$
$$= 0.5 \text{ μm}$$

Use Equation 5.3 to determine the rate of emission for Earth.

$$E = (5.67 \times 10^{-8} \text{ W·m}^{-2}\text{·K}^{-4}) (288 \text{ K})^4$$
$$= 390.1 \text{ W/m}^2$$

Use Equation 5.5 to determine the wavelength of maximum emission for Earth.

$$\lambda_{max} = \frac{2897 \text{ μm·K}}{288 \text{ K}}$$
$$= 10.1 \text{ μm}$$

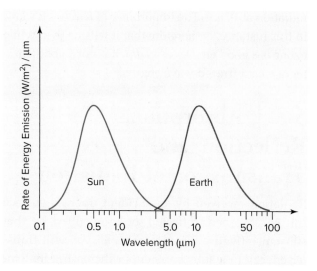

FIGURE 5.6 | Planck curves for the sun and Earth. The vertical scale has been normalized so that the two curves can be shown on the same graph.

Notice that there is little overlap between the range of wavelengths emitted by the sun and the range of wavelengths emitted by Earth (Figure 5.6). The sun's emission drops to almost nothing at about 3.0 µm, which is roughly where Earth's emission begins. Atmospheric scientists take advantage of this division by calling the sun's radiation **shortwave radiation** and Earth's radiation **longwave radiation**.

5.2.5 | Kirchhoff's Law

Kirchhoff's law makes an important connection between the *emission* of radiation and the *absorption* of radiation. It states that the emissivity of a substance at a given wavelength, ε_λ, is equal to its absorptivity at that wavelength, a_λ, as shown in Equation 5.7.

$$\varepsilon_\lambda = a_\lambda \qquad (5.7)$$

In other words, a good absorber at a certain wavelength is a good emitter at that wavelength, and vice versa. For example, fresh snow, with an emissivity of 0.99, is an almost perfect *emitter* of infrared radiation. It follows from Kirchhoff's law that fresh snow has an *absorptivity* of 0.99 for infrared wavelengths. Thus, because snow is an almost perfect *absorber* of infrared radiation, it is a poor *reflector* of these wavelengths. At the same time, snow is a poor absorber, but very good reflector, of *solar* radiation.

Kirchhoff's law can sometimes cause confusion. For example, ozone gas is a good absorber of both ultraviolet radiation from 0.2 to 0.3 µm and infrared radiation at 10 µm. Based on Kirchhoff's law, one might conclude that ozone is also a good *emitter* of these wavelengths. At temperatures typical of the Earth-atmosphere system, however, ozone gas is not hot enough to emit ultraviolet radiation, but it will emit infrared

SHORTWAVE RADIATION
The radiation emitted by the sun, which includes ultraviolet, visible, and infrared radiation.

LONGWAVE RADIATION
The radiation emitted by Earth, which includes only infrared radiation.

KIRCHHOFF'S LAW
A radiation law stating that the emissivity of a substance at a given wavelength is equal to the absorptivity of that substance at the same wavelength.

radiation at 10 μm. The importance of Kirchhoff's law, in this instance, is therefore that it tells us that, while ozone is a good emitter at 10 μm, it is not a good emitter of *other* infrared wavelengths.

5.3 | Absorption, Reflection, and Transmission of Radiation

Radiation received by, or incident upon, a surface can be absorbed, reflected, or transmitted by that substance (Figure 5.7). *Absorbed radiation* will transfer energy to a substance, either increasing its temperature or causing a phase change. Because our bodies *absorb* radiation, a person standing outside in the sun will feel much warmer than a person standing in the shade nearby. *Reflected radiation* "bounces off" a surface. All visible objects *reflect* a certain amount of radiation— in fact, it is the reflection of visible radiation that allows us to see things that are not light sources. For example, Earth is visible from space, not because it *emits* visible radiation, but because it *reflects* it. *Transmitted radiation* travels right through a substance. We can see through glass, for example, because it *transmits* most visible radiation.

REFLECTIVITY
A measure of a substance's ability to reflect incident radiation.

TRANSMISSIVITY
A measure of a substance's ability to transmit incident radiation.

ALBEDO
The proportion of the sun's incident radiation that is reflected by a surface.

The relative proportions of incident radiation that are absorbed, reflected, or transmitted depend to a large extent on the properties of the surface itself. Also important are the wavelengths of the incident radiation. Because energy is conserved, all the radiation received by a surface must be accounted for by some combination of absorption, reflection, and transmission (Equation 5.8).

$$a_\lambda + \alpha_\lambda + t_\lambda = 1 \quad or \quad a_\lambda + \alpha_\lambda + t_\lambda = 100\% \qquad (5.8)$$

In this equation, *a* indicates a substance's absorptivity—the fraction (or percentage) of incident radiation that the substance absorbs; α indicates a substance's **reflectivity**—the fraction (or percentage) of incident radiation that the substance reflects; and *t* indicates a substance's **transmissivity**—the fraction (or percentage) of incident radiation that the substance transmits. Thus, we can use Equation 5.8 to show, for example, that an opaque surface ($t = 0$) with a reflectivity of 0.1 must have an absorptivity of 0.9. In other words, 10 per cent of the incident radiation will be reflected, and 90 per cent will be absorbed. The subscripts (λ) indicate that absorptivity, reflectivity, and transmissivity are all wavelength dependent—for example, while glass *transmits* most visible radiation, it *absorbs* most ultraviolet and infrared radiation.

5.4 | Albedo

Albedo refers to the reflectivity of a surface to the sun's radiation. Table 5.2 gives typical albedo values for some surfaces. In general, these values depend on

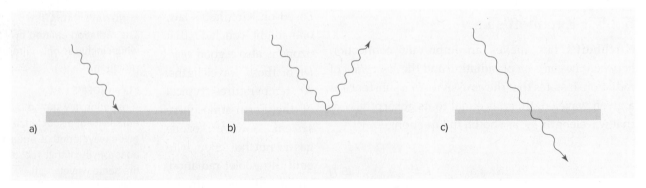

a) b) c)

FIGURE 5.7 | A surface will either a) absorb, b) reflect, or c) transmit incident radiation.

the colour and roughness of a surface, but they are also somewhat dependent on the angle of the sun.

High albedos are typical of clouds, snow and ice, and deserts. The bright glare you may have noticed when flying in a jet above clouds is due to the high albedo of clouds. As clouds get thicker, their albedos get higher because there are more water droplets to reflect light. Once clouds reach a thickness of about 1000 m, they *transmit* very little light and may even look black when viewed from below. The albedo of fresh snow is about the same as that of thick clouds. (As snow gets older, it gets dirty, and the albedo typically decreases.) With albedos up to 0.40, deserts are also quite reflective surfaces. It follows that deserts are relatively poor absorbers of solar radiation, so it may seem strange that they can get very hot. But, as you will see in Section 6.6.3, properties other than albedo also influence surface temperature.

Vegetated surfaces generally have lower albedos than non-vegetated surfaces. Forests and areas of long grass have low albedos because of internal reflection within the trees or the long blades of grass, which prevents much of the radiation from reflecting back out. Clear-cutting a forest will increase albedo, as will agricultural practices that leave the soil bare

for extended periods. It follows that desertification increases albedo as well. Because higher albedos can cause cooling, these land-use changes can influence climate (Section 17.3.2).

Of the natural surfaces on Earth, water has, on average, the lowest albedo of all (Figure 5.8). This often surprises people who think of the way water sparkles in the sunlight or produces a blinding glare at sunset. As it turns out, the albedo of water is strongly dependent on the sun's angle. When the sun is low in the sky, as it is in the high latitudes and at sunrise and sunset, the albedo of water is very high, at times approaching 1.0. However, with higher sun angles, the albedo of water drops below 0.10, and water becomes a very good absorber of solar radiation. When a water surface is disturbed by waves or ripples, or when the sky is cloudy, the sun's rays will meet the water at many different angles, thus increasing the water's albedo. In such cases, the water will often appear to sparkle, and it may reflect enough sunlight to produce sunburn. Through a similar process, the wakes of large ships have recently been shown to increase the albedo of the oceans very slightly. In fact, researchers have estimated that these wakes are causing a slight decrease in the solar energy absorbed by the oceans (Gatebe, Wilcox, Poudyal, & Wang, 2011).

TABLE 5.2 | Albedos of various surfaces.

Surface	Albedo
Fresh Snow	0.75–0.95
Old Snow	0.40–0.70
Water (Low Sun Angle)	0.10–1.00
Water (High Sun Angle)	0.03–0.10
Sea Ice	0.30–0.45
Glacier Ice	0.20–0.40
Thick Clouds	0.70–0.95
Thin Clouds	0.40–0.60
Deserts	0.20–0.40
Wet Soil	0.05–0.15
Dry Soil	0.25–0.35
Coniferous Forest	0.05–0.15
Deciduous Forest	0.15–0.25
Grass	0.15–0.25
Asphalt	0.05–0.20
Concrete	0.10–0.35
Urban Area	0.15 (average)

FIGURE 5.8 | An iceberg near Twillingate, Newfoundland, shows the high albedo of glacial ice relative to the low albedo of the ocean in high sun.

Figure 5.9 shows the variation in the albedo of Earth's *surface*. Only polar ice caps and deserts stand out as having relatively high albedos. The rest of Earth's surface, about 70 per cent of which is ocean, has a relatively low albedo. Based on data collected by satellites, researchers estimate that the albedo of the Earth-atmosphere system as a whole is currently about 30 per cent. This value is higher than the albedo of the surface alone because it includes the effects of the atmosphere and, in particular, the clouds. In comparison, Mars, with its thin atmosphere, has an albedo of only 17 per cent; Venus, with its dense atmosphere, has an albedo of 78 per cent; and the moon, with no atmosphere at all, has a very low albedo of only 7 per cent. From space, Earth appears much brighter than the moon.

DIFFUSE RADIATION
The sun's radiation that reaches Earth's surface after being scattered.

DIRECT BEAM RADIATION
The sun's radiation that reaches Earth's surface without first being scattered.

SCATTERING
The process by which atmospheric gases and aerosols reflect radiation in multiple directions.

RAYLEIGH SCATTERING
Scattering of radiation by particles smaller than the wavelengths they scatter.

MIE SCATTERING
Scattering of radiation by particles bigger than the wavelengths they scatter.

5.5 | Radiation in the Atmosphere: Scattering and Absorption

As radiation from the sun and from Earth travels through the atmosphere, it is absorbed, reflected (or scattered), and transmitted by gas molecules and aerosols, including cloud droplets. If the radiation is absorbed, it contributes to heating the atmosphere. If it is reflected or scattered, it may eventually reach earth as **diffuse radiation**. If the radiation is not reflected, scattered, or absorbed, it reaches us as **direct beam radiation**. On average, just over half of the sunlight reaching Earth's surface is in the form of diffuse radiation; when the sky is overcast, all light is diffuse.

5.5.1 | Scattering

In the context of the atmosphere, we usually use the term **scattering** in addition to, or in place of, *reflection*. Whereas reflection sends the radiation off in *one* direction, scattering sends it off in *many* directions (Figure 5.10). Atmospheric scattering is of two main types: **Rayleigh scattering** and **Mie scattering**. The type of scattering that occurs depends on the size of the particle responsible for the scattering.

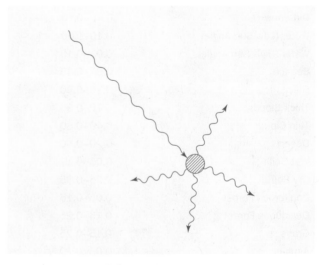

0	50	100	150	200	250	300

Flux (W/m²)

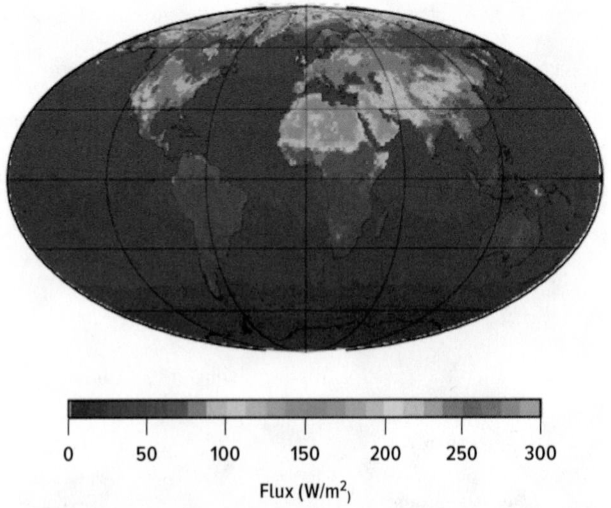

FIGURE 5.9 | The global variation in Earth's clear-sky albedo in July 2004

FIGURE 5.10 | In the atmosphere, gas molecules and aerosols scatter radiation in many directions.

5.5.1.1 | Rayleigh Scattering

Rayleigh scattering is named after Lord Rayleigh, who first described it in 1871. This form of scattering is also known as *selective scattering* because it scatters some wavelengths more than others. Rayleigh scattering occurs when the particles responsible for the scattering are considerably smaller than the wavelengths of light being scattered. Gas molecules, with diameters of about 0.0001 μm, fit this description in relation to the sun's visible radiation and are, thus, responsible for Rayleigh scattering.

As Expression 5.9 shows, the amount of scattering that occurs is inversely proportional to the wavelength of the radiation raised to the fourth power.

$$\text{Rayleigh Scattering} \sim \lambda^{-4} \qquad (5.9)$$

As wavelength increases, the amount of scattering quickly decreases, meaning that the shorter wavelengths of solar radiation are scattered considerably more than the longer ones. The result is that, of the three primary colours of light, blue is scattered most, and red, least, making the sky appear blue. Further, the sun appears yellow because the red and green light that are left coming directly from the sun combine to make yellow (Figure 5.11a). When the sun is setting, its rays travel such a long distance through the atmosphere that much more blue light, and even some green light, is scattered, making the sun appear red (Figure 5.12). If there are clouds in the sky at sunset, they too appear red, as they reflect the red light from the sun. Without an atmosphere to scatter the solar rays, the sun would appear white because all three colours would travel directly from it to our eyes (Figure 5.11b). (Even without scattering, the sun might still appear somewhat yellow, rather than pure white, because it emits most strongly in the green part of the spectrum.)

Question for Thought

What would the sun and the "sky" look like from the moon?

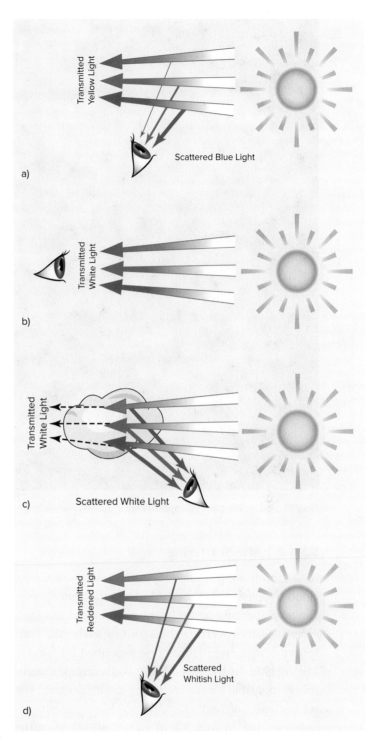

FIGURE 5.11 | The colours we see are a result of the scattering of sunlight. a) With an atmosphere, gas molecules scatter blue light, producing a blue sky; red and green light are left to come directly from the sun, making the sun appear yellow. b) With no atmosphere, the sun would appear white. c) Clouds scatter all wavelengths of light equally, so clouds appear white. d) When the atmosphere is hazy, aerosols are scattering all the light, making the sky appear whitish. However, since blue light is still preferentially scattered, haze can make the sun appear red.

Source: Adapted from Turco, 2002, p. 62.

FIGURE 5.12 | The sun often appears red at sunset.

5.5.1.2 | Mie Scattering

In addition to gas molecules, aerosols—including cloud droplets—also scatter light in the atmosphere. However, because aerosols are much bigger than gas molecules, they scatter light according to the Mie scattering process, which was first described by Gustav Mie in 1908. In the case of Mie scattering, the aerosols responsible for the scattering are bigger than the wavelengths of light. In addition, Mie scattering is non-selective, in that it will scatter all wavelengths nearly equally, so that objects appear white. Mie scattering allows us to see clouds, fogs, and haze.

The water droplets in clouds are very effective at scattering all wavelengths of light; as a result, clouds appear white (Figure 5.11c). It is the *total* effect of this scattering by the many water droplets in clouds that causes clouds to have such high albedos. The water

droplets in fogs are, of course, just as effective at scattering visible light. In fact, that's why it is usually more difficult to drive in fog with headlights on than with headlights off—the fog scatters and reflects the beam of the headlights back at the driver.

Haze occurs when the amount of aerosols in the atmosphere is greater than normal. Salt crystals released into the air by breaking waves can produce haze, as can smoke from forest fires, and other particulate matter from urban and industrial areas. Such aerosols will scatter all wavelengths of light, but they tend to scatter more blue and green than red. Haze, therefore, is white, and the sun, seen through the haze, can appear red (Figure 5.11d). As a consequence of this scattering process, the skies in polluted cities normally appear as a much paler, almost white, blue compared to the deep blue of non-polluted skies.

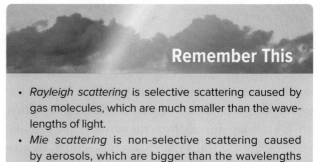

Remember This

- *Rayleigh scattering* is selective scattering caused by gas molecules, which are much smaller than the wavelengths of light.
- *Mie scattering* is non-selective scattering caused by aerosols, which are bigger than the wavelengths of light.

5.5.2 | Absorption

Whereas atmospheric *scattering* of the sun's shortwave radiation is quite strong, scattering of Earth's longwave radiation is almost negligible. The opposite is true for *absorption*: longwave radiation is more strongly absorbed in the atmosphere than is shortwave radiation (Figure 5.13). This is because gases are *selective absorbers*; they absorb some wavelengths very effectively, some not at all, and some partially. For example, Figure 5.13 shows that methane gas is a good absorber of radiation with wavelengths of just over $3100\mu m$ and of radiation with wavelengths of about $8\ \mu m$.

The gases that are most effective at absorbing the *sun's radiation* are oxygen, ozone, and water vapour. Oxygen and ozone effectively absorb almost

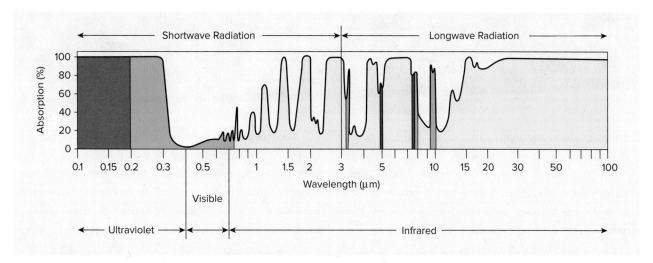

FIGURE 5.13 | Per cent absorption of radiation, by wavelength, by atmospheric gases. Oxygen absorbs the shortest wavelengths of ultraviolet radiation (red); ozone absorbs longer wavelengths of ultraviolet radiation, as well as infrared radiation at wavelengths of about 10 μm (green); and methane (blue) and nitrous oxide (purple) absorb infrared radiation in a few narrow bands. Water vapour and carbon dioxide are responsible for the remaining absorption (yellow); of this absorption, water vapour absorbs more than carbon dioxide does. This graph shows that the atmosphere is much more effective at absorbing Earth's longwave radiation than it is at absorbing the sun's shortwave radiation.

all radiation with wavelengths shorter than about 0.3 μm. Recall that this is how oxygen and ozone help protect life on Earth from the harmful effects of this radiation (Section 2.7). Water vapour partially absorbs the sun's radiation between 0.8 and 3 μm. However, as we saw in Section 5.2.4, most of the sun's radiation falls between these two regions of the spectrum, in the visible portion. Figure 5.13 shows that much of this visible radiation passes through the atmosphere without being absorbed.

The gases that are most effective at absorbing *Earth's radiation* are water vapour, carbon dioxide, methane, nitrous oxide, and ozone—the greenhouse gases (Figure 5.13). In combination, the greenhouse gases absorb almost all of Earth's radiation, with the exception of radiation with wavelengths from 8 to 11 μm, which is known as the **atmospheric window**. (Ozone absorbs a small amount of the radiation in this window.) Radiation with wavelengths in this range will travel from Earth's surface right through the atmosphere and escape to space.

Notice that the vertical line in the middle of Figure 5.13 corresponds to a wavelength of 3.0 μm, thus marking the division between shortwave radiation from the sun and longwave radiation from Earth. To the left of this line, the only significant absorption is of ultraviolet radiation by oxygen and ozone; there is virtually no absorption of visible light. In contrast, to the right of the line, absorption is almost complete, with the exception of the wavelengths that fall in the atmospheric window. In summary, Figure 5.13 shows that while the atmosphere is almost transparent to the sun's shortwave radiation, it is almost opaque to Earth's longwave radiation.

The atmospheric absorption of longwave radiation is increased further by the presence of clouds. Clouds absorb all wavelengths of longwave radiation, even those in the atmospheric window. Aerosols also absorb both shortwave and longwave radiation, but their exact effects are not as well understood.

> **ATMOSPHERIC WINDOW**
> The band of wavelengths of radiation, from 8 to 11 μm, that is not absorbed by gases in the atmosphere.

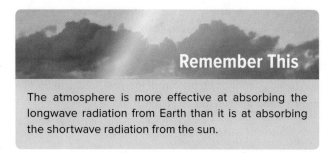

Remember This

The atmosphere is more effective at absorbing the longwave radiation from Earth than it is at absorbing the shortwave radiation from the sun.

In the Field

A Global Network to Measure Atmospheric Aerosols Using Sunlight

Glen Lesins, Dalhousie University

Sunlight provides the energy that drives the atmospheric and oceanic circulations that result in our ever-changing weather and climate. Solar energy arrives in the form of electromagnetic waves, or radiation, in the ultraviolet, visible, and near infrared portions of the spectrum; these waves interact with air molecules, atmospheric aerosols, and clouds before reaching Earth's surface. There are two types of interactions with the atmosphere: scattering and absorption. The sum of scattering and absorption is called extinction, which is a measure of the total loss of radiation from the solar beam. These interactions differ as the composition of the atmosphere varies, resulting in changes in the amount of solar heating at Earth's surface.

I am a participant in a global network of sun photometers called AERONET (Aerosol Robotic Network) that measures the changing aerosol content of the atmosphere by measuring how much solar energy reaches Earth's surface. We study aerosols because they affect climate by scattering solar energy back to space and by acting as centres of condensation for cloud droplets, which also scatter solar energy back to space. As a result, aerosols cool our climate, partially offsetting the effects of global warming.

In our research, we use sun photometers to measure the amount of solar energy reaching Earth's surface. The telescope portion of the photometer is automatically pointed toward the sun on a clear day, and it focuses the direct solar beam from the sun's disk onto a semiconductor detector that is sensitive to solar energy. By comparing the measured solar energy with the theoretical amount of solar energy that would reach the surface if the atmosphere were completely transparent, we can calculate the optical depth of the aerosols. We calculate this theoretical value based on the elevation angle of the sun above the horizon (which depends on the latitude and local time at the photometer's location) and the distance between Earth and the sun (which depends on the day of the year, because Earth's orbit around the sun has an elliptical shape.)

Aerosol optical depth is a dimensionless number that measures the amount of aerosols in the atmospheric column above the ground, based on the total extinction of sunlight caused by those aerosols. We have to correct for the loss of direct-beam sunlight caused by scattering from the atmospheric molecules, primarily nitrogen and oxygen, which is responsible for making the clear skies blue. Since we are measuring only the aerosols, we do not collect data when clouds are obscuring the sun.

The sun photometer contains a series of filters that are used to sample specific wavelengths of the solar spectrum, from the ultraviolet to the near infrared, allowing us to compute an aerosol optical depth for each solar wavelength. This is useful because the smaller aerosol particles will scatter the shorter wavelengths more efficiently than the larger particles will. By comparing the optical depths at the different wavelengths, we can make a rough estimate of the size distribution of the aerosols. This remote-sensing method is not as accurate as measuring the particles directly with a very sophisticated particle-sizing instrument, but it does provide a relatively inexpensive way to determine whether the aerosol column is dominated by smaller or larger particles; this information, in turn, gives us a clue about where the aerosols originated. For example, small aerosols are more likely to be caused by human activities, whereas larger ones are probably natural.

Since the photometer that I manage in Halifax, Nova Scotia, is part of a global network operated by NASA AERONET, and its Canadian partner AEROCAN, we are able to compare our measurements with other sites to monitor how aerosol levels are changing and impacting climate around the world. This is an exciting example of how the work of individual scientists from different countries can be combined to increase our knowledge on a global scale. We are using our measurements to learn about the role of aerosols in cooling the climate, and we are providing observations that can be used to test global climate models that predict how much warming will take place in the coming decades. Our measurements are also used to provide a "ground truth" for satellites that use remote sensing to determine aerosol optical depths from space.

For more information visit the following websites:
http://aeronet.gsfc.nasa.gov/new_web/index.html
www.aerocanonline.com/templates/nature/index.html

GLEN LESINS is an associate professor in the Department of Physics and Atmospheric Science at Dalhousie University. His current research involves studying the role that clouds, water vapour, and aerosols play in determining the transfer of solar and terrestrial radiation in the Arctic atmosphere in order to further our understanding of climate-change processes.

5.6 | The Greenhouse Effect

The fact that the atmosphere easily transmits the sun's radiation but almost completely absorbs Earth's radiation is the basis of the greenhouse effect. Because the atmosphere is relatively transparent to solar radiation, much of the sun's radiation is able to travel to Earth's surface and warm it. Earth's surface, in turn, emits longwave radiation, most of which is absorbed in the atmosphere by greenhouse gases and clouds. Of the gases, water vapour and carbon dioxide are the most important absorbers; they are responsible for 60 per cent and 26 per cent, respectively, of the total absorption under clear-sky conditions (Kiehl & Trenberth, 1997). The only wavelengths of terrestrial radiation not absorbed by the greenhouse gases are those in the range from 8.0 to 11.0 μm (Section 5.5.2). When clouds

are present, however, even these wavelengths will be absorbed, because clouds absorb all wavelengths of longwave radiation.

In turn, greenhouse gases and clouds will emit the same wavelengths of radiation that they absorb, according to Kirchhoff's law. Without clouds, the emissivity of the atmosphere will be quite a bit less than 1.0. But, since clouds are very good absorbers of all longwave radiation, it follows that they are also very good emitters of this radiation; as a result, they will bring the emissivity of the atmosphere closer to 1.0. This is particularly true of low- and middle-height clouds.

With or without clouds, the atmosphere emits radiation. This longwave radiation is emitted in all directions, but most of it ends up at Earth's surface where it is absorbed, making Earth 33 K warmer than it would be without an atmosphere (Figure 5.14). If the

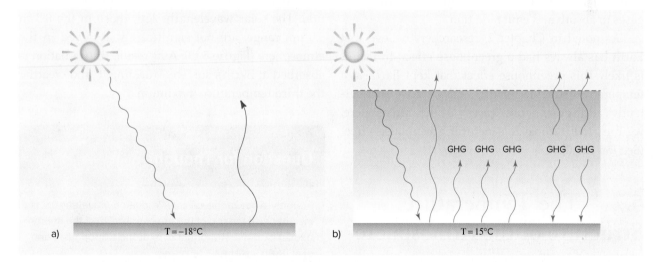

FIGURE 5.14 | The greenhouse effect. a) Without an atmosphere, Earth's average temperature would be about −18°C because Earth's longwave radiation would escape to space. b) With an atmosphere, Earth's average temperature is 15°C because much of Earth's longwave radiation is absorbed by greenhouse gases (GHG). This warms the atmosphere and the atmosphere emits longwave radiation back to Earth and out to space.

atmosphere were as transparent to longwave radiation as it is to shortwave radiation, Earth would be much colder.

The strength of the greenhouse effect can change at a location due to the variability of clouds and water vapour in the atmosphere. Because clouds are such effective absorbers and emitters of Earth's radiation, cloudy nights are usually warmer than clear nights. In addition, in winter, when the sun's radiation is weak, cloudy days can be warmer than clear days. Further, deserts often cool off rapidly at night because the skies are both clear *and* dry. In a similar way, high altitude locations cool off rapidly at night because, as the atmosphere thins with height, there are fewer gases to absorb and emit radiation.

As noted in Chapter 2, researchers believe that Earth has always had a greenhouse effect. In fact, it is likely this greenhouse effect that kept Earth at a temperature suitable for life. Today, as we inadvertently add greenhouse gases to the atmosphere, we are *strengthening* this natural greenhouse effect (Section 17.3.2).

5.7 | The Temperature Structure of the Atmosphere

The absorption characteristics of atmospheric gases can also explain the temperature profile of the atmosphere, first described in Section 1.6.2. Figure 1.7c shows three maximums in the temperature profile: one in the upper thermosphere, one in the upper stratosphere, and one at Earth's surface. Each one

represents a region of the atmosphere in which the sun's radiation is strongly absorbed (Figure 5.15).

At the "top" of Earth's atmosphere, the shortest wavelengths of the sun's radiation—those that range from 0.1 to 0.2 μm—are absorbed by molecules of oxygen and nitrogen. This process splits the oxygen and nitrogen molecules into atoms, and it warms the air, creating the temperature maximum at the top of the thermosphere. As the sun's radiation continues down through the thermosphere, these short wavelengths are quickly used up. Eventually there is little of this radiation left, so that there is only weak absorption through the mesosphere.

The second temperature maximum occurs in the upper stratosphere. At this altitude, the continued weak absorption of the sun's shortest wavelengths by oxygen becomes significant because the atmosphere is dense enough that collisions are far more frequent. It is here that oxygen atoms collide with oxygen molecules to form the ozone layer, as described in Section 2.7. Ozone then quite effectively absorbs the remaining longer wavelengths of ultraviolet radiation—particularly those in the 0.2 to 0.3 μm range. The warming associated with this absorption is responsible for the second temperature maximum.

Below the ozone layer, the sun's spectrum is almost completely depleted of ultraviolet radiation. The solar wavelengths left, those in the 0.3 to 3.0 μm range, are not significantly absorbed in the atmosphere (Figure 5.13). As a result, this radiation is absorbed at Earth's surface, warming it, and creating the third temperature maximum.

The air of the troposphere is then heated by Earth's surface. There are three ways in which this heat transfer occurs (Section 4.4). First, the longwave radiation emitted by Earth is strongly absorbed by the water vapour

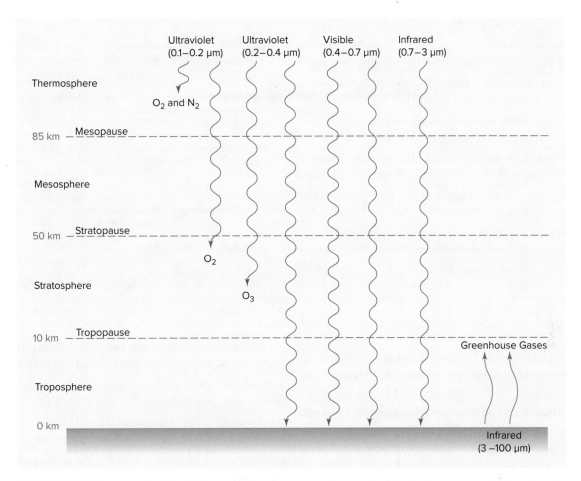

FIGURE 5.15 | The absorption of radiation in Earth's atmosphere. Where solar radiation is most strongly absorbed,—in the upper thermosphere, the upper stratosphere, and at Earth's surface—temperature maximums occur (Figure 1.7c).

and carbon dioxide in the atmosphere (Figure 5.15). Second, air heated at Earth's surface by conduction will rise, transferring sensible heat to the atmosphere by convection. Third, heat is transferred from Earth's surface to the atmosphere through phase changes. Latent heat is required in order for water at Earth's surface to evaporate; when the water vapour rises, it will eventually condense into clouds and release this latent heat.

Question for Thought

The temperature profiles for the atmospheres of Venus and Mars have only two temperature maximums: one at the surface, and one at the top of the atmosphere. Why do you think these planets' atmospheres have only two maximums, while Earth's has three?

5.8 | Solar, or Shortwave, Radiation

The **solar constant** is a measure of the amount of solar radiation reaching the top of Earth's atmosphere. Using data collected by satellites, researchers have estimated that the solar constant is currently about 1365 W/m² (Trenberth, Fasullo, & Kiehl, 2009). We can also estimate the solar constant of a planet mathematically, by applying the **inverse square law** to radiation from the sun. As with any radiation leaving

SOLAR CONSTANT
The amount of energy that strikes the top of the atmosphere, on a surface perpendicular to the solar beam, when Earth is at an average distance from the sun.

INVERSE SQUARE LAW
A general mathematical law used to determine the amount of any physical quantity propagating from a point source at a given distance from that source.

a point source, the radiation leaving the sun spreads out more and more the farther it travels from its starting point. (Think of how the light from a light bulb at the centre of a room is dimmer at the edges of the room.) We can imagine that radiation leaving the sun is travelling through successively larger spheres. Because the surface area of a sphere is $4\pi r^2$, each time the distance, r, from the source doubles, the same amount of radiation will be spread over a sphere four times bigger. This means that the radiation received *per unit of surface area* gradually decreases with distance from the source, as shown in Equation 5.10, which is the mathematical expression of the inverse square law.

$$E_2 = E_1 \left(\frac{R_1}{R_2} \right)^2 \tag{5.10}$$

In this equation, E_1 is the flux of energy at radius R_1, and E_2 is the flux of energy at radius R_2 (Figure 5.16). For the specific case of the sun and Earth, R_1 is the sun's radius, R_2 is the distance from the sun to Earth, E_1 is the rate of flow of radiation from the sun, and E_2 is the solar constant.

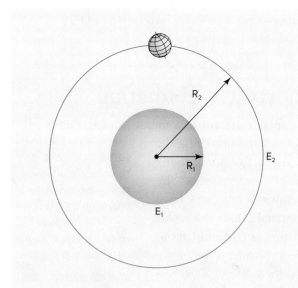

FIGURE 5.16 | We can use the inverse square law to determine the amount of radiation received on surfaces at increasing distances from a point source. This diagram shows the relationship between the variables used in Equation 5.10, using the sun and Earth as examples.

Example 5.5

Calculate the solar constant for Earth. The average distance between the sun and Earth is 1.5×10^8 km. The radius of the sun is 6.96×10^5 km. As we discovered in Example 5.4, the radiation emitted by the sun is 6.4×10^7 W/m^2.

Use Equation 5.10, and define the variables as shown in Figure 5.16.

$$E_2 = 6.4 \times 10^7 \text{ W/m}^2 \left(\frac{6.96 \times 10^5 \text{ km}}{1.5 \times 10^8 \text{ km}} \right)^2$$

$$= 1378 \text{ W/m}^2$$

This value is similar to the solar constant estimated by satellites.

The solar constant is the maximum amount of solar radiation that can be received anywhere in the Earth-atmosphere system. This value might be *approached* at the surface—for instance, on a very clear day in the tropics, when the sun is directly overhead—but it can never be *exceeded*. If the sun's rays are less than perpendicular to Earth's surface, the amount of solar radiation will be reduced. Further, passage through the atmosphere will reduce the amount of radiation received at the surface. In short, the amount of solar radiation reaching Earth's surface is controlled by the sun angle and by the depletion of the sun's rays in the atmosphere.

Remember This

The inverse square law shows that the solar constant of a planet depends on

- the output of the sun and
- the distance of the planet from the sun.

The amount of solar radiation received at Earth's surface depends on

- the solar constant,
- the sun angle, and
- depletion of the sun's rays in the atmosphere.

5.8.1 | Sun Angle

Sun angle can be defined in two possible ways: as the **altitude angle**, or as the **zenith angle** (Figure 5.17). When the sun is directly overhead, the altitude angle is 90° and the zenith angle is 0°. Altitude angle will be used in this book.

The amount of solar radiation a surface receives decreases as the altitude angle decreases because of the processes of **beam spreading** and **beam depletion**. Beam spreading occurs because, as the sun angle decreases, the beam is *spread* over a larger surface area, decreasing its intensity (Figure 5.18). We can use the **sine law of illumination** (Equation 5.11) to determine how beam spreading affects the amount of radiation received on a surface.

$$S = S_i \sin \theta \qquad (5.11)$$

In this equation, θ is the sun's altitude angle, S_i is the amount of radiation that the surface receives when the sun is directly overhead, and S is the amount of radiation

> ### Example 5.6
>
> If 950 W/m² of radiation is received on a horizontal surface when the sun is directly overhead, how much radiation will be received when the altitude angle is a) 30° and b) 60°?
>
> Use Equation 5.11.
>
> a) $S = (950 \ \text{W/m}^2) \sin 30°$
>
> $= 475.0 \ \text{W/m}^2$
>
> b) $S = (950 \ \text{W/m}^2) \sin 60°$
>
> $= 822.7 \ \text{W/m}^2$
>
> As the sun angle increases, the amount of solar radiation received on a horizontal surface will increase because the solar beam will be more concentrated.

that the surface receives when the sun is at an altitude angle of θ (Figure 5.19). Notice that when the altitude angle is 90°, $\sin \theta$ will be one, and S will equal S_i. For all other angles, S will be less than S_i.

Like beam spreading, beam depletion also increases with decreasing sun angle. This is because as the sun becomes lower in the sky, its **path length** through the atmosphere increases (Figure 5.20). With a greater path length, there is more scattering, reflection, and absorption of the solar beam by atmospheric components. Equation 5.12 shows that path length, *PL*, is inversely proportional to the sine of the sun's altitude angle, θ.

ALTITUDE ANGLE
The angle of the sun above the horizon.

ZENITH ANGLE
The angle of the sun from the zenith.

BEAM SPREADING
The spreading of the solar beam over an increasing surface area as the sun's angle decreases.

BEAM DEPLETION
The increasing depletion of the solar beam by atmospheric constituents as the sun's path length through the atmosphere increases.

SINE LAW OF ILLUMINATION
An equation used to calculate the amount of radiation incident on a surface, based on altitude angle.

PATH LENGTH
The distance that the sun's rays must travel through the atmosphere to reach Earth's surface.

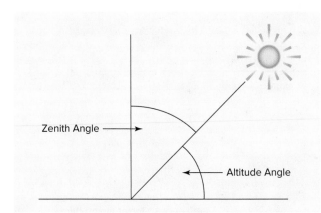

FIGURE 5.17 | The sun's altitude angle compared to the sun's zenith angle.

> ## Question for Thought
>
> How would the amount of solar radiation received at the surface change in Example 5.6 if the surface was not horizontal?

$$PL = \text{Depth of Atmosphere} \left(\frac{1}{\sin \theta} \right) \qquad (5.12)$$

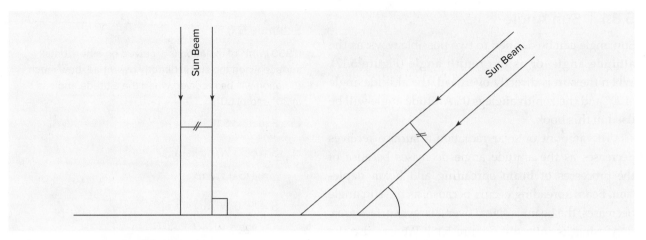

FIGURE 5.18 | As the sun's altitude angle decreases, the sun's beam is spread over an increasingly larger area, so that any one point on the surface receives less solar radiation.

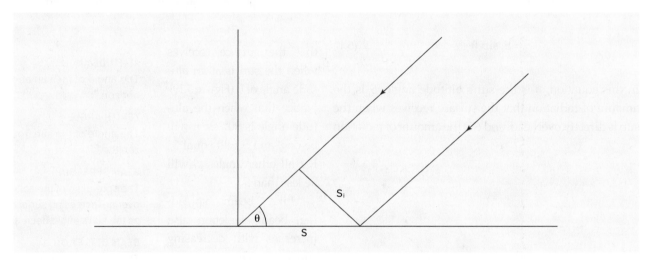

FIGURE 5.19 | **The sine law.**

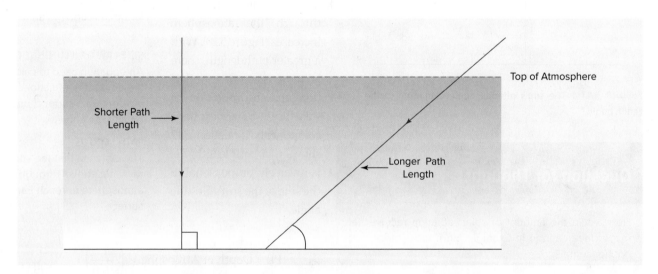

FIGURE 5.20 | **The sun's path length through the atmosphere increases with decreasing sun angle.**

Because the atmosphere does not have a well-defined "top," we cannot provide an accurate number for the "depth of the atmosphere." However, this equation is still useful for studying the relationship between sun angle and path length, as shown in Example 5.7.

Example 5.7

How much greater is the sun's path length when the sun's altitude angle is a) 30° and b) 10°, compared to when the sun is directly overhead?

Use Equation 5.12 and set the value of "Depth of Atmosphere" to be 1. This means that when the sun is directly overhead, the path length will be 1.

a)
$$PL = 1 \times \frac{1}{\sin 30°} = \frac{1}{0.5}$$
$$= 2.0$$

b)
$$PL = 1 \times \frac{1}{\sin 10°} = \frac{1}{0.17}$$
$$= 5.9$$

When the sun angle is 30°, the path length is double what it is when the sun is overhead; and when the sun angle drops to 10°, the path length is almost six times what it is when the sun is overhead.

Remember This

As sun angle decreases, the amount of solar radiation striking a surface will decrease for two reasons:

- a given amount of radiation will be *spread* over an increasingly larger area, and
- the solar beam will experience greater *depletion* due to a longer path length.

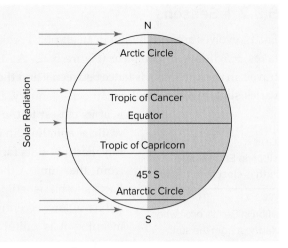

FIGURE 5.21 | The effect of Earth's sphericity on sun angle.

Example 5.8

When the sun is directly overhead at the equator, the amount of solar radiation received at the top of the atmosphere above the equator is the solar constant, 1365 W/m^2. Under these conditions, how much solar radiation is received at noon at latitudes a) 30°, b) 60°, and c) 90°? When the noon sun is overhead at the equator, the noon sun angles at these latitudes are 60°, 30°, and 0°, respectively.

Use Equation 5.11, and set S_i equal to the solar constant.

a) $S = (1365 \text{ W/m}^2) \sin 60°$
    $= 1182.1 \text{ W/m}^2$

b) $S = (1365 \text{ W/m}^2) \sin 30°$
    $= 682.5 \text{ W/m}^2$

c) $S = (1365 \text{ W/m}^2) \sin 0°$
    $= 0.0 \text{ W/m}^2$

These calculations show that the amount of solar radiation received at the top of the atmosphere decreases rapidly with latitude.

We have seen that sun angle affects the amount of solar radiation a surface receives. In turn, sun angle varies with latitude, time of day, and time of year. First, sun angle varies with *latitude* because Earth is a sphere; as such, when the sun is directly overhead at the equator, it will be on the horizon at the poles (Figure 5.21). As shown in Example 5.8, this large change in sun angle—from 90° at the equator to 0° at the poles—results in a considerable change in the amount of solar radiation received. Second, sun angle varies over a *day* because of the rotation of Earth about its axis. The sun begins and ends each day on the horizon, and it will be highest in the sky at noon. Third, the reasons for the *annual* variation of sun angles are somewhat more complex and are associated with the seasons.

5.8.2 | Seasons

Earth revolves around the sun in an elliptical orbit that takes 365¼ days to complete (Figure 5.22). As Earth travels in this orbit, the distance between it and the sun varies slightly. At **perihelion**, in early January, Earth is at its closest to the sun, while at **aphelion**, in early July, Earth is at its farthest from the sun. Although Earth's orbit is elliptical, it has a low eccentricity, meaning it is quite close to being a circle. The result is that the difference between the Earth–sun distance at perihelion and that at aphelion is very small—only about 3 per cent. (It is because the distance between Earth and the sun varies over the course of a year that the solar constant is defined as applying only when Earth is at an *average* distance from the sun.)

As Earth revolves around the sun, it is tilted on its axis at an angle of 23½° to a perpendicular through the plane of the orbit. In addition, because the north end of Earth's axis *always* points in the same direction (toward the North Star), the orientation of Earth to the sun changes over the course of a year (Figure 5.22). The result is that, at a given latitude, the noon sun angle and the hours of daylight will change from one day to the next. In turn, these changes in noon sun angle and day length result in seasons.

PERIHELION
The position in Earth's orbit when Earth is closest to the sun.

APHELION
The position in Earth's orbit when Earth is farthest from the sun.

JUNE SOLSTICE
The date on which the sun is directly overhead at the Tropic of Cancer. This is the first day of summer in the northern hemisphere and the first day of winter in the southern hemisphere.

DECEMBER SOLSTICE
The date on which the sun is directly overhead at the Tropic of Capricorn. This is the first day of winter in the northern hemisphere and the first day of summer in the southern hemisphere.

MARCH EQUINOX
The date on which the sun is directly overhead at the equator. This is the first day of spring in the northern hemisphere and the first day of fall in the southern hemisphere.

SEPTEMBER EQUINOX
The date on which the sun is directly overhead at the equator. This is the first day of fall in the northern hemisphere and the first day of spring in the southern hemisphere.

CIRCLE OF ILLUMINATION
A circular boundary between Earth's light half and its dark half.

Remember This

Seasons occur because

- Earth revolves around the sun,
- the axis is tilted, and
- the north end of the axis always points to the North Star.

Together, these conditions cause sun angle and day length to vary throughout the year, at a given latitude.

Question for Thought

Why might Earth be less suitable for life if a) the axis of rotation had a greater tilt or b) the speed of rotation was slower?

The first day of each season is marked by a *solstice* (for summer and winter) or an *equinox* (for spring and fall) (Figure 5.23). The **June solstice** occurs around June 21. On this date, the north end of the axis points *toward* the sun. In this position, the sun is directly overhead at the Tropic of Cancer, 23½° N, and all latitudes north of 23½° N will experience their highest noon sun angles of the year. The **December solstice** occurs around December 21. On this date, the north end of the axis points *away* from the sun; this puts the sun directly overhead at the Tropic of Capricorn, 23½° S, and all latitudes south of 23½° S will experience their highest noon sun angles of the year. The **March** and **September equinoxes** fall around March 21 and September 21, respectively. At these positions, Earth is not tilted with respect to the sun; this puts the sun directly overhead at the equator.

Day lengths, or the hours of daylight at a given latitude, also change as Earth revolves around the sun. To get a rough idea of day length, we can begin by drawing a **circle of illumination** around the globe (Figure 5.23). When a latitude line is cut in half by the circle of illumination, then day and night are equal in length at that latitude. This is always the case at

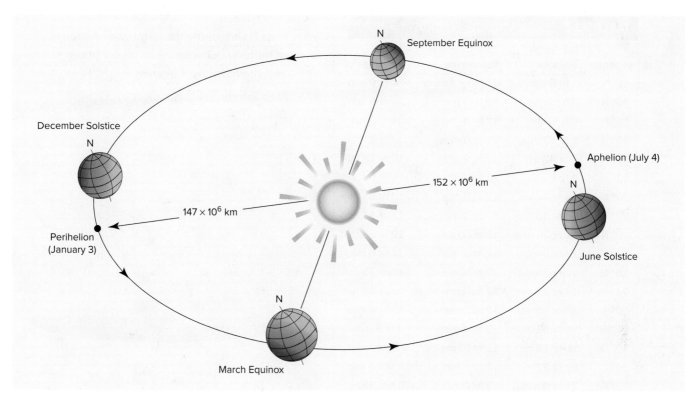

FIGURE 5.22 | Earth's orbit around the sun. The positions of Earth represent the first day of each season.

the equator. Otherwise, the proportion of the line on the light side of the circle represents the length of the day at that location (Table 5.3). In the northern hemisphere, the June solstice is the longest day and the December solstice is the shortest day; in the southern hemisphere, this relationship is reversed. On the

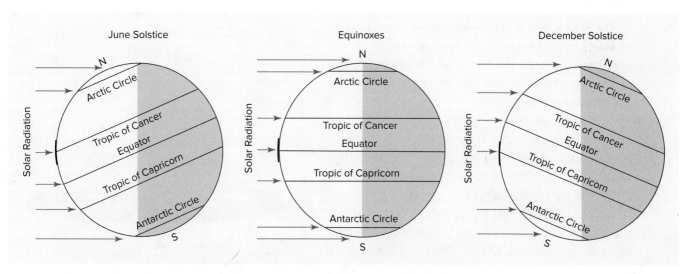

FIGURE 5.23 | The Earth–sun positions associated with the solstices and equinoxes. Notice that the north end of the axis is pointed toward the sun on the June solstice and away from the sun on the December solstice. On the equinoxes, Earth is not tilted with respect to the sun. These diagrams also show the circle of illumination, which separates Earth's dark half from its light half.

TABLE 5.3 | Day lengths by latitude and date.

Latitude	December Solstice	Equinoxes	June Solstice
90° N	0 hr 0 min	24 hr on horizon	24 hr 0 min
80° N	0 hr 0 min	12 hr 0 min	24 hr 0 min
70° N	0 hr 0 min	12 hr 0 min	24 hr 0 min
60° N	5 hr 33 min	12 hr 0 min	18 hr 27 min
50° N	7 hr 42 min	12 hr 0 min	16 hr 18 min
40° N	9 hr 8 min	12 hr 0 min	14 hr 52 min
30° N	10 hr 4 min	12 hr 0 min	13 hr 56 min
20° N	10 hr 48 min	12 hr 0 min	13 hr 12 min
10° N	11 hr 25 min	12 hr 0 min	12 hr 38 min
0°	12 hr 0 min	12 hr 0 min	12 hr 0 min
10° S	12 hr 38 min	12 hr 0 min	11 hr 25 min
20° S	13 hr 12 min	12 hr 0 min	10 hr 48 min
30° S	13 hr 56 min	12 hr 0 min	10 hr 4 min
40° S	14 hr 52 min	12 hr 0 min	9 hr 8 min
50° S	16 hr 18 min	12 hr 0 min	7 hr 42 min
60° S	18 hr 27 min	12 hr 0 min	5 hr 33 min
70° S	24 hr 0 min	12 hr 0 min	0 hr 0 min
80° S	24 hr 0 min	12 hr 0 min	0 hr 0 min
90° S	24 hr 0 min	24 hr on horizon	0 hr 0 min

Step 1: Determine the number of degrees of latitude between the subsolar point and the latitude of interest.

Step 2: Subtract the answer in Step 1 from 90°. (5.13)

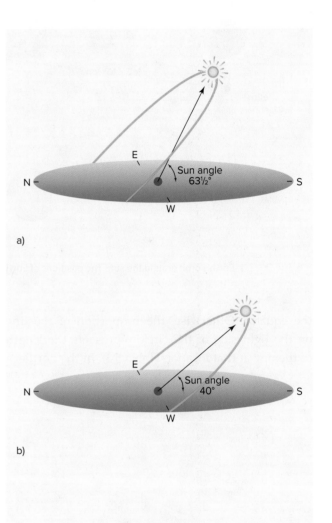

a)

b)

c)

FIGURE 5.24 | **The sun's path through the sky at 50°N, from the point of view of an observer at the centre of the circle, for a) the June solstice, b) the equinoxes, and c) the December solstice.**

equinoxes, days and nights are of equal length everywhere on Earth, except at the poles, where the sun is on the horizon for 24 hours. On the June solstice, the Arctic Circle, 66½° N, and all latitudes north of it experience 24 hours of day light, while the Antarctic Circle, 66½° S, and all latitudes south of it experience 24 hours of darkness; on the December solstice, the opposite occurs. At the North and South poles, six-month "days" and "nights" are experienced from one equinox to the other.

Just as the noon sun angle and day length will be different from day to day at a given location, the sun's path through the sky will also be different each day. That is, each day the exact location of the sun's rise and set, as well as the height it reaches in the sky, will be different (Figure 5.24). It is possible to calculate the sun angle at any time of day, but such calculations are quite complicated. On the other hand, it is quite simple to calculate the *noon* sun angle. To do so, all you need to know is the **subsolar point** and the latitude of interest (Equation 5.13).

SUBSOLAR POINT
The latitude at which the sun is directly overhead at noon.

For the equinoxes and solstices, the subsolar points are easy to remember; for all other days of the year, you can use an **analemma graph** to determine these points (Figure 5.25a). The shape of this graph represents the path of the sun in the sky over the course of a year—if a camera were set to take a picture of the sun at noon every day for a year, this is the shape that the sun's path in the sky would take (Figure 5.25b). Each planet has its own unique analemma because an analemma's shape depends on the shape of the planet's orbit around the sun and the tilt of the planet's axis. If Earth's orbit were perfectly circular, then the analemma would be a *symmetrical* figure eight. In addition, if Earth's axis were not tilted, the analemma would be a dot because the noon sun would always be overhead at the equator, and the noon sun angle at any latitude would always be the same. In other words, the sun would appear at the same place in the sky at the same time every day.

ANALEMMA GRAPH
A graph giving the latitude at which the sun is directly overhead for any day of the year.

The graph in Figure 5.26 shows the variation in solar radiation received *at the top of the atmosphere* for a variety of northern-hemisphere latitudes. This graph reveals two important patterns. The first pattern is that the difference in solar radiation received between the equator and the poles is much greater in winter than it is in summer. In winter, both sun angle and day length decrease with distance from the equator; this progressive decrease explains the large drop in solar radiation received at this time. On the other hand, in summer, sun angle decreases but day length increases with distance from the equator. The increase in day length compensates for the decrease in sun angle, so that the amount of solar radiation received at the North Pole is *greater* than the amount received at the equator. At Earth's surface, the added effect of beam depletion means that more solar radiation will be received at the equator than at the North Pole, but there is still much less difference between these two values at this time than there is in winter. An important result of this is that the temperature gradient between the equator and the poles is much less in summer than it is in winter (see Figure 16.2).

The second pattern evident from Figure 5.26 is that not all latitudes experience seasonal variation, or seasonality, in solar radiation received to the same degree; the *least* variation occurs in the tropics, while

---

Example 5.9

Calculate the noon sun angle at 50° N for a) the equinoxes, b) the June solstice, and c) the December solstice.

Use Equation 5.13.

a) For the equinoxes, the subsolar point is 0°.

Step 1: $50° - 0° = 50°$

Step 2: noon sun angle $= 90° - 50° = 40°$

b) For the June solstice, the subsolar point is 23½° N.

Step 1: $50° - 23½° = 26½°$

Step 2: noon sun angle $= 90° - 26½° = 63½°$

c) For the December solstice, the subsolar is 23½° S.

Step 1: $50° - 23½° = 73½°$

Step 2: noon sun angle $= 90° - 73½° = 16½°$

At 50° N, the noon sun angle varies from 16½° on the December solstice to 63½° on the June solstice.

Do similar calculations for the equator and the North Pole. What do you notice?

Remember This

- At the June solstice, the subsolar point is 23½° N.
- At the December solstice, the subsolar point is 23½° S.
- At the equinoxes, the subsolar point is 0°.

Question for Thought

If Earth's axis were not tilted, why would we not experience seasons?

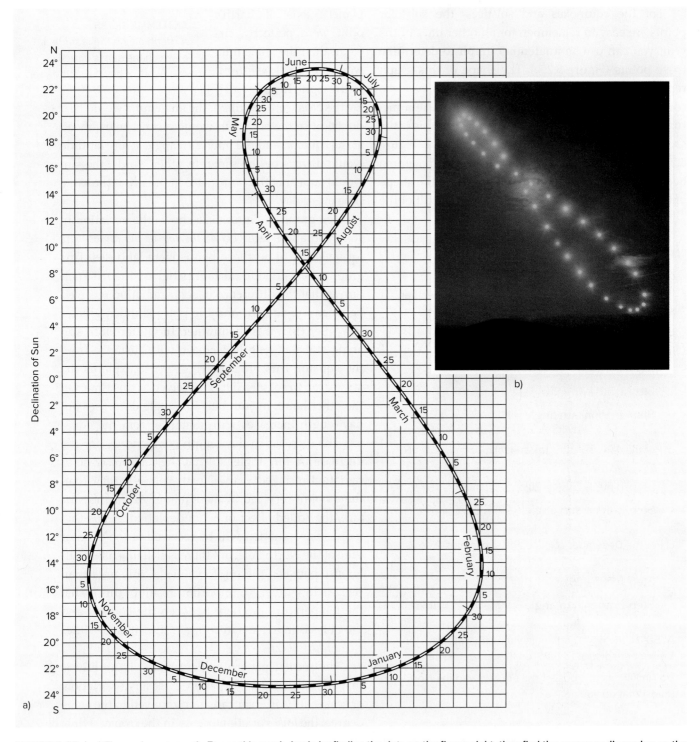

FIGURE 5.25 | a) The analemma graph. To use this graph, begin by finding the date on the figure eight, then find the corresponding value on the vertical axis. This value is the subsolar point for that date. The vertical axis runs from 23½° N to 23½° S because the sun is never overhead beyond this range. The graph also shows that at all locations within this range the sun is overhead twice a year. b) A photograph showing the shape of the sun's path in the sky over the course of a year.

Source: a) Adapted from Hidore, John J., John E. Oliver, Mary Snow, and Rich Snow. (2010). *Climatology: An atmospheric science* (3rd ed.). Upper Saddle River, New Jersey: Prentice Hall, p. 331.

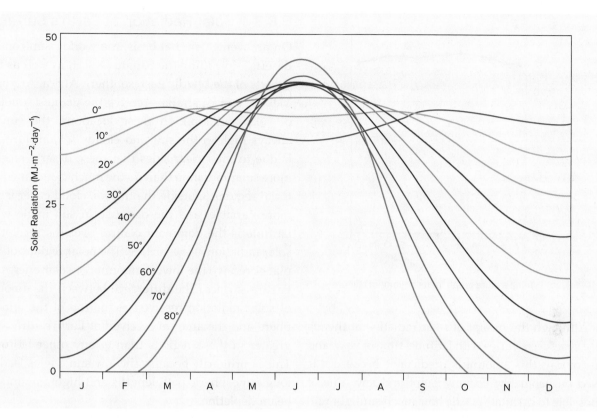

FIGURE 5.26 | Solar radiation, in MJ·m⁻²·day⁻¹, received at the top of the atmosphere throughout the year for a variety of northern-hemisphere latitudes. (Remember from Section 1.5 that 1 W = 1 J/s. Here we are giving the amount of solar radiation received over a day, so we must use J/day, instead of watts. Further, megajoules [MJ] are used, where 1 MJ = 1 x 10⁶ J, because of the large amount of energy involved.) The variation shown at each latitude is a result of the combined effects of the variation of day length and sun angle. Why does the least annual variation of solar radiation occur in the tropical latitudes? Why does the greatest annual variation of solar radiation occur in the higher latitudes? Why does 80° N latitude receive more solar radiation than any other location on the June solstice?

the *greatest* variation occurs in the polar latitudes. Near the equator, neither sun angle nor day length changes significantly. The sun is always high in the sky, and days and nights are of roughly equal length all year. As a result, there is no cold season, and most equatorial climates are characterized by small annual temperature ranges (Figure 5.27). In the polar latitudes, there is a small annual variation in sun angle, but a very large annual variation in day length. Due to the latter, the polar regions experience a greater variation in solar radiation received at the top of the atmosphere than does anywhere else on Earth. Here there are long hours of light or long hours of darkness. The annual variation in temperature is very large, but it is *always* cold (Figure 5.28).

The mid-latitudes are the only regions that experience true seasons. Here, there is a large variation in both day length *and* sun angle. Throughout the mid-latitudes, the annual variation in sun angle is 47°, and the annual variation in day length increases with latitude. The result is that mid-latitude locations experience both warm seasons and cold seasons (Figure 5.29). Within the mid-latitudes, annual temperature variation increases with latitude and is strongly affected by **continentality**, with the result that inland locations experience larger daily and annual temperature ranges than do coastal locations (Section 6.6.3).

CONTINENTALITY
The degree to which a climate is affected by its distance from a body of water.

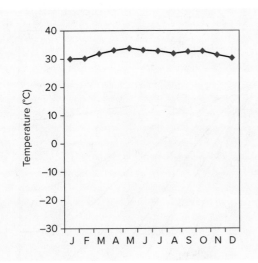

FIGURE 5.27 | Temperature graph for an equatorial climate.

5.8.3 | Solar Radiation at Earth's Surface

On an average annual basis, the world's subtropical deserts—including the Sahara Desert, as well as the deserts of the Middle East, southern Africa, Arizona, Mexico, and Australia—receive the greatest amount of solar radiation. In these locations, the sun is always high in the sky, and clouds seldom form. It is due to their clear skies that these deserts receive more solar radiation than do the much cloudier equatorial regions. Outside the tropics, cloud cover is far more variable, and the decrease in sun angle with latitude is the dominant control over the receipt of solar radiation at the surface. The polar regions of the planet receive the lowest amounts of solar energy at Earth's surface. The difference between the amount of solar radiation received at the top of the atmosphere and the amount received at Earth's surface is greater in these regions than at any other latitude. This is primarily because the low sun angles lead to long atmospheric path lengths and high amounts of beam depletion.

Although the receipt of solar radiation at the *top of the atmosphere* varies with latitude, time of year, and time of day, this amount is predictable because it is based on geometry. For any point and any time, it is possible to calculate *exactly* how much sunlight will be received at the top of the atmosphere if you know the latitude, the date, and the time of day. It is far more difficult, however, to make accurate predictions about the amount of solar radiation received, and ultimately absorbed, at Earth's surface, because more than geometry is involved.

The amount of solar radiation received at the surface on the *local scale* is further controlled by *shadiness* and *slope*. Because shady areas receive diffuse rather than direct beam radiation, they tend to be a little cooler than areas that receive direct sunlight. The slope of a surface, on the other hand, does not affect

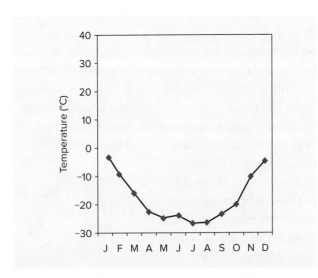

FIGURE 5.28 | Temperature graph for a southern-hemisphere polar climate.

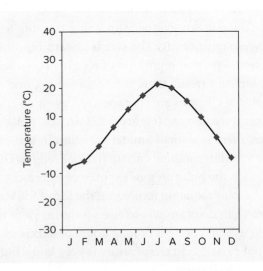

FIGURE 5.29 | Temperature graph for a continental mid-latitude climate in the northern hemisphere.

the receipt of *diffuse* beam radiation, but it does affect the angle at which *direct* beam radiation strikes it. In the tropics, when the sun is overhead, a horizontal surface will receive more sunshine than will a sloping surface. In the mid-latitudes, where the sun is never directly overhead, certain combinations of sun angle and slope angle can increase the amount of solar radiation received over what would be received on a horizontal surface. For example, if the sun is 70° above the horizon, its rays will be perpendicular to the surface of a south-facing slope of 20° (Figure 5.30). Grapes are often grown on steep slopes to take advantage of such effects. For the same reason, the angle of solar panels can be changed to maximize the amount of solar radiation the panels receive (Figure 5.31).

Shadiness and slope angle determine the amount of solar radiation a surface *receives*, but it is the albedo of a surface that ultimately determines the amount of radiation that the surface will *absorb* (Table 5.2). For simple surfaces, then, once the solar radiation reaches the surface, the amount absorbed is dependent on the slope angle, the amount of shade, and the albedo. However, most of Earth's surfaces are far from simple. For water bodies, vegetated surfaces, and even snow and ice, the sun's radiation is absorbed through a *volume* rather than at the somewhat imaginary boundary between the ground and the atmosphere that represents a simple surface. Whatever the surface characteristics, the amount of solar radiation absorbed is an important—though not the sole—determinant of climate, as we shall see in Chapter 6.

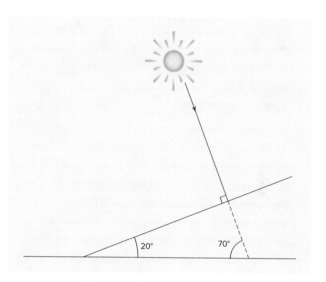

FIGURE 5.30 | When the sun angle is 70° relative to Earth's surface, the sun's rays will be received on a 20° slope as though the sun were directly overhead.

FIGURE 5.31 | Solar panels, such as these on top of Exhibition Place in Toronto, are positioned on an angle to maximize the amount of sunlight they receive.

5.9 | Chapter Summary

1. Radiation is energy that travels in the form of electromagnetic waves; it can transfer heat and travel in a vacuum. Electromagnetic radiation is grouped and named according to wavelength. In atmospheric science, only ultraviolet, visible, and infrared wavelengths are significant.

2. Radiation is emitted according to the radiation laws. As temperature increases, the rate of radiation emission increases, and the wavelengths of the emitted radiation decrease. Most of the radiation emitted by the sun is at wavelengths between 0.15 and 3.0 μm; we call this shortwave radiation. Most of the radiation emitted by Earth is at wavelengths between 3.0 and 100.0 μm; we call this longwave radiation.

3. All radiation incident on a substance is accounted for through some combination of absorption, reflection, and transmission.

4. Albedo is the reflectivity of surfaces to the sun's radiation. Surfaces with high albedos—such as snow or deserts—will reflect much of the sun's radiation; surfaces with low albedos—such as wet soil and vegetative coverings—will absorb much of the sun's radiation.

5. The sun's radiation is scattered by gas molecules and aerosols in the atmosphere. Selective—or Rayleigh—scattering by gas molecules makes the sky blue. Non-selective—or Mie—scattering by aerosols, including cloud droplets, makes clouds and fog white.

6. The gases of the atmosphere selectively absorb radiation. The atmosphere is much more effective at absorbing longwave radiation than it is at absorbing shortwave radiation. The gases most responsible for absorbing longwave radiation are water vapour, carbon dioxide, methane, nitrous oxide, and ozone. These are greenhouse gases. The gases most responsible for absorbing shortwave radiation are oxygen and ozone. Atmospheric gases do not effectively absorb visible radiation (0.4 to 0.7 μm) or wavelengths within the atmospheric window (8 to 11 μm). Clouds, however, absorb all longwave radiation.

7. The greenhouse effect occurs because the atmosphere allows the transmission of shortwave radiation, but greenhouse gases and clouds absorb longwave radiation. Earth is currently 33°C warmer than it would be without this greenhouse effect.

8. We can use the selective absorption of radiation by gases to explain the temperature structure of the atmosphere. There are three points at which temperature reaches a maximum due to strong absorption of solar radiation: at the "top" of the atmosphere, in the upper stratosphere, and at Earth's surface.

9. The maximum amount of solar radiation that can be received anywhere in the Earth-atmosphere system is the solar constant. Solar radiation received at Earth's surface will be less than the solar constant due to the effects of sun angle and atmospheric attenuation. Sun angle varies with latitude, time of year, and time of day. Atmospheric attenuation varies with the content of the atmosphere, combined with the sun angle. Once solar radiation reaches Earth's surface, the amount absorbed largely depends on surface shadiness, slope, and albedo.

Review Questions

1. What do the radiation laws tell us about the emission of radiation? How and why is the radiation emitted by Earth different from that emitted by the sun?

2. What determines the colours we see around us? What, specifically, determines the colour of a) sky, b) clouds, and c) stars?

3. How is Rayleigh scattering similar to and different from Mie scattering?

4. What are the differences between the atmosphere's radiative properties for shortwave radiation and its radiative properties for longwave radiation?

5. How do the radiative properties of gases (Figure 5.13) help explain a) the greenhouse effect and b) the temperature structure of the atmosphere?

6. Distinguish between beam spreading and beam depletion. How is each related to sun angle? Why does sun angle vary with latitude?

7. What causes seasons?

8. How would you describe and account for the relationship between latitude and seasonality?

9. What factors influence the amount of solar radiation received at the top of the atmosphere?

10. Why is very little solar radiation received at Earth's surface at 60° N on the December solstice?

11. What determines the distribution of shortwave radiation received at Earth's surface?

12. How do clouds influence the receipt of a) shortwave radiation and b) longwave radiation at Earth's surface?

Problems

1. Determine roughly how much more blue light than red light is scattered in the atmosphere. Use Expression 5.9, and make the simplification that the wavelength of blue light is 0.4 μm and the wavelength of red light is 0.7 μm.

2. Use the data in Table 2.4 to determine the following for Venus.
   a) the solar constant (show that the solar constant for Venus can also be determined using Earth's solar constant)
   b) the amount of radiation emitted
   c) the wavelength of maximum emission

3. Compare the annual variation in noon sun angle at 55° N to that at 15° N.

4. a) How much more solar radiation is received on a surface when the sun angle is 70° compared to when the sun angle is 40°?
   b) How much shorter is the sun's path length through the atmosphere when the sun angle is 70° compared to when it is 40°?

5. Calculate the noon sun angle for each of the following.
   a) 57° S on 9 January
   b) 12° N on 30 April
   c) 39° N on 15 November

6. a) A soil surface has a temperature of 31°C and an emissivity of 0.92. How much radiation is emitted from this surface?
   b) What is the wavelength of maximum emission for this surface?

7. a) Calculate the noon sun angles on the June solstice for each of the following latitudes: 0°, 50° N, and 90° N. Why are the noon sun angles at the equator and at 50° N so similar?
   b) For the same three latitudes, calculate the amount of solar radiation received at noon at the top of the atmosphere on the June solstice. Use a solar constant of 1365 W/m².

Suggestions for Research

1. Provide a detailed analysis of the radiative properties of aerosols, and investigate how the types and amounts of aerosols in the atmosphere are changing. Suggest how we might work to slow or even reverse the environmental impact of anthropogenic aerosols.

2. Investigate how the sun is being used as a source of renewable energy. Describe the advantages and disadvantages related to using solar energy in place of energy produced by burning fossil fuels.

References

Gatebe, C.K., E. Wilcox, R. Poudyal, & J. Wang. (2011). Effects of ship wakes on ocean brightness and radiative forcing over ocean. *Geophysical Research Letters*, *38*, L17702, 1–6. doi: 10.1029/2011GL048819

Kiehl, J.T., & K.E. Trenberth. (1997). Earth's annual global mean energy budget. *Bulletin of the American Meteorological Society, 78*(2), 197–208.

Trenberth, K.E., J.T. Fasullo, & J. Kiehl. (2009). Earth's global energy budget. *Bulletin of the American Meteorological Society, 90*, 311–23.

Turco, R.P. (2002). *Earth under siege* (2nd ed.). New York, NY: Oxford University Press.

Large cities such as Toronto radically alter Earth's surface. The varieties of surface types and geometries associated with urbanization have a dramatic impact on the energy balance—and, thus, the climate—at the local scale.

6

Energy Balance

Learning Goals

After studying this chapter, you should be able to

1. *explain* the concept of effective radiating temperature as it applies to a planet;

2. *describe* and *account* for the flows of energy that make up the planetary energy balance;

3. *understand* the significance of the latitudinal radiative imbalances;

4. *describe* the radiative and non-radiative heat flows at a surface;

5. *account* for the variation of radiative and non-radiative heat flows at a surface; and

6. *explain* how radiative and non-radiative heat flows influence microclimates.

In Chapter 4, you learned that heat is transferred by the emission and absorption of radiation, and also by the non-radiative processes of conduction and convection. Radiation, as a form of energy, was described in detail in Chapter 5. In this chapter, you will learn about how radiative and non-radiative heat transfers relate to the **energy balance** concept.

ENERGY BALANCE
An accounting system for the flows of energy in a system.

Because the energy balance concept provides an accounting method for the energy flows in the Earth-atmosphere system, we can use it as a tool to understand climate, and climate change, at a variety of scales. In its simplest form, the energy balance concept is based on the conservation of energy, and it can be expressed as a word equation.

$$\text{Energy Input} - \text{Energy Output} = \text{Change in Energy Storage} \qquad (6.1)$$

Equation 6.1 shows that when energy inputs and outputs are balanced, energy storage won't change. However, when energy inputs are greater than energy outputs, there will be an *increase* in energy storage; conversely, when energy outputs are greater than energy inputs, there will be a *decrease* in energy storage. When inputs and outputs remain balanced in a climate system, climate remains unchanged. When an imbalance occurs in a climate system, the result is often a temperature change, but there are additional possibilities. In the Earth-atmosphere system, for example, an excess of energy might be used as latent heat to melt ice or evaporate water, or it might be stored in the oceans. (Recall from Chapter 4 that the oceans have a tremendous capacity to store energy because of water's high specific heat.)

RADIATIVE FORCING
A change in a climate system's energy balance that will ultimately lead to climate change.

EFFECTIVE RADIATING TEMPERATURE
The temperature at which a system radiates away as much energy as it receives.

A change in the flows of energy in a climate system is called a *forcing* because it will *force* the climate to change. If the change influences the flow of radiation, it is called a **radiative forcing**. An increase in atmospheric carbon dioxide concentrations, such as that associated with human activity, can cause a radiative forcing. This change creates a *positive* forcing because carbon dioxide, being a greenhouse gas, *increases* the amount of energy stored within a system, thus reducing energy output. In fact, recent energy balance studies show that, as a result of the increase of greenhouse gases in Earth's atmosphere, the input of energy *to* the Earth-atmosphere system is now greater than the output of energy *from* the Earth-atmosphere system. Thus, there is an imbalance that is causing more energy to be stored in the system and, as a result, global climate is changing (Trenberth, Fasullo, & Kiehl, 2009; Hansen, et al., 2005) (Section 17.3.2).

At the smaller scale, analysis of the energy balance of a *surface* allows us to account for the energy exchanges between the surface and the air layer just above it. These exchanges are a direct result of the *properties* of both the surface and the air layer. A practical application of the surface energy balance concept is the analysis of how land-use changes influence small-scale climates. For example, we can apply this concept to help us understand how urbanization affects climate. Cities are a radical alteration of Earth's surface and, since so many of us live in cities, it is important to not only understand how these alterations affect climate, but also consider how we might modify these built environments to our advantage.

Ultimately, the surface energy balance is important because the surface is the source of moisture, and most of the heat, for the atmosphere as a whole. In turn, the state of the atmosphere influences surface climates by determining the amount of solar and longwave radiation that the surface receives. To understand these complex relationships, we will begin by considering the simple energy balance at the top of the atmosphere.

6.1 | Effective Radiating Temperature

We can use the energy balance at the top of the atmosphere to determine the **effective radiating temperature** of the Earth-atmosphere system. Since energy can travel through space only as radiation, the energy balance at the top of the atmosphere is a simple *radiation* balance rather than a *total energy* balance; there are no non-radiative terms (Equation 6.2).

$$\text{Solar Radiation Absorbed} =$$
$$\text{Terrestrial Radiation Emitted} \qquad (6.2)$$

In this equation, the solar radiation absorbed at the top of the atmosphere is the difference between the incoming solar radiation and the solar radiation reflected by the Earth-atmosphere system.

Originally, the balance described in Equation 6.2 might have been achieved in the following way. When Earth was young, it was so cold that it emitted very little energy. As heat was transferred from the sun to Earth, Earth began to warm up, and it gradually emitted more and more energy in accordance with the Stefan–Boltzmann law (Section 5.2.1). For as long as Earth continued to emit less energy than it received from the sun, its temperature increased. Eventually, Earth was hot enough to emit as much radiation as it was receiving from the sun; balance was achieved, and Earth's temperature stopped increasing.

This balance was achieved, and is maintained, due to a rather simple negative feedback mechanism (Section 1.2) that prevents Earth's temperature from continually rising or continually falling. This feedback works to *stabilize* the system: for example, if the sun's output increases, the warmer Earth will respond by emitting more radiation, thus preventing further warming; if the sun's output decreases, the opposite will occur. Because we know that such negative feedbacks operate, it follows that, for the most part, an energy balance should exist. Over Earth's history, however, changes have continually upset this balance, as evidenced by the fact that the planet has experienced warmer and cooler periods (Section 17.3.1).

Question for Thought

Earth is currently warming due to increasing concentrations of greenhouse gases in the atmosphere. Do you think a negative feedback such as that described in this section might operate to reduce the warming?

To obtain an equation for the *effective radiating temperature*, we first need to determine an expression for the left side of Equation 6.2, the amount of solar radiation Earth and its atmosphere absorb. To do this, imagine that there is a wall on the other side of Earth from the sun so that, as the sun shines on Earth, Earth's shadow is cast on the wall. This shadow represents the amount of the sun's radiation *intercepted* by Earth (Figure 6.1). Thus, the expression for the total amount of the sun's radiation *incident* on Earth is $S \pi r_E^2$, where S is the solar constant (Section 5.8) and r_E is equal to Earth's radius. It follows that the amount

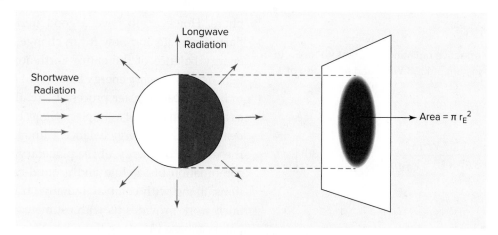

FIGURE 6.1 | When Earth intercepts solar energy, it acts as though it were a circle with an area of πr_E^2. In turn, it emits longwave radiation over its entire surface area, $4 \pi r_E^2$.

Source: Adapted from Marshall & Plumb, 2007, p. 12.

of solar radiation that the Earth-atmosphere system *absorbs* is $S(1 - \alpha)\pi r_E^2$, where α is the albedo. To determine an expression for the right side of Equation 6.2, we use the Stefan–Boltzmann law (Equation 5.3). Because Earth emits radiation over its entire surface area, $4\pi r_E^2$, an expression for the total amount emitted is $(\sigma T_E^4)(4\pi r_E^2)$, where T_E is the effective radiating temperature.

Using the above, we can now express Equation 6.2 mathematically; then, rearranging it to solve for T_E, we can arrive at an equation for the effective radiating temperature (Equation 6.3).

$$S(1 - \alpha)\,\pi\,r_E^2 = (\sigma\,T_E^4)\,(4\,\pi\,r_E^2)$$

$$T_E = \left[\left(\frac{S}{4\sigma}\right)(1 - \alpha)\right]^{0.25} \tag{6.3}$$

Given that the albedo of the Earth-atmosphere system is 30 per cent, we can solve Equation 6.3, as shown in Example 6.1, to give an effective radiating temperature for Earth of 255 K, or −18°C. This temperature is 33°C below Earth's actual temperature of about 15°C. The reason for this difference is that Equation 6.3 treats Earth as having an atmosphere that reflects shortwave radiation, but that does not absorb or emit longwave radiation. In other words, it does not include the greenhouse effect, which is the mechanism responsible for the extra 33°C.

Example 6.1

Calculate the effective radiating temperature for Earth. Use a solar constant of 1365 W/m² and an albedo of 30 per cent.

Use Equation 6.3.

$$T_E = \left[\left(\frac{1365\ \text{W/m}^2}{4 \times 5.67 \times 10^{-8}\ \text{W}\cdot\text{m}^{-2}\cdot\text{K}^{-4}}\right)(1 - 0.30)\right]^{0.25}$$

$$= 255\ \text{K}$$

$$= -18°\text{C}$$

Equation 6.3 is an example of a very simple *model* that we can use to understand the temperature of a planet. As a model, it allows us to make predictions—in this case, predictions about how changes in the planetary albedo or the output of the sun might influence a planet's temperature. In addition, if we know a planet's *actual* temperature, we can use the equation to estimate the size of that planet's greenhouse effect.

Remember This

Four factors determine a planet's temperature:

- solar output,
- distance from the sun,
- albedo, and
- greenhouse effect.

Question for Thought

Venus has a lower effective radiating temperature and a larger greenhouse effect than Earth has. Why do you think this is so? (Hint: Consult Section 2.11, which provides information about Venus's atmosphere.)

The model represented by Equation 6.3 is based on a simple radiation balance at the top of the atmosphere. However, to have a good understanding of Earth's climate and how it can change, we study the energy balance of the entire Earth-atmosphere system, or the *planetary* energy balance. In this case, non-radiative heat transfer processes are also important; as a result, instead of a simple *radiation* balance, we are dealing with an *energy* balance. Climatologists study these flows of energy, at the planetary scale, using a combination of satellite and ground-based observations, along with complex radiative transfer models. Such work provides us with estimates of the energy balance, based on long-term averages, for the entire Earth-atmosphere system.

6.2 | Flows of Solar (Shortwave) Radiation

Figure 6.2 summarizes the flows of solar radiation in the Earth-atmosphere system. As you can see from this figure, the radiation that enters the system is either absorbed, reflected, or scattered in the atmosphere, or reflected or absorbed at Earth's surface. The amounts given are annual averages for the entire Earth; they will, of course, vary over Earth's surface throughout a year, and from year to year. The total amount of solar radiation received at the top of the atmosphere, averaged over Earth's total surface area, is about 341.3 W/m² (Example 6.2). For the purpose of simplifying our discussion of the planetary energy balance, we will make 341.3 W/m² equal to 100 units. That is, 1 unit on figures 6.2, 6.3, and 6.6 equals 3.41 W/m².

Example 6.2

Calculate the total amount of solar radiation received at the top of the atmosphere, averaged over Earth's surface.

Remember that the *total* amount of solar radiation intercepted by Earth is

$$S \pi r_E^2$$

Averaged over Earth's spherical surface, this becomes

$$\frac{S \pi r_E^2}{4 \pi r_E^2} = \frac{S}{4}$$

Substitute the solar constant of 1365 W/m² into this equation.

$$\frac{1365 \text{ W/m}^2}{4} = 341.3 \text{ W/m}^2$$

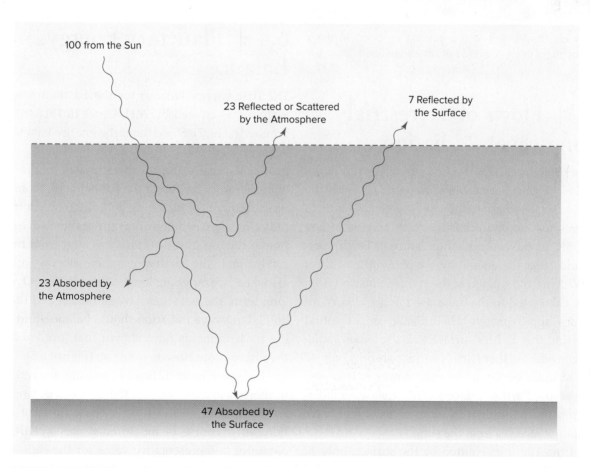

100 from the Sun

23 Reflected or Scattered by the Atmosphere

7 Reflected by the Surface

23 Absorbed by the Atmosphere

47 Absorbed by the Surface

FIGURE 6.2 | The flows of solar radiation. (Note that the amounts given are annual averages, and the units have been adjusted so that 1 unit in the diagram equals 3.41 W/m².)

Of the 100 units entering the top of the atmosphere, 23 are absorbed by water vapour, ozone, and clouds. Another 23 units are lost to space due to reflection and scattering by gas molecules, aerosols, and clouds. The solar radiation that is not absorbed or reflected in the atmosphere reaches Earth's surface. Of this, 47 units are absorbed by the surface, and 7 are reflected back to space. In total, 30 units of solar radiation are thus reflected back out to space. This value represents Earth's albedo of 30 per cent (Section 5.4). Figure 6.2 shows that this relatively high albedo is due mostly to reflection by the atmosphere; the surface itself reflects only 7 units of the radiation incident upon it.

Question for Thought

How might changes in the amount of cloud in the atmosphere influence flows of solar radiation?

6.3 | Flows of Terrestrial (Longwave) Radiation

Figure 6.3 summarizes the flows of terrestrial radiation in the Earth-atmosphere system. As with the flows of solar radiation, the values given are annual averages for the entire Earth. Using a value of just over 15°C as the average temperature of Earth's surface and 1.0 as the emissivity, we obtain the value of 396 W/m^2 as an estimate of the *average* emission from Earth. Converted to the units used above, 396 W/m^2 becomes approximately 116 units. It might sound surprising that Earth's surface emits so much radiation, considering that Earth's surface absorbs only 47 units of radiation from the sun. However, it is possible because Earth's surface receives longwave radiation from the atmosphere in addition to the shortwave radiation it receives from the sun.

Of the 116 units emitted by the surface, only 12 travel through the atmosphere to space, while 104 are absorbed by the greenhouse gases and clouds in the atmosphere (Figure 6.4). The atmosphere, in turn, emits 156 units of longwave radiation, 58 of which are emitted upward to space, and 98 of which are emitted back down to Earth. (Some of the 156 units of radiation come from sensible and latent heat transfers, as discussed in the following section.) Notice that less radiation is emitted upward than downward because the temperature of the atmosphere decreases with height. Notice also that Earth receives less longwave radiation *from* the atmosphere than it emits because, on average, the atmosphere is cooler than Earth.

Question for Thought

How might changes in the amount of cloud in the atmosphere influence the flows of longwave radiation?

6.4 | Planetary Energy Balance

The first energy balance for the Earth-atmosphere system was created by W.H. Dines in 1917 (Kiehl & Trenberth, 1997). Since then, the energy balance has been revised many times. Recently, satellites have provided greatly improved estimates of the flows of radiation at the top of the atmosphere. From their measurements of incoming and outgoing shortwave radiation, we can determine the net shortwave radiation at the top of the atmosphere, and from this we can estimate the Earth-atmosphere albedo. Satellites also measure outgoing longwave radiation. Over the long term, the net shortwave radiation and the outgoing longwave radiation should balance, and satellite measurements have shown that, until recently, they have come close to doing so (Figure 6.5).

It is much more difficult to measure the reflection of shortwave radiation and the emission of longwave radiation from Earth's *surface*. These flows of radiation can, of course, be measured at sites on Earth, but obtaining a representative value for the *entire* Earth would require measurement at a tremendous number of sites. To further complicate things, well over half of Earth's surface is water! In addition, in order to come

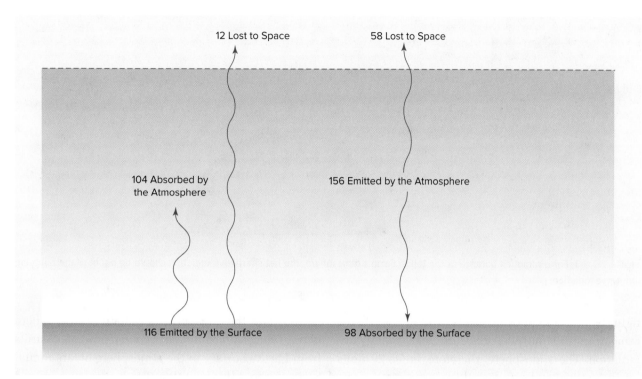

12 Lost to Space

58 Lost to Space

104 Absorbed by
the Atmosphere

156 Emitted by the Atmosphere

116 Emitted by the Surface

98 Absorbed by the Surface

FIGURE 6.3 | **The flows of terrestrial radiation. (Note that the units have been adjusted so that 1 unit in the diagram equals 3.41 W/m².)**

up with a complete planetary energy balance based on measurements, we would need to take on the complex task of measuring radiative transfer *within* the

FIGURE 6.4 | **Clouds seen from space. Clouds reflect and absorb shortwave radiation from the sun, and they absorb longwave radiation from Earth.**

atmosphere. Because it is so difficult to make such measurements, we generally estimate the surface and atmospheric flows of radiation by using *models* that are based on our best understanding of radiative transfer processes. In turn, these modelled results are checked with observations whenever possible (Kiehl & Trenberth, 1997).

The complete planetary energy balance is depicted in Figure 6.6. This diagram combines the flows of solar radiation and terrestrial radiation illustrated in figures 6.2 and 6.3. It also includes *convective* flows of sensible heat and latent heat from Earth's surface to the atmosphere.

These convective flows develop because there is a radiative *surplus* for Earth's surface and a radiative *deficit* for the atmosphere. As shown in Figure 6.6, Earth *absorbs* 47 units of shortwave radiation and 98 units of longwave radiation, for a total of 145 units of radiation. At the same time, Earth *emits* 116 units of longwave radiation. Earth, therefore, absorbs 29 units of radiation more than it emits, so that there is a radiative *surplus* for Earth. The atmosphere *absorbs* 23 units of shortwave radiation and 104 units of longwave

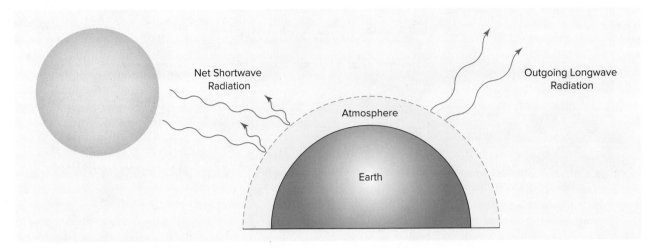

FIGURE 6.5 | For a radiation balance at the top of Earth's atmosphere, the net shortwave radiation should be equal to the outgoing longwave radiation.

radiation, for a total of 127 units of radiation. At the same time, the atmosphere *emits* 156 units of longwave radiation. The atmosphere, therefore, emits 29 units more than it absorbs, so that there is a radiative *deficit* for the atmosphere. As a result of Earth's radiative surplus and the atmosphere's radiative deficit, there is on average a *net* convective heat flow of 29 units from Earth's surface to the atmosphere on an annual basis. Without this heat flow, Earth would warm and the atmosphere would cool.

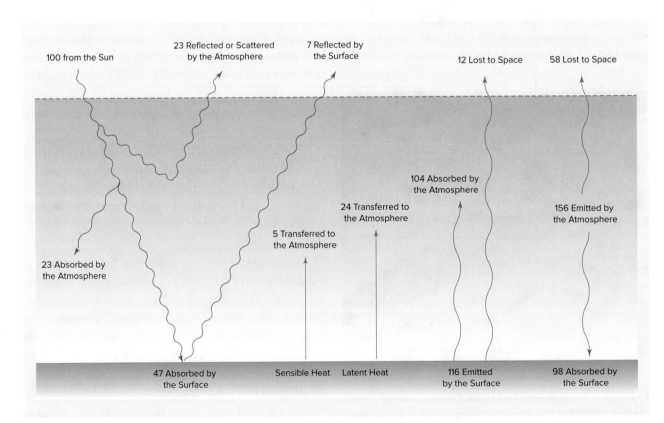

FIGURE 6.6 | The planetary energy balance. Convective flows of sensible heat and latent heat have been added to the flows depicted in figures 6.2 and 6.3.

Of the 29 units of convective heat transfer, 24 are transferred as latent heat. This means that a little more than 80 per cent of the surplus radiation at Earth's surface is used to *evaporate* water. As the water is evaporated at the surface, it absorbs latent heat. Upon rising in the atmosphere, the water vapour may eventually condense into clouds, releasing latent heat; in the process, heat is transferred from the surface to the atmosphere. The remaining 5 of the 29 units of convective heat transfer are transferred as sensible heat in the thermals created as warm air rises.

Question for Thought

How would you use the energy balance diagram to show that the atmosphere is heated mostly by Earth's surface?

Researchers estimate the apportionment of the 29 units between sensible heat and latent heat using observations of total annual precipitation for the planet. Since evaporation is the process that carries water into the atmosphere and precipitation is the process that returns water to the surface, they use knowledge of precipitation amounts to estimate evaporation amounts. In turn, they can calculate the amount of energy required for this amount of evaporation; the result of the calculation represents the convective flow of latent heat, and the remainder is the convective flow of sensible heat.

The radiative properties of clouds also play very important roles in the planetary energy balance. As we saw in Chapter 5, clouds both *reflect* solar radiation and *absorb* and *emit* longwave radiation (Figure 6.7). The result is that clouds can both cool and warm Earth; their *net* effect is a subject of ongoing research. It seems that low, thick clouds are better at reflecting solar radiation than they are at absorbing longwave radiation. These clouds, therefore, are believed to have a net cooling effect. High, thin clouds, on the other hand, are not very effective at reflecting solar radiation. Because of this, these clouds appear to have a net warming effect. Satellite measurements have allowed us to compare the amount of radiation leaving the top of the atmosphere under clear-sky conditions to that under cloudy-sky conditions. With cloudy skies, 50 W/m² *more* solar radiation leaves the system than with clear skies. With cloudy skies, 30 W/m² *less* longwave radiation leaves the system than with clear skies. This means that, on average, clouds cause a net loss of 50 W/m² of shortwave radiation and a net gain of 30 W/m² of longwave radiation (Kiehl & Trenberth, 1997). Therefore, on an average annual basis, clouds *cool* the planet.

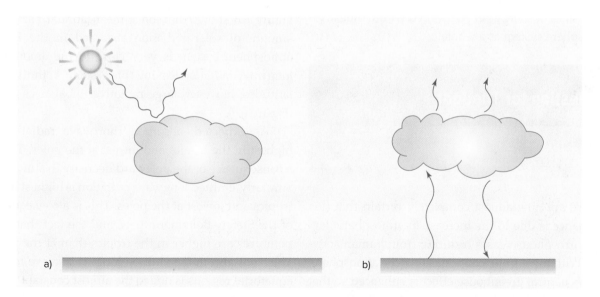

FIGURE 6.7 | The radiative properties of clouds. a) Because clouds reflect solar radiation, they can have a cooling effect on Earth's surface. **b)** Because clouds absorb and emit longwave radiation, they can have a warming effect on Earth's surface.

Finally, Figure 6.6 shows that there is a radiative balance between absorbed solar radiation and outgoing longwave radiation: 70 units of solar radiation are absorbed in the Earth-atmosphere system (23 by the atmosphere and 47 by Earth's surface), while 70 units of longwave radiation leave the top of the atmosphere. Under such balanced conditions, there should be no change in the storage of energy within the Earth-atmosphere system, and temperatures should remain more or less constant. However, as mentioned at the beginning of this chapter, current energy balance research shows a radiative *imbalance* at the top of the atmosphere (see Trenberth, Fasullo, & Kiehl, 2009). Figure 6.6 doesn't show this imbalance because the numbers have been rounded off, thus obscuring the relatively small imbalance.

Though small, the imbalance is still significant, and it becomes evident when we consider *actual* values. Recall that we estimate incoming solar radiation to be 341.3 W/m² at the top of the atmosphere. Additionally, our best estimate for the amount of solar radiation reflected at the top of the atmosphere is 101.9 W/m². Thus, the value for absorbed solar radiation at the top of the atmosphere is 239.4 W/m². In contrast, our best estimate of outgoing longwave radiation at the top of the atmosphere is 238.5 W/m² (Trenberth, Fasullo, & Kiehl, 2009). Together, these values show that 0.9 W/m² more energy is entering the Earth-atmosphere system than is leaving. This is equivalent to an extra 1.45×10^{22} J of energy per year stored within the Earth-atmosphere system as a whole!

> ## Question for Thought
>
> How do you think the value of 1.45×10^{22} J of energy per year was determined?

We are now almost completely certain that this imbalance is due to the increase in atmospheric levels of greenhouse gases resulting from human activities. With more greenhouse gases in the atmosphere, Earth's natural greenhouse effect is enhanced so that less terrestrial radiation leaves the system. Researchers have long predicted that increasing levels of greenhouse gases could warm the planet and, in fact, we have observed a temperature increase of about 0.8°C over the last 100 years. As a response to this warming, Earth is emitting more radiation in accordance with the negative feedback mechanism described in Section 6.1. Despite this attempt to regain radiative equilibrium, an imbalance of 0.9 W/m² remains. We are now observing and measuring the effects of this increase in stored energy on a global scale. Some of it is being used as latent heat to melt ice and evaporate water. The rest of it is being stored in the oceans, resulting in thermal expansion of the water and a rise in sea level. In fact, sea level has risen about 17 cm over the last century as a result of the *combined* effects of thermal expansion and melting ice.

The discussion in this section has illustrated the value of the energy balance concept as an accounting tool. When an energy imbalance is discovered, it must be accounted for. In a complex system, such as the Earth-atmosphere system, this is not easy. However, by trying to account for this imbalance, we hope to gain a greater understanding of the effects of human activities on climate (Section 17.3.2).

6.5 | Latitudinal Radiative Imbalances

Figure 6.6 shows that, on a mean annual basis, the amount of solar radiation absorbed in the Earth-atmosphere system is very close to the amount of longwave radiation leaving this system. At individual latitudes, however, large radiative imbalances occur (Figure 6.8).

As expected, absorbed shortwave radiation is highest in the tropics and lowest at the poles. This is a consequence of the poleward decrease in sun angle. Similarly, emitted longwave radiation is highest in the tropics and lowest at the poles. This is a consequence of the Stefan–Boltzmann Law, and the fact that temperatures are higher in the tropics than at the poles. The small dip in the emitted longwave curve for the equatorial regions is due to the almost constant cloud cover in this region. Taken together, the two curves show that the amount of absorbed solar radiation is *greater* than the amount of emitted longwave radiation between roughly 40° N and 40° S, and *less* than the amount of emitted longwave radiation beyond these latitudes. This trend is reflected in the curve for *net* radiation; the positive values of net radiation for lower latitudes indicate a radiative surplus, while the negative values at higher latitudes indicate a radiative deficit.

Acting alone, this radiative imbalance would cause temperatures at the poles to become colder until the amount of longwave radiation emitted decreased to balance the smaller amount of solar radiation absorbed. In the tropics, it would cause temperatures to become warmer until the amount of longwave radiation emitted increased to achieve balance. This doesn't happen, however, because the fluids of our planet—the atmosphere and the oceans—help to transport sensible and latent heat from the equator to the poles through **advection**. Without fluids to transfer heat, the difference in temperature between the equator and the poles would be much greater than it is. This heat flow is carried out by the large-scale winds and ocean currents that make up what we know as the **general circulation** of our planet.

The driving force behind the general circulation is the **meridional**, or north–south, temperature gradient itself. As the warm air at the equator rises, it flows toward the colder poles; at the same time, the cold air at the poles sinks and flows back toward the warmer equator (Figure 6.9). Of course, this is a very simple model for explaining the origins of the heat flow on Earth; as you will discover in Chapter 12, the general circulation is far more complex in reality.

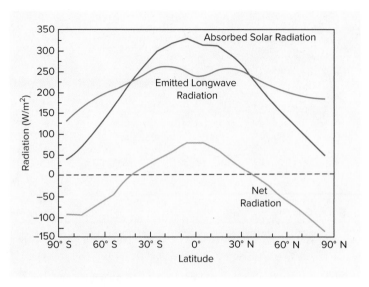

FIGURE 6.8 | The absorbed solar radiation, emitted longwave radiation, and net radiation at the top of the atmosphere by latitude.

While the atmosphere is responsible for about three-quarters of the latitudinal transfer of heat, the oceans are responsible for the rest. Like heat flow in the atmosphere, heat flow in the oceans is driven by density differences—which are influenced by salinity as well as temperature—and by winds. The circulation system known as the *global ocean conveyor belt* (Figure 6.10) is an example of how density differences and winds work together to transport heat toward the polar regions. As the cold, salty water of the North Atlantic Ocean sinks, it draws warmer water northward in currents known as the Gulf Stream and the North Atlantic Drift. The winds also help support these currents. As a result of this circulation system, the climates of northern Europe are much warmer than locations at similar latitudes in North America (Figure 6.11).

The result of the general circulation is that it keeps both the poles and the tropics within a relatively comfortable range of temperatures. There is an important

ADVECTION
Horizontal transfer across a fluid, caused by movement within the fluid.

GENERAL CIRCULATION
The average patterns of pressure and wind over Earth's surface.

MERIDIONAL
Relating to or varying along a meridian, or in the north–south direction.

Remember This

Because heat is transferred from areas of high temperature to areas of low temperature, convection transports heat from the warmer surface to the cooler atmosphere, and advection transports heat from the warmer equatorial regions to the cooler polar regions.

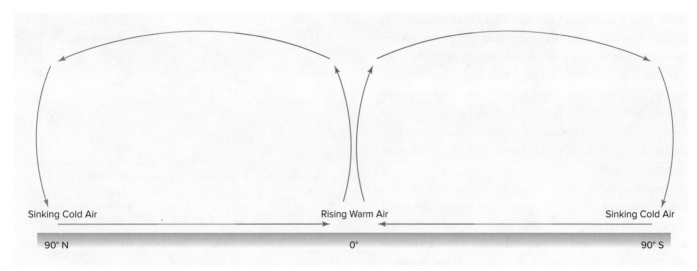

FIGURE 6.9 | Hypothetical convection cells due to Earth's meridional temperature gradient.

negative feedback mechanism involved here. If the temperature gradient were to increase, the circulation would become more vigorous. Then, as more heat was carried poleward, the poles would get warmer, the tropics would get cooler, and the temperature gradient would decrease. However, heat transfer can never completely eliminate the temperature gradient because the sun will always warm the tropical regions more than it warms the polar regions.

Studies of past climates show the importance of this link between the planetary circulation and the meridional temperature gradient. For example, the

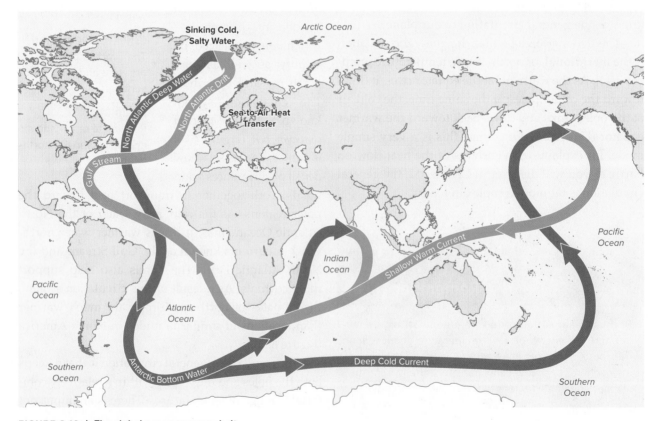

FIGURE 6.10 | The global ocean conveyor belt.

Source: Smithsonian Institution.

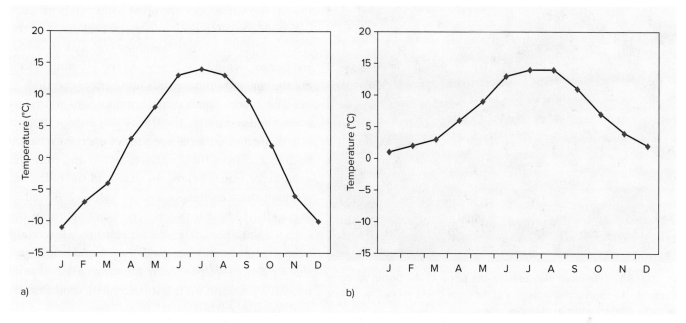

FIGURE 6.11 | Temperature graphs for a) Anchorage, Alaska, and b) Bergen, Norway. These coastal cities are both located at approximately 60° N latitude but, due to the effects of the North Atlantic Drift, winter temperatures are significantly higher in Bergen than they are in Anchorage.

movement of continents has, in the past, altered ocean currents and, in turn, caused climates to change. This is what happened when Antarctica became a separate continent; with no major land masses near enough to deflect warm currents its way, it became much colder than it was. As a consequence, the temperature gradient between the equator and the South Pole strengthened.

It is particularly important for us to understand the links between the meridional temperature gradient and the planetary circulation because anthropogenic climate change is causing the temperature gradient to decrease. As the planet warms, the polar regions are warming more than the tropical regions—thus the decrease in the temperature gradient. Because mid-latitude weather systems are largely driven by this temperature gradient, it is expected that a change in the temperature gradient will influence weather systems.

6.6 | Surface Energy Balance

We can also apply the energy balance model on a smaller scale, based on a surface *type*. Just as the *planetary* energy balance determines Earth's climate, the *surface* energy balance determines microclimates. A microclimate occurs in a small area, ranging anywhere from a few square metres to several square kilometres in size, in which climate is different than it is in the surroundings. Within an area where the overall larger scale atmospheric conditions are similar, it is the *characteristics of surfaces* that create microclimates.

The differences in surface type are greatest between land surfaces and water surfaces, but much variety exists even among land surfaces. Consider gardens, which often contain many small microclimates. Sometimes conditions in one spot within a garden can be just right to grow an exotic plant that might not otherwise grow in the climatic region as a whole. In vineyards, for example, grape growers often take advantage of differences in slope steepness and aspect to grow several varieties of grapes in a relatively small area (Figure 6.12).

On a somewhat larger scale, cities can create climatic conditions considerably different from their rural surroundings. Cities as a whole tend to create **urban heat islands**, which can occur anywhere in the world, no matter what the *regional* climate type. Within cities, many different microclimates can develop (Figure 6.13). Parks can create cool, moist escapes from the often hot, dry conditions of downtown cores. Fountains in vast concrete plazas can significantly chill

URBAN HEAT ISLAND
A microclimate created by a city, in which temperatures are higher than they are in the surrounding region.

FIGURE 6.12 | Terraced vineyards on the slope of the Douro Valley in Portugal. The varying characteristics of the slope create microclimates that allow for different varieties of grapes to be grown.

the air. A courtyard, on the other hand, might produce welcoming warmth on an otherwise cool day. And the "canyons" between tall buildings can lead to excessively cool, windy conditions, as the buildings block out the sun and funnel winds into narrow corridors.

The energy balance concept allows us to understand and quantify the flows of energy within microclimates. Surfaces are sites of energy exchange both to and from the air above, and to and from the ground (or water) below. As shown in the following sections, these exchanges involve radiative as well as non-radiative heat transfers. The exact nature of the exchange, at a particular site, is determined by certain properties of the surface itself, the ground below, and the air above—of particular importance are albedo, emissivity, roughness, moisture content, conductivity, and specific heat.

FIGURE 6.13 | *Top left:* Hikers enjoy a cool stroll through Mount Royal Park in downtown Montreal. *Top right:* The fountain in front of the Vancouver Art Gallery chills the air amid concrete buildings. *Bottom left:* An inner courtyard at the University of Toronto provides warmth on a spring day. *Bottom right:* Skyscrapers, such as these ones in Calgary, block out much of the sky, reducing the nighttime loss of longwave radiation from the surface, thus contributing to the urban heat-island effect. In addition, they cast shadows on the streets below while funnelling the wind.

6.6.1 | Radiative Heat Transfer

The radiation balance of a surface is made up of incoming and outgoing shortwave and longwave radiation. In this context, the surface properties of albedo and emissivity are most important.

The difference between the incoming shortwave radiation for a surface, $K\!\downarrow$, and that reflected by the surface, $K\!\uparrow$, is the *net* shortwave radiation, K^*, or the amount of shortwave radiation *absorbed* by the surface. We can represent this relationship in a simple balance equation.

$$K^* = K\!\downarrow - K\!\uparrow \tag{6.4}$$

It follows that we can calculate the surface albedo, α, as the ratio of reflected shortwave radiation to incoming shortwave radiation.

$$\alpha = \frac{K\!\uparrow}{K\!\downarrow} \tag{6.5}$$

The amount of incoming shortwave radiation, $K\!\downarrow$, varies with latitude, season, time of day, and atmospheric attenuation (Section 5.8). It will be highest when sun angles are high and skies are clear. At the local scale, the amount of sunlight incident upon a surface is further dependent on slope angle and aspect. Finally, the amount of solar radiation that is absorbed, K^*, depends on surface albedo.

We can write a similar balance equation for longwave exchange. The *net* longwave radiation at the surface, L^*, is the difference between the longwave radiation coming from the atmosphere, $L\!\downarrow$, and the longwave radiation emitted by the surface, $L\!\uparrow$.

$$L^* = L\!\downarrow - L\!\uparrow \tag{6.6}$$

We can then expand Equation 6.6 so that it expresses the net longwave radiation as a function of atmospheric and surface temperatures, T_a and T_s, and atmospheric and surface emissivities, ε_a and ε_s. We can do this using the Stefan–Boltzmann law (Equation 5.4).

$$L^* = L\!\downarrow - L\!\uparrow = \varepsilon_a \sigma T_a^4 - \varepsilon_s \sigma T_s^4 \tag{6.7}$$

In Equation 6.7, the longwave radiation leaving the surface, $L\!\uparrow$, is shown as *emitted* radiation only, but it is possible that a small amount of this radiation could be longwave radiation from the atmosphere that is *reflected* by the surface. To derive an expression for this reflected longwave radiation, we can use Kirchhoff's law (Section 5.2.5). For longwave radiation, the transmissivity of most natural surfaces is zero, so we can set t_λ equal to zero; thus we reduce Equation 5.8 to Equation 6.8.

$$a_{\text{long}} + \alpha_{\text{long}} = 1 \tag{6.8}$$

Since Kirchhoff's law tells us that, for a given wavelength, emissivity is equal to absorptivity, we can substitute emissivity for absorptivity in Equation 6.8. This gives an expression for the reflectivity of a surface, α_{long}, to longwave radiation.

$$\alpha_{\text{long}} = 1 - \varepsilon_s \tag{6.9}$$

Because the emissivity of most natural surfaces is usually above 0.9, Equation 6.9 shows that the reflection of longwave radiation by such surfaces will normally be quite low. But, however small the amount, any reflected longwave radiation will contribute to the total amount of longwave radiation leaving Earth's surface. This reflected longwave radiation is included in Equation 6.10.

$$L^* = L\!\downarrow - L\!\uparrow = \varepsilon_a \sigma T_a^4 - [\varepsilon_s \sigma T_s^4 + (1 - \varepsilon_s) L\!\downarrow] \tag{6.10}$$

For most surfaces, the term $(1 - \varepsilon_s) L\!\downarrow$, which represents reflection, will be very small; if we ignore it, Equation 6.10 reduces to Equation 6.7.

Equation 6.10 shows that longwave exchange for a surface depends on the *difference* between atmospheric and surface temperatures as well as the *difference* between atmospheric and surface emissivities. First, the bulk temperature of the atmosphere is almost always lower than that of the surface. This is particularly true when the skies are clear. However, under a heavy overcast of low clouds, the temperature of the atmosphere can be as warm as, or almost as warm as, the temperature of the surface. Second, with thick low or middle-height cloud cover, the emissivity of the atmosphere is close to 1.0, but with clear skies or thin clouds, atmospheric emissivity is considerably lower.

Under clear skies, then, the temperature and emissivity of the surface are usually quite a bit higher than the temperature and emissivity of the atmosphere. With cloudy skies, the differences are often less, but the surface temperature is still generally higher than the temperature of the atmosphere. Therefore, whether the skies are cloudy or clear, the net longwave radiation, L^*, will usually be negative. That is, Earth's surface will usually emit more radiation than the atmosphere, so longwave *loss* from the surface is the norm. It follows that *longwave exchange usually cools the surface*, although this cooling effect is less under cloudy skies than it is under clear skies. On the other hand, net shortwave radiation, K^*, is always positive, so *shortwave exchange always warms the surface*.

SKY-VIEW FACTOR
A measure of the amount of sky that can be "seen" from a point on the ground.

Longwave exchange is also influenced by **sky-view factor**, which can be approximated using a fish-eye lens photograph of the sky. As you might expect, open spaces have much higher sky-view factors than do more urban or enclosed spaces (Figure 6.14). A lower sky-view factor results in a smaller longwave loss, because the buildings or trees blocking the sky from view are considerably warmer than the sky itself, thus increasing the amount of longwave radiation coming from the atmosphere, $L\downarrow$.

Taken together, the net shortwave radiation and the net longwave radiation comprise the net radiation, Q^*, at the surface. Naturally, the equation for Q^* during the day (Equation 6.11a) will differ from the equation for Q^* at night (Equation 6.11b).

$$Q^* = K^* + L^* = (K\downarrow - K\uparrow) + (L\downarrow - L\uparrow) \qquad \text{(6.11a)}$$

$$Q^* = L^* = L\downarrow - L\uparrow \qquad \text{(6.11b)}$$

FIGURE 6.14 | The sky-view factor for an open field (*left*) is much higher than that for an urban area (*right*).

rock. As warm air rises by convection, Q_H, heat is transferred to the air above. At the same time, sensible heat will flow downward through the rock by conduction, Q_G. The simple energy balance described above is depicted in Figure 6.18, which shows that the net radiation, Q^*, is divided between two sensible heat fluxes: one into the air, Q_H, and one into the rock, Q_G.

The exact way in which heat is divided between the two sensible heat fluxes depends on the *relative* ease at which the two flows operate. The conductive sensible heat flux depends to a large extent on the conductivity of the ground. For most natural Earth surfaces, conductivity is relatively low (Table 4.2), although conductivity increases with moisture content. In most cases, the conduction of heat into the ground is a far less efficient process than is convection of heat into the air. As a result, most of the radiative heat input to the surface is usually transferred upward into the air. Recall from Section 4.4.2 that convection, or turbulence, in the atmosphere has both thermal and mechanical origins; thermal convection arises due to surface heating, while mechanical convection arises due to wind shear and surface roughness. Conditions that increase such turbulence will increase the flow of sensible heat upward from the surface.

In the above example, the simple rock surface was *dry*, so all the available energy produced a temperature increase. For a surface containing moisture, such as a wet soil, some energy will be used in evaporation, which leaves less energy to raise the temperature. This will reduce the transfer of sensible heat and introduce a transfer of latent heat. The evaporation of water at the surface will cause a water vapour gradient to develop, with more water vapour near the surface and less water vapour higher up. Just as temperature gradients drive flows of sensible heat away from the surface, this gradient will drive a flow of water vapour away from the surface, thus increasing the water vapour content of the air. The energy associated with this flow is the convective latent heat flux, Q_E. Like the convective sensible heat flux, the convective latent heat flux depends on turbulence to transfer heat to the air above the surface. Recall that this heat is transferred to the atmosphere when water vapour condenses (Section 6.4).

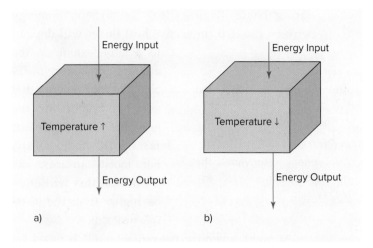

FIGURE 6.17 | a) Flux convergence occurs when the input of energy to a volume is *greater* than the output of energy from that volume. b) Flux divergence occurs when the input of energy to a volume is *less* than the output of energy from that volume.

Plants also control the latent heat flux through **transpiration**, which is controlled by the same atmospheric factors that control evaporation. When the air is warm and dry, and it is windy, transpiration will be high. However, plants need to replace the moisture they lose by drawing up more water through their roots; if there is too little water in the soil, the plant's stomata will close, transpiration will cease, and the latent heat flux, Q_E, will decrease.

> **TRANSPIRATION**
> The process by which water vapour is released through a plant's stomata.

Thus, unless the ground is dry, there will be three fluxes of heat transferring energy away from a surface

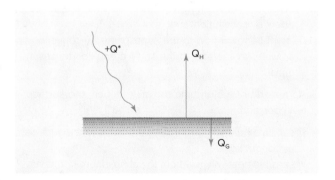

FIGURE 6.18 | For a bare rock surface, the radiative surplus during the day drives flows of sensible heat downward into the rock and upward into the air.

during the day (Figure 6.16a). The division of energy between the two convective heat fluxes will depend to a large extent on the amount of moisture available. The more energy that goes into evaporation, the less that is available for raising the temperature. For moist surfaces, the latent heat flux will often be higher than the sensible heat flux.

DEW
Atmospheric water vapour that has condensed onto a cool surface during the night.

FROST
Atmospheric water vapour that has collected on a surface as ice crystals.

At night, when the net radiation, Q^*, is negative, the surface begins to cool because the loss of radiation is greater than the gain. The surface becomes cooler than the air and the ground, the temperature gradients are reversed, and sensible heat is transferred toward the surface (Figure 6.16b). Now the conductive sensible heat flux is upward, and the convective sensible heat flux is downward. The latter will operate only under windy conditions, as it requires *mechanical* convection to *force* warm air from above toward the surface. Thus, the surface will get much cooler under

calm conditions than it would under windy conditions. In some cases, a water vapour gradient will be directed toward the surface at night, so that the latent heat flux will flow from the air to the surface. When this happens, water vapour can condense on the cool surface to form **dew**. If temperatures are below freezing, **frost** will form instead.

The typical daily variation of net radiation, Q^*, is shown in Figure 6.15. This curve becomes the starting point for the energy balance diagram shown in Figure 6.19, which also includes curves representing the three heat fluxes, Q_H, Q_E, and Q_G. Figure 6.19 shows that the three heat fluxes are of roughly equal magnitude in the morning. However, as the surface warms, the atmospheric temperature and water vapour gradients strengthen, and the winds pick up. The result is that the convective heat fluxes become the dominant terms in the afternoon. Of these two, the latent heat flux is largest because, in this case, the surface is fairly moist. In mid-afternoon, the conductive heat flux becomes negative, indicating that heat is flowing toward the surface instead of away, as it was earlier. This flow of heat adds to the amount of energy available at the surface, so that evaporation continues into the evening. Notice that, in the evening, the latent heat flux exceeds the net radiation. At night, both convective heat fluxes are almost zero. With a temperature gradient directed toward the surface, and with light winds, heat transfer by the air is negligible. As a

Remember This

As a result of a daytime radiative surplus, the surface will *warm*, and

- sensible heat will flow from the surface to the ground by conduction,
- sensible heat will flow from the surface to the atmosphere by convection, and
- latent heat will flow to the atmosphere by convection as water evaporates at the surface and condenses in the air.

As a result of a nighttime radiative deficit, the surface will *cool*, and

- sensible heat will flow from the ground to the surface by conduction,
- sensible heat will flow from the atmosphere to the surface by convection, and
- latent heat will flow from the atmosphere by convection as water vapour condenses on the surface as dew.

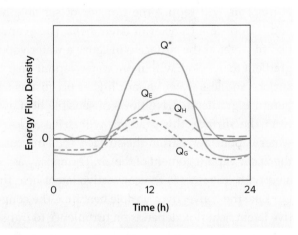

FIGURE 6.19 | Idealized energy balance for a moist vegetated surface on a clear day in summer.

Source: Adapted from Bailey, Oke, & Rouse, 1997, p. 29.

result, the radiative heat loss from the surface is balanced almost solely by a flow of sensible heat from the ground. This example applies to a fairly typical moist surface. However, just as surface types and atmospheric conditions can vary, so too can the energy balance. The possibilities are endless.

In fact, an added complication is the possibility that heat, moisture, or both could be carried *horizontally* over a surface by advection. Equation 6.12 assumes that there is no advection, but this is not to say that advection is not important to the surface energy balance. In actuality, advection can influence the temperature and moisture gradients by bringing in warmer or colder, or drier or wetter, air. For example, if cold air blows over a warm surface, the temperature gradient will increase and drive a greater convective sensible heat flux. Similarly, if dry air blows over a moist surface, the water vapour gradient will increase, increasing the flow of latent heat. Advection can also change the *direction* of the heat flows. For example, if warm, dry air flows over a cool, moist surface, sensible heat might begin to flow *toward* the surface. In addition, this flow of heat will greatly enhance evaporation from the moist surface.

Question for Thought

Which surface characteristics influence the surface energy balance? How does each characteristic affect the radiative, convective, and/or conductive heat fluxes?

6.6.3 | Examples of Two Surface Climates

To illustrate the way in which the energy balance influences surface climates, we will consider the energy balance of a desert surface and that of an ocean surface, each located in the subtropics (Figure 6.20). These two surfaces are both relatively simple: they are smooth, extensive, and non-vegetated. In addition, they represent extreme opposites in terms of climate.

Being located in the subtropics, both surfaces will receive high amounts of solar radiation, $K\downarrow$. Because deserts have high albedos of 30 to 40 per cent and

FIGURE 6.20 | *Top:* A sea surface in the Bahamas and *bottom:* a desert surface in Mauritania. What accounts for the difference in daily temperature ranges between the two surfaces?

oceans have low albedos of less than 10 per cent (Table 5.2), the ocean will absorb considerably more of the incident solar radiation than will the desert. This means that K^* will be higher for the ocean. Further, net longwave radiation, L^*, will be at a greater loss for the desert because the hotter desert surface will emit more longwave radiation, $L\uparrow$, than will the cooler ocean surface, and the drier desert air will result in less incoming longwave radiation, $L\downarrow$. With a lower absorption of shortwave radiation and a higher loss of longwave radiation, the net radiation, Q^*, will be less for the desert surface than it will be for the ocean surface (Figure 6.21).

Despite absorbing less radiation, the desert surface can reach much higher daytime temperatures

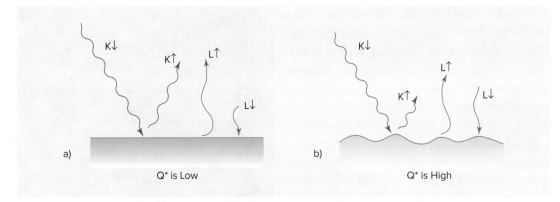

FIGURE 6.21 | Schematic daytime radiation balance for a) a desert surface, and b) an ocean surface.

than can the ocean surface. Because the desert is dry, the latent heat flux will likely be zero or, at least, very close to it. Dry sand is a poor conductor (Table 4.2), so heat will not easily be transferred down into the ground. In addition, dry sand has a low specific heat (Table 4.1). The combination of little to no evaporation, poor conductivity, and low specific heat will cause the dry sand of the desert surface to get very hot. In fact, surface temperatures of deserts can reach up to 70°C, while temperatures just below the surface are considerably lower. With such high surface temperatures, thermal turbulence is greatly enhanced, and there is, thus, a large convective transport of sensible heat into the atmosphere. At night, the clear skies, along with the low water-vapour content of the air, allow for strong radiative cooling. This, combined with the poor conductivity of the ground, means the surface will rapidly cool. Deserts, therefore, represent thermally extreme environments in which *daily* temperature ranges are very large. Mid-latitude deserts also exhibit large *annual* temperature ranges (Figure 6.22).

Oceans, on the other hand, are thermally conservative. Despite absorbing much more radiation than deserts absorb, ocean surfaces remain cool during the day for several reasons. First, water is somewhat transparent, so solar radiation is absorbed through a large volume rather than just at the surface. Second, evaporation from ocean surfaces is high, and this evaporation utilizes energy that would otherwise be available for raising the temperature. Third, although water is not a particularly good conductor of heat (Table 4.2), its fluid nature allows it to *mix*, and mixing effectively transfers heat by convection. (Because of this mixing,

the energy balance of a water body is more complicated than Equation 6.12 implies.) Mixing can be *enhanced* by evaporation—which causes the surface water to cool and sink while the warmer water from below rises to take its place—and *forced* by wind. Finally, water has a very high specific heat (Table 4.1). To produce a given temperature increase, about three times more heat is required for water than for an equivalent mass of most land materials. At night, the high specific heat of water has the opposite effect; water must *lose* a lot of heat to cool. In addition, radiative cooling of surface waters leads to mixing. Thus, just as water does not warm significantly during the day, it does not cool significantly at night. This leads to the low *daily* temperature ranges observed for water bodies. *Annual* temperature ranges for water bodies, and coastal locations, are also small (Figure 6.23).

The above examples show that the amount of radiation absorbed and what happens to this energy once it is absorbed are *both* determinants of surface climate.

Remember This

Water bodies heat and cool more slowly than land does because

- water is transparent,
- water is a fluid,
- a large portion of the available energy is used in evaporation, and
- water has a high specific heat.

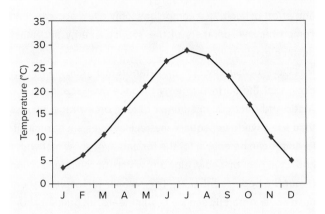

FIGURE 6.22 | Temperature graph for a mid-latitude desert location.

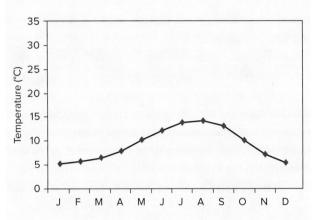

FIGURE 6.23 | Temperature graph for a mid-latitude coastal location.

In the Field

Measuring the Energy Balance of Sea Ice in the Arctic and Antarctic

Christian Haas, York University

Sea ice forms when sea water is exposed to very cold air. Once it forms, this ice floats as a thin layer, 0.5 to 3 m thick, on top of the water. Sea ice, along with its snow cover, plays an important role in atmosphere–ocean processes in the polar regions, and its high surface albedo significantly affects the polar regions' radiation balance.

As a consequence of climate warming, the areal extent of sea ice is currently retreating in the Arctic. In contrast, Antarctic sea-ice coverage has changed little in the past 30 years. The sea-ice decline in the Arctic is particularly strong during summer, when the snow ablates and the resulting meltwater forms ponds on the ice surface. These melt ponds have a lower albedo than do snow and ice, and they therefore absorb more solar radiation, which leads to accelerated melt via albedo-feedback mechanisms. Melt ponds therefore play an important role in the Arctic's sea-ice mass and energy balance. Interestingly, melt-pond formation is uncommon on sea ice in the Antarctic, although the ice generally resides at much lower latitudes and experiences higher levels of solar shortwave radiation there than it does in the Arctic. However, sporadic summer-thaw events in the Antarctic highly metamorphose the ice's snow cover, which

results in significantly different optical and microwave properties, as seen by Earth-observing satellites that monitor the ice cover of the polar regions.

In order to better understand the differences between the surface energy balance of Arctic and Antarctic sea ice and their importance for climate processes, scientists install weather and radiation stations on the ice. They use these stations to observe the downward and upward fluxes of latent and sensible heat and of short- and longwave radiation, the transmission of radiation and the conductance of heat through the ice, and heat fluxes from the water to the ice. All these variables affect the thickness, growth, and melt of the ice. Typical sensors include up- and down-looking pyranometers and pyrgeometers, anemometers, hygrometers, thermometers, thermistor strings, and sonic rangefinders. Radiometers must be mounted such that they are not influenced by radiation reflected or emitted from masts or other supports, and such that they are not shadowed, even with the sun moving 360 degrees over the horizon during the polar day. Ideally, process studies should be accompanied by careful in-situ measurements of changes of snow and ice properties resulting from variable energy

(continued)

fluxes. It is challenging to perform such measurements in a small homogenous and representative region around the energy-balance station without disturbing the original snow surface.

Energy and mass-balance studies require extended observational periods to monitor concurrent changes of snow and ice thicknesses. The required measurements are significantly more challenging to make in the Arctic and Antarctic than in populated regions because of the remoteness of these polar areas and the lack of infrastructure like transportation, accommodation, and electricity. Additionally, long-term observations are difficult to maintain because sea ice drifts—typically several kilometres per day—and may fracture or raft, causing damage or loss of sensors, and because adverse weather can lead to condensation or icing on sensors. Therefore, automatic stations rarely provide reliable measurements in polar regions; instead, researchers must operate ice camps based on aircraft or icebreaker support. Even with continuous supervision, however, instruments and the snow are regularly disturbed or destroyed by local wildlife like polar bears or penguins, which have little interest in or respect for human endeavours. Due to the cost and effort of such field campaigns, only a few accurate and comprehensive datasets of the sea-ice energy and mass balance exist.

Our measurements and theoretical considerations show that the surface energy balance over sea ice in the Arctic and Antarctic in summer differs, on average, by less than a few watts per square metre. However, such small differences can be critical for the formation of snow meltwater and changes of snow albedo, which accelerate melting through albedo feedbacks. The smaller atmospheric heat fluxes over Antarctic sea ice in summer result primarily from the Southern Ocean's specific climate conditions, which are dominated by cold, dry air originating from the Antarctic ice sheet.

CHRISTIAN HAAS is a professor and Canada research chair for Arctic sea ice geophysics in the Department of Earth and Space Science and Engineering at York University. He uses in-situ measurements, airborne surveys, and satellite remote sensing to observe and understand the impacts of climate change on Arctic and Antarctic sea ice, and the consequences for the ecosystem and for human activities in ice-covered seas.

6.7 | Chapter Summary

1. The effective radiating temperature of a planet is the temperature at which that planet radiates away as much energy as it receives from its sun. We can derive an equation for calculating this temperature based on the radiation balance at the top of the atmosphere. This equation shows that the effective radiating temperature of a planet depends on the planet's solar constant and the planet's albedo. The *actual* temperature of a planet is also affected by the planet's greenhouse effect.

2. Of the solar radiation that reaches the top of the atmosphere, just under half of it is absorbed by Earth's surface. The rest is mostly reflected, scattered, or absorbed by clouds, gases, and aerosols in the atmosphere, while a small amount is reflected by Earth's surface. On an annual average, 30 per cent of the solar radiation that reaches the top of the atmosphere is reflected back to space, and the remaining 70 per cent is absorbed in the Earth-atmosphere system.

3. Earth's surface emits longwave radiation; almost 90 per cent of this radiation is absorbed in the atmosphere. The atmosphere also emits longwave radiation; just over 60 per cent of this radiation is absorbed at Earth's surface, while the rest escapes to space.

4. The planetary energy balance is an accounting system for flows of energy, by both radiation and convection, within the Earth-atmosphere system. It applies on an average annual basis. As a result of increases in greenhouse gases due to human activity, there is currently a small imbalance, at the top of the atmosphere, between absorbed solar radiation and outgoing longwave radiation.

5. Large radiative imbalances occur at individual latitudes. Between roughly 40° N and 40° S, absorbed solar radiation is greater than outgoing longwave radiation. Beyond these latitudes, absorbed solar radiation is less than outgoing longwave radiation. Heat flow from the equator to the poles prevents tropical regions from getting continually warmer, and it prevents polar regions from getting continually colder; it also reduces, but does not eliminate, the meridional temperature gradient. This heat flow is accomplished by winds and ocean currents.

6. We can apply the energy balance concept to a surface. A radiative surplus during the day drives a flow of sensible heat into the ground by conduction, and it drives flows of sensible and latent heat into the air by convection. These flows typically carry heat and water vapour away from the surface during the day. At night, a radiative deficit normally causes heat flow to operate in the reverse, carrying heat and water vapour toward the surface. Certain characteristics of the surface determine the relative magnitude of these flows. These characteristics are albedo, emissivity, conductivity, specific heat, moisture content, and surface roughness. The last characteristic influences winds.

Review Questions

1. Since Earth is continually receiving radiation from the sun, why doesn't it keep getting hotter? What causes Earth to re-establish a balance when the energy balance is disturbed?

2. Why is a planet's effective radiating temperature likely to be different from the planet's actual temperature?

3. How can we use Equation 6.3 as a model to help us understand a planet's temperature?

4. Why is only about 47 per cent of the sun's radiation absorbed at Earth's surface? How would changes to Earth's surface and the atmosphere change the amount of solar radiation absorbed at Earth's surface?

5. Over the course of a year, why is there a net transfer of heat by convection from the surface to the atmosphere?

6. How is latent heat transferred from Earth's surface to the atmosphere?

7. What role do clouds play in the planetary energy balance? What is the net effect of clouds on Earth's temperature?

8. How and why is heat transferred from tropical regions to polar regions? Why would this heat transfer be greater in the winter than in the summer?

9. What factors influence the net radiation, Q^*, at a surface? How does each of these factors influence net radiation?

10. What is the driving force behind the three surface heat fluxes: Q_H, Q_E, and Q_G? What factors determine the relative size of these fluxes? How are winds important?

11. Why can deserts get very hot during the day and very cold at night?

12. Why do water bodies heat and cool much more slowly than land surfaces?

Problems

1. Given the following information for a planet in another solar system, calculate the planet's radiative equilibrium temperature. In this solar system, the sun has a radius of 500,000 km, and it emits 25,000,000 W/m². The planet has an albedo of 21 per cent and it is 76,000,000 km from its sun.

2. Calculate the greenhouse effect for Venus. To do this, you will need information from Table 2.4. Use Equation 5.10 to calculate the solar constant for Venus, and use Equation 6.3 to calculate the effective radiating temperature of Venus.

3. a) In Section 2.10, you learned that the sun was producing about 30 per cent less energy when the solar system first formed than it is today. Given these conditions, what would have been Earth's effective radiating temperature? (Assume that Earth's albedo was the same as it is today.)

   b) The temperature you calculated in a) should fall below the freezing point of water, yet there is evidence that liquid water existed on Earth as early as 3.8 billion years ago. This apparent contradiction has been called *the faint young sun paradox*. Provide two solutions to this paradox, given the following:

   - your understanding of the factors affecting the temperature of a planet;
   - the greater amount of both carbon dioxide and methane in Earth's early atmosphere (Section 2.10); and
   - the greater area of ocean surface compared to ↓ the area of land surface on the young Earth.

4. At noon, incoming solar radiation ($K\downarrow$) is 625 W/m², and incoming longwave radiation ($L\downarrow$) is 345 W/m². Given that the surface temperature is 17°C, the surface albedo is 12 per cent, and the surface emissivity is 0.94, what is the net radiation? (Ignore surface reflection of longwave radiation.)

Suggestions for Research

1. Investigate the various ways in which people around the world have designed their homes to be suited to the climate of the region in which they live. Design a home that would be well suited to the climate in your region, and comment on how you could keep energy requirements for heating and cooling to a minimum.

2. Work with a partner to find two microclimates within your university campus. Gather pairs of whatever weather-measuring instruments you can find—at minimum, you will need thermometers. (Psychrometers would also be valuable to this investigation. These simple instruments measure atmospheric moisture and are described in Chapter 7.) Take your half of the instruments to one location, and have your partner go to the other location with her or his instruments. At a pre-agreed upon time, take your measurements. Compare notes with your partner, and suggest how variations in the surface types and immediate surroundings could have contributed to any differences in your observations.

References

Bailey, W.G., T.R. Oke, & W.R. Rouse, eds. (1997). *The surface climates of Canada.* Montreal, QC: McGill-Queen's University Press.

Hansen, J., et. al. (2005, 3 June). Earth's energy imbalance: Confirmation and implications. *Science, 308*, 1431–5.

Kiehl, J.T., & K.E. Trenberth. (1997). Earth's annual global mean energy budget. *Bulletin of the American Meteorological Society, 78*(2), 197–208.

Marshall, J., & R. A. Plumb. (2007). *Atmosphere, ocean, and climate dynamics: An introductory text.* Boston, MA: Elsevier Academic Press.

Trenberth, K.E., J.T. Fasullo, & J. Kiehl. (2009). Earth's global energy budget. *Bulletin of the American Meteorological Society, 90*, 311–23.

This fog around islands in Lake Superior Provincial Park has formed because the air is saturated with water vapour. When it occurs higher in the atmosphere, saturation results in the formation of clouds.

7 Water Vapour

Learning Goals

After studying this chapter, you should be able to

1. *explain* the concept of saturation as it applies to water vapour in the atmosphere;

2. *describe* the relationship between saturation and temperature;

3. *appreciate* the significance of saturation to the processes of evaporation and condensation;

4. *explain* how absolute measures of water vapour are different from relative measures of water vapour, and *distinguish* between the various absolute and relative measures;

5. *describe* the relationships between the various measures of water vapour; and

6. *explain* how psychrometers work and, in particular, the meaning of wet-bulb temperature.

In Chapter 2, we saw that water vapour is the most variable gas in the atmosphere (Figure 7.1) and of fundamental importance to our weather. In Chapter 6, we discussed the energy flows that transfer not only water vapour but also energy between the surface and the atmosphere. Under certain conditions, to be explored in *this* chapter, water vapour in the atmosphere reaches **saturation**; when it does, it begins to condense, producing clouds and precipitation, fog, or dew (Chapter 9). In the process, latent heat is released (Section 4.1.2); this form of energy is particularly important to storms (Chapter 14). Water vapour also affects the buoyancy of rising air parcels (Section 8.1) and the character of air masses (Chapter 13). In addition, water vapour influences our comfort levels; hot, humid days feel much warmer than hot, dry days. For these reasons, the amount of water vapour in the atmosphere is routinely measured as part of weather reports. In this chapter, we will not only explore the concept of saturation but also the variety of indicators that are used to express **humidity**, the relationships between these indicators, and some of the methods used for measuring them.

SATURATION
The maximum amount of water vapour that can exist at a given temperature.

HUMIDITY
The amount of water vapour in a quantity of air.

PARTIAL PRESSURE
The pressure contributed by a single gas in a mixture of gases.

VAPOUR PRESSURE
Partial pressure exerted by water vapour.

7.1 | Vapour Pressure

Each atmospheric gas exerts a **partial pressure** that contributes to the total atmospheric pressure. The partial pressure exerted by atmospheric water vapour is known as **vapour pressure**. In Earth's atmosphere, vapour pressure varies from close to nothing to about 4 kPa.

As an air parcel rises or sinks in the atmosphere, its vapour pressure will change as it adjusts to the surrounding pressure. This is misleading because, although vapour pressure is changing, the air parcel's actual water vapour content may or may not be changing. Therefore, vapour pressure's usefulness in meteorology is limited. As we shall see, there are other moisture variables that *don't* change with vertical motions. However, as you will discover in the following sections, vapour pressure *is* useful in explaining the concept of *saturation*.

7.2 | Evaporation and Condensation

Evaporation requires energy: for a water molecule to evaporate, it must be moving quickly enough to break the strong molecular bonds that characterize liquid water. The energy needed is transferred as latent heat from the surroundings; as a result, the surroundings cool (Section 4.1.2). This is what makes evaporation a cooling process.

Evaporation can occur at any temperature, but the *rate* of evaporation tends to increase as temperature increases. To understand why this is, remember that temperature is a measure of the *average* kinetic energy of the molecules. This means that at any given temperature, the molecules will not all have the same amount of energy. Thus, when temperature is low, only a few liquid water molecules will have enough energy to escape the water surface and become vapour. As

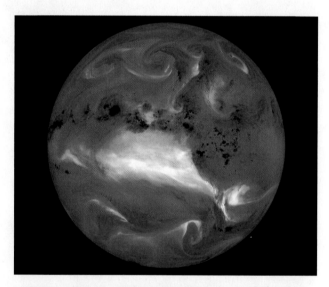

FIGURE 7.1 | This infrared satellite image of Earth shows the spatial variability of atmospheric water vapour on a global scale. The white areas of the image show the presence of water vapour, while the dark areas are regions of dry air. The large white area in the centre indicates the relatively large amount of water vapour that has been evaporated in warm equatorial regions; the black areas indicate the drier air of the subtropics; and the wave patterns, which lie toward the poles, indicate the presence of mid-latitude storms.

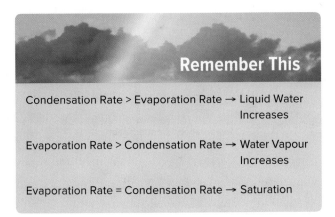

Remember This

Condensation Rate > Evaporation Rate → Liquid Water Increases

Evaporation Rate > Condensation Rate → Water Vapour Increases

Evaporation Rate = Condensation Rate → Saturation

temperature increases, more and more liquid molecules will have enough energy to become vapour.

Conversely, the *rate* of condensation tends to increase as temperature *decreases*. At lower temperatures in the atmosphere, not all water vapour molecules will have enough energy to remain in their gaseous state. As a result, these molecules will be pulled toward other water molecules and collect as liquid water. The energy lost by the slower moving molecules as they condense is transferred as latent heat to the surroundings. Thus, condensation is a warming process.

Over any water surface, whether it be the ocean or a tiny water droplet in a cloud, water molecules are both evaporating and condensing all the time. It is the dominance of one process over the other that

determines whether the amount of *liquid* water will increase or whether the amount of water *vapour* will increase. When the rate of condensation exceeds the rate of evaporation, there will be net condensation. When the rate of evaporation exceeds the rate of condensation, there will be net evaporation. When the rate of condensation equals the rate of evaporation, the air will be saturated.

7.3 | Saturation Vapour Pressure

To illustrate the concept of saturation, imagine that we are doing a simple experiment involving a closed container that is partially filled with water (Figure 7.2). Initially, an impermeable barrier separates the water from the air above, and the air contains no water vapour. When we remove the barrier, evaporation will begin to add water vapour molecules to the dry air. In turn, some of these water vapour molecules will condense, becoming liquid once again. As long as the *rate* of evaporation exceeds the *rate* of condensation there will be net evaporation, or an increase in the amount of water vapour in the container. However, as the amount of water vapour increases, so too will the *rate* of condensation. Eventually, the closed environment will reach a point

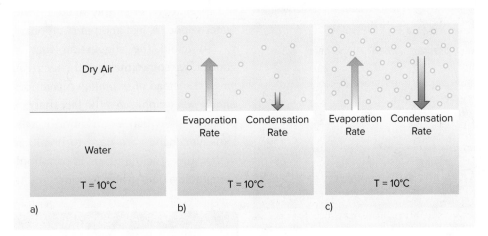

FIGURE 7.2 | a) Dry air is separated from the pure water below. b) When the barrier is removed, evaporation occurs at a rate proportional to the temperature. Once there is water vapour in the air, condensation can occur as well. c) Eventually, the rate of condensation equals the rate of evaporation, and the air has reached its *saturation vapour pressure*.

Source: Adapted from Aguado & Burt, 2010, p. 135.

where the *rate* of condensation equals the *rate* of evaporation, and the amount of water vapour will stop increasing. At this point, the air is *saturated*—it has reached its **saturation vapour pressure**.

Now imagine that, as part of our experiment, we increase the temperature. As a result of this change, the *rate* of evaporation will increase. Once again there will be more water evaporating than condensing, and the amount of water vapour in the container will increase. As the amount of water vapour increases, the rate of condensation will increase until it equals the new evaporation rate. At this point, the air has reached a new, *higher* saturation vapour pressure that corresponds to the new *higher* temperature. The *rates* of evaporation and condensation will also be greater (Figure 7.3). This simple experiment shows that *as temperature increases, saturation vapour pressure increases*. If the temperature had decreased, the evaporation and condensation rates would have decreased until a new, *lower* saturation vapour pressure was reached. Saturation vapour pressure, therefore, is defined as the maximum vapour pressure that can exist at a given temperature.

SATURATION VAPOUR PRESSURE
The maximum water vapour pressure that can exist at a given temperature.

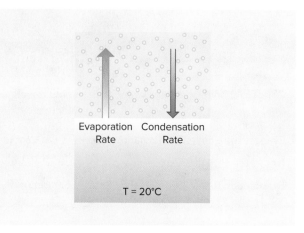

FIGURE 7.3 | With a higher temperature than in the initial experiment (Figure 7.2), the evaporation rate increases and, at saturation, there is more water vapour in the air.

vapour that the air can *hold* at a given temperature. Although this is a common, and rather convenient, way of defining the concept of air saturation, it is not literally correct. Air does not *hold* water vapour any more than it *holds* oxygen, or nitrogen, or any other gas. The idea that air *holds* water vapour, as well as the use of the term *saturation* in relation to air, arose because, prior to the nineteenth century, people believed that water dissolved in air, just as salt dissolves in water. In the same way that water can become saturated with salt, it was thought that air could become saturated with water vapour. After much experimentation, John Dalton published a paper, in 1802, in which he concluded that all the gases together simply make up the "air" and that water vapour is just another component.

To avoid the suggestion that air *holds* water vapour, we sometimes use the term *equilibrium vapour pressure* instead of *saturation vapour pressure*. The word *equilibrium* emphasizes the fact that this is the vapour pressure at which evaporation *equals* condensation. However, habits are hard to break, so the term *saturation* has stuck, and air is commonly described as being able to *hold* water vapour.

Remember This

Saturation vapour pressure

- is the vapour pressure when evaporation equals condensation;
- is the maximum vapour pressure that can exist at a given temperature; and
- it increases with increasing temperature.

Question for Thought

In the experiment illustrated in figures 7.2 and 7.3, what would have happened if the container had not been closed?

Saturation vapour pressure is also often incorrectly described as the maximum amount of water

Question for Thought

When working in a scientific context, why should we avoid saying that the air "holds" water vapour?

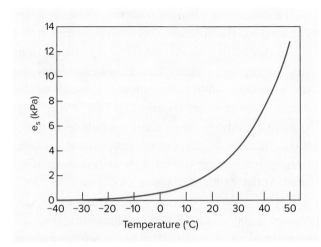

FIGURE 7.4 | Saturation vapour pressure over a flat surface of pure liquid water as a function of temperature. The values depicted on this graph are those listed in Table 7.1.

As shown in our above experiment, saturation vapour pressure increases as temperature increases. This relationship is approximated by Equation 7.1, which is a form of the Clausius–Clapeyron equation.

$$e_s = e_o \exp\left[\frac{L}{R_v}\left(\frac{1}{T_o} - \frac{1}{T}\right)\right] \tag{7.1}$$

In this equation, e_s is saturation vapour pressure, T is temperature, and e_o is the saturation vapour pressure at T_o. Since T_o is 273 K (0°C), e_o is the saturation vapour pressure at 0°C, which is 0.611 kPa (Table 7.1).

R_v is the gas constant for water vapour, which is equal to 461.5 J·kg^{-1}·K^{-1}. Finally, L is the latent heat. For a liquid water surface, this is the latent heat of vaporization; for an ice surface, this is the latent heat of deposition (Section 4.1.2). Equation 7.1 shows that the relationship between temperature and saturation vapour pressure is *exponential* (Figure 7.4), meaning that the amount of water vapour that can exist in warm conditions is considerably greater than that which can exist in colder conditions. This explains why heating an object is a very effective way to dry it. Values of saturation vapour pressure, calculated using Equation 7.1, are given in Table 7.1.

For temperatures below 0°C, Table 7.1 gives two values: one for the saturation vapour pressure over liquid water, and one for the saturation vapour pressure over ice. Because we normally consider 0°C to be the temperature at which water freezes, this may seem strange. However, in the atmosphere, tiny cloud droplets will often remain **supercooled** until temperatures drop to about −40°C (Section 9.2). The table shows that, for any given temperature, the saturation vapour pressure over ice is lower than that over liquid water (Example 7.1) (Figure 7.5). This is because it requires more energy for ice to become vapour than it does for water to become vapour.

> **SUPERCOOLED**
>
> Cooled below the temperature at which a substance would normally freeze, while remaining in a liquid state.

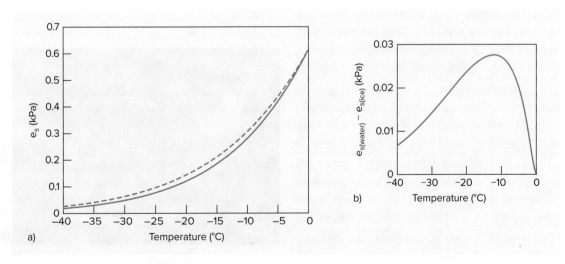

FIGURE 7.5 | a) Saturation vapour pressure over flat surfaces of pure liquid water (dashed line) and ice (solid line) at temperatures between 0°C and −40°C. b) The difference between saturation vapour pressure over liquid water and that over ice. Note that this difference is greatest at temperatures just below −10°C.

Example 7.1

Determine the saturation vapour pressure for −5°C for a) liquid water and b) ice.

Use Equation 7.1.

a) Use the latent heat of vaporization, L_v, which is approximately 2.5×10^6 J/kg for a temperature of −5°C.

$$e_s = 0.611 \exp\left[\left(\frac{2.5 \times 10^6 \text{ J/kg}}{461.5 \text{ J} \cdot \text{kg}^{-1} \cdot \text{K}^{-1}}\right)\left(\frac{1}{273 \text{ K}} - \frac{1}{268 \text{ K}}\right)\right]$$

$$= 0.42 \text{ kPa}$$

b) Use the latent heat of deposition, L_D. Recall from Section 4.1.2 that this is simply the sum of the latent heat of vaporization and the latent heat of fusion, or 2.8×10^6 J/kg.

$$e_s = 0.611 \exp\left[\left(\frac{2.8 \times 10^6 \text{ J/kg}}{461.5 \text{ J} \cdot \text{kg}^{-1} \cdot \text{K}^{-1}}\right)\left(\frac{1}{273 \text{ K}} - \frac{1}{268 \text{ K}}\right)\right]$$

$$= 0.40 \text{ kPa}$$

This example confirms that the saturation vapour pressure over liquid water is greater than that over ice.

The saturation vapour pressure values given in Table 7.1 apply only for *flat* surfaces of *pure* water. Most of the time, water surfaces can be considered to be flat, but for the very tiny water droplets in clouds, the curvature becomes significant enough that it influences the relationship between temperature and saturation vapour pressure. Because water molecules are not held as tightly to a curved surface as they are to a flat surface, it is *easier* for water to evaporate from a curved surface. As a result, the saturation vapour pressure above a curved surface is *higher* than it would be above a flat surface at the same temperature. Conversely, impurities in the water tend to have the opposite effect—they make it *harder* for water to evaporate, so the saturation vapour pressure above a solution is less than it is above pure water at the same temperature. These effects on saturation vapour pressure are known as the *curvature effect* and the *solute effect*; in the atmosphere, they are important to the formation and growth of water droplets in clouds (Section 9.1).

The concept of saturation vapour pressure helps us understand the difference between evaporation and boiling. Although both involve the phase change from liquid to vapour, evaporation occurs only from the water's surface, while boiling occurs throughout the entire volume of water (Figure 7.6). This is why bubbles form in boiling water—these are bubbles of water vapour. In addition, evaporation can occur at any temperature, but boiling occurs only at certain temperatures. At the standard sea level pressure of 101.3 kPa, water will boil at 100°C because the saturation vapour pressure at 100°C is 101.3 kPa. It follows that the boiling point of any liquid occurs at the temperature at which the liquid's saturation vapour pressure equals atmospheric pressure. Thus, changes in the surrounding pressure can change the boiling point of a liquid.

When the surrounding pressure is lower, water will boil at a lower temperature; when the surrounding pressure is higher, water will boil at a higher temperature. It follows that water boils at lower temperatures at higher altitudes. For example, at an elevation of 1000 m, where atmospheric pressure is roughly 90 kPa, water will boil at about 96.5°C because the saturation vapour pressure at 96.5°C is 90 kPa. Since boiling occurs at lower temperatures at higher altitudes, foods need to be boiled longer at higher altitudes in order to cook fully. On the other hand, pressure cookers allow foods to cook more quickly because the pressure inside the pressure cooker is higher than atmospheric pressure and, therefore, the boiling point is higher.

FIGURE 7.6 | As the bubbles in this pot of boiling water show, liquid water is changing to vapour throughout the entire volume of water.

TABLE 7.1 | Saturation vapour pressure at different air temperatures, or vapour pressure at different dew-point temperatures,[a] over flat surfaces of pure liquid water or ice.[b]

Air Temperature/Dew-Point Temperature (°C)	Saturation Vapour Pressure/Vapour Pressure (kPa)	Air Temperature/Dew-Point Temperature (°C)	Saturation Vapour Pressure/Vapour Pressure (kPa)
−40	0.013 (0.019)	0	0.611
−39	0.014 (0.021)	1	0.657
−38	0.016 (0.023)	2	0.705
−37	0.018 (0.026)	3	0.758
−36	0.020 (0.028)	4	0.813
−35	0.022 (0.031)	5	0.872
−34	0.025 (0.035)	6	0.935
−33	0.028 (0.038)	7	1.001
−32	0.031 (0.042)	8	1.072
−31	0.034 (0.046)	9	1.147
−30	0.038 (0.051)	10	1.227
−29	0.042 (0.056)	11	1.312
−28	0.047 (0.061)	12	1.401
−27	0.052 (0.067)	13	1.497
−26	0.057 (0.074)	14	1.598
−25	0.063 (0.081)	15	1.704
−24	0.070 (0.088)	16	1.817
−23	0.077 (0.097)	17	1.937
−22	0.085 (0.105)	18	2.063
−21	0.094 (0.115)	19	2.196
−20	0.103 (0.125)	20	2.337
−19	0.114 (0.137)	21	2.486
−18	0.125 (0.149)	22	2.643
−17	0.137 (0.162)	23	2.809
−16	0.151 (0.176)	24	2.983
−15	0.165 (0.191)	25	3.167
−14	0.181 (0.208)	26	3.361
−13	0.198 (0.225)	27	3.565
−12	0.217 (0.244)	28	3.780
−11	0.238 (0.264)	29	4.006
−10	0.260 (0.286)	30	4.243
−9	0.284 (0.310)	31	4.493
−8	0.310 (0.335)	32	4.755
−7	0.338 (0.362)	33	5.031
−6	0.369 (0.391)	34	5.320
−5	0.402 (0.421)	35	5.624
−4	0.437 (0.455)	36	5.942
−3	0.476 (0.490)	37	6.276
−2	0.517 (0.528)	38	6.626
−1	0.562 (0.568)	39	6.993
0	0.611 (0.611)	40	7.378

[a] Dew-point temperatures will be discussed in Section 7.5.1.3.
[b] Values over liquid water at subfreezing temperatures appear in parentheses.

Question for Thought

How would you determine the boiling point of water at an altitude of 1500 m?

7.4 | The Concept of Saturation

The above has shown that when the amount of water vapour in a sample of air is the *same* as the saturation vapour pressure at the temperature of the air, the air is saturated. An example of a saturated air sample might be one with a temperature of 10°C and a vapour pressure of 1.2 kPa. This air sample would fall on the saturation vapour pressure curve, as shown in Figure 7.7 (point A). When the air is saturated, *net* evaporation cannot occur. But, if there is a vapour pressure gradient directed away from the surface and/or winds to generate turbulence (Section 6.6.2), water vapour will be carried away from the surface, and evaporation can continue.

Remember This

Evaporation depends on

- the availability of water;
- the availability of energy for the phase change;
- a vapour pressure gradient directed away from the evaporating surface; and
- turbulence in the atmosphere.

Question for Thought

As water evaporates from a swimming pool, the water in the pool will cool. What are some ways that you could decrease the rate of evaporation, in order to keep the water at a temperature comfortable for swimming? (List all possibilities you can think of, no matter how impractical they might be.)

Of course, air is not always in a state of saturation. Most commonly, air is **unsaturated**. When the air is

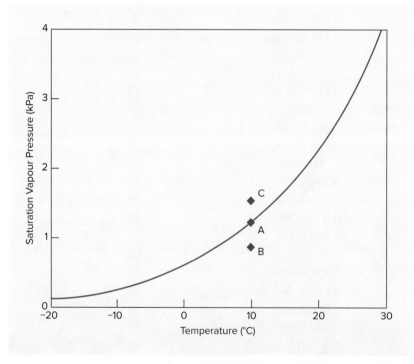

FIGURE 7.7 | On this saturation vapour pressure curve, point A represents a *saturated* air sample, point B represents an *unsaturated* air sample, and point C represents a *supersaturated* air sample.

unsaturated and water is available, evaporation will occur. An example of an unsaturated air sample could be one with a temperature of 10°C and a vapour pressure of 0.9 kPa (point B in Figure 7.7). It is also conceivable that air could be **supersaturated**. An example of a supersaturated air sample could be one with a temperature of 10°C and a vapour pressure of 1.5 kPa (point C in Figure 7.7). Such supersaturations rarely occur in our atmosphere; when they do, they are usually no more than 1 per cent above the saturation value. This is because when air becomes saturated, the water vapour begins to condense to form clouds, fogs, or dew (Figure 7.8).

Three processes can make air saturated and, thus, lead to condensation: cooling, adding water vapour, or mixing. For example, the air sample represented by point B in Figure 7.7 could be cooled to bring it to saturation (Figure 7.9). In this case, the vapour pressure of the air would remain constant at 0.9 kPa, but the temperature would drop until it reaches about 6°C. Cooling is the most common way for air to become saturated in the atmosphere, and the most common way for air to cool is by rising (Section 8.1). Rising air produces almost all clouds.

The air sample represented by point B could also become saturated if water vapour is added to it (Figure 7.9). In this case, the temperature of the air would remain constant at 10°C, but the water vapour content would rise from 0.9 to 1.2 kPa. In the atmosphere, water vapour is added to the air through evaporation. For example, when rain falls through unsaturated air, its evaporation may saturate the air. If it does, the water vapour will condense, forming fog.

Finally, the mixing of unsaturated air parcels can produce saturation (Figure 7.10). When two parcels that are close to saturation are mixed, they may form a parcel of air that is supersaturated; as a result, the water vapour will begin to condense. (Example 7.2 shows this result mathematically.) We most commonly observe this process when we breathe out on a cold day—as our warm, moist breath mixes with the cold air, it produces saturation, and condensation will occur (Figure 7.11). A few cloud types, and some fogs, are formed by mixing (Section 9.3 and 9.6).

UNSATURATED
The condition that occurs when the amount of water vapour in the air is lower than the saturation value.

SUPERSATURATED
The condition that occurs when the amount of water vapour in the air is higher than the saturation value.

FIGURE 7.8 | **This fog in New Glasgow, Prince Edward Island, has formed because the air is saturated.**

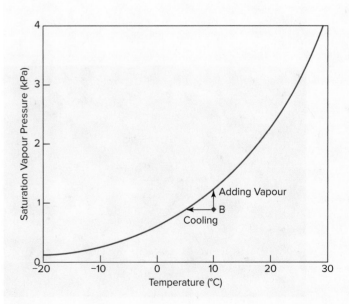

FIGURE 7.9 | **The air parcel represented by point B in Figure 7.7 can become saturated if it cools or if water vapour is added to it.**

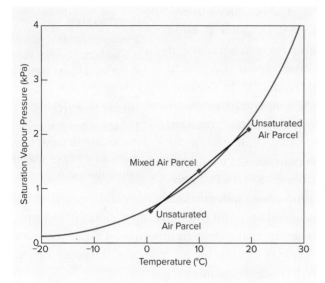

FIGURE 7.10 | When the two air parcels that are close to saturation are mixed, the newly formed air parcel will be supersaturated; as a result, the water vapour will begin to condense.

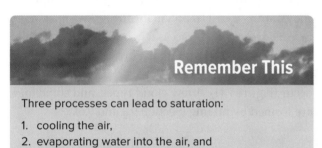

Remember This

Three processes can lead to saturation:

1. cooling the air,
2. evaporating water into the air, and
3. mixing two air parcels that are both close to saturation.

FIGURE 7.11 | We can see the exhaled breath of this soccer player as it mixes with the cool air.

Example 7.2

One air parcel has a temperature of 19°C and a vapour pressure of 2.1 kPa. A second air parcel, with the same mass as the first, has a temperature of 5°C and a vapour pressure of 0.8 kPa. If these two air parcels are mixed together, will the newly formed air parcel be saturated?

Because these two air parcels have the same mass, you can find the properties of the new air parcel by simply taking averages.

$$T = \frac{19°C + 5°C}{2}$$

$$= 12°C$$

$$e = \frac{2.1 \text{ kPa} + 0.8 \text{ kPa}}{2}$$

$$= 1.5 \text{ kPa}$$

Since the saturation vapour pressure for 12°C is 1.4 kPa, this new air parcel will be saturated.

Question for Thought

Do temperatures need to be below freezing for us to "see" our breath? Why, or why not? (Hint: See Problem 7 at the end of the chapter.)

7.5 | Absolute and Relative Measures of Humidity

In addition to vapour pressure, there are several other measures for indicating the amount of water vapour in the atmosphere (Table 7.2). These measures can be broadly classified into two groups: *absolute* measures of water vapour, and *relative* measures of water vapour. Absolute measures give the *actual* amount of water vapour in the air, while relative measures compare the actual amount of water vapour in the air to the maximum amount possible, or the saturation value. The latter are valuable because they provide a measure of the degree of saturation of the air.

7.5.1 | Absolute Measures

As the name suggests, absolute measures of water vapour tell us the *absolute* amount of water vapour in the air. As shown in Table 7.2, there are five such measures: vapour pressure, absolute humidity, mixing ratio, specific humidity, and dew-point temperature. The first of these, vapour pressure, has been covered in previous sections; the following sections cover the other four.

7.5.1.1 | Absolute Humidity

Imagine you wanted to determine the amount of water vapour in a contained sample of air. One way of doing this would be to pass a beam of infrared radiation through the sample and measure the amount by which the beam is reduced when it reaches the other side. This technique, which works because water vapour is a good absorber of infrared radiation

TABLE 7.2 | Indicators of atmospheric water vapour.

Absolute Measures	Equation	Comments
Vapour Pressure	$e = P\left(\dfrac{r}{\varepsilon + r}\right)$ $\varepsilon = 0.622$	not conserved in vertical motion
Absolute Humidity	$\rho_v = \dfrac{e}{R_v\, T}$	not conserved in vertical motion; can be measured directly
Mixing Ratio	$r = \varepsilon\left(\dfrac{e}{P - e}\right)$	conserved in vertical motion; used on thermodynamic diagrams
Specific Humidity	$q = \varepsilon\left(\dfrac{e}{P}\right)$	conserved in vertical motion
Dew-Point Temperature	$T_d = \left[\left(\dfrac{1}{T_o}\right) - \dfrac{R_v}{L}\ln\left(\dfrac{e}{e_o}\right)\right]^{-1}$	not conserved in vertical motion; can be measured directly; reported as part of routine weather observations
Relative Measures		
Relative Humidity	$RH = \dfrac{e}{e_s} \times 100\%$ $RH = \dfrac{r}{r_s} \times 100\%$ $RH = \dfrac{\rho_v}{\rho_{vs}} \times 100\%$	a measure of how close the air is to being saturated; can be directly measured; reported as part of routine weather observations
Vapour Pressure Deficit	$VPD = e_s - e$	an indicator of the drying power of the air
Dew-Point Depression	$T_{dd} = T - T_d$	an indicator of how much the air needs to cool to reach saturation; labelled on 700 hPa charts
Wet-Bulb Temperature		a measure of evaporative cooling; can be measured directly

(Section 5.5.2), provides us with a way of measuring **absolute humidity**. This measure is sometimes referred to as *vapour density* because it has the same units as density—grams per cubic metre.

Because the pressure and density of any gas are related through the gas law (Section 3.3), we can use Equation 3.6 (the form of the gas law most applicable to the atmosphere) to create an equation for water vapour. Pressure becomes *vapour* pressure, e, and the gas law constant becomes the constant for water vapour, R_v. (Recall from above that this value is 461.5 $J \cdot kg^{-1} \cdot K^{-1}$.) Thus, we obtain Equation 7.2 for absolute humidity, ρ_v.

ABSOLUTE HUMIDITY

The mass of water vapour in a unit volume of air.

MIXING RATIO

The ratio of the mass of water vapour to the mass of dry air.

SPECIFIC HUMIDITY

The ratio of the mass of water vapour to the total mass of air.

$$\rho_v = \frac{e}{R_v\, T} \qquad (7.2)$$

Example 7.3 shows how vapour pressure can be used to calculate absolute humidity.

Example 7.3

For a vapour pressure of 1.1 kPa and a temperature of 23°C, what is the absolute humidity?

Convert 1.1 kPa to 1100 Pa, and 23°C to 296 K, then use Equation 7.2.

$$\rho_v = \frac{1100\ kg \cdot m^{-1} \cdot s^{-2}}{461.5\ J \cdot kg^{-1} \cdot K^{-1}\ x\ 296\ K}$$

$$= 0.008\ kg/m^3$$

$$= 8\ g/m^3$$

Absolute humidity is a useful indicator of the amount of infrared radiation that will be absorbed by the water vapour in a *volume* of air, as described above. It is less useful in meteorological applications, because it depends on volume and, therefore—like vapour pressure—does not remain constant during vertical motions. For example, if there is 1 g of water vapour in 1 m^3 of air at the surface, the absolute humidity is 1.0 g/m^3. If this air rises and expands to 2 m^3, its absolute humidity will drop to 0.5 g/m^3, assuming that no condensation or evaporation took place. The amount of water vapour won't have changed, but the change in the value of the absolute humidity will imply that it has.

7.5.1.2 | Mixing Ratio and Specific Humidity

Unlike vapour pressure and absolute humidity, **mixing ratio** and **specific humidity** do not change when an air parcel rises or sinks in the atmosphere. They are measures of the mass of water vapour in a *mass* of air and, because a mass of air won't change as an air parcel rises or sinks, these two measures are conserved in rising or sinking air.

There is only a subtle difference between mixing ratio and specific humidity. The mixing ratio, r, is the ratio of the mass of water vapour to the mass of *dry* air, as shown in Equation 7.3.

$$r = \frac{\text{Mass of Water Vapour}}{\text{Mass of Dry Air}} \qquad (7.3)$$

In contrast, the specific humidity, q, is the ratio of the mass of water vapour to the *total* mass of air, as shown in Equation 7.4.

$$q = \frac{\text{Mass of Water Vapour}}{\text{Mass of Dry Air + Mass of Water Vapour}} \qquad (7.4)$$

Mixing ratio and specific humidity are usually expressed in units of grams per kilogram. The units don't cancel because they represent masses of different substances.

It is far more useful to express Equations 7.3 and 7.4 in terms of pressure, P, and vapour pressure, e. We can do this by using the gas law.

$$r = \varepsilon \left(\frac{e}{P - e} \right) \qquad (7.5)$$

$$q = \varepsilon \left(\frac{e}{P} \right) \qquad (7.6)$$

In equations 7.5 and 7.6, ε is equal to 0.622; this value represents the ratio of the gas law constant for dry air, R_d, to the gas law constant for water vapour, R_v. When we solve Equation 7.5 for vapour pressure, the result is Equation 7.7.

$$e = P\left(\frac{r}{\varepsilon + r}\right)$$ (7.7)

Equations 7.5 and 7.6 show why mixing ratio and specific humidity don't vary with height. As air rises, both vapour pressure and air pressure decrease but, because they decrease at the *same rate*, their *ratio* does not change. Equation 7.7, on the other hand, shows why vapour pressure *does* change with height. It is dependent on air pressure; as air rises or sinks, air pressure changes, and so too does vapour pressure.

In Example 7.4, mixing ratio and specific humidity are calculated using Equations 7.5 and 7.6. This example shows that vapour pressure is so small compared to total atmospheric pressure that there is little difference between the value of mixing ratio and the value of specific humidity. Of the two, mixing ratio is normally used in atmospheric applications; a typical value might be about 10 g/kg.

Example 7.4

If the vapour pressure is 1.1 kPa and the atmospheric pressure is 101.3 kPa, what are the mixing ratio and the specific humidity?

Use Equation 7.5 to calculate the mixing ratio.

$$r = 0.622\left(\frac{1.1 \text{ kPa}}{101.3 \text{ kPa} - 1.1 \text{ kPa}}\right)$$

$$= 0.00683 \text{ g/g}$$

$$= 6.83 \text{ g/kg}$$

Use Equation 7.6 to calculate the specific humidity.

$$q = 0.622\left(\frac{1.1 \text{ kPa}}{101.3 \text{ kPa}}\right)$$

$$= 0.00675 \text{ g/g}$$

$$= 6.75 \text{ g/g}$$

Mixing ratio and specific humidity are almost the same.

Example 7.5

Determine the saturation vapour pressure, saturation absolute humidity, and saturation mixing ratio at 10°C for both sea level and an altitude of 5 km. Assume that the pressure at sea level is 100 kPa and the pressure at 5 km is 50 kPa.

First, use Equation 7.1, or Table 7.1, to determine saturation vapour pressure.

$$e_s = 0.611 \exp\left[\left(\frac{2.5 \times 10^6 \text{ J/kg}}{461.5 \text{ J} \cdot \text{kg}^{-1} \cdot \text{K}^{-1}}\right)\left(\frac{1}{273 \text{ K}} - \frac{1}{283 \text{ K}}\right)\right]$$

$$= 1.23 \text{ kPa}$$

Equation 7.1 indicates that saturation vapour pressure does not depend on atmospheric pressure; therefore, 1.23 kPa is the saturation vapour pressure at both sea level and 5 km altitude.

Next, rewrite equations 7.2 and 7.5 to apply to saturation absolute humidity and saturation mixing ratio, respectively.

$$\rho_{vs} = \frac{e_s}{R_v T}$$ (7.8)

$$r_s = \varepsilon\left(\frac{e_s}{P - e_s}\right)$$ (7.9)

Use Equation 7.8 for saturation absolute humidity.

$$\rho_{vs} = \frac{1230 \text{ Pa}}{461.5 \text{ J} \cdot \text{kg}^{-1} \cdot \text{K}^{-1} \times 283 \text{ K}}$$

$$= 0.0094 \text{ kg/m}^3$$

Equation 7.8 indicates that saturation absolute humidity does not depend on atmospheric pressure; therefore, 0.0094 kg/m³ is the saturation absolute humidity at both sea level and 5 km altitude.

Use Equation 7.9 for saturation mixing ratio.

For sea level: $$r_s = 0.622\left(\frac{1.23 \text{ kPa}}{100 \text{ kPa} - 1.23 \text{ kPa}}\right)$$

$$= 7.7 \text{ g/kg}$$

For 5 km: $$r_s = 0.622\left(\frac{1.23 \text{ kPa}}{50 \text{ kPa} - 1.23 \text{ kPa}}\right)$$

$$= 15.7 \text{ g/kg}$$

These calculations show that the saturation mixing ratio increases with height. Therefore, while saturation vapour pressure and saturation absolute humidity *do not* vary with pressure, saturation mixing ratio *does* vary with pressure.

7.5.1.3 | Dew-Point Temperature

A final *absolute* measure of atmospheric water vapour is **dew-point temperature**. It may seem strange that a temperature is used to represent humidity, but it is not difficult to see why this is. Imagine two samples of air: the first has a temperature of 10°C and a vapour pressure of 0.71 kPa; the second also has a temperature of 10°C but a vapour pressure of 1.1 kPa. Table 7.1 shows that the air sample containing 0.71 kPa of water vapour will be saturated at a temperature of 2°C, while the air sample containing 1.1 kPa of water vapour will be saturated at a temperature of 8°C. These two temperatures—2°C and 8°C, respectively—are the dew-point temperatures of the two air samples. This example shows that it is the amount of moisture in the air that determines dew-point temperature. Air with a higher vapour pressure will have a higher dew-point temperature, and air with a lower vapour pressure will have a lower dew-point temperature.

Because the relationship between dew-point temperature and vapour pressure is the same as that between air temperature and saturation vapour pressure, we can use Table 7.1 to determine dew-point temperature given vapour pressure. Alternatively, we can use Equation 7.1 to derive an equation for dew-point temperature. To do this, we substitute dew-point temperature, T_d, for temperature, and we substitute vapour pressure for saturation vapour pressure.

> **DEW-POINT TEMPERATURE**
> The temperature to which the air must be cooled, at constant pressure, to reach saturation.

$$T_d = \left[\frac{1}{T_o} - \frac{R_v}{L} \ln\left(\frac{e}{e_o}\right) \right]^{-1}$$

(7.10)

In Section 7.4, we explored the ways in which air can become saturated. Now we see that when air becomes saturated due to cooling, it is cooling to its dew-point temperature. Clouds form when rising air cools to its dew-point temperature. In the same way, water droplets form on a surface that is at, or below, the dew-point temperature of the air. This is how dew forms on the grass or condensation forms on a glass containing a cold drink (Figure 7.12). In fact, one way to measure the dew-point temperature of the air is to cool a surface until condensation forms on it. The temperature at which condensation forms is, of course, the dew-point temperature. Remember that condensation can also occur when water evaporates into the air. In this case, instead of the air cooling to the dew-point temperature, evaporation is increasing the dew-point temperature until it reaches the air temperature. For example, when rain evaporates into the air, the dew-point temperature increases; if enough water evaporates, fog will form.

> **Example 7.6**
>
> It is common for condensation to form on eyeglasses when the person wearing them comes into a warm house on a cold day. If the inside air has a vapour pressure of 1.2 kPa, will condensation form on a pair of glasses that have cooled to 5°C?
>
> Consult Table 7.1.
>
> For a vapour pressure of 1.2 kPa, Table 7.1 shows that the dew-point temperature is about 10°C. Since the glasses are only 5°C, condensation will form on them.

> **Remember This**
>
> Using the values in Table 7.1, we can
>
> - find saturation vapour pressure based on air temperature, and vice versa; and
> - find dew-point temperature based on vapour pressure, and vice versa.

FIGURE 7.12 | The dew on this bunchberry flower in British Columbia has formed because the flower is at, or below, the dew-point temperature of the air.

Where surfaces are moist and there is sufficient energy, evaporation rates will be high and, as a result, so will dew-point temperatures. It follows that dew-point temperatures will tend to be highest over vegetated surfaces and warm water bodies. Some of the highest dew-point temperatures, about 25°C, are reached over warm tropical oceans. It is difficult for dew-point temperatures to get much higher than 25°C. This is because the exponential relationship between temperature and saturation vapour pressure means that for higher dew-point temperatures to be reached, there must be increasingly greater amounts of water vapour in the air. For example, as Table 7.1 shows, a dew-point temperature of 20°C requires a vapour pressure of 2.3 kPa, while a dew-point temperature of 30°C requires that vapour pressure almost doubles to 4.2 kPa. It is rare for vapour pressures to get this high.

Question for Thought

Why might it be possible to estimate the amount of moisture in the air over the ocean, or over an ice surface, by simply measuring the air temperature? Why would this not work over a desert?

7.5.2 | Relative Measures

Relative measures of water vapour are derived from the absolute measures described above, and they provide information about how close the air is to being saturated. As shown in Table 7.2, there are four relative measures: relative humidity, vapour pressure deficit, dew-point depression, and wet-bulb temperature. The following sections discuss the first three; the fourth will be discussed in Section 7.6.

7.5.2.1 | Relative Humidity

Relative humidity is probably the most commonly used relative measure. This measure is expressed as a percentage, and we can calculate it using vapour pressure, mixing ratio, or absolute humidity, as shown in Equation 7.11.

$$RH = \frac{e}{e_s} \times 100\% = \frac{r}{r_s} \times 100\% = \frac{\rho_v}{\rho_{vs}} \times 100\% \qquad (7.11)$$

Example 7.7

Air near the Earth's surface has a temperature of 6°C and a vapour pressure of 0.7 kPa. What is the relative humidity?

Use Equation 7.11.

Table 7.1 shows that the saturation vapour pressure at 6°C is 0.94 kPa.

$$RH = \frac{0.7 \text{ kPa}}{0.94 \text{ kPa}} \times 100\%$$

$$= 74\%$$

A relative humidity of 74 per cent means that the amount of water vapour in the air is 74 per cent of the maximum it could be at 6°C.

When relative humidity is 100 per cent, air temperature and dew-point temperature will be the same; the air will be saturated and, as a result, no net evaporation can occur. It follows that if relative humidity rises even slightly above 100 per cent, condensation is likely. In the atmosphere, this condensation will take the form of fog or clouds (Chapter 9). During a fog, relative humidity near the surface is 100 per cent. On the other hand, when relative humidity is below 100 per cent, the dew-point temperature will be less than the air temperature, and the air will not be saturated. The lower the relative humidity the less saturated is the air, and the greater is the difference between the air temperature and the dew-point temperature. Under such conditions, net evaporation can occur.

RELATIVE HUMIDITY
The ratio of the actual amount of water vapour in the air to the saturation value at the air's temperature.

Question for Thought

Imagine that it is raining and the relative humidity at the surface is 100 per cent. How would you explain why 100 per cent relative humidity is not likely to be a *cause* of the rain, but rather a *consequence* of it?

Equation 7.11 shows that there are two ways relative humidity can change: either the amount of water vapour in the air can change, or the temperature can change (see Example 7.8). The latter changes the saturation vapour pressure of the air. Because of its dependence on temperature, relative humidity often changes through the day, even if the air's water vapour content does not change. Increasing temperatures—generally in the afternoon—cause relative humidity to fall, and decreasing temperatures—particularly overnight—cause relative humidity to rise (Figure 7.13).

Remember This

Relative humidity depends on

- the water vapour content of the air and
- the temperature of the air.

Example 7.8

Example 7.7 showed that with a temperature of 6°C and a vapour pressure of 0.7 kPa, the relative humidity is 74 per cent. a) If the temperature of the air stays constant at 6°C, but the amount of water vapour decreases to 0.5 kPa, what is the relative humidity? b) If the amount of water vapour in the air stays constant at 0.7 kPa, but the air temperature increases to 12°C, what is the relative humidity?

a) Use Equation 7.11.

$$RH = \frac{0.5 \text{ kPa}}{0.94 \text{ kPa}} \times 100\%$$

$$= 53\%$$

b) Use Equation 7.11.

Table 7.1 shows that the saturation vapour pressure at 12°C is 1.40 kPa.

$$RH = \frac{0.7 \text{ kPa}}{1.40 \text{ kPa}} \times 100\%$$

$$= 50\%$$

This example shows that relative humidity will decrease if the amount of water vapour decreases or the air temperature increases.

Because relative humidity is dependent on both water vapour content *and* temperature, we can properly interpret relative humidity only when we know the temperature. Example 7.9 shows how important this is. In this example, the air sample with the greatest relative humidity has the lowest water vapour content, and the air sample with the lowest relative humidity has the highest water vapour content. This occurs because the former air sample is very cold and the latter sample is very warm. Another reason for providing temperature alongside relative humidity is that when we know both, we can calculate other humidity variables. This is shown in Example 7.10.

Heating and cooling of buildings can cause relative humidity extremes that lead to unhealthy conditions. Heating dry winter air can produce very low indoor relative humidities, causing scratchy noses, chapped lips, dry skin, and, for some, breathing problems. When relative humidity is below about 25 per cent, the air is too dry to be healthy. In the summer time, the

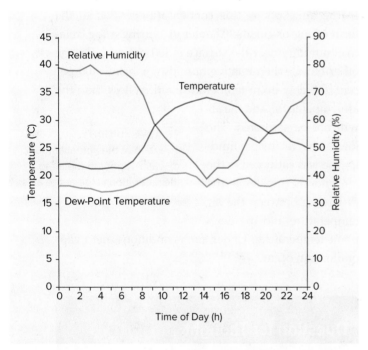

FIGURE 7.13 | This graph shows the variation of temperature, dew-point temperature, and relative humidity on a hot summer day in Toronto. Use the information on the graph to account for the change in relative humidity over this day.

Source: Data from Environment Canada, Weather office, www.climate.weatheroffice.gc.ca/climateData/canada_e.html, 10 August 2011.

Example 7.9

How much water vapour is in a sample of air with a) a temperature of −2°C and a relative humidity of 95 per cent and b) a temperature of 27°C and a relative humidity of 49 per cent? Express the answers as vapour pressure.

a) Use Equation 7.11.

At −2°C, the saturation vapour pressure is 0.52 kPa.

$$e = \frac{95}{100} \times 0.52 \text{ kPa}$$

$$= 0.49 \text{ kPa}$$

b) Use Equation 7.11.

At 27°C, the saturation vapour pressure is 3.57 kPa.

$$e = \frac{49}{100} \times 3.57 \text{ kPa}$$

$$= 1.75 \text{ kPa}$$

The air sample with the lower relative humidity contains more water vapour than the air parcel with the higher relative humidity contains.

Example 7.10

The air temperature in a sample of air is 19°C, and the relative humidity is 65 per cent. a) What is the vapour pressure? b) What is the dew-point temperature?

a) Use Equation 7.1.

At 19°C, the saturation vapour pressure is 2.2 kPa.

$$e = \frac{65}{100} \times 2.2 \text{ kPa}$$

$$= 1.4 \text{ kPa}$$

b) Using Table 7.1 and a vapour pressure of 1.4 kPa, the dew-point temperature is found to be 12°C.

Example 7.11

The air outside has a temperature of 0°C and a dew-point temperature of −1°C. Inside air is heated to 22°C. What are the inside and outside relative humidities? It is reasonable to assume that the vapour pressure inside is the same as that outside.

Use Table 7.1 to determine that the vapour pressure of the outside air is about 0.56 kPa, the saturation vapour pressure of the outside air is 0.61 kPa, and the saturation vapour pressure of the inside air is 2.64 kPa.

Use Equation 7.11.

$$RH_{(outside)} = \frac{0.56 \text{ kPa}}{0.61 \text{ kPa}} \times 100\%$$

$$= 92\%$$

$$RH_{(inside)} = \frac{0.56 \text{ kPa}}{2.64 \text{ kPa}} \times 100\%$$

$$= 21\%$$

Outside, the relative humidity is high because the air is so cold. Once the air is heated inside the house, its relative humidity drops to unhealthy levels.

Question for Thought

What activities could increase the relative humidity in a home?

7.5.2.2 | Vapour Pressure Deficit

The second relative measure of atmospheric water vapour—**vapour pressure deficit** (VPD)—tells us the difference between the saturation vapour pressure and the actual vapour pressure.

VAPOUR PRESSURE DEFICIT
The difference between the saturation vapour pressure and the actual vapour pressure.

$$VPD = e_s - e \qquad (7.12)$$

In other words, vapour pressure deficit indicates the amount of water vapour that must be added to the

opposite problem occurs. Cooling moist summer air can increase the indoor relative humidity, often leading to condensation problems. When relative humidity is consistently over 60 per cent, unhealthy molds and mildews can form. Recommended indoor relative humidity is between 30 per cent and 50 per cent. To control humidity levels, people often use humidifiers in the winter and dehumidifiers in the summer.

air to make it saturated. As such, we can use it as a measure of the drying power of the air. For example, a high vapour pressure deficit in the air above a moist surface will drive evaporation from the surface; as a result, the surface will dry. Thus, vapour pressure deficit can influence the convective latent heat flux, Q_E, described in Section 6.6. Example 7.12 shows that an increase in temperature will increase the vapour pressure deficit, increasing the drying power of the air.

Vapour pressure deficit is also a useful measure of humidity in greenhouses (Figure 7.14). Through the process of transpiration, plants lose water through the stomata in their leaves and pull water, and nutrients, up from the soil. When it is hot and humidity is either high or low, plant transpiration will decrease

FIGURE 7.14 | The vapour pressure deficit of the air, such as that in this organic greenhouse, is an important consideration for healthy plant transpiration.

Example 7.12

Air has a temperature of 14°C and a relative humidity of 45 per cent. If the temperature of the air is increased to 22°C, how much is its drying power increased?

Use Table 7.1 to determine that the saturation vapour pressure is 1.6 kPa at 14°C and 2.6 kPa at 22°C. Then use Equation 7.11 to determine the vapour pressure of the air.

$$e = \frac{45}{100} \times 1.6 \text{ kPa}$$

$$= 0.72 \text{ kPa}$$

Next, calculate the vapour pressure deficit of this air using Equation 7.12.

$$\text{VPD} = 1.6 \text{ kPa} - 0.72 \text{ kPa}$$

$$= 0.9 \text{ kPa}$$

Now, calculate the vapour pressure deficit of the warmer air.

$$\text{VPD} = 2.6 \text{ kPa} - 0.72 \text{ kPa}$$

$$= 1.9 \text{ kPa}$$

Finally, calculate the *change* in the drying power of the air.

$$1.9 \text{ kPa} - 0.9 \text{ kPa} = 1.0 \text{ kPa}$$

Example 7.13

a) The temperature in a greenhouse is 20°C. Determine the vapour pressure deficit for relative humidities of 60 per cent and 80 per cent.

The saturation vapour pressure for 20°C is 2.3 kPa.

At a relative humidity of 60 per cent, the vapour pressure is 1.4 kPa (60 per cent of 2.3 kPa), and the vapour pressure deficit is 0.9 kPa (2.3 kPa – 1.4 kPa).

At a relative humidity of 80 per cent, the vapour pressure is 1.8 kPa (80 per cent of 2.3 kPa) and the vapour pressure deficit is 0.5 kPa (2.3 kPa – 1.8 kPa).

The vapour pressure deficits indicate that both conditions are conducive to healthy plants.

b) The temperature in another greenhouse is 25°C. Determine the vapour pressure deficit for relative humidities of 60 per cent and 80 per cent.

The saturation vapour pressure for 25°C is 3.2 kPa.

At a relative humidity of 60 per cent, the vapour pressure is 1.9 kPa (60 per cent of 3.2 kPa) and the vapour pressure deficit is 1.3 kPa (3.2 kPa – 1.9 kPa).

At a relative humidity of 80 per cent, the vapour pressure is 2.6 kPa (80 per cent of 3.2 kPa) and the vapour pressure deficit is 0.6 kPa (3.2 kPa – 2.6 kPa).

The vapour pressure deficits indicate that, under these warmer conditions, a relative humidity of 60 per cent is too low for healthy plants.

to dangerous levels. Studies have shown that plants grow best when the vapour pressure deficit is between 0.45 and 1.25 kPa, regardless of temperature (Example 7.13). Thus, vapour pressure deficit provides a more consistent measure of ideal conditions than does relative humidity, because the optimal relative humidity varies with temperature.

7.5.2.3 | Dew-Point Depression

The third relative measure of humidity is **dew-point depression**, T_{dd}, which is the difference between the air temperature and the dew-point temperature.

$$T_{dd} = T - T_d \qquad (7.13)$$

Dew-point depression is, therefore, a measure of how much the air needs to cool to become saturated. It is often plotted on 700 hPa upper-air charts (Section 3.7) because areas with dew-point depressions of less than 5°C are likely to be cloudy (Figure 7.15). The 700 hPa chart is used for this because it represents the average height of most cloud cover.

> **DEW-POINT DEPRESSION**
> The difference between the air temperature and the dew-point temperature.

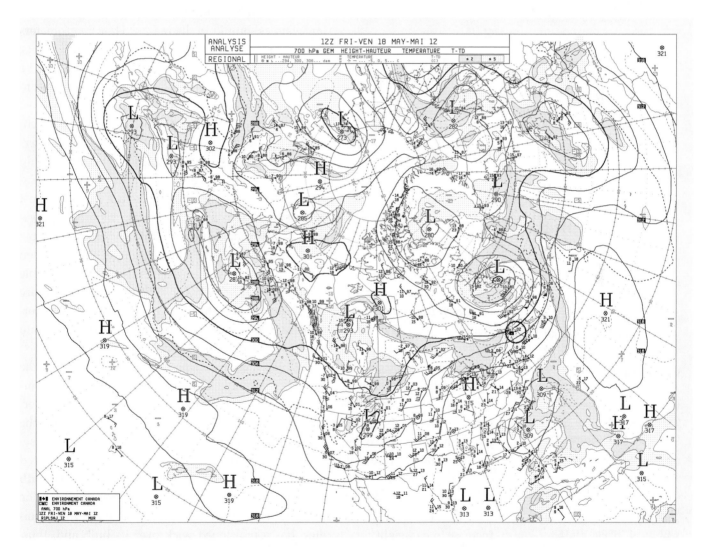

FIGURE 7.15 | A 700 hPa chart. The solid lines are heights of the 700 hPa surface. The dashed lines are isotherms. All shaded areas have dew-point depressions of less than or equal to 5°C; the darker shaded areas have dew-point depressions of less than or equal to 2°C. As a result of their low dew-point depressions, these shaded areas are likely to be cloudy.

Source: Environment Canada, Weather Office, www.weatheroffice.gc.ca/analysis/index_e.html, 18 May 2012.

Example 7.14

Determine the vapour pressure deficit and the dew-point depression for the air sample described in Example 7.7.

Use Equation 7.12 to find the vapour pressure deficit.

$$VPD = 0.94 \text{ kPa} - 0.7 \text{ kPa}$$
$$= 0.24 \text{ kPa}$$

Table 7.1 gives a dew-point temperature of 2°C for this air sample. Use Equation 7.13 to find the dew-point depression.

$$T_{dd} = 6°C - 2°C$$
$$= 4°C$$

For this air sample, the relative humidity is 74 per cent (see Example 7.7). The vapour pressure deficit indicates that 0.24 kPa of water vapour is needed to saturate it. The dew-point depression indicates that it must cool 4°C to be saturated.

Question for Thought

What does a vapour pressure deficit of 0.24 kPa tell you that a relative humidity of 74 per cent does not? What does a relative humidity of 74 per cent tell you that a vapour pressure deficit of 0.24 kPa does not?

7.6 | Methods of Measuring Atmospheric Humidity

HYGROMETER
An instrument used to directly measure the amount of water vapour in the air, based on a variety of different methods.

PSYCHROMETER
An instrument used to measure the amount of water vapour in the air, based on the cooling produced by evaporation.

DRY-BULB THERMOMETER
The thermometer in a psychrometer that is used to measure air temperature.

WET-BULB THERMOMETER
The thermometer in a psychrometer that is used to measure the wet-bulb temperature.

Many interesting methods have been devised to measure humidity. The earliest methods made use of the tendency for certain substances to absorb water vapour. These methods worked on the principle that as substances such as sponges, wool, or paper absorbed water vapour, they got heavier. When using such methods, researchers hung the absorbing substance on one side of a balance and observed the increase in its weight.

It was soon realized that materials that changed *dimensions* after absorbing water vapour provided more reliable results. The *hair **hygrometer**,* invented in 1783, is based on this principle. It measures the length of either human hair or horse hair, both of which respond to humidity by lengthening in a regular way as humidity increases. Hair hygrometers measure relative humidity, but they have drawbacks: they are not particularly accurate, they need to be calibrated regularly, and they are slow to respond to humidity changes.

Other hygrometers use more sophisticated techniques to measure humidity. *Electrical hygrometers* measure relative humidity based on the relationship between changes in electrical resistance and relative humidity. These instruments are the ones most commonly used in radiosondes. *Infrared hygrometers* measure absolute humidity by recording how much infrared radiation an air sample absorbs when a beam of this radiation passes through the sample (Section 7.5.1.1). Finally, *dew-point hygrometers* measure dew-point temperature by measuring the temperature at which water vapour begins to condense onto a surface as the surface is cooled (Section 7.5.1.3).

While hygrometers provide direct readings of the humidity variable they are measuring, **psychrometers** provide indirect readings that must be interpreted, through tables or calculations, to obtain the desired measure. These readings come from the psychrometer's two thermometers, which are mounted side by side. One thermometer, called the **dry-bulb thermometer**, measures the air temperature. The other thermometer, called the **wet-bulb thermometer**, has a moistened wick over its bulb and, thus, measures the **wet-bulb temperature** (Figure 7.16). The cooling of the wet-bulb thermometer is similar to the cooling people experience when their sweat evaporates as a breeze blows over their skin.

FIGURE 7.16 | Wet- and dry-bulb thermometers, shown here hanging vertically, are standard components of a weather station. The thermometers shown here are shielded from direct sunlight, wind, and rain—all of which could cause false readings—by a screened box known as a *Stevenson screen*. Note that the wick attached to the wet-bulb thermometer, on the right, is connected to a reservoir of water, which keeps the wick damp.

The wet-bulb thermometer will cool as water evaporates from its wick. Therefore, since the amount of water vapour in the air determines the amount of evaporation that will occur, it also determines how much the wet-bulb thermometer will cool. If the air is saturated, the wet-bulb thermometer will not cool at all, and the wet-bulb temperature will be equal to the dry-bulb, or air, temperature. As the amount of water vapour in the air decreases, the amount of cooling of the wet-bulb thermometer will increase, and the difference between the wet-bulb temperature and the dry-bulb temperature will get larger. (Psychrometers generally use a mechanism such as a fan to increase the turbulence around the thermometer bulb; this turbulence helps to prevent the air around the bulb from becoming saturated due to evaporation from the wick.)

Question for Thought

Under saturated conditions, how would the wet-bulb temperature compare to the dew-point temperature? How would these temperatures compare under unsaturated conditions?

The difference between the dry-bulb temperature and the wet-bulb temperature thus provides a measure of atmospheric water vapour. This difference is known as the **depression of the wet bulb**. Once the two thermometers have been read, and the depression of the wet bulb has been obtained, other measures of humidity can be determined using the psychrometric equations (equations 7.15 and 7.16), the psychrometric chart (Figure 7.17), or psychrometric tables (Table 7.3).

The psychrometric equation is based on the law of conservation of energy (Chapter 4). Because energy is conserved, the energy used to evaporate water is equal to the loss of energy from the surroundings. More precisely, the sensible heat energy lost from the surroundings becomes the latent heat energy involved in evaporating the water from the wet-bulb thermometer. In Equation 7.14, the left-hand side represents the energy lost as the surroundings cool, and the right-hand side represents the energy used to evaporate the moisture from the wet-bulb thermometer.

$$C\,(T - T_w) = L_v\,(\rho_{vws} - \rho_v) \qquad (7.14)$$

In this equation, C is the **heat capacity** of the air in units of $J \cdot m^{-3} \cdot K^{-1}$. This term is similar to specific heat (Section 4.1.1), in that it is used to convert a change in energy into a change in temperature, but it is for a unit *volume* of air. T_w is the wet-bulb temperature, so that $T - T_w$ is the wet-bulb depression. The term ρ_{vws} is the saturation absolute humidity at the wet-bulb temperature, while ρ_v is the absolute humidity of the air. The closer these last two values are, the less evaporation will occur.

We can rearrange Equation 7.14 to produce an equation for absolute humidity based on the measured wet-bulb depression. The result is the *psychrometric equation.*

$$\rho_v = \rho_{vws} - \gamma\,(T - T_w) \qquad (7.15)$$

WET-BULB TEMPERATURE
The temperature to which air will cool by evaporating water into it.

DEPRESSION OF THE WET BULB
The difference between the dry-bulb temperature and the wet-bulb temperature.

HEAT CAPACITY
The amount of heat required to raise a unit volume of a substance by 1 K.

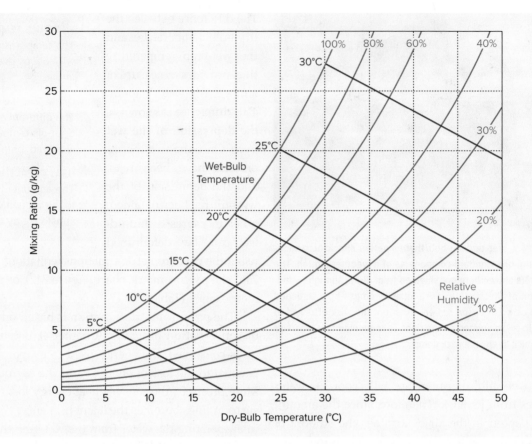

FIGURE 7.17 | **A simplified psychrometric chart for 101.325 kPa. Dry-bulb temperatures are indicated by the black vertical lines; wet-bulb temperatures are indicated by the red diagonal lines; and relative humidity is indicated by the curved blue lines. For any pair of dry-bulb and wet-bulb temperatures, the two lines will meet at the corresponding relative humidity. For example, in Example 7.15, the dry-bulb temperature is 16°C and the wet-bulb temperature is 14°C. For these two values, the chart shows that the relative humidity is close to 80 per cent. In a similar way, the mixing ratio can be determined using the black horizontal lines.**

In Equation 7.15, γ is the psychrometric constant, which is equal to C/L_v, or $0.5 \ \text{g} \cdot \text{m}^{-3} \cdot \text{K}^{-1}$. The psychrometric equation can also be written for vapour pressure.

$$e = e_{ws} - \gamma \, (T - T_w) \qquad (7.16)$$

In Equation 7.16, e_{ws} is the saturation vapour pressure at the wet-bulb temperature, and the psychrometric constant is equal to 65 Pa/K.

7.7 | Humidity and Human Comfort

Humidity is an important determinant of our own comfort, and even safety, when it is warm. When humidity is high, it is more difficult for our bodies to regulate our internal temperature because our sweat will not easily evaporate. At first, we may just feel sticky and uncomfortable, but as temperature and humidity rise, heat stroke becomes a possibility. Because the amount of water vapour in the air determines how effectively our sweat will evaporate, hot and humid conditions *feel* hotter than hot and dry conditions.

To quantify our perceptions of just how hot it feels under conditions of high humidity, indices have been developed. These indices are used to warn people of the potential dangers of high temperatures combined with high humidity. In Canada, we use an index known as the **humidex**. In the United States, the *heat index*, or *apparent temperature*, is used. These indices are calculated using regression equations, developed based on peoples' perceptions of how hot

TABLE 7.3 | A psychrometric table, giving relative humidity values for different combinations of dry-bulb temperature and depression of the wet bulb.

Dry Bulb (°C)	Depression of Wet Bulb (°C)																																							
	0.5	1.0	1.5	2.0	2.5	3.0	3.5	4.0	4.5	5.0	5.5	6.0	6.5	7.0	7.5	8.0	8.5	9.0	9.5	10	11	12	13	14	15	16	17	18	19	20	21	22	23	24	25	26	27	28	29	30
−1	90	79	69	59	49	39	30	20	10	1																														
0	90	81	71	61	52	44	34	25	16	7																														
1	90	81	73	64	55	47	38	29	20	13	4																													
2	91	82	73	64	57	49	41	33	24	17	9	1																												
3	91	83	74	65	57	49	43	36	28	21	14	7																												
4	91	83	75	67	59	51	43	35	32	25	18	11	4																											
5	92	84	76	68	61	53	46	38	31	24	21	15	8	2																										
6	92	85	77	70	62	55	48	41	34	27	20	14	6																											
7	93	85	78	71	64	57	50	44	37	30	24	17	11	5	1																									
8	93	86	79	72	65	59	52	46	39	33	27	21	15	9	3																									
9	93	86	80	73	67	60	54	48	42	36	30	24	18	12	7	3																								
10	93	87	81	75	69	62	56	50	44	38	33	27	21	16	10	5	1																							
11	94	87	81	75	69	63	58	52	46	41	35	30	24	19	14	9	5																							
12	94	88	82	76	70	65	59	54	48	43	37	32	27	21	16	12	7	2																						
13	94	88	83	77	71	66	60	55	50	45	40	35	30	25	20	15	11	6	1																					
14	94	89	83	78	72	67	62	57	52	47	42	37	33	28	23	18	14	9	4																					
15	94	89	84	78	73	68	63	58	53	48	43	39	34	30	25	21	17	12	8	4																				
16	95	89	84	79	74	69	64	59	55	50	45	41	37	32	28	24	19	15	11	7	1																			
17	95	90	85	80	75	70	65	61	56	52	47	43	39	34	30	26	22	18	14	10	3																			
18	95	90	85	80	76	71	66	62	57	53	49	45	40	36	32	28	24	21	17	13	6																			
19	95	90	86	81	76	72	67	63	59	54	50	46	42	38	34	30	26	23	19	16	9	2																		
20	95	91	86	82	77	73	68	64	60	56	52	48	44	40	36	32	29	25	22	18	12	5																		
21	95	91	86	82	78	73	69	65	61	57	53	49	45	41	38	34	30	27	24	20	14	7	1																	
22	95	91	87	82	78	74	70	66	62	58	54	50	47	43	40	36	32	29	26	23	16	10	4																	
23	96	91	87	83	79	75	71	67	63	59	55	52	48	45	41	38	34	31	28	25	19	13	7	1																
24	96	92	88	83	79	75	71	68	64	60	57	53	50	46	43	39	36	33	30	27	21	15	9	4																
25	96	92	88	83	80	76	72	68	65	61	58	54	51	47	44	41	38	34	31	28	22	17	11	6	1															
26	96	92	88	84	80	76	73	69	65	62	58	55	52	48	45	42	39	36	33	30	24	19	14	9	3															
27	96	92	89	84	81	77	73	70	66	63	59	56	53	50	47	44	40	37	34	32	26	21	16	11	7	3														
28	96	92	89	85	81	77	74	70	67	64	60	57	54	51	48	45	42	39	36	33	28	22	18	13	8	3														
29	96	93	89	85	82	78	75	71	68	65	61	58	55	52	49	46	43	40	38	35	29	24	20	15	11	7	3													
30	96	93	89	85	82	78	75	72	68	65	62	59	56	53	50	47	45	42	39	37	31	26	22	18	13	10	5	1												
32	96	93	90	86	83	79	76	73	70	66	63	60	57	54	51	49	46	43	40	38	32	28	24	20	16	13	10	7	3											
34	96	93	90	86	83	80	77	74	71	68	65	62	59	57	54	51	48	45	43	41	35	31	27	24	21	18	15	13	10	7										
36	96	94	90	87	84	81	78	75	72	69	66	64	61	58	55	53	50	48	45	43	38	34	30	27	24	22	19	16	13	11	7	3								
38	96	94	91	87	85	82	79	76	73	71	68	65	63	60	58	55	52	50	48	45	41	37	34	31	28	25	23	20	18	16	13	10	7							
40	96	94	91	88	85	82	79	77	74	72	69	67	64	62	59	57	54	52	50	47	43	40	36	34	32	29	27	24	21	19	17	15	12	10	7	4	1			
42	97	94	91	88	85	82	80	77	75	72	70	67	65	63	61	59	57	55	53	50	47	43	40	37	34	32	29	27	24	21	19	16	14	12	10	7	4	2	1	
44	97	94	91	88	86	83	80	78	76	73	71	68	66	64	62	60	58	56	54	52	49	45	42	39	37	34	32	29	27	24	23	20	18	15	13	11	9	7	5	3
46	97	94	91	89	86	83	81	78	76	74	72	69	67	65	63	61	59	57	55	54	51	47	44	41	39	36	34	32	29	27	25	22	20	18	15	13	11	9	7	5
48	97	95	92	89	86	84	81	79	77	75	72	70	68	66	64	62	60	58	57	55	52	49	46	43	41	38	36	34	31	29	27	24	22	20	18	15	13	11	9	7
50	97	95	92	89	87	84	82	80	77	75	73	71	69	67	65	63	61	59	58	56	53	50	47	45	42	39	37	35	33	31	29	26	24	21	19	16	14	12	10	9
52	97	95	93	90	87	85	82	80	78	76	74	71	69	67	66	64	62	60	59	57	54	51	49	46	44	41	39	37	35	33	31	28	26	23	21	18	16	14	11	9
54	97	95	93	90	88	85	83	81	79	76	74	72	70	68	66	65	63	61	60	58	55	52	49	47	45	42	40	38	36	34	32	30	27	24	22	19	17	15	13	11
56	97	95	93	90	88	86	83	81	79	77	75	73	71	69	67	66	64	62	61	59	56	54	51	48	46	44	42	39	37	35	33	31	28	25	23	21	18	16	15	13
58	97	95	93	90	87	86	83	81	79	77	76	74	72	70	68	67	65	63	61	59	57	55	52	49	47	45	42	40	38	37	34	32	30	27	25	23	20	18	16	13
60	98	96	93	90	87	85	83	81	79	77	75	72	70	68	67	65	64	62	60	57	54	51	48	45	42	39	36	34	32	29	27	25	23	20	18	16	13	11	9	

HUMIDEX
An index used in Canada to provide a measure of how warm it feels due to a combination of high temperature and high humidity.

WIND CHILL
A measure of how cold it feels due to a combination of low temperature and high wind.

it is. They are analogous to the **wind chill**, a measure of how cold we *feel* when it is windy as well as cold. Because they are based on our perceptions, these indices are rather abstract. We all respond quite differently to our environment; what is hot to one person might be comfortable to another.

The humidex is calculated using the air temperature and the dew-point temperature; the number obtained is dimensionless, but it is meant to approximate temperature (Table 7.4). Table 7.5 is used to determine the humidex, given air temperature and relative humidity. For example, if the temperature is 35°C and the relative humidity is 50 per cent, the table shows that the humidex is 45. So, whereas a relative humidity of 50 per cent may not sound like anything to be concerned about, a humidex of 45 indicates that it might be.

Although indices such as the humidex are widely used, some scientists believe that an index is not really necessary. Instead, they argue, we should use dew-point temperature, which is often cited as a much more straightforward measure of human comfort. As mentioned above, even with a relative humidity of only 50 per cent, air at 35°C can still contain enough water vapour to adversely affect us. This air will have a dew-point temperature of about 23°C.

In general, most people begin to feel uncomfortable when dew-point temperatures reach about 18°C. Once they go above about 22°C, the heat becomes oppressive. Consider that with a dew-point temperature of 22°C and an air temperature of 25°C, relative humidity will be 83 per cent. With a dew-point temperature of 22°C and an air temperature of 35°C, relative humidity will be only 47 per cent. However, in both cases, the combination of heat and humidity will cause discomfort. Relative humidity would not tell us this, but dew-point temperature will. Dew-point temperature measures the amount of moisture in the air and, regardless of the relative humidity, the more moisture there is, the more uncomfortable we become.

Example 7.15

A psychrometer reading gives a dry-bulb temperature of 16°C and a wet-bulb temperature of 14°C. Determine the vapour pressure, relative humidity, and dew-point temperature for this air.

Use Table 7.1 to obtain the saturation vapour pressure at the wet-bulb temperature, then use Equation 7.16 to calculate vapour pressure.

$$e = 1500 \text{ Pa} - 65 \text{ Pa/K } (16°C - 14°C)$$
$$= 1370 \text{ Pa}$$
$$= 1.4 \text{ kPa}$$

Use Table 7.1 to obtain the saturation vapour pressure for 16°C, then use Equation 7.11 to calculate relative humidity.

$$RH = \left(\frac{1.4 \text{ kPa}}{1.8 \text{ kPa}} \right) \times 100\%$$
$$= 78\%$$

Note that you can also obtain relative humidity from Table 7.3 or Figure 7.17.

Use Table 7.1, and the vapour pressure of 1.3 kPa obtained above, to determine that the dew-point temperature is about 11°C.

TABLE 7.4	Humidex and degree of comfort.	
Humidex	**Degree of Comfort**
20–29	no discomfort
30–39	some discomfort
40–45	great discomfort; avoid exertion
46 and over	dangerous; possible heat stroke

Source: Environment Canada, www.ec.gc.ca/meteo-weather/default.asp?lang= En&n=86C0425B-1#h2.

TABLE 7.5 | Humidex table.

Temperature (C°)	Relative Humidity (%)																
	100	95	90	85	80	75	70	65	60	55	50	45	40	35	30	25	20
21°C	29	29	28	27	27	26	26	24	24	23	23	22					
22°C	31	29	29	28	28	27	26	26	24	24	23	23					
23°C	33	32	32	31	30	29	28	27	27	26	25	24	23				
24°C	35	34	33	33	32	31	30	29	28	28	27	26	26	25			
25°C	37	36	35	34	33	33	32	31	30	29	28	27	27	26			
26°C	39	38	37	36	35	34	33	32	31	31	29	28	28	27			
27°C	41	40	39	38	37	36	35	34	33	32	31	30	29	28	28		
28°C	43	42	41	41	39	38	37	36	35	34	33	32	31	29	28		
29°C	46	45	44	43	42	41	39	38	37	36	34	33	32	31	30		
30°C	48	47	46	44	43	42	41	40	38	37	36	35	34	33	31	31	
31°C	50	49	48	46	45	44	43	41	40	39	38	36	35	34	33	31	
32°C	52	51	50	49	47	46	45	43	42	41	39	38	37	36	34	33	
33°C	55	54	52	51	50	48	47	46	44	43	42	40	38	37	36	34	
34°C	58	57	55	53	52	51	49	48	47	45	43	42	41	39	37	36	
35°C		58	57	56	54	52	51	49	48	47	45	43	42	41	38	37	
36°C			58	57	56	54	53	51	50	48	47	45	43	42	40	38	
37°C					58	57	55	53	51	50	49	47	45	43	42	40	
38°C							57	56	54	52	51	49	47	46	43	42	40
39°C									56	54	53	51	49	47	45	43	41
40°C										57	54	52	51	49	47	44	43
41°C											56	54	52	50	48	46	44
42°C												56	54	52	50	48	46
43°C													56	54	51	49	47

Source: Environment Canada, www.ec.gc.ca/meteo-weather/default.asp?lang=En&n=86C0425B-1#h2.

7.8 | Chapter Summary

1. Water vapour enters the atmosphere by evaporation and leaves by condensation. Whereas evaporation is a cooling process that absorbs energy, condensation is a warming process that releases energy.

2. There is a limit to how much water vapour can exist in the air at a given temperature. At this limit, the rate of evaporation is equal to the rate of condensation, and the vapour pressure is the saturation vapour pressure.

As temperature increases, saturation vapour pressure increases. This relationship is an exponential one.

3. Once saturation is reached, condensation can occur. Saturation can be reached by cooling the air, adding water vapour to it, or mixing together two air parcels that are close to saturation. The most common process for producing condensation in the atmosphere is cooling, which is responsible for the formation of most clouds and fogs.

4. We use many methods to express the amount of water vapour in the air. The absolute measures include vapour pressure, absolute humidity, mixing ratio, specific humidity, and dew-point temperature. The relative measures are derived from the absolute ones and include relative humidity, vapour pressure deficit, dew-point depression, and wet-bulb temperature.

5. We can measure absolute humidity and dew-point temperature directly. With a psychrometer, we can measure wet-bulb temperature. Evaporation lowers the wet-bulb temperature so that the less water vapour there is in the air, the greater is the difference between the wet-bulb temperature and the dry-bulb (air) temperature. We can use this difference to determine other humidity measures.

6. Certain combinations of heat and humidity can cause heat stress. In Canada, an index known as the humidex has been developed to make people aware of potentially stressful conditions.

Review Questions

1. What is vapour pressure? Why does it change with vertical motion?

2. What is meant by the term *saturation*? What does saturation vapour pressure depend on?

3. What factors influence evaporation? Provide an explanation for each.

4. What is the difference between evaporation and boiling? Why does boiling point decrease with altitude?

5. What are the three processes by which air can reach saturation? Which is most common in the atmosphere?

6. What is absolute humidity? Why does it change with vertical motion?

7. What are mixing ratio and specific humidity? Why are they almost the same in the atmosphere?

8. Why is mixing ratio (specific humidity) more useful to meteorologists than either vapour pressure or absolute humidity?

9. What is dew-point temperature? How can dew-point temperature increase?

10. What is relative humidity? How is it different from absolute measures of humidity? What can cause relative humidity to change?

11. Why would it be very unlikely for the relative humidity to be 90 per cent if the temperature was 34°C?

12. Why is the relative humidity inside buildings usually low in the winter and high in the summer?

13. What does vapour pressure deficit tell us? What does dew-point depression tell us?

14. How does a psychrometer work? What is the relationship between the amount of water vapour in the air and the depression of the wet bulb?

15. When it is hot, why do we *feel* much hotter if it is also humid?

Problems

1. The temperature is 24°C, the air pressure is 101.3 kPa, and the vapour pressure is 2.4 kPa.
   a) What is the dew-point temperature?
   b) What is the relative humidity?
   c) What is the absolute humidity?
   d) What is the mixing ratio?
   e) What is the saturation mixing ratio?
   f) Use your answers to d) and e) to recalculate the relative humidity.

2. The temperature is −8°C, the air pressure is 85 kPa, and the vapour pressure is 0.2 kPa.
   a) What is the dew-point temperature?
   b) What is the relative humidity?
   c) What is the absolute humidity?
   d) What is the mixing ratio?
   e) What is the saturation mixing ratio?
   f) Use your answers to d) and e) to recalculate the relative humidity.

3. Use the psychrometric equation to determine the vapour pressure, relative humidity, and dew-point temperature for each of the following sets of observations. (Compare the calculated relative humidity in each case to that found in Table 7.3 or estimated from Figure 7.17.)
   a) $T = 30°C$, $T_w = 20°C$
   b) $T = 14°C$, $T_w = 12°C$
   What do you notice about the relationship between T, T_w, and T_d?

4. What is the humidex in each of the following cases?
   a) $T = 30°C$, $RH = 65\%$
   b) $T = 30°C$, $RH = 30\%$
   c) $T = 25°C$, $RH = 65\%$
   Calculate the dew-point temperature in each case above. Does there appear to be a relationship between dew-point temperature and how warm the air *feels*?

5. Find the saturation vapour pressure and the saturation mixing ratio for the following parcels of air.
   a) $T = 17°C$, $P = 100$ kPa

   b) $T = 17°C$, $P = 80$ kPa
   Comment on your findings.

6. The temperature is 26°C and the dew-point temperature is 19°C. Determine vapour pressure, saturation vapour pressure, vapour pressure deficit, and relative humidity. If air temperature rose, how would each of these variables be affected?

7. a) Assume your breath has a temperature of 32°C and a dew-point temperature of 29°C. If the outside air temperature is 2°C with a dew-point temperature of 0°C, will your breath form a "cloud" when you exhale? Assume that your breath mixes in equal proportions with the outside air.
   b) Do some calculations to determine the importance of the outside air temperature and dew-point temperature in influencing whether or not your breath will form a cloud.

8. a) If the air temperature in a kitchen is 24°C, the temperature of the single-paned windows is 14°C, and the relative humidity is 40 per cent, what is the highest amount of water vapour that can be added to the air before condensation forms on the windows? How would double-paned windows change the conditions in this problem?
   b) If the temperature of a glass containing a cold drink is 4°C and the air temperature is 20°C, what is the highest possible relative humidity that can occur before condensation will form on the glass?

Suggestions for Research

1. Explore the different ways people have measured humidity throughout history, and list practical reasons for obtaining such measurements.

2. Investigate practical uses for information about

atmospheric water vapour. Consider applications for weather forecasting and for other realms, such as public health. Choose one or two applications that interest you and research them further.

References

Aguado, E., and J.E. Burt. (2010). *Understanding weather and climate* (5th ed.). Upper Saddle River, NJ: Prentice Hall.

This dust devil in Etosha National Park, Namibia, is a sign of an unstable atmosphere. In this case, the instability is a result of the intense heating of Earth's surface.

8

Adiabatic Lapse Rates and Atmospheric Stability

Learning Goals

After studying this chapter, you should be able to

1. *apply* the concept of an adiabatic process to the atmosphere;

2. *distinguish* between the dry adiabatic lapse rate and the saturated adiabatic lapse rate;

3. *distinguish* between actual temperature and potential temperature;

4. *use* a thermodynamic diagram to follow the change in state of a rising or sinking air parcel;

5. *use* a thermodynamic diagram to assess stability conditions indicated by an atmospheric sounding;

6. *distinguish* between the various stability types; and

7. *describe* conditions that influence the stability of an air layer.

This chapter will build on the introduction to adiabatic processes given in Chapter 4. As you will discover, these processes are fundamental to an understanding of **atmospheric stability**, because they control temperature changes in rising and sinking air parcels. To explain the concept of atmospheric stability, we will begin by applying the adiabatic process to derive two adiabatic lapse rates: one for dry air, the other for saturated air. We will also consider the temperature structure of the surroundings, as illustrated on **thermodynamic diagrams**, through which these air parcels move. By comparing the temperature of rising air to the temperature of surrounding air, we can determine atmospheric stability.

We will continue to apply the concept of atmospheric stability in subsequent chapters. In Chapter 9, we will explore its importance in the formation of different cloud types. In Chapter 13, we will consider how air masses can be characterized by their stability, and how that stability can change as the air masses move. Then, in Chapter 14, we will see how stability conditions influence the occurrence and intensity of thunderstorms. Finally, in Chapter 17, we will examine the important role of atmospheric stability in the dispersal of air pollutants.

ATMOSPHERIC STABILITY
A measure of the tendency for a parcel of air, once disturbed, to move vertically in the atmosphere due to temperature differences.

THERMODYNAMIC DIAGRAMS
Diagrams, or graphs, that show thermodynamic processes.

DRY ADIABATIC LAPSE RATE (DALR)
The rate of change of temperature of a rising or sinking unsaturated air parcel.

8.1 | Adiabatic Processes

Recall that an adiabatic process is one in which a temperature change occurs *without* adding or removing heat (Section 4.3). In the atmosphere, adiabatic processes occur as air parcels expand and contract in response to the pressure changes they encounter as they rise and sink. As a rising air parcel expands, it performs work on its surroundings, causing it to cool. As a sinking air parcel is compressed, the surroundings are performing work on it, causing it to warm.

Remember This

Recall the following equations.

- The hydrostatic equation:

$$\frac{\Delta P}{\Delta z} = -\rho\, g \qquad (3.9)$$

- The thermodynamic equation for an adiabatic process:

$$c_p\, \Delta T = \frac{\Delta P}{\rho} \qquad (4.9)$$

- Poisson's equation:

$$\frac{T_2}{T_1} = \left(\frac{P_2}{P_1}\right)^{\frac{R_d}{c_p}} \qquad (4.10)$$

8.1.1 | The Dry Adiabatic Lapse Rate (DALR)

To derive the **dry adiabatic lapse rate (DALR)**, let's begin by reconsidering Poisson's equation (Equation 4.10). In Chapter 4, you saw how this equation can be used to determine the temperature of a rising air parcel based on the change in pressure the parcel experienced as it rose. Yet such calculations are much easier to make if we use the DALR. We can find this constant by substituting the hydrostatic equation (Equation 3.9) into the thermodynamic equation for an adiabatic process (Equation 4.9), then rearranging.

$$c_p\, \Delta T = -\frac{\rho\, g\, \Delta z}{\rho}$$

$$\frac{\Delta T}{\Delta z} = -\frac{g}{c_p} = \Gamma_d \qquad (8.1)$$

Substituting 9.8 m/s^2 for g and 1004 J·kg^{-1}·K^{-1} for c_p into Equation 8.1 gives a value for the DALR, Γ_d, of 0.98°C/100 m, or 9.8°C/km. These values are normally rounded off to 1°C/100 m, or 10°C/km. (The negative sign in Equation 8.1 simply indicates that temperature decreases as height increases.)

Example 8.1

At sea level, the temperature of an air parcel is 23°C. If this parcel rises from sea level to a height of 2000 m, what will be its temperature?

Use the DALR.

$$23°C - \left(\frac{1°C}{100\ m}\right)(2000\ m) = 3°C$$

Recall that this problem was solved using Poisson's equation in Example 4.6. Why do both methods yield the same result?

Question for Thought

If Earth were much larger and, therefore, had a much larger value for *g*, how would the dry adiabatic lapse rate be different? How do you account for this difference?

The DALR tells us that if an unsaturated air parcel rises from Earth's surface to the tropopause, a distance of roughly 10 km, it will cool by about 100°C. Air experiences such large temperature changes as it ascends or descends through the atmosphere because it is compressible (Section 3.3). As air sinks, it is compressed and, as a result, its temperature increases; as air rises, it is decompressed and, as a result, its temperature decreases. On the other hand, water, an incompressible fluid, will not experience significant temperature changes as it rises and sinks in the oceans.

Air's compressibility has another interesting result. Imagine a layer of air at Earth's surface in which the temperature decreases with height. This air will mix as warm surface air rises and colder air from above sinks. It might also be mixed by a wind. In an incompressible fluid, like water, such mixing would eventually cause temperature to become constant with height. In the air, however, vertically moving air parcels will warm and cool at the DALR, so the temperature profile for a layer of air that is being mixed will eventually approach the DALR (Figure 8.1).

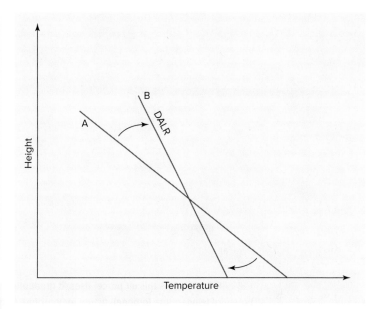

FIGURE 8.1 | The initial temperature profile for a layer of air is represented by line A. As mixing occurs, rising and sinking air parcels will warm and cool. Eventually, the temperature profile will approach the DALR, which is represented by line B.

8.1.2 | Potential Temperature

Adiabatic temperature changes can greatly complicate some atmospheric analyses; so, to simplify things, we use the theoretical concept of **potential temperature**. As long as no heat is transferred between an air parcel and its surroundings, the *potential* temperature of the parcel will not change as the parcel rises or sinks in the atmosphere. Therefore, while the *actual* temperature of an air parcel will drop by approximately 100°C as it rises to the tropopause, the *potential* temperature will remain constant (Figure 8.2). In the following discussion, you will find two definitions of potential temperature. The first is meant only to help you understand the concept; the second is more precise.

For conceptual purposes, we can loosely define potential temperature as the temperature an unsaturated air parcel would have if it were brought adiabatically back to the surface. Based on this preliminary definition, we can calculate the potential

> **POTENTIAL TEMPERATURE**
> The temperature an unsaturated air parcel would have if brought adiabatically to a pressure of 100 kPa.

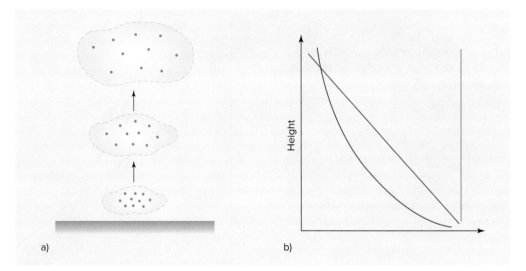

FIGURE 8.2 | a) As this air parcel rises, it gradually expands. **b)** The changes in this air parcel—in terms of potential temperature (orange), actual temperature (red), and density (blue)—are shown graphically.

Source: Adapted from Salby, Murry L. *Fundamentals of Atmospheric Physics*. Academic Press—Elsevier Science, San Diego, 1996. Page 74.

temperature, θ, of an air parcel using Equation 8.2, when its actual temperature, T, and height, z, above sea level are known.

$$\theta = T + \Gamma_d z \qquad (8.2)$$

Example 8.2

A parcel of air at a height of 2000 m above sea level has a temperature of 11°C. What is its potential temperature?

Use Equation 8.2.

$$\theta = 11°C + \left(\frac{1°C}{100\ m}\right) 2000\ m$$

$$= 31°C$$

The potential temperature is greater than the actual temperature because potential temperature removes the effects of adiabatic cooling.

This preliminary definition, and the equation that it suggests, is not entirely accurate because the change in *height* does not directly affect the temperature of a rising air parcel; instead, the change in *pressure* does. Potential temperature is, therefore, more precisely defined as the temperature an unsaturated air parcel would have if brought adiabatically to a *pressure* of 100 kPa. Based on this more accurate definition, we can use Poisson's equation (Equation 4.10) to obtain Equation 8.3 for potential temperature.

$$\theta = T \left(\frac{100}{P}\right)^{\frac{R_d}{c_p}} \qquad (8.3)$$

In this equation, T is the actual temperature and θ is the potential temperature of air at pressure P. Recall that R_d is the gas law constant for dry air (Section 3.3), and c_p is the specific heat of air at constant pressure (Section 4.3).

Equation 8.3 shows that, as would be expected, the potential temperature is equal to the actual temperature at 100 kPa. When the pressure of air is less than 100 kPa, which is generally the case for air that is above sea level, potential temperature will be greater than actual temperature. This is because when pressure is below 100 kPa, the air must be compressed to bring it to 100 kPa, and this compression will warm the air.

The concept of potential temperature is needed because air is compressible. If air were incompressible, it would not expand upon rising, nor would it be compressed upon sinking; as a result, vertical

Example 8.3

A parcel of air at a height of 2000 m above sea level and a pressure of 79.5 kPa has a temperature of 11°C. Determine its potential temperature using Equation 8.3.

$$\theta = 284\ \text{K} \left(\frac{100\ \text{kPa}}{79.5\ \text{kPa}} \right)^{0.286}$$

$$= 303.3\ \text{K}$$

$$= 30.3°\text{C}$$

This answer is close to the answer calculated in Example 8.2, using Equation 8.2.

Question for Thought

How do equations 8.2 and 8.3 compare to one another?

motions would not cause temperature changes. By using the concept of potential temperature, we are eliminating the effect of pressure on the density of air. Therefore, using potential temperature in certain atmospheric applications is simpler than using actual temperature. For example, it was shown above that when a layer of air is mixed, its temperature profile is likely to approach the DALR (Figure 8.1). On the other hand, when air is mixed, its *potential* temperature will be constant with height.

8.1.3 | Saturated Adiabatic Lapse Rate (SALR)

As air rises and cools at the DALR, chances are good that it will reach its dew-point temperature (Section 7.5.1.3). When it does, the water vapour contained in the air will begin to condense, forming a cloud. As the water vapour condenses, it will release latent heat (Section 4.1.2), thus slowing the cooling rate of the rising air so that the dry adiabatic lapse rate no longer applies. Instead, the rate of cooling in the cloudy air is the **saturated adiabatic lapse rate (SALR)**.

Unlike the DALR, the SALR is *not* constant, because it depends on the amount of latent heat being released. The more latent heat that is released, the slower the rate of cooling will be. This relationship is shown by Equation 8.4, which is used to calculate the SALR, Γ_s.

$$\Gamma_s = \Gamma_d - \left(\frac{L_v\, r}{c_p\, \Delta z} \right) \tag{8.4}$$

Recall that L_v is the latent heat of vaporization (Section 4.1.2), r is the mixing ratio (Section 7.5.1.2), and c_p is the specific heat of air at constant pressure (Section 4.3). In its entirety, the term enclosed in round brackets represents the change in temperature with height, Δz, due to the release of latent heat. As would be expected, this term increases with the mixing ratio. The more water vapour in the air, the greater the difference between the DALR and the

> **SATURATED ADIABATIC LAPSE RATE (SALR)**
> The rate of change of temperature of a rising or sinking saturated air parcel.

Example 8.4

a) A parcel of air is rising and, as it does so, 1 g/kg of water vapour is condensing. What is the saturated adiabatic lapse rate under these conditions?

Use Equation 8.4.

$$\Gamma_s = 10\ \text{K/km} - \left[\frac{(2.26 \times 10^6\ \text{J/kg})\,(1 \times 10^{-3}\ \text{kg/kg})}{(1004\ \text{J} \cdot \text{kg}^{-1} \cdot \text{K}^{-1})\,(1\ \text{km})} \right]$$

$$= 7.8\ \text{K/km} = 7.8°\text{C/km}$$

b) For another rising air parcel, 2 g/kg of water vapour is condensing. What is the saturated adiabatic lapse rate in this case?

Again, use Equation 8.4.

$$\Gamma_s = 10\ \text{K/km} - \left[\frac{(2.26 \times 10^6\ \text{J/kg})\,(2 \times 10^{-3}\ \text{kg/kg})}{(1004\ \text{J} \cdot \text{kg}^{-1} \cdot \text{K}^{-1})\,(1\ \text{km})} \right]$$

$$= 5.5\ \text{K/km} = 5.5°\text{C/km}$$

The SALR is closer to the DALR in the first case because less water vapour is condensing in that case than in the second.

SALR. Near Earth's surface, where the air is fairly warm and moist, the SALR is generally about 0.4°C/100 m, or 4°C/km. The SALR then tends to increase with height in the troposphere until it is nearly equal to the DALR at higher altitudes, where the air is cold and dry. In calculations an average value of 0.6°C/100 m, or 6°C/km, is often used for the SALR.

8.1.4 | Lifting Condensation Level (LCL)

The height at which water vapour in air rising from the surface begins to condense is called the **lifting condensation level** (LCL) (Figure 8.3). The LCL can often be "seen" in the atmosphere as the base of a cumulus cloud (Figure 8.4).

> **LIFTING CONDENSATION LEVEL**
> The height at which a rising parcel of air will reach its dew-point temperature and the water vapour it contains will begin to condense.

As an air parcel rises, not only does its temperature drop, its dew-point temperature also drops. The rate of decrease of the dew-point temperature in a rising air parcel is about 0.18°C/100 m. Because the air temperature drops faster than the dew-point temperature, the two temperatures will eventually

FIGURE 8.4 | The flat, distinct bases of these cumulus clouds clearly show the height of the lifting condensation level.

be equal; this occurs at the LCL (Figure 8.5). Based on this equality, we can derive a simple equation for estimating the height of the LCL, z_{LCL}. To do this, we set the temperature at the LCL as equal to the dew-point temperature at the LCL, and then we rearrange the terms until we arrive at Equation 8.5. (Note that the exact value of the DALR has been used in deriving Equation 8.5.)

$$T - \left(\frac{0.98°C}{100\ m}\right)(z_{LCL}) = T_d - \left(\frac{0.18°C}{100\ m}\right)(z_{LCL})$$

$$T - T_d = \left(\frac{0.98°C}{100\ m}\right)(z_{LCL}) - \left(\frac{0.18°C}{100\ m}\right)(z_{LCL})$$

$$(z_{LCL}) = 125\ (T - T_d) \qquad (8.5)$$

In Equation 8.5, z_{LCL} is in metres, and T and T_d are the *surface* temperature and dew-point temperature, respectively.

Equation 8.5 provides the best results when it is applied to air that is rising from the surface due to convection and the air layer from the surface to cloud height is well-mixed; this occurs most commonly on a warm, sunny afternoon (Section 8.6). This equation cannot be applied to air that is rising at a front because, in this case, the air rises on a slanted path

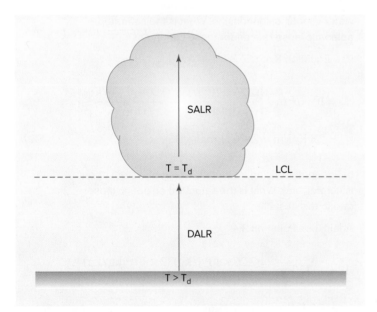

FIGURE 8.3 | Air rising due to surface heating will cool at the DALR until it reaches its dew-point temperature, T_d. At this height, the air has reached its lifting condensation level, and the water vapour it contains will begin to condense. The rising air will continue to cool at the SALR.

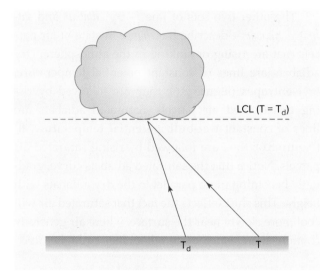

FIGURE 8.5 | As an air parcel rises, both its temperature and its dew-point temperature drop. When they are equal, the water vapour in the air will begin to condense, forming a cloud. How will the difference between the temperature and dew-point temperature at the surface affect the height of the cloud base?

Example 8.5

A warm air parcel with a temperature of 29°C and a dew-point temperature of 17°C rises from the surface.

a) What is the LCL of this air parcel?

b) If this air rises to 2000 m, what will be its temperature? Assume that the SALR is 0.5°C/100 m.

a) Use Equation 8.5.

$$z_{LCL} = 125 \ (29°C - 17°C)$$

$$= 1500 \ m$$

This rising air parcel will begin to condense and form a cloud at a height of 1500 m.

b) The air parcel will cool at the DALR until it reaches the LCL, then it will cool at the SALR.

$$29°C - (1500 \ m) \left(\frac{1°C}{100 \ m} \right) = 14°C$$

$$14°C - (500 \ m) \left(\frac{0.5°C}{100 \ m} \right) = 11.5°C$$

At 2000 m, the temperature of the air parcel will be 11.5°C.

Remember This

- The dry adiabatic lapse rate is 9.8°C/km (generally rounded to 10°C/km).
- The average value of the saturated adiabatic lapse rate is 6°C/km.
- The dew-point temperature lapse rate is approximately 1.8°C/km.

---

from hundreds or even thousands of kilometres away. As a result, the temperature and dew-point temperature at the surface directly below the clouds will have no relationship to the air forming the clouds.

8.2 | Environmental Lapse Rates (ELR)

The temperature of the air surrounding a rising or sinking parcel of air also changes with height. The rate of this change is often referred to as the **environmental lapse rate (ELR)**. We can estimate the *average* ELR by consulting the standard atmosphere (Table 1.3), and we can measure *actual* ELRs with information gathered from atmospheric soundings. Recall that soundings are made as radiosondes ascend through the atmosphere measuring temperature, pressure, dew-point temperature, and wind (Section 3.7).

ENVIRONMENTAL LAPSE RATE (ELR)
The change in temperature with height in the atmosphere.

On average, the ELR in mid-latitudes, as given by the standard atmosphere, is about 6.5°C/km through the troposphere. Actual atmospheric soundings, however, show considerable variability in both time and space. There are layers of the atmosphere where temperatures increase with height, layers where temperatures decrease with height, and layers where temperatures remain relatively constant with height. In addition, sometimes the *rate* of change of temperature through a layer is high, while at other times it is low. This variability is a result of many different processes—operating separately or together, at different levels in the atmosphere—that can influence temperature structure.

8.3 | Thermodynamic Diagrams

Atmospheric soundings are commonly plotted on *thermodynamic diagrams*. Several different types of thermodynamic diagrams are used when studying the atmosphere. In Canada, forecasters use **tephigrams**. In the following discussion, we will begin by examining another type of thermodynamic diagram, known as an *emagram* (Figure 8.6), which is easier to use; then we will move on to examine a tephigram in brief.

An emagram, like all thermodynamic diagrams, has five sets of lines. Three of these sets—isobars, **isotherms**, and **isohumes**—describe the *state* of the air (Figure 8.6). The isobars, or lines of constant pressure, run horizontally across the graph and are labelled in kilopascals along the left vertical axis. These lines are used to represent height in the atmosphere, and the scale is logarithmic because the decrease in pressure with height is exponential. (Height, in kilometres, is often plotted on the right vertical scale, as shown in Figure 8.6.) The isotherms, or lines of constant temperature, run vertically and are measured in degrees Celsius along the horizontal axis. Thus, temperature and pressure create the axes of emagrams. The third set of lines that describe the state of the air are **isohumes**, which, in this case, are lines of constant mixing ratio. These lines slope downward from left to right, and they are labelled along the top of the graph in grams per kilogram.

TEPHIGRAM
The thermodynamic diagram used in Canada.

ISOTHERMS
Lines of constant temperature.

ISOHUMES
Lines of constant atmospheric moisture.

ISENTROPES
Lines of constant potential temperature or entropy.

WET-BULB POTENTIAL TEMPERATURE
The wet-bulb temperature that an air parcel would have if brought adiabatically back to 100 kPa.

The other two sets of lines—*dry adiabats* and *saturated adiabats*—describe the *change* in state of air parcels that are rising or sinking in the atmosphere. Dry adiabats are lines of constant potential temperature, or **isentropes** (Figure 8.6); they are followed by rising unsaturated air parcels. Saturated adiabats are lines of constant **wet-bulb potential temperature**, θ_L (Figure 8.6); they are followed by rising saturated air parcels. Notice that the saturated adiabats curve, gradually becoming more parallel to the dry adiabats with height. This shift reflects the fact that saturated air will cool more slowly near the surface, where air generally contains more water vapour than does air higher up.

Example 8.6

This example will show how you can use thermodynamic diagrams in place of calculations.

First, use Equation 7.9 to calculate the saturation mixing ratio, r_s, for an air parcel with a temperature of 15°C and a pressure of 90 kPa. For a temperature of 15°C, the saturation vapour pressure from Table 7.1 is 1.7 kPa.

$$r_s = 0.622 \; \frac{(1.7 \text{ kPa})}{(90 \text{ kPa} - 1.7 \text{ kPa})}$$

$$= 0.012 \text{ kg/kg}$$

$$= 12 \text{ g/kg}$$

Second, find the point on Figure 8.6 corresponding to a temperature of 15°C and a pressure of 90 kPa. Use the isohumes to determine that the saturation mixing ratio for this air parcel is roughly 12 g/kg, just as calculated above.

Because 15°C is the *air* temperature in this example, 12 g/kg is the *saturation* mixing ratio, r_s. If 15°C had been the *dew-point temperature*, then 12 g/kg would be the *actual* mixing ratio, r.

As shown in Example 8.7, a rising air parcel follows the dry adiabat corresponding to its temperature and the isohume corresponding to its dew-point temperature. Because the dew-point temperature decreases more slowly than the air temperature, these two lines will eventually meet at the LCL. This is part of a rule

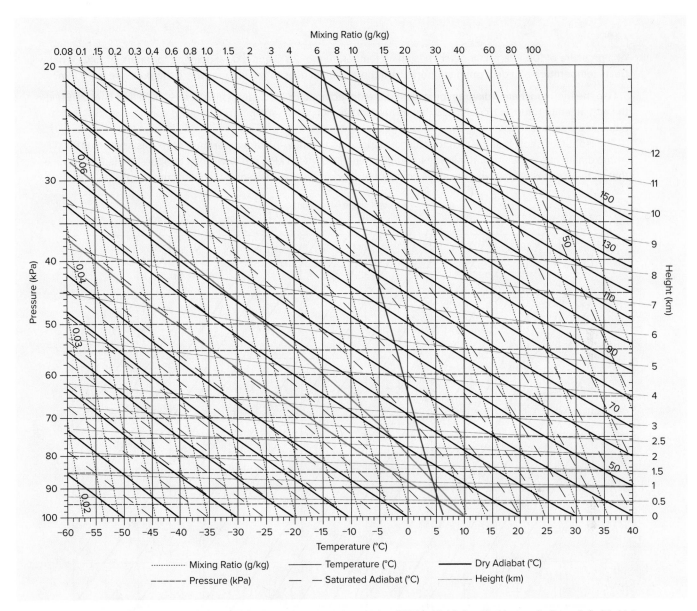

FIGURE 8.6 | A blank emagram. An isotherm (°C) is highlighted in red, an isobar (kPa) is highlighted in blue, an isohume (g/kg) is highlighted in purple, a dry adiabat (°C) is highlighted in green, and a saturated adiabat (°C) is highlighted in orange. Notice that the base of the grid pattern for the emagram is created by the isotherms and the isobars.

Source: Adapted from R.B. Stull. (2000). *Meteorology for scientists and engineers* (2nd ed.). Brooks/Cole Thomson Learning. Retrieved from http://en.wikipedia.org/wiki/File:Emagram.GIF

known as *Normand's rule*, which states that, for a given air parcel, the dry adiabat corresponding to the parcel's temperature, T, the isohume corresponding to the parcel's dew-point temperature, T_d, and the saturated adiabat corresponding to the parcel's wet-bulb temperature, T_w, intersect at the parcel's LCL (Figure 8.7). This is true because the air is saturated at the LCL; therefore, all three temperatures should be equal at this point.

An important result of Normand's rule is that if we know the temperature and dew-point temperature of an air parcel, we can determine its wet-bulb temperature. Because the saturated adiabat follows a parcel's wet-bulb temperature, we can trace this line back down from the LCL to obtain the wet-bulb temperature at any height, and tracing it back to 100 kPa will give the parcel's wet-bulb *potential* temperature.

Example 8.7

At a pressure of 100 kPa, an air parcel has an air temperature of 25°C and a dew-point temperature of 15°C. What will be its temperature and dew-point temperature if the air parcel rises to an altitude where the pressure is 90 kPa?

Use the thermodynamic diagram below. Find the dry adiabat for 25°C, and follow it to 90 kPa (red line). At 90 kPa, the actual temperature is 16°C. Notice that as the air rises, its potential temperature stays constant at 25°C, but its actual temperature drops.

The isohumes show that the mixing ratio corresponding to a dew-point temperature of 15°C is about 11 g/kg. Approximate this isohume, and follow it to 90 kPa (blue line). At 90 kPa, the dew-point temperature is 13°C. Notice that as the air rises, its mixing ratio stays constant at 11 g/kg, while its dew-point temperature drops.

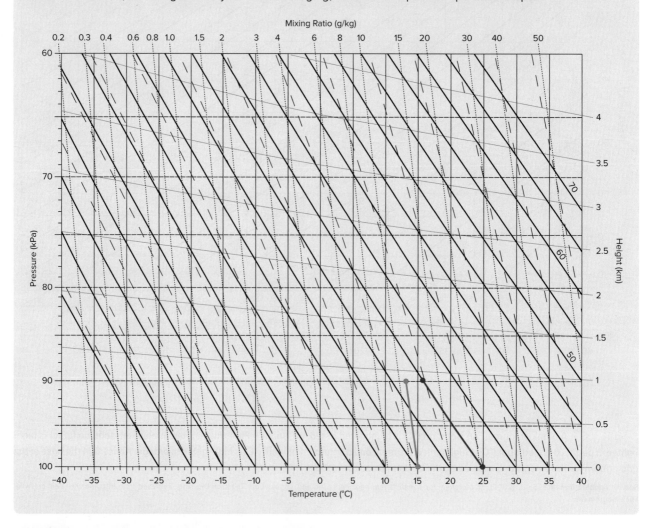

Question for Thought

As air rises, what happens to the mixing ratio? What happens to the saturation mixing ratio? (Hint: Use a thermodynamic diagram to help you answer this question.)

Once the LCL is reached, the saturated air parcel will continue to rise and cool, and the water vapour it contains will start to condense, forming a cloud. At this point, the rising air will begin to follow a *saturated* adiabat. This indicates that the temperature and dew-point temperature are now equal to each other and decreasing together, and that the mixing ratio and

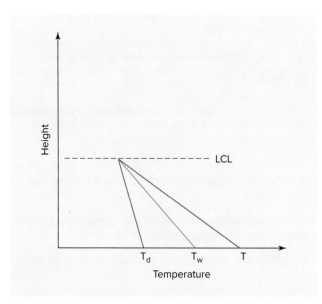

FIGURE 8.7 | A depiction of Normand's rule. Why does the wet-bulb temperature fall between the air temperature and the dew-point temperature until the LCL is reached?

saturation mixing ratio are also equal to each other and decreasing together. The dew-point temperature and mixing ratio are decreasing because the water vapour is condensing. At any height, the amount of water vapour in the cloud, r_s, is given by the intersection of the saturated adiabat with an isohume. As the amount of water vapour decreases, the amount of liquid water, r_L, will increase. However, as long as no precipitation falls, the total amount of water in the cloud, r_T, does not change, so we can write a simple water balance for the cloud.

$$r_T = r_L + r_s \qquad (8.6)$$

The total amount of *liquid* water in the cloud is an indicator of the amount of precipitation that is possible.

Air that has risen can eventually descend. For example, air can rise up the windward side of a mountain and sink down the leeward side. As the air sinks, it will be compressed and, thus, it will adiabatically warm. If *all* the liquid water precipitates from a cloud on the windward side of the mountain, the air descending on the leeward side will warm at the DALR (Figure 8.8a). (On a thermodynamic diagram, the air will follow a dry adiabat as it descends.) The loss of water represents a gain of heat because the latent heat

that was released when the water vapour condensed on the windward side of the mountain was left behind when the water precipitated. As a result, the air will be both warmer and drier on the leeward side than it was, at the same height, on the windward side. Such warm, dry winds descending the leeward side of mountains are called **chinooks** in Canada and *foehns* in the Alps. These winds are common in Calgary, Alberta, and can sometimes raise winter temperatures as much as 15°C in just a few hours.

CHINOOK
The Canadian name for a warm, dry wind that blows down the leeward side of a mountain range.

If none of the water precipitates from the cloud on the windward side of the mountain, the air descending on the leeward side will initially warm at the SALR (Figure 8.8b). At the LCL, all the water will have evaporated, and the air will continue to warm at the DALR. (On a thermodynamic diagram, it will follow a saturated adiabat back to the LCL, then it will follow a dry adiabat.) As a result, the air on the leeward side will be the same temperature that it was, at the same height, on the windward side. In this scenario, there is no gain of heat because the latent heat that was released in condensation is absorbed again during evaporation.

The scenarios described in the previous two paragraphs and depicted in figures 8.8a and 8.8b represent extremes. It is most likely that *some* of the water will precipitate out of the cloud, leaving the rest of it to be evaporated during descent (Figure 8.8c). In this case, temperatures on the leeward side will be somewhere between those in the first scenario and those in the second.

Until this point, we have worked with *emagrams* but, because we use *tephigrams* in Canada, it is important for you to be familiar with this form of thermodynamic diagram as well. The tephigram uses the same five sets of lines as the emagram, but they are oriented differently (Figure 8.9 and 8.10). This is because the tephigram uses temperature and potential temperature as axes, but rotates them 45° so that the isotherms slope up to the right and the dry adiabats slope up to the left. Tephigrams are named for these axes: *T* is for temperature, and the Greek letter phi, ϕ, is for entropy, which is related to potential

Example 8.8

For the air parcel in Example 8.7, determine

- the LCL, the temperature at the LCL, and the dew-point temperature at the LCL;

- what the wet-bulb temperature was at the surface;

- what the temperature, dew-point temperature, and liquid water content will be if the parcel continues to rise to a height of 70 kPa.

Use the diagram provided below.

a) Follow the dry adiabat for 25°C (red line) and the isohume for 11 g/kg (blue line) past 90 kPa, to where they intersect at 86 kPa; this is the LCL of the air parcel. At the LCL, the temperature and dew-point temperature are both 12.5°C.

b) Approximate the saturated adiabat that intersects with the dry adiabat and isohume at the LCL (green line). Follow this adiabat back to the surface to reveal that the wet-bulb temperature at the surface was 17°C.

c) Follow the same saturated adiabat you used in b) up to 70 kPa. At 70 kPa, the temperature and dew-point temperature are both 3°C, and the mixing ratio is about 7 g/kg.

Use Equation 8.6.

$$r_L = 11 \text{ g/kg} - 7 \text{ g/kg}$$
$$\quad = 4 \text{ g/kg}$$

Because some of the water has condensed, the amount of liquid water is 4 g/kg.

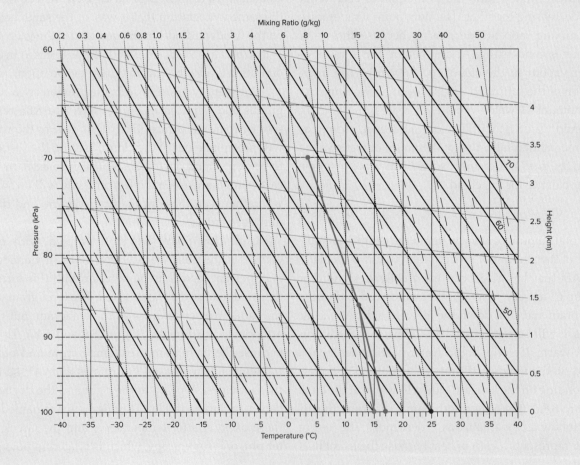

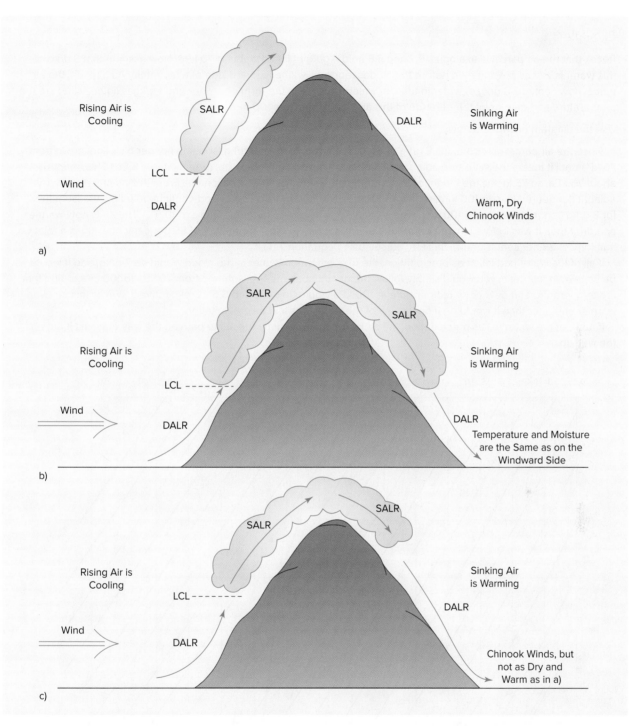

FIGURE 8.8 | As air rises on the windward side of a mountain, it cools at the DALR until it reaches its dew-point temperature at the LCL. As it continues to rise, water vapour begins to condense, and the air cools at the SALR. a) When all the liquid water in the cloud falls as precipitation on the windward side, air descending on the leeward side will warm at the DALR. On the leeward side, the air will be warmer and drier than it was on the windward side. b) When *no* precipitation falls from the cloud on the windward side, the air descending on the leeward side will warm at the SALR until the liquid water of the cloud has evaporated. As it continues to descend, it will warm at the DALR. Thus, on the leeward side, the air will have the same temperature and moisture content as it had on the windward side, at the same height. c) When *some* precipitation falls from the cloud on the windward side, the rest of the water will be evaporated as the air descends down the leeward side. With less water, the cloud base will be higher here than it was on the windward side. At the base of the mountain, the state of the air will be somewhere between what it was in a) and what it was in b).

Example 8.9

Recall that the air parcel in examples 8.7 and 8.8 held 4 g/kg of liquid water at 70 kPa. Now assume that 3 g/kg of this water is lost as precipitation before the air descends down the leeward side of a mountain. As a result, the air parcel now contains 8 g/kg of water in total, of which 7 g/kg is in the form of vapour and 1 g/kg is liquid. What are the temperature and dew-point temperature of this air when it reaches 100 kPa?

Use the diagram provided below.

Since the air parcel now contains 8 g/kg of water in total, follow the 8 g/kg isohume (longer blue line) down from 70 kPa until it meets the saturated adiabat that the air followed on the way up. (See Example 8.8c.) This happens at about 75 kPa, so 75 kPa is the height of the bottom of the cloud as the air descends. From this point, follow a dry adiabat (longer red line) to 100 kPa, where it indicates a temperature of 30°C, and continue to follow the isohume for 8 g/kg (longer blue line) to 100 kPa, where it indicates a dew-point temperature of 10.5°C. The air is now warmer and drier than it was before it rose. (Before the air rose, it had a relative humidity of 55 per cent; now it has a relative humidity of 28 per cent. Confirm this for yourself using equation 7.11.)

If all of the water had fallen as precipitation, the total amount of water in the cloud would be 7 g/kg, and it would all be vapour. In this case, you would simply follow the dry adiabat and the isohume from 70 kPa to 100 kPa to find that the temperature and dew-point temperature at 100 kPa would be 34°C and 8.5°C, respectively. Why is this air even warmer and drier than it would be if only some of the water was lost as precipitation?

If none of the water fell as precipitation, the air would follow exactly the same path on the way down as it did on the way up.

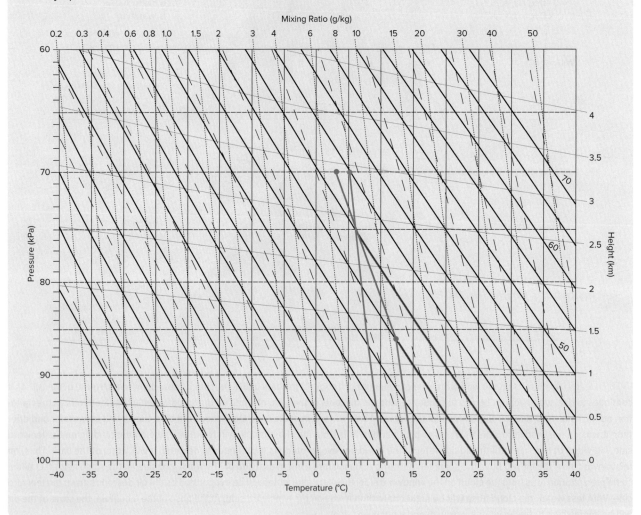

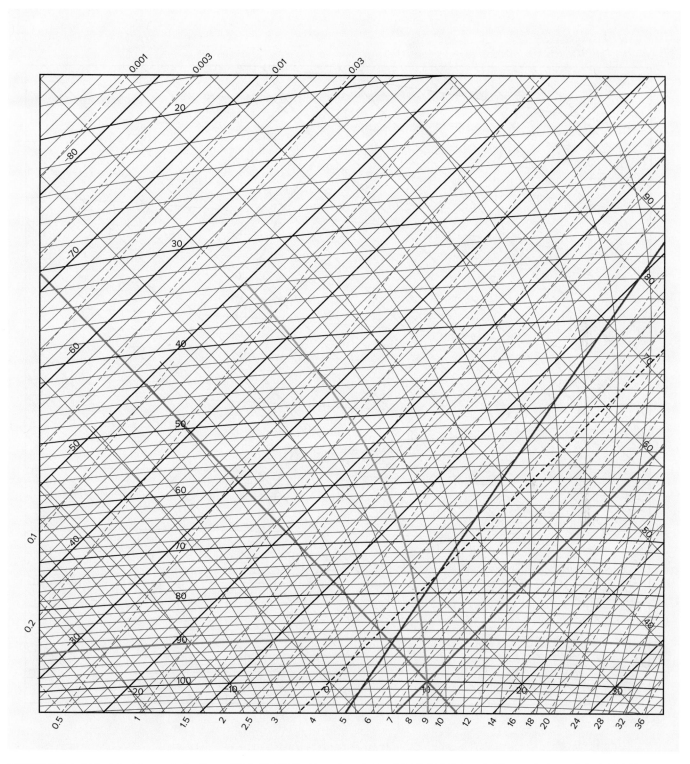

FIGURE 8.9 | A blank tephigram. An isotherm (°C) is highlighted in red, an isobar (kPa) is highlighted in blue, an isohume (g/kg) is highlighted in purple, a dry adiabat (°C) is highlighted in green, and a saturated adiabat (°C) is highlighted in orange. Notice that the base of the grid pattern for the tephigram is created by the isotherms and the dry adiabats thus, they meet at right angles. Use this diagram to help you interpret the sounding in Figure 8.10.

Source: Tephigram adapted from Maarten Ambaum's tephigram, retrieved from http://www.met.reading.ac.uk/sws97mha/Tephigram/tephi_b_colour.pdf.

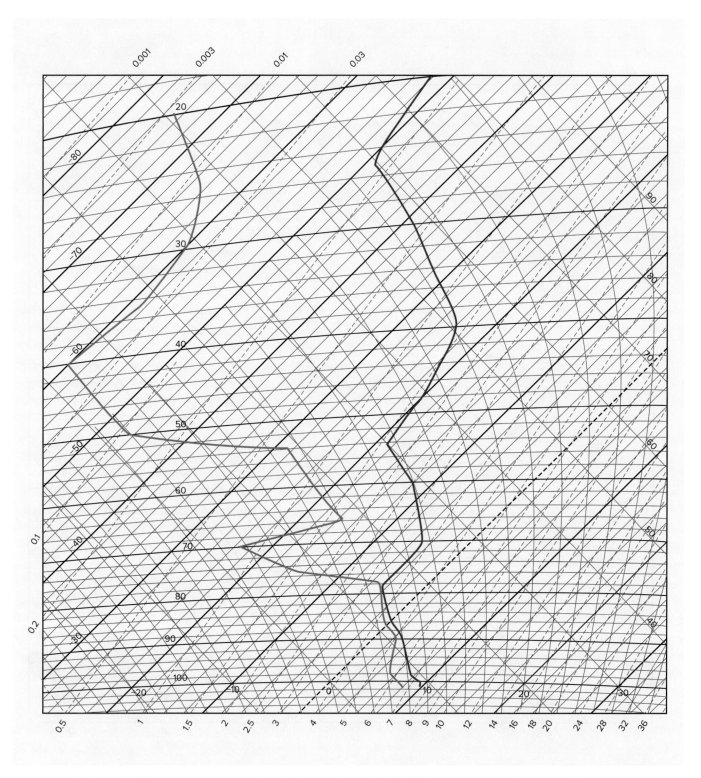

FIGURE 8.10 | An atmospheric sounding for Port Hardy on Vancouver Island, as observed on 7 June 2011 at 1200 UTC. These observations were made using a radiosonde and have been plotted on a tephigram. The red line shows the change in temperature with height, while the blue line shows the change in dew-point temperature with height. A cloudy layer is indicated where the air temperature and dew-point temperature lines merge. Above this layer, the temperature and dew-point temperature lines diverge, indicating that the air becomes drier with height.

Source: Data from University of Wyoming, Department of Atmospheric Science, http://weather.uwyo.edu/upperair/sounding.html, 4 June 2012. Graph adapted from Maarten Ambaum's tephigram, retrieved from http://www.met.reading.ac.uk/~sws97mha/Tephigram/tephi_b_colour.pdf.

temperature. Notice that these two sets of lines are perpendicular to each other and straight. All other lines are based on these axes. The isobars run horizontally across the graph but, instead of being straight as they are on the emagram, they curve slightly down toward the left. The isohumes slope up to the right, but at a greater angle than the isotherms. Finally, the saturated adiabats are almost vertical in the lower portion of the graph but slope toward the top left with height. Despite these differences, the tephigram is applied as explained above for the emagram.

Question for Thought

If air continued to follow the dry adiabat and the isohume *above* the LCL, what would the thermodynamic diagram indicate? Could this happen?

Remember This

As an unsaturated air parcel rises, it follows the dry adiabat corresponding to its air temperature at the surface, and the isohume corresponding to its dew-point temperature at the surface. This shows that as *unsaturated* air rises, its temperature and dew-point temperature decrease, while its potential temperature and mixing ratio remain constant. The LCL of this air parcel is the point where these two lines intersect.

Above the LCL, the air parcel is saturated. As saturated air rises, it follows a saturated adiabat. This shows that as *saturated* air rises, and the water vapour in it condenses, its temperature and dew-point temperature are equal to each other and decrease together, and its saturation mixing ratio and actual mixing ratio are also equal to each other and decrease together. As the water *vapour* mixing ratio decreases, the amount of *liquid* water in the cloud increases.

8.4 | Stability Types

So far, we have seen how thermodynamic diagrams can be used to study the changes in state of ascending and descending air parcels. When atmospheric soundings are plotted on these diagrams, as shown in Figure 8.10, they can be used to analyze *atmospheric stability*.

Atmospheric stability is a measure of the tendency for air—once it has been disturbed by turbulence (Section 4.4.2)—to move vertically due to temperature differences. We can characterize an air parcel as either *unstable, stable,* or *neutral,* based on the way in which it reacts when it is disturbed. If air that is disturbed continues in the direction of the disturbance, vertical motion is favoured, and the air is said to be *unstable.* If the disturbed air parcel returns to its original position, vertical motion is suppressed, and the air is said to be *stable.* If the disturbed air parcel remains in its new position, the air is said to be *neutral.*

The following analogy should help clarify these three stability types. Imagine a marble balanced on top of a dome (Figure 8.11a). As long as the marble is undisturbed, it will remain there, but the slightest disturbance will cause it to roll off the dome. This marble is in an *unstable* position. Now, imagine a marble sitting in the bottom of a bowl (Figure 8.11b). A slight disturbance could cause the marble to roll up the side of the bowl, but the marble will naturally roll right back to its starting position. This marble is in a *stable* position. Finally, imagine a marble on a flat surface (Figure 8.11c). If this marble is disturbed, it will roll away from its starting point and stop. This marble is in a *neutral* position; it neither keeps moving away from nor returns to its starting point.

In the atmosphere, stability is associated with temperature differences. An undisturbed air parcel will be at the same temperature as its surroundings. If the air parcel is vertically displaced by turbulence, it will likely find itself in a situation where its temperature is different from that of its surroundings (Figure 8.12). There are two possible reasons for this difference in temperature. First, the air parcel itself will experience an adiabatic temperature change resulting from its displacement. Second, the temperature of the new surroundings may be different from the temperature of the original surroundings; this difference is determined by the ELR of the air layer and the direction in which the parcel travelled. Once in its new position, if the air parcel is warmer than its surroundings, it will be less dense and, therefore, have a tendency to rise.

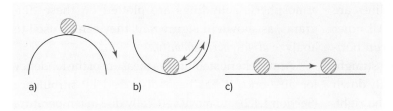

FIGURE 8.11 | a) A marble on a dome is unstable. b) A marble at the bottom of a bowl is stable. c) A marble on a flat surface is neutral.

On the other hand, if it is colder than its surroundings, it will be more dense and have a tendency to sink.

To determine atmospheric stability, we compare the temperature of a displaced air parcel, using the DALR or the SALR, to that of the surroundings, using the ELR. First, let's consider the conditions that produce an *unstable* air layer. Look at Figure 8.13a, and imagine that an unsaturated air parcel is initially at z. If the air parcel is displaced upward from z, it will follow the dry adiabat and become warmer than the surroundings. Being warmer, the air parcel will also be less dense than the surroundings and will continue to rise. If the air parcel is displaced downward from z, it will become colder than the surroundings and continue to sink. Under either of these conditions, vertical motion is favoured; the air layer is unstable. Like the marble on the top of the dome, this air parcel will remain in its initial position only if it is left undisturbed; if disturbed, it will continue to move in the direction of the displacement. Thus, when the temperature change with height in the surroundings (ELR) is greater than the temperature change of a rising or sinking *unsaturated* air parcel (DALR), the air layer is *unstable* for unsaturated air. In the same way, when the ELR is greater than the temperature change of a rising or sinking *saturated* air parcel (SALR), the air layer is *unstable* for saturated air (Figure 8.13b).

Next, let's consider the conditions that would produce a *stable* air layer. In Figure 8.13c, an unsaturated air parcel that is displaced upward from z will become colder than its surroundings and sink back to z. Similarly, if the air parcel is displaced downward from z, it will become warmer than its surroundings and rise back to z. In either case, vertical motion is suppressed; the air layer is stable. Like the marble at the bottom of the bowl, this air has a tendency to return to its original position if disturbed. Thus, when the temperature change with height in the surroundings (ELR) is less than the temperature change of a rising or sinking unsaturated air parcel (DALR), the air layer is *stable* for unsaturated air; similarly, when the ELR is less than the SALR, the air layer is *stable* for saturated air (Figure 8.13d). Furthermore, an inversion, the condition in which the temperature of the surroundings *increases* with height, would result in extremely stable conditions (Figure 8.14).

It is also possible that the ELR could be *equal* to the DALR or the SALR. Under these conditions, displaced air parcels will always find themselves at the same temperature as their surroundings and will, therefore, remain in their new positions after a disturbance, just as the marble on the flat surface remains in its new position after it has been displaced. If the ELR is equal to the DALR, the air layer is *neutral* for unsaturated air, and if the ELR is equal to the SALR, the air layer is *neutral* for saturated air.

The three stability types have been described above in terms of *actual* air temperatures. It can be much simpler, however, to use *potential* temperature. This way, we can depict the DALR graphically as a vertical line because the potential temperature of a rising or sinking air parcel does not change (Figure 8.15).

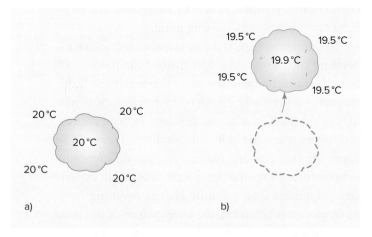

FIGURE 8.12 | a) This undisturbed air parcel is in thermal equilibrium with its surroundings. b) After the air parcel has been displaced from its position in a), its temperature is lower than before because the parcel has experienced an adiabatic temperature change resulting from its displacement. In addition, the temperature of the surroundings has changed. Because the parcel is now warmer than its surroundings, it will have a tendency to rise.

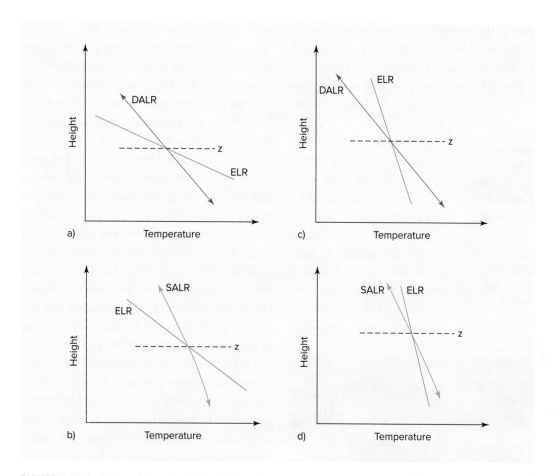

FIGURE 8.13 | a) When the ELR is greater than the DALR, the air layer is unstable for unsaturated air. b) When the ELR is greater than the SALR, the air layer is unstable for saturated air. c) When the ELR is less than the DALR, the air layer is stable for unsaturated air. d) When the ELR is less than the SALR, the air layer is stable for saturated air.

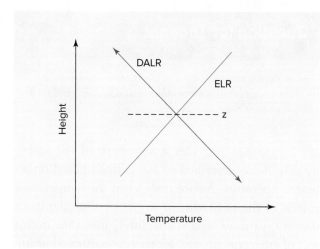

FIGURE 8.14 | An inversion represents an extremely stable air layer because unsaturated air that is displaced upward will be much colder than the surroundings, and unsaturated air that is displaced downward will be much warmer than the surroundings. Such an air layer would also be extremely stable for saturated air.

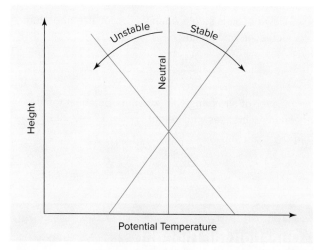

FIGURE 8.15 | Stability types for unsaturated air, where the ELR of the air layer has been expressed in terms of potential temperature. In a neutral air layer, potential temperature is constant with height. In an unstable air layer, potential temperature decreases with height. In a stable air layer, potential temperature increases with height.

It follows that if the potential temperature of the air layer is constant with height, that air layer is neutral. If the potential temperature *decreases* with height, the air layer is *unstable*, and if the potential temperature *increases* with height, the air layer is *stable*.

The use of potential temperature provides a somewhat more intuitive view of atmospheric stability. When graphed using potential temperature, a stable layer of air appears to have warm air over cold. This is clearly a stable situation because warm air is less dense than cold air, so the layer will have a tendency to stay this way. On the other hand, when graphed using potential temperature, an unstable layer of air appears to have cold air over warm. This stratification is clearly unstable because cold air is denser than warm air; the cold air will have a tendency to sink, while the warm air will have a tendency to rise. This air layer is top-heavy.

Potential temperature also helps us understand the temperature structure of the troposphere. Remember that the standard atmosphere shows temperature decreasing with height in the troposphere. This puts warm air below cold air and, since warm air rises, it would seem that air should be continuously and spontaneously rising from Earth's surface. This isn't the case, however, because the density of air is controlled not only by temperature but also by pressure (Section 3.3). As a result, even though surface air is warmer, it is denser because of the great weight of the air above. Recall that when we consider potential temperature, we eliminate the effect of pressure on density. It follows that, under standard conditions, potential temperature *increases* with height in the troposphere; thus, this interpretation puts air with a lower potential temperature below air with a higher potential temperature, resulting in a more stable situation. Air with a higher potential temperature doesn't *feel* warmer, but it *acts* warmer.

ABSOLUTELY UNSTABLE
The condition in which an air layer is unstable for saturated and unsaturated air parcels; this occurs when the ELR is greater than the DALR.

ABSOLUTELY STABLE
The condition in which an air layer is stable for saturated and unsaturated air parcels; this occurs when the ELR is less than the SALR.

Remember This

The following is true for stability based on potential temperature:

- a layer of air is neutral when its potential temperature is constant with height

$$\frac{\Delta\theta}{\Delta z} = 0$$

- a layer of air is stable when its potential temperature increases with height

$$\frac{\Delta\theta}{\Delta z} > 0$$

- a layer of air is unstable when its potential temperature decreases with height

$$\frac{\Delta\theta}{\Delta z} < 0$$

Question for Thought

For a *small* decrease in actual temperature with height, an air layer is stable, even though this means that cold air is lying above warm air. Why is this so?

Question for Thought

How could you use the standard atmosphere to confirm that potential temperature increases with height in the troposphere?

Figure 8.16 provides a summary of the stability conditions described so far, defined based on actual temperature. Notice that, when the temperature profile of the environment is greater than the DALR (ELR > Γ_d), the air layer is **absolutely unstable**. In this sort of environment, any air parcel—saturated or unsaturated—that is displaced upward will be warmer than the surrounding air, so vertical motion will be favoured (Figure 8.17). Conversely, when the temperature profile of the environment is less than the SALR (ELR < Γ_s), the air layer is **absolutely stable**. In this sort

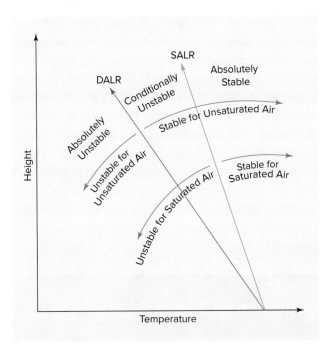

FIGURE 8.16 | A layer of air is either *absolutely unstable*, *conditionally unstable*, or *absolutely stable*, depending on the relationship between the ELR and the DALR or the SALR.

of environment, any air parcel—saturated or unsaturated—that is displaced upward will be colder than the surrounding air, so vertical motion will be suppressed (Figure 8.18).

In addition, notice that there is an area of Figure 8.16, between the DALR and the SALR, that is *stable* for unsaturated conditions but *unstable* for saturated conditions. Thus, an air layer characterized by an ELR that falls within this area of the graph is **conditionally unstable**. In other words, this air

Remember This

The ELR of an air layer determines the stability of that air layer.

- If the ELR is greater than the DALR (ELR > Γ_d), the air will be *absolutely unstable*.
- If the ELR is greater than the SALR but less than the DALR (Γ_s < ELR < Γ_d), the air will be *conditionally unstable*.
- If the ELR is less than the SALR (ELR < Γ_s)—and this includes the possibility that the ELR is an inversion—the air will be *absolutely stable*.

layer is unstable based on the *condition* that it becomes saturated. Such a possibility arises because there is a greater chance that an air layer will be unstable when it is saturated than when it

CONDITIONALLY UNSTABLE
The condition in which an air layer is unstable on the condition that it becomes saturated; this occurs when the ELR is greater than the SALR but less than the DALR.

FIGURE 8.17 | A California condor takes advantage of the rising currents of warm air associated with an unstable atmosphere.

FIGURE 8.18 | Under stable conditions, the accumulation of aerosols can create haze such as that seen here along the Toronto waterfront. Under unstable conditions, these aerosols would be easily dispersed.

What is the stability type if potential temperature is constant with height? What is the stability type if actual temperature is constant with height? Why?

LEVEL OF FREE CONVECTION (LFC)
The height at which air can rise due to its own buoyancy.

is unsaturated. The reason for this is that the release of latent heat in a rising saturated air parcel increases the likelihood that a displaced *saturated* air parcel will be warmer than its environment.

Conditional instability is depicted in Figure 8.19. This diagram shows that if saturated air is displaced upward from z, it will follow the saturated adiabat and become warmer than the surroundings; in this case, the layer of air is *unstable* for saturated air. On

the other hand, if dry air is displaced upward from z, it will follow the dry adiabat and become colder than the surroundings; in this case, the layer of air is *stable* for unsaturated air. If, however, the dry air is displaced enough to reach its LCL, it will begin to condense and cool more slowly, following a saturated adiabat. If it continues to be forced far enough beyond its LCL, this air will eventually reach its **level of free convection** (LFC), at which point it will become warmer than the surroundings and continue to rise on its own. Notice that this instability is possible only because the initially unsaturated air became saturated.

What factors could influence the height of the level of free convection? (Refer to Figure 8.19.)

We can also use the marble analogy to understand conditional instability (Figure 8.20). In this case, the marble begins in a "valley." If it is displaced to point A, it will simply roll back to its starting point. This would be analogous to air that remains unsaturated. If, however, the marble is forced to point B, it will continue on its own. This is analogous to what would happen if the air is forced to its LCL, so that it begins to cool more slowly, then beyond to its LFC.

Conditional instability is the most common state in the troposphere. This is clear from the standard atmosphere, which indicates that the *average* lapse rate through the troposphere is 6.5°C/km—a rate that falls between the DALR and the SALR. However, without strong forcings, such as a cold front or a rise in topography, conditionally unstable air will not produce the vigorous convection characteristic of instability. (In Section 14.4.3, we will examine the importance of conditional instability in the formation of thunderstorms.)

As mentioned in Section 8.2, the ELR is very unlikely to be constant throughout the troposphere. Instead, ELRs will exhibit a great deal of variability with height, as shown in Figure 8.10. This means

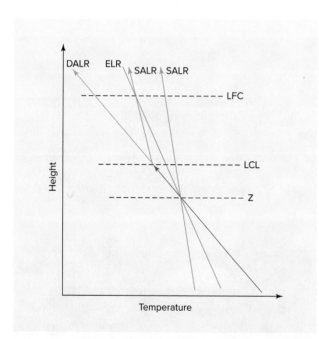

FIGURE 8.19 | If *saturated* air is displaced slightly from z, it will behave as though the air layer is unstable. If *unsaturated* air is displaced slightly from z, it will behave as though the air layer is stable. If *unsaturated* air is forced to its LCL, it will follow a dry adiabat. Beyond this point, the air that was initially unsaturated will cool more slowly, following a saturated adiabat. If the air reaches its LFC, it will continue to rise on its own.

that the determination of stability is not as simple as the rules given above might imply (Figure 8.16). Figure 8.21 depicts an atmospheric sounding showing some variability. In the lowest part of the sounding, the ELR is greater than the DALR; in the middle part of the sounding, the ELR is equal to the DALR; and in the upper part of the sounding, the ELR is an inversion. The rules from above would imply that the lowest layer is unstable, the middle layer is neutral, and the upper layer is very stable. Notice, however, that if air were to rise from the surface, it would continue to be warmer than the surroundings until well into the inversion layer. Therefore, the whole layer through which air from the surface can rise is unstable.

To determine stability for an atmospheric sounding plotted on a thermodynamic diagram, we start by finding unstable layers. For this we imagine that an air parcel is rising from each local maximum, or sinking from each local minimum. If the air is unsaturated, we use a dry adiabat; if the air is saturated, we use a saturated adiabat. For the entire depth through which this air parcel could conceivably rise, or sink, the layer is designated unstable. Once unstable layers are identified, we look for neutral layers by finding layers where the adiabats are parallel to the sounding. Finally, any layers not identified as either unstable or neutral are stable. This technique is illustrated in Example 8.10.

8.5 | Factors Influencing Stability

8.5.1 | Unstable Air

Unstable conditions occur when temperature decreases rapidly with height in the atmosphere. Such conditions lead to a well-mixed, turbulent atmosphere. When the atmosphere is unstable, rising air tends to favour the formation of **cumuliform clouds** (Section 9.3.4). Winds are most often gusty, as strong mixing carries faster moving winds from higher above down toward the surface. Under unstable conditions, pollutants are easily dispersed; smoke from smokestacks can be seen looping up and down (Section 17.1.2). In addition, strong heat transfer away

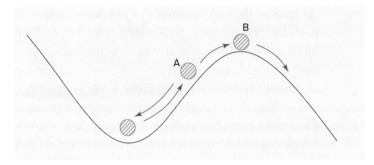

FIGURE 8.20 | If the marble is displaced to A, it will roll back to where it started. If the marble is displaced to B, it will keep rolling away from its starting position.

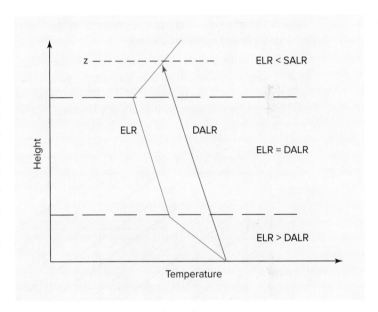

FIGURE 8.21 | A temperature profile in which the ELR varies with height. Because air rising from the surface would be warmer than the surrounding air all the way to z, this layer is unstable, although the ELRs of the two upper layers might suggest otherwise.

from Earth's surface can cause optical phenomena, like mirages or shimmering. It can also cause dust devils to rise and swirl in dry conditions (Figure 8.22).

One way in which an air layer can become unstable is through the warming of the air near the surface (Figure 8.23a). Heating of the surface by the sun can make the lowest layer of the atmosphere unstable. These conditions are best met during days when the sun angle is high and the skies are clear.

CUMULIFORM CLOUD
Cloud that is heaped in form and often exhibits strong vertical development.

As long as the sun continues to provide a source of heat for the surface, unstable conditions will be maintained. Otherwise, a negative feedback operates, through which warm air rising from the surface eventually reduces the instability. Over warm tropical oceans, the air is almost always unstable because the oceans provide a continuous source of heat from below (Section 13.4). Warming of near-surface air can also occur as air moves over a warmer surface. This often happens in the winter as cold stable air moves southeastward over the Great Lakes, creating what are known as *lake-effect snows* (Section 13.3).

Another way in which an air layer can become unstable is through the cooling of the air aloft (Figure 8.23b). This can occur due to the advection of colder air aloft, or the cooling of cloud tops as they emit longwave radiation to space. Such processes can create deep unstable layers that are conducive to the formation of thunderstorms (Section 14.4). In fact, while thunderstorms are most common

Example 8.10

An atmospheric sounding has been plotted on the thermodynamic diagram shown below. Identify unstable, neutral, and stable layers on this sounding.

First, find unstable layers by following dry adiabats upward from each large dot. Wherever the dry adiabat shows that a rising air parcel will be warmer than the surroundings, the air is unstable. The only portion of the sounding for which this is true is the layer from 100 to 85 kPa. Next, find neutral layers by looking for sections where the slope of the sounding is the same as that of a dry adiabat. This is true from 66 to 60 kPa. All the other layers are stable.

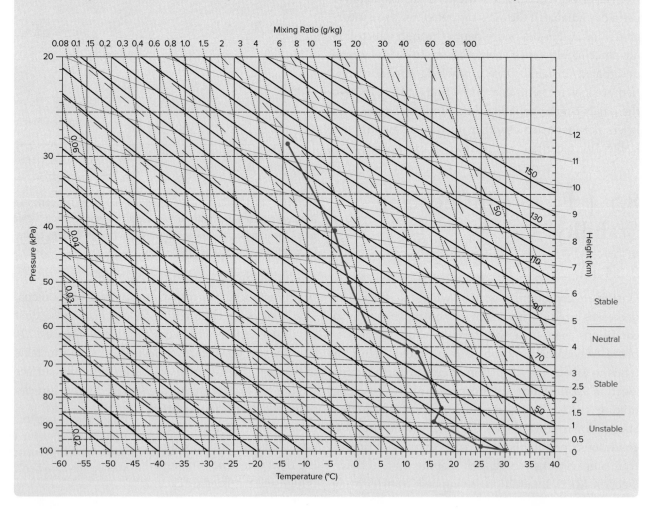

FIGURE 8.22 | This dust devil—spiralling up from a flat, dry surface in Namibia—indicates an unstable atmosphere.

Example 8.11

Suppose that an air layer at the surface has a depth of 1 km. The temperature at the bottom of the air layer is 28°C, and the temperature at the top of the air layer is 20°C. How will the ELR of this air layer change if the surface air warms 3°C?

Initially, the ELR is 8°C/km. This makes the air conditionally unstable because the ELR falls between the SALR and the DALR. After warming, the ELR will be 11°C/km, and the air will be absolutely unstable because the ELR will be greater than the DALR.

on warm afternoons due to the instability created when the sun heats the surface, the instability created by the cooling of cloud tops can be a cause of nighttime thunderstorms.

An air layer may also be destabilized if it is lifted (Figure 8.24). For example, an air layer might rise at fronts, or it might rise due to topography. Just as a tiny *parcel* of air will expand as it rises in the atmosphere, a much larger *layer* of air will also expand as it rises. Essentially, the air layer *stretches*. It follows that air at the top of the layer will rise further than air at the bottom of the layer will. Therefore, although the entire air layer will be cooling at the DALR as it rises, the top of the layer will cool more than the bottom will, simply because it rises farther. As a result, the ELR of the

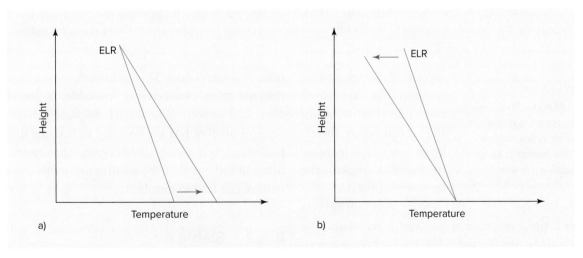

FIGURE 8.23 | a) Warming of air near the surface can destabilize the air layer. b) Cooling of air aloft can also destabilize the air layer. In both cases, the destabilization is represented by an increase in the ELR.

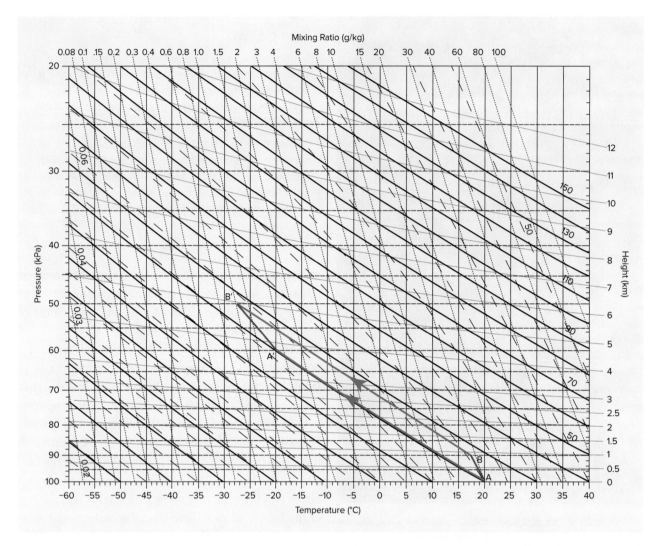

FIGURE 8.24 | An air layer at the surface extends from 100 to 90 kPa. The ELR of this air layer is indicated by the line AB. As the air layer is lifted, the bottom of the layer cools along the dry adiabat AA′, while the top of the layer cools along the dry adiabat BB′. After lifting, the air layer extends from 60 to 50 kPa. Notice that the air layer is now thicker, indicating that the top of the layer rose, and cooled more, than the bottom of the layer did. The new ELR, represented by the line A′B′, shows a greater change in temperature with height, thus a less stable air layer.

CONVECTIVELY UNSTABLE
The condition of an air layer in which the lower air is moist and the upper air is dry, so that the layer has the potential to become unstable if it is lifted.

layer will change so that temperature decreases more quickly with height than it did before the air layer rose. This increase in the ELR makes the layer more unstable.

The destabilization of a lifted air layer will be even greater if the lower part of the layer is moist and the upper part is dry (Figure 8.25). This is because once the moist lower part of the layer becomes saturated, it will cool even more slowly than the top due to the

release of latent heat. This condition, most commonly referred to as **convectively unstable**, is sometimes described as *potentially unstable* because it refers to a layer of air that has the *potential* to become unstable because of its moisture distribution; it just needs to be lifted. It follows that the addition of moisture at the surface can lead to instability.

8.5.2 | Stable Air

Stable conditions occur when temperature decreases very gradually with height in the atmosphere.

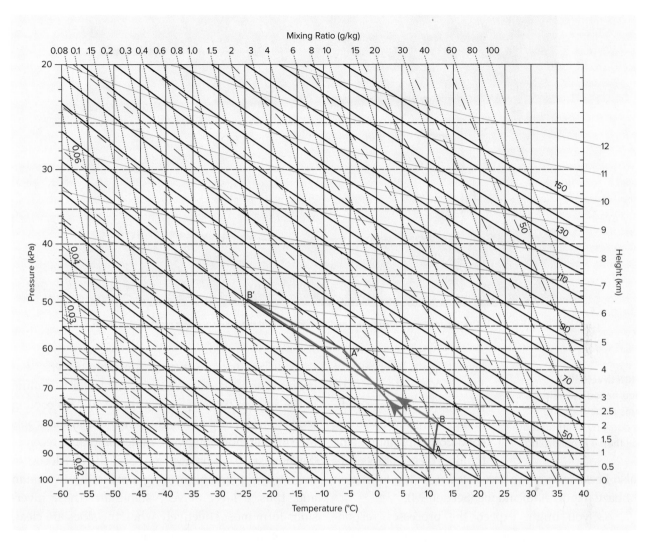

FIGURE 8.25 | The air layer **AB**, extending from 90 to 80 kPa, is absolutely stable, and the bottom of the layer is saturated. As it is lifted, the bottom of the layer cools along the saturated adiabat **AA'**, while the top of the layer cools along the dry adiabat **BB'**. As a result, the bottom of the layer cools much more slowly than the top does, and the air layer is absolutely unstable after lifting.

Remember This

The following processes can destabilize an air layer:

- radiative heating of the surface;
- passage of air over a warmer surface;
- advection of warm air below cold air;
- radiative cooling of cloud tops;
- cold air advection aloft; and
- lifting of an air layer, especially if the bottom of the layer is moist.

However, the strongest stability occurs with inversions, or increases in temperature with height in the atmosphere. These conditions inhibit thermal turbulence—in fact, when it is stable, the atmosphere is unlikely to be turbulent at all, unless there is strong wind shear. With stable conditions, **stratiform** clouds are favoured (Section 9.3.5). In addition, winds are usually light, and pollutants are not easily dispersed—smoke from smokestacks can be seen to rise a little and then spread in a horizontal

STRATIFORM CLOUD
Cloud that is layered in form.

FIGURE 8.26 | The horizontal spread of this smoke indicates that the atmosphere is most likely stable.

RADIATION INVERSION
A surface-based inversion that forms as air near the surface cools by emitting more radiation than it is absorbing.

layer (Figure 8.26; Section 17.1.2). Visibility can be poor with stable conditions, partly because these conditions promote the buildup of pollutants, but also because they are associated with the formation of haze (Figure 8.18) and fog (Section 9.4).

As you might expect, the processes responsible for forming stable layers are opposite to those responsible for producing unstable layers. For instance, an air layer can become stable due to cooling of the air near the surface (Figure 8.27a). This cooling tends to take place overnight, when the surface cools by emitting more longwave radiation than it receives (Section 6.6). This process will cool the air closest to the surface, possibly forming a **radiation inversion** and, thus, a layer of very stable air. These inversions form most effectively when the skies are clear, because with clear skies longwave radiation from

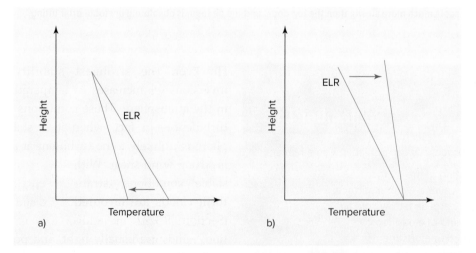

FIGURE 8.27 | a) Cooling of air near the surface can make the air layer more stable. b) Warming of air aloft can also make the air layer more stable. In both cases, the stabilization is represented by a decrease in the ELR. How could the processes depicted here eventually produce inversions?

Example 8.12

Recall that the bottom of the air layer described in Example 8.11 warmed during the day to 31°C. In the late afternoon, it begins to cool. If the surface air cools to 15°C by sunrise, how will the stability change? (Assume that the temperature at a height of 1 km remains at 20°C.)

Initially, the ELR is 11°C/km, and the air is unstable. After cooling, the temperature will *increase* with height at a rate of 5°C/km, and the air will be very stable because the ELR will be an inversion.

the atmosphere, $L\downarrow$, is small. It also helps if winds are light, preventing forced convection from mixing warm air down from above. Stable surface air created by radiative cooling over snow- and ice-covered surfaces can even be maintained during the day; this is why cold air masses that form in polar latitudes are usually very stable (Section 13.3). In addition, a stable air layer can form if warm air is advected over a cold surface and then cooled from below. Both radiative cooling of surface air and the advection of warm air over cold surfaces are processes that can produce fogs (Section 9.4).

Warming aloft can also stabilize an air layer (Figure 8.27b). Remember that the inversion of the stratosphere results from radiative warming in its upper portions due to the presence of the ozone layer. In the troposphere, warm air can be advected over cold air at fronts. Whether the front is warm or cold, the warm air always moves over the colder air below. This creates a **frontal inversion** above the surface.

Another type of upper-air inversion, the **subsidence inversion**, can form when a layer of air sinks, as happens in areas of high pressure (Figure 8.28). While a rising layer of air will *stretch*, a sinking layer of air will *shrink* so that the top of the layer will sink more than the bottom of the layer will sink. This means that the top of the layer will warm more than will the bottom, potentially forming an inversion. This type of inversion usually occurs at heights of about 1 to 2 km above the surface, where the air from above meets with warm air rising from below.

Whether they form at the surface or aloft, inversions influence atmospheric stability. Upper-air inversions can produce "caps." A cap can prevent

surface-based convection from reaching higher into the atmosphere and, as a result, it tends to keep heat, moisture, and pollutants trapped below it. In contrast, surface-based inversions prevent convection from the surface completely.

Remember This

The following processes can stabilize an air layer:

- radiative cooling of the surface;
- passage of air over a colder surface;
- advection of cold air below warm air;
- warm air advection aloft; and
- sinking of an air layer.

Question for Thought

How is the stability of the troposphere different from the stability of the stratosphere?

8.5.3 | Neutral Stability

For a layer of the atmosphere to be of **neutral stability**, the ELR for that layer must be the same as, or at least very close to, the DALR (if the air is unsaturated) or the SALR (if the air is saturated). This can happen when skies are overcast or when it is windy. With overcast skies, the surface will not be strongly cooled or heated. Remember that cooling and heating of the surface are two of the conditions responsible for creating stable and unstable layers. With winds there will be mixing and, as shown in Section 8.1.1, mixing can

FRONTAL INVERSION
An upper-air inversion that forms at a front because warm air rides over colder air.

SUBSIDENCE INVERSION
An upper-air inversion that commonly forms in the subsiding air of a high-pressure area, as air sinks from above and warms.

NEUTRAL STABILITY
The condition of an air layer in which the ELR is equal to the DALR when the air is dry or to the SALR when the air is saturated, so that an air parcel that is displaced will be the same temperature as the surrounding air.

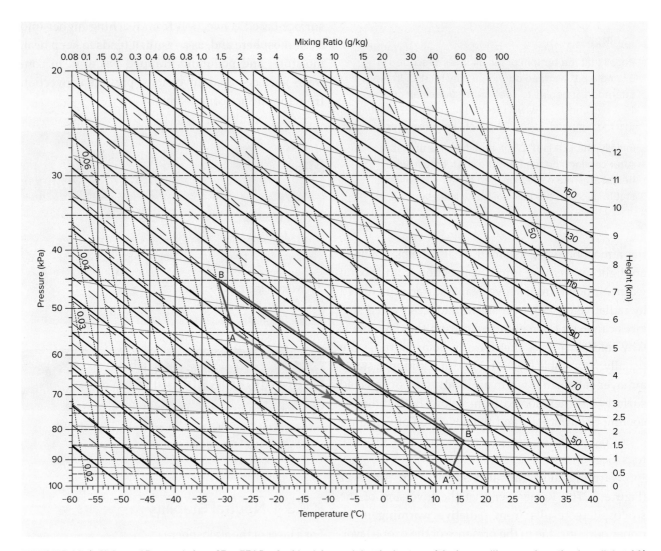

FIGURE 8.28 | Air layer **AB** extends from 65 to 55 kPa. As this air layer sinks, the bottom of the layer will warm along the dry adiabat **AA′**, while the top of the layer will warm along the dry adiabat **BB′**. After sinking, the air layer extends from 95 to 85 kPa, and the ELR of the air layer is now an inversion.

produce an ELR that is the same as the DALR, a condition indicative of neutral stability. It follows that strong stability or instability will not develop when it is overcast or windy. Under neutral or stable conditions, only mechanical convection can occur, but under unstable conditions, both thermal and mechanical convection can occur.

Question for Thought

If the ELR equals the DALR, why will the ELR not be affected by mixing?

8.6 | Stability Conditions over a Clear Day

The above discussion shows that, under *clear-sky conditions*, surface warming by day can lead to unstable conditions in the lowest layer of the atmosphere, while surface cooling by night can lead to stable conditions in the lowest layer of the atmosphere. In addition, instability during the day can lead to the formation of a **mixed layer** that can extend

MIXED LAYER
A turbulent air layer that extends from near the surface to an upper-air inversion; the temperature lapse rate through most of this layer is close to the DALR.

through the lowest 1 or 2 km of the atmosphere. As its name suggests, this is a turbulent layer characterized by strong mixing.

Under clear-sky conditions, the surface radiation inversion generally reaches its maximum depth, probably no more than 100 m, just before sunrise. As the day progresses, this inversion gradually disappears, from the bottom up, as the sun's heating replaces it with an unstable layer in which the ELR is greater than the DALR (Figure 8.29a). On a warm day, this unstable layer may reach up to about 100 m in depth, and heat rising from this layer will warm the overlying air and make it turbulent. The *mixing* created by this turbulence produces a layer with a neutral temperature profile lying directly above the unstable surface layer. This is the *mixed layer*, throughout which the potential temperature, humidity, and wind speed remain constant with height. The upper boundary of the mixed layer is a subsidence inversion that creates a cap, keeping pollutants trapped below. Therefore,

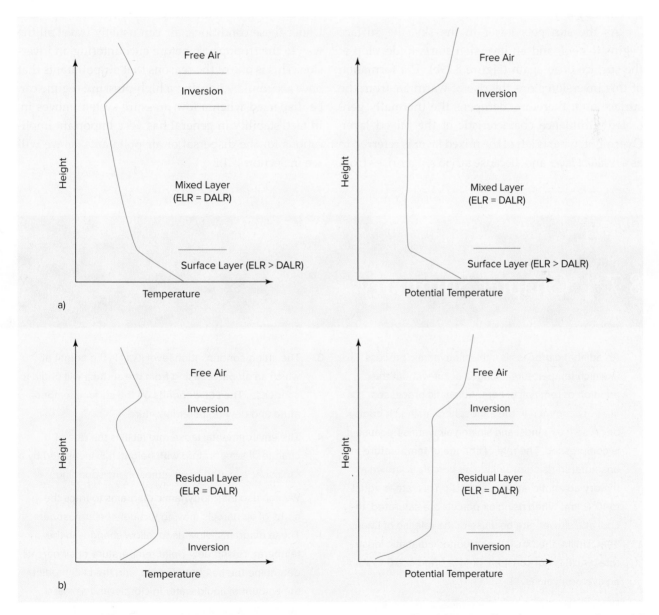

FIGURE 8.29 | a) The daytime profiles of temperature and potential temperature. **b)** The nighttime profiles of temperature and potential temperature. Notice that both the mixed layer and the residual layer have neutral temperature profiles (ELR = DALR). Why, then, is the mixed layer described as being unstable, while the residual layer is described as being neutral?

the deeper is the mixed layer, the lower will be pollutant concentrations. The depth of the mixed layer can grow throughout the day as the air warms and as *free* air is mixed downward from *above* the inversion.

During the day, therefore, a typical fair-weather profile consists of a shallow unstable surface layer, overlain by a mixed layer, and topped by an inversion. Although the ELR of the mixed layer would suggest neutral stability, air parcels can rise all the way from the surface before being stopped in the inversion. Thus, the entire air layer below the inversion is unstable by day.

As the sun gets lower in the sky, the surface begins to cool, and an inversion starts to develop at the surface once again (Figure 8.29b). The formation of this inversion stops the rise of warm air from the surface and, therefore, dampens the thermally generated turbulence characteristic of the mixed layer. Overnight, what is left of the mixed layer is referred to as a *residual layer* and, because air no longer rises from

the surface, it now acts as a neutral layer. The characteristic nighttime temperature profile is one in which there are two inversions: one at the surface below the residual layer, the other atop the residual layer.

The above description is most representative of summertime, high-pressure conditions. In winter, the nighttime inversion is deeper, and the daytime mixed layer is shallower. With low-pressure conditions, things are quite different. The rising air associated with low-pressure areas is strong enough to push through the capping inversion, destroying it and carrying heat, moisture, and pollutants upward. Under these conditions, air can usually travel all the way to the tropopause before encountering an inversion. This is one of the reasons that air pollutants that have accumulated during a high-pressure regime can be dispersed when a low-pressure system moves in. In fact, stability in general has very important implications for the dispersal of air pollutants, as we will see in Section 17.1.2.

8.7 | Chapter Summary

1. An adiabatic process is a thermodynamic process in which temperature changes occur without the addition or removal of heat. Adiabatic processes are very common in the atmosphere; rising air cools because it expands, and sinking air warms because it is compressed. The rate of change of temperature in unsaturated rising or sinking air parcels is known as the dry adiabatic lapse rate (DALR). This rate is equal to 10°C/km. When rising air parcels are saturated, they cool at a slower rate because of the release of latent heat. Unlike the DALR, the saturated adiabatic lapse rate (SALR) is not constant, but we often use 6°C/km as an average value.

2. Potential temperature is the temperature an air parcel would have if it were brought to a pressure of 100 kPa. Potential temperature remains constant for vertical motions.

3. The lifting condensation level (LCL) is the height at which an air parcel rising from the surface will begin to condense. The LCL depends on the surface temperature and dew-point temperature.

4. The environmental lapse rate (ELR) is the rate of change of temperature with height. It is measured by a radiosonde as part of an atmospheric sounding.

5. We can use thermodynamic diagrams to trace the paths of air parcels moving through the atmosphere. These diagrams allow us to follow changes in the temperature and dew-point temperature of air parcels, determine the height of the LCL and the LFC, predict the amount of liquid water in clouds, and assess atmospheric stability.

6. The environmental lapse rate (ELR) of an air layer determines the stability of that air layer. In an *absolutely*

unstable layer, the ELR is greater than the DALR, and vertical motions are favoured. In an *absolutely stable* layer, the ELR is less than the SALR, and vertical motions are suppressed. We can use potential temperature to describe the stability of unsaturated air: if potential temperature increases with height, the air is stable; if potential temperature decreases with height, the air is unstable; and if potential temperature is constant with height, the air is neutral.

7. If the ELR falls between the DALR and the SALR, the atmosphere is *conditionally unstable*. In this sort of atmosphere, unsaturated air is stable, but saturated air is unstable. This is because if air is forced to the LCL, it will begin to cool more slowly and, if it is forced further, will eventually be warmer than the surroundings. At this point, known as the level of free convection (LFC), the air can rise on its own.

8. Any processes that cause the temperature to decrease more rapidly with height will make an air layer more unstable. These processes include any mechanisms that will warm the air near the surface or cool the air aloft. In addition, an air layer can be destabilized by lifting because the top will cool more than the bottom will cool. This is particularly true for air layers that contain more moisture at the bottom than they do at the top. The potential instability of such an air layer is called *convective instability*.

9. Any processes that cause the temperature to decrease less rapidly with height, or produce an inversion, will make an air layer more stable. These processes include any mechanisms that will cool the air near the surface or warm the air aloft. In addition, an air layer can become more stable by sinking because the top will warm more than the bottom will warm. The formation of inversions will make the air strongly stable. Surface radiative cooling can produce radiation inversions overnight. Upper-air inversions can form at fronts as warm air rides up and over cold air, or they can form as air sinks in regions of high pressure.

10. Under warm and clear conditions, the lowest 1 or 2 km of the atmosphere acts as an unstable layer by day. The lowest 100 m or so of this layer is characterized by an ELR that is greater than the DALR. Above that is the *mixed layer*, where the ELR is equal to the DALR. This layer is capped by an inversion. On clear, calm nights, a surface inversion forms that prevents air from rising from the surface. As a result, mixing stops, and the mixed layer becomes a residual layer.

Review Questions

1. Why does an air parcel cool as it rises and warm as it sinks? Why does saturated air cool more slowly than unsaturated air?

2. What is the lifting condensation level? How is it related to the surface temperature and the dew-point temperature?

3. If an *unsaturated* air parcel rises adiabatically, how will *temperature, vapour pressure, relative humidity, dew-point temperature, absolute humidity, mixing ratio*, and *potential temperature* be affected? Will these properties increase, decrease, or stay constant? How will the same properties be affected if a saturated air parcel rises adiabatically? (You may want to draw a chart to compare your answers for the unsaturated parcel to those for the saturated parcel.)

4. What is the relationship between the different stability types and the environmental lapse rate (ELR)?

5. How can we use potential temperature to distinguish between stability types?

6. In a conditionally unstable atmosphere the ELR falls between the DALR and the SALR. Under what conditions will thermal turbulence develop in such an air layer?

7. How are stable atmospheric conditions different from unstable atmospheric conditions?

8. How might an air layer become a) unstable or b) stable?

9. How can winds and cloud cover influence stability?

10. How are stability conditions in the day similar to and different from stability conditions at night?

Problems

1. Calculate the potential temperature in each case. Check your answer using the blank emagram in Figure 8.6. (You may want to scan or photocopy the emagram so that you can mark the points on the graph.)
   a) $T = 21°C$, $P = 94$ kPa
   b) $T = 0°C$, $P = 76$ kPa

2. Air at the surface has a temperature of 7°C. The LCL of this air is 1.2 km. If this air rises to a height of 3 km, what will be its temperature? Assume the SALR is 0.6°C/100 m.

3. Use Equation 8.5 to calculate the LCL for the air parcel in examples 8.7 and 8.8. Use the height scale on the thermodynamic diagram used in these examples to compare your answer to that determined in Example 8.8.

4. Use the thermodynamic diagram in Figure 8.6 to find the missing values in each case below.
   a) $T = 12°C$, $P = 92$ kPa, $T_d = 7°C$, $r_s = $ _____, $r = $ _____, $RH = $ _____, $LCL = $ _____
   b) $T = 24°C$, $P = 100$ kPa, $r = 5$ g/kg, $T_d = $ _____, $r_s = $ _____, $RH = $ _____, $LCL = $ _____

5. At 88 kPa, the temperature is 9°C and the dew-point temperature is –1°C. Use the thermodynamic diagram in Figure 8.6 to answer the following questions.
   a) If this air rises to 80 kPa, what will be its temperature, dew-point temperature, and relative humidity?
   b) What is the LCL for this air? What are the temperature and dew-point temperature at the LCL? Check these values by calculating them.
   c) What is the wet-bulb temperature at 88 kPa?
   d) If this air continues to rise to 55 kPa, what will be its temperature, dew-point temperature, and liquid water content?
   e) If 1 g/kg of water falls as precipitation and the air then descends, what will be the temperature and dew-point temperature at 90 kPa? What will be the relative humidity?
   f) If all the water had precipitated out, what would be the temperature and dew-point temperature in e)?

6. Air at 90 kPa on the windward side of the Rockies has a temperature of 10°C and a dew-point temperature of 5°C. This air rises to the summit, the pressure at which is about 60 kPa. If all the water precipitates out on the windward side of the mountains, what will be the temperature, dew-point temperature, and relative humidity at Calgary, where the pressure is about 90 kPa? Use the thermodynamic diagram in Figure 8.6 to answer this question.

Suggestion for Research

1. Find examples of types of thermodynamic diagrams other than those discussed in this chapter. Compare the various types, and suggest reasons for the variety of different diagrams.

2. Search the Internet for examples of atmospheric soundings for a variety of times and locations. Look for relationships between these soundings and the season, the large-scale atmospheric conditions, and the location.

9

Condensation

Learning Goals

After studying this chapter, you should be able to

1. *distinguish* between homogeneous nucleation and heterogeneous nucleation;

2. *distinguish* between the curvature effect and the solute effect, and *explain* how together these effects influence the formation of water droplets;

3. *appreciate* the significance of atmospheric aerosols to the formation of water droplets and ice crystals in the atmosphere;

4. *list* the five main ways in which air can rise to form clouds;

5. *describe* the classification system used to name clouds, and *apply* this system to *recognize* major cloud types;

6. *distinguish* between cumuliform clouds and stratiform clouds; and

7. *explain* how fogs form.

Clouds—the visible result of condensation in the atmosphere—can take a variety of forms. This aerial view shows the heaped tops of cumuliform clouds as well as the very different thin, fibrous streaks of higher cirriform clouds in the distance.

In Chapter 7, you learned what it means for the air to be saturated with water vapour. In theory, when saturation is reached, the water vapour should *condense*. This chapter is about condensation in the atmosphere, where water condenses on aerosols to form the water droplets that make up clouds or fogs. Throughout the first part of this chapter, we will study the rather complex factors that operate at the microscopic scale to influence the formation of water droplets in the atmosphere. In the second part, we will focus on the clouds and fogs that are the visible result of condensation. First you will learn about the classification system used to name clouds; then, you will discover some of the general conditions under which clouds and, finally, fogs form.

HOMOGENEOUS NUCLEATION
The formation of water droplets by the chance collision of water vapour molecules.

HETEROGENEOUS NUCLEATION
The formation of water droplets on a nucleus; in the atmosphere, these nuclei are aerosols.

9.1 | Formation of Cloud Droplets

In Section 7.3, it was shown that saturation refers to a state of equilibrium between water vapour and a surface of liquid water. Saturation vapour pressures, as given in Table 7.1, were for *flat* surfaces of *pure* water. In the atmosphere, liquid water is in the form of droplets. Such droplets have curved, rather than plane, surfaces. In addition, the water in these droplets is not always pure; instead, it is often in the form of a solution. This is because, in the atmosphere, water often condenses on soluble substances. For curved surfaces of water containing dissolved substances, saturation vapour pressures are different than those given in Table 7.1. Because of this, the formation of water droplets in the atmosphere is not straightforward; water droplets don't just spontaneously form when the relative humidity reaches 100 per cent.

CURVATURE EFFECT
The effect in which increased curvature of a droplet's surface, as the droplet gets smaller, increases the relative humidity required for the droplet to be in equilibrium with its surroundings.

Once water droplets form, they may grow into cloud droplets. On average, cloud droplets are about 10 μm in radius. Considering the difficulty of observing microscopic processes at work in the atmosphere, it is not easy to study the formation of cloud droplets. Despite this, studies have been done; the methods used include direct measurement in the atmosphere, observations of the processes at work in cloud chambers, and computer modelling. These studies have revealed that water droplets can form in two ways: through **homogeneous nucleation**, which results in droplets made up of water only, and through **heterogeneous nucleation**, which results in droplets made up of water *and* the substance upon which the water condensed. Of these two processes, only heterogeneous nucleation is responsible for the formation of water droplets in the atmosphere.

9.1.1 | Homogeneous Nucleation

Recall from Section 7.3 that water can evaporate more easily from a curved surface than it can from a plane surface because water molecules are not held as tightly to a curved surface. Because of this, the saturation vapour pressure over a curved surface is greater than that over a flat surface. It follows that in order for air to be saturated for, or in equilibrium with, a curved surface, it must be supersaturated for a flat surface. As water droplets get smaller, their curvature increases, so that there is a relationship between a droplet's size and the relative humidity at which it will be in equilibrium (Figure 9.1). This relationship is known as the **curvature effect**; it was first expressed mathematically by Lord Kelvin. Figure 9.1 shows that as the size of a droplet *decreases*, the relative humidity required for it to be in equilibrium with its surroundings will *increase*. This is because as water droplets get smaller and their curvature increases, they evaporate more easily. Consider a droplet with a radius of 0.05 μm. Figure 9.1 shows that if the relative humidity is 102 per cent—or, put another way, if the supersaturation is 2 per cent—this droplet will be in equilibrium with its surroundings; it will neither evaporate nor grow. If the relative humidity is below 102 per cent, this droplet will evaporate, and if the relative humidity is above 102 per cent, the droplet will grow by condensation. Notice that once a droplet reaches about 1 μm in radius, it begins to act like a plane surface and will be in equilibrium with its surroundings at a relative humidity of 100 per cent. The curvature effect, therefore, applies only for very small droplets.

It is unlikely that droplets formed by the chance collisions associated with homogeneous nucleation will be larger than about 0.01 μm in radius. In fact, they will usually be considerably smaller. The graph shows that a droplet with a radius of 0.01 μm requires a relative humidity of just over 112 per cent to survive. Because the relative humidity in most clouds rarely goes over about 101 per cent, such tiny droplets will quickly evaporate. Thus, water droplets forming by homogeneous nucleation will not survive in the atmosphere. This has been shown in cloud chambers. Once all aerosols have been removed from the air in these chambers, droplets will not survive except at high supersaturations. Therefore, cloud droplets must form by heterogeneous nucleation.

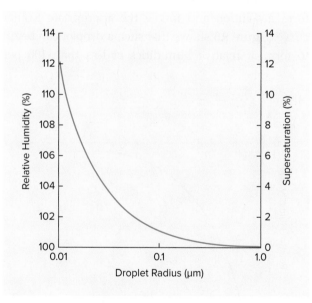

FIGURE 9.1 | As the size of a droplet increases, the relative humidity at which it is in equilibrium with its environment will decrease because the curvature of the droplet's surface decreases with its size. A droplet that falls on the curve is in equilibrium with its environment. A droplet that falls below the curve will shrink by evaporation, while a droplet that falls above the curve will grow by condensation. This relationship is a consequence of the *curvature effect*.

9.1.2 | Heterogeneous Nucleation

In Section 2.9, it was shown that the atmosphere is composed of not only gases but also tiny particles, or aerosols. As it turns out, some of these aerosols are **cloud condensation nuclei (CCN)**. To be a CCN, an aerosol needs to be hydrophilic, or **wettable**. Sometimes, CCN are referred to as "cloud seeds" because it is on CCN that cloud droplets begin to *grow* by the process of heterogeneous nucleation. Unlike water droplets formed by homogeneous nucleation, water droplets formed by heterogeneous nucleation are quite likely to survive and grow in the atmosphere.

> **CLOUD CONDENSATION NUCLEI (CCN)**
> Atmospheric aerosols on which water vapour can condense to form water droplets.
>
> **WETTABLE**
> Able to allow water to form a film.

When water condenses on a wettable aerosol, the water will spread over the aerosol surface so that the droplet resembles a pure water droplet, but such a droplet will be much bigger than one formed through homogeneous nucleation. Imagine that water begins to condense on a wettable aerosol with a radius of about 0.2 μm. Right from the start, this droplet will be larger than one formed by homogeneous nucleation. Figure 9.1 shows that a droplet of this size will need a relative humidity of only about 100.5 per cent to be in equilibrium with its surroundings. If the relative humidity in the cloud is below 100.5 per cent, any water condensing on this aerosol will quickly evaporate, and the aerosol will remain dry. However, since relative humidities of around 100.5 per cent are quite common in clouds, there is a good chance that a droplet of 0.2 μm will not evaporate. In fact, with a relative humidity in the cloud of over 100.5 per cent, this droplet will not only survive, it will also grow.

In some cases, when water condenses on a CCN, a solution will form because the CCN dissolves in the water. This happens with CCN that are **hygroscopic nuclei**. An example is sea salt (NaCl), which is an abundant atmospheric aerosol. Sulphuric acid and nitric acid are

> **HYGROSCOPIC NUCLEI**
> CCN that both attract water and dissolve in it.

also examples of hygroscopic aerosols. Recall from Section 7.3 that, since it is more difficult to evaporate water from a solution than from pure water, the saturation vapour pressure over a solution will be less than it is for pure water. It follows that when the air is saturated for pure water, it will be supersaturated for a solution. This is known as the **solute effect**.

Whereas the curvature effect tends to *inhibit* the growth of very small droplets, the solute effect tends to *promote* the growth of very small droplets. This is because water vapour can condense on solution droplets when the relative humidity is *below* 100 per cent. For the same size droplets, higher relative humidities are required for those formed on insoluble CCN to be in equilibrium than for those formed on soluble CCN to be in equilibrium.

> **SOLUTE EFFECT**
> The effect in which a dissolved substance reduces the relative humidity required for a droplet to be in equilibrium with its surroundings.
>
> **KOHLER CURVE**
> A graph showing the relationship between the size of a solution droplet and the relative humidity required for it to be in equilibrium with its surroundings.

As a solution droplet grows, the solution will become diluted, and greater relative humidities will be required for further growth (Figure 9.2). Things get complicated, though, because once the droplet begins to act like a pure water droplet, it might still be small enough that the curvature effect takes over. If it does, relative humidities *above* 100 per cent will be required for further growth. Figure 9.2, therefore, does not tell the whole story of the growth of a solution droplet. This is because both the curvature effect *and* the solute effect play a role in the growth of tiny solution droplets.

We can get a more accurate sense of how solution droplets grow by consulting a **Kohler curve**, which takes both the curvature effect and the solute effect into account. Different Kohler curves can be drawn for different *types* and *amounts* of solutes, as shown in Figure 9.3. As a droplet grows, it will follow its Kohler curve. A Kohler curve thus shows that the equilibrium relative humidity of a droplet depends not only on its size, but also on the type and amount of solute that it contains.

If a droplet forms on a *hygroscopic* aerosol, it will form a solution and follow the appropriate Kohler curve. Figure 9.3 shows that such a droplet can begin to form at relative humidities of less than 100 per

Remember This

- The *curvature effect* states that a curved surface requires a higher saturation vapour pressure than does a flat surface in order to be in equilibrium with the surroundings.
- The *solute effect* states that a solution requires a lower saturation vapour pressure than does pure water in order to be in equilibrium with the surroundings.

Remember This

CCN encourage the formation of water droplets because

- they make them bigger and
- they can form solutions.

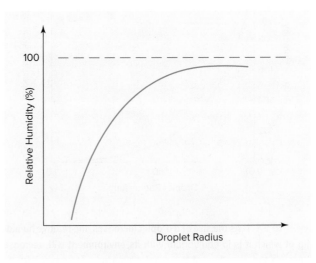

FIGURE 9.2 | As the size of a solution droplet increases, the relative humidity at which it is in equilibrium with its environment will increase because the solution becomes diluted as the droplet grows. This relationship is a consequence of the *solute effect*.

cent. However, as the droplet grows and the solution becomes diluted, greater and greater relative humidities are required for it to be in equilibrium. Eventually, the solution will have been diluted to the point where the droplet acts as a pure water droplet. At this point, the Kohler curve must rise to meet the pure water curve, thus forming the *peak* of the Kohler curve. From this point, the droplet will be influenced by the curvature effect until its curvature is no longer significant.

It is easiest to illustrate the use of Kohler curves through examples. For the first example, we will compare the two curves for droplets containing sodium chloride on Figure 9.3, and we will assume that the relative humidity in the cloud is 100.3 per cent. If water begins to condense on a sodium chloride aerosol with a mass of 10^{-19} kg, the droplet will grow according to its curve until it reaches an equilibrium relative humidity of 100.3 per cent, which is the same as the relative humidity of the surroundings. At this point, the droplet will be just a little bigger than 0.1 µm in radius. The droplet will stay this tiny and form *haze* (Section 2.9).

If, however, water begins to condense on a sodium chloride aerosol with a mass of 10^{-18} kg, the droplet will grow according to *its* curve. In this case, the droplet *will* reach the peak of the curve because, even at the peak, the relative humidity of the surroundings remains higher than the equilibrium relative humidity of the droplet. Because any growth beyond the peak requires *lower* relative humidities, the droplet will grow very rapidly beyond the peak of its curve. Droplets that reach the peak of their Kohler curve are described as *activated*. An activated droplet can continue to grow, and by the time it reaches a size of just over 1 µm in radius it will behave like a plane surface of pure water. This comparison shows that larger hygroscopic aerosols are more favourable to cloud droplet growth than are smaller ones.

We can obtain similar results by comparing the curve for the droplet containing 10^{-19} kg of sodium chloride to the curve for the droplet containing the same mass of ammonium sulphate. This time, we will assume that the relative humidity in the cloud is 100.5 per cent. Now the Kohler curves show that a droplet forming on sodium chloride will become

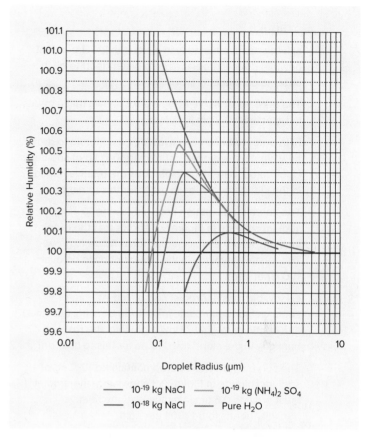

FIGURE 9.3 | Kohler curves for different types and amounts of solutes: one for a solution drop containing 10^{-19} kg of NaCl (sodium chloride), one for a solution drop containing 10^{-18} kg of NaCl, and one for a solution drop containing 10^{-19} kg of $(NH_4)_2 SO_4$ (ammonium sulphate). Also shown is the curve for pure water, which represents the curvature effect alone (Figure 9.1).

activated and continue to grow, while a droplet forming on ammonium sulphate will grow only until it reaches an equilibrium relative humidity that matches the relative humidity of the surroundings, at which point it will form haze. This comparison shows that certain hygroscopic substances are more favourable to cloud droplet growth than are others.

As we have seen, Kohler curves show that when water forms a solution, droplets can begin to form and grow even when relative humidities are *below* 100 per cent. For example, water will condense on salt aerosols when relative humidities are as low as 70 to 80 per cent. Recall that as waves break, they add salt to the air. Water vapour will condense on the salt at relative humidities below 100 per cent, forming a haze over the breaking waves.

Example 9.1

a) If the relative humidity in a cloud is 100.4 per cent, how big does a droplet of pure water need to be in order to continue to grow?

The pure water curve in Figure 9.3 shows that it needs to be at least 0.3 μm in radius.

b) In order to become activated and continue to grow, how big must a droplet containing 10^{-19} kg of sodium chloride become? How high does the relative humidity in the cloud need to be in order for this to happen?

The curve for a solution drop containing 10^{-19} kg of sodium chloride in Figure 9.3 shows that the droplet must grow to 0.2 μm in radius to continue to grow. The relative humidity in the cloud needs to be over 100.4 per cent.

c) How big must a droplet containing 10^{-18} kg of sodium chloride become in order to become activated and continue to grow? How high does the relative humidity in the cloud need to be for this to happen?

The curve for a solution drop containing 10^{-18} kg of sodium chloride in Figure 9.3 shows that the droplet must grow to 0.6 μm in radius, and the relative humidity needs to be 100.1 per cent.

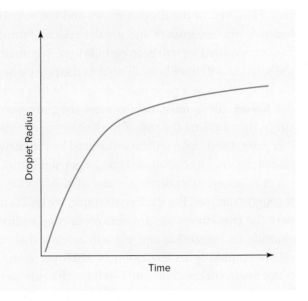

FIGURE 9.4 | When the relative humidity is constant, droplets growing by condensation grow very quickly at first, then much more slowly.

Once a droplet becomes activated, it will continue to grow by **diffusion**. In conditions of *constant* relative humidity, an individual droplet will grow very quickly at first, but its growth will slow as it gets bigger (Figure 9.4). In a cloud, however, the relative humidity will *decrease* as droplets grow because their growth uses up the available water. This creates a limit to how big droplets in a cloud can grow by condensation alone. Since smaller droplets grow faster than bigger droplets, the size of droplets in a cloud tends to become more uniform with time. Remember that the radius of a typical cloud droplet is about 10 μm. For continued growth to raindrop size—which is about 1 mm—other processes take over, as will be shown in Chapter 10.

The *average* size of a cloud droplet may be 10 μm, but the actual size of droplets ultimately depends on the amount of water in the cloud and the number of CCN available. Given the same amount of

DIFFUSION
The movement of water vapour molecules toward a water droplet or ice crystal upon which they condense or are deposited, respectively.

water, more CCN will mean smaller water droplets because with more CCN, there is less water for each. Measurements have shown that there are more, and smaller, CCN over continents than over oceans. It follows that, if continental and marine clouds have roughly the same water content, cloud droplets will be smaller and more numerous over the continents than over the oceans (Figure 9.5).

As noted above, only some aerosols can be CCN. In fact, studies have shown that far less than half of the aerosols in the atmosphere can be CCN (Wallace & Hobbs, 2006, p. 215). Some aerosols are more *favoured* as CCN than are others. Favoured aerosols are those that are wettable, large or giant, and hygroscopic. In the case of those that are hygroscopic, the more soluble they are, the better. Figure 9.3 shows that, because relative humidities in clouds rarely go above 101 per cent, insoluble aerosols will need to be at least 0.1 μm in radius in order to be CCN, while those that are soluble can be smaller. The larger hygroscopic aerosols are favoured for CCN because as air rises and cools, these are the aerosols that water will condense on first.

The exact sources of CCN remain somewhat uncertain and are, therefore, an area of active research. In general, it seems that aerosols produced by combustion are a particularly important source of CCN over

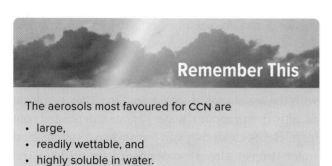

the continents. It is known, for example, that forest fires increase the number of CCN in the atmosphere. Combustion associated with human activity is also a significant, and increasing, source for CCN. In addition, hydrocarbons emitted by vegetation may be an important source of CCN. Finally, the continents are a significant source of dust, but it is less certain how important dust is as CCN because dust is generally non-hygroscopic.

Over the oceans, a source of CCN is sea salt. Salt aerosols are both large and soluble. Recently, it has been recognized that dimethyl sulphide (DMS) gas, produced by phytoplankton, may be an even more important source of CCN over the oceans than salt is. In this case, the CCN are droplets of sulphuric acid

formed from DMS. In general, the conversions to acid droplets of both sulphate and nitrate gases are major natural and anthropogenic sources of CCN over both the continents and the oceans.

In the late 1980s, a group of scientists hypothesized that DMS might produce a negative feedback that would partially counter global warming (see Charlson, Lovelock, Andreae, & Warren, 1987). These scientists thought that as the oceans warmed, there would be an increase in phytoplankton and, thus, an increase in the production of DMS. With more DMS, there would be more CCN and, with more CCN, cloud droplets would be more numerous and, therefore, smaller. As a result, cloud albedo and, thus, the cooling effect of clouds might increase. This hypothesis became known as the CLAW hypothesis, named using the initials of its four authors. One of these authors, James Lovelock (2006), has recently suggested that DMS may, in fact, be part of a positive rather than a negative feedback. Warmer oceans, it turns out, are *less* productive than cooler ones. This is mostly because a warmer surface layer prevents the upwelling of cold, nutrient-containing water from below. With fewer nutrients, there will be less phytoplankton, less DMS, and, ultimately, lower cloud albedos.

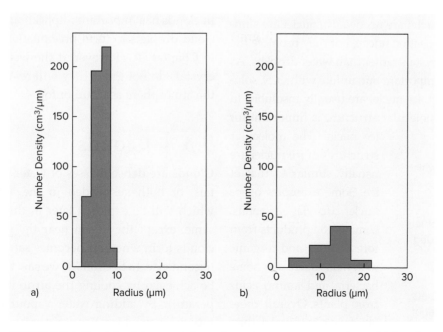

FIGURE 9.5 | Cloud droplets are smaller and more numerous over continents (a) than they are over oceans (b). These graphs show that there are about 10 times more droplets in continental clouds than in marine clouds. Why do you think this is?

9.2 | Nucleation below Freezing

When the temperature drops below 0°C, it might be expected that ice crystals, rather than water droplets, should form. However, just as the formation of water droplets in the atmosphere does not follow expected rules, neither does the formation of ice crystals—water does not necessarily freeze at 0°C. In fact, temperatures must drop to below about −40°C before water droplets will spontaneously freeze to form ice crystals. The *exact* temperature at which they will freeze depends upon their size; larger droplets will freeze at slightly higher temperatures than will smaller droplets. The spontaneous freezing of water droplets is known as *homogeneous nucleation* because it is similar to the homogeneous nucleation of water droplets. Between temperatures of 0°C and −40°C, however, ice crystals grow by heterogeneous nucleation.

For this, **ice nuclei** are needed. Ice nuclei are similar to cloud condensation nuclei, but they promote the formation of ice crystals rather than water droplets. As with CCN, size is important but, unlike with CCN, solubility is not; in fact, ice nuclei are usually insoluble. In addition to size, molecular structure is important for ice nuclei. The molecular structures of ice nuclei are usually similar to that of ice. Some examples of ice nuclei are clay minerals, combustion products from forest fires, and organic materials such as some bacteria and amino acids from plants. Overall, there are fewer ice nuclei than CCN in the atmosphere.

ICE NUCLEI
Atmospheric aerosols on which ice crystals can form.

CLOUD
A dense mass of water droplets and/or ice crystals suspended in the atmosphere.

FOG
Suspensions of water droplets and/or ice crystals in a layer of air at Earth's surface.

Ice nuclei become more effective as temperatures decrease. This means that as temperatures drop from 0°C to −40°C, more and more ice crystals will form. There are three ways in which ice crystals can form with the assistance of ice nuclei. First, they can form directly from vapour if water vapour is deposited onto ice nuclei. Second, they can form when a supercooled water droplet already contains an ice nucleus and temperatures get cold enough. Third, they can form when supercooled droplets collide with ice nuclei.

In clouds with temperatures just below 0°C, there are usually no ice crystals, only supercooled water droplets. At these temperatures, the air is full of water droplets cold enough to freeze; they just need something to freeze onto. Thus it is at temperatures just below 0°C that the danger of ice forming on aircraft becomes greatest. As temperatures drop, ice nuclei become more and more effective, and the proportion of ice crystals to water droplets gradually increases. At first, there are usually on the order of thousands of times more water droplets than ice crystals. Once the temperature drops below about −40°C, however, there will be only ice crystals. It follows that in clouds or portions of clouds where the temperature is between 0°C and −40°C, supercooled water droplets and ice crystals will coexist (Figure 9.6).

The coexistence of water droplets and ice crystals in clouds has important implications for the *growth* of cloud droplets to form precipitation, as we shall see in Chapter 10. However, if the water droplets or ice crystals do not grow, they will remain suspended in the atmosphere as cloud or fog.

9.3 | Clouds

Clouds are dense masses of water droplets, ice crystals, or both suspended in the atmosphere. **Fogs**, which will be explored later in this chapter, are the same, except they form near the ground. Fogs and clouds form when air becomes saturated and begins to condense. In Section 7.4, we saw that saturation can be achieved by cooling the air to its dew-point temperature, by adding water vapour to the air, or by mixing air samples. Various combinations of the three can also produce saturation.

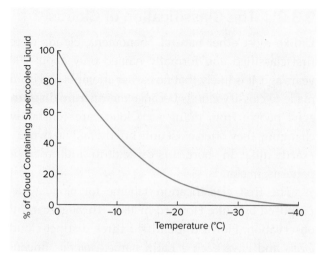

FIGURE 9.6 | **The relationship between the temperature of a cloud and the percentage of the cloud that is supercooled liquid.**

Source: Adapted from Salby, 1996, p. 275.

9.3.1 | Cloud Formation

In almost all cases, clouds form due to the cooling produced when air rises. Far less commonly, clouds form due to mixing; clouds formed by mixing include jet stream cirrus, jet contrails, and some stratocumulus.

There are five basic mechanisms by which air rises to form clouds (Figure 9.7). First, air can rise from the surface by *convection* (Example 9.2). Second, air can be forced to rise up the side of a hill or mountain. Such **orographic lift** explains why some high mountain peaks are often shrouded in cloud, and why the windward sides of mountains are so wet. Third, air can rise due to **convergence**,

OROGRAPHIC LIFT
The process by which air is forced to rise up a slope.

CONVERGENCE
The net inflow of air to a region.

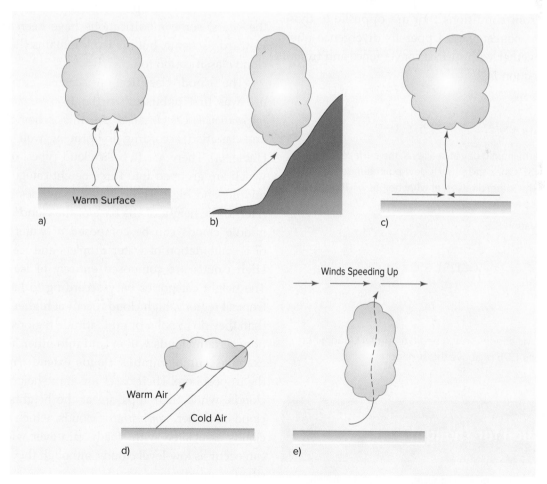

FIGURE 9.7 | **The five basic mechanisms by which air can rise to form clouds: a) convection, b) orographic lift, c) lifting due to surface convergence, d) frontal lifting (this diagram shows a warm front), and e) divergence aloft.**

which generally occurs as air streams flow together (*directional convergence*) or as air slows down along the direction of flow (*speed convergence*). Areas of low pressure are associated with convergence because air flows into them from areas of higher pressure. Fourth, air can rise when it meets colder air at a *front*. Frontal lift can also be considered as a type of convergent lifting. In the mid-latitudes, low-pressure systems develop along fronts. As a result, such systems, in which air is rising due to both low pressure and the presence of fronts, are often referred to as **frontal systems**. Finally, air can be *pulled* up from the surface when **divergence** occurs in the middle to upper troposphere. Divergence results from conditions that are opposite to those that cause convergence. Upper-air divergence often occurs together with surface convergence and frontal lifting (Section 14.2).

FRONTAL SYSTEM
A large area of low pressure that forms in the mid-latitudes due to fronts.

DIVERGENCE
The net outflow of air from a region.

Example 9.2

On a warm, humid day at sea level, the surface temperature is 35°C and the surface dew-point temperature is 24°C. If this warm air rises, at what height will it begin to condense and form a cloud?

Use Equation 8.5

$$z_{LCL} = 125\,(35°C - 24°C)$$

$$= 1375\text{ m}$$

The air will condense and begin to form a cloud when it reaches 1375 m above the surface.

Question for Thought

How does dew-point temperature affect the height of the base of a convective cloud?

9.3.2 | The Classification of Clouds

Unlike most other natural phenomena, clouds were first classified and formally named only about 200 years ago. It is likely that no earlier attempts had been made to classify clouds because their nature discouraged people from doing so. Clouds are constantly changing, they often exist only briefly, and no two are exactly alike. In short, it is difficult to find order, or patterns, in clouds.

The first classification scheme for clouds was proposed by Luke Howard in 1802. Through careful observation, Howard identified three distinct cloud *forms* and gave each a Latin name: *cumulus* (meaning "heap"), *stratus* (meaning "layer"), and *cirrus* (meaning "curl of hair"). He added a fourth Latin name, *nimbus* (meaning "rain"), to use for clouds actively producing precipitation. He then combined these terms to come up with seven cloud types. Over the years, some modifications have been made to Howard's scheme, but his work remains the basis of cloud classification today.

The cloud classification scheme currently in use was first published in the 1956 version of the *International Cloud Atlas*. In this scheme, clouds are classified according to form as well as *height* (Table 9.1). There are 10 basic cloud types, or genera, which are grouped into three height categories: low, middle, and high. These height categories generally refer to the height of the base of the cloud. Low and middle clouds can be composed of water droplets or a combination of water droplets and ice crystals. High clouds are composed entirely of ice crystals. The height categories vary according to latitude; in tropical regions, high clouds occur at higher altitudes than they do in polar or mid-latitude regions. In addition, certain clouds will extend into other levels. For example, cumulonimbus clouds extend throughout the troposphere. Their bases are at the heights of low clouds, while their tops are at the heights of high clouds. Further, *nimbostratus* clouds, which are layer clouds associated with steady rain over wide areas, can occur as low-level clouds, although they are classified as middle-level clouds.

Cloud genera are further subdivided into species and varieties. These sub-categories make use of additional Latin terms to provide a more detailed description of the cloud. For example, the name *cumulus humilis* refers to the small fair-weather cumulus clouds that often dot the sky on a summer afternoon.

Here, clouds will be considered in three main groups based on their form: cirriform, cumuliform, and stratiform. *Stratocumulus* clouds, the most common of all cloud types, have characteristics of both the cumuliform and stratiform types. In fact, Luke Howard originally named these clouds *cumulo-stratus*, but the name was later changed to *stratocumulus* to highlight their *layered* character, as opposed to their *heaped* character. In the following discussion, you will see that the *form* of clouds is closely linked to the conditions under which the clouds develop.

TABLE 9.1 | Cloud classification.[a]

Low-Level Clouds (0–2 km above sea level)

Stratus (St)	• a featureless sheet of cloud, often with ragged edges • similar to fog, but not at the ground • can be deep enough to produce drizzle or light snow
Stratocumulus (Sc)	• an almost continuous sheet of cloud, with a well-defined base that resembles cumuliform cloud and can show strong variations in tone • can produce light precipitation
Cumulus (Cu)	• detached heaps of cloud with mounded tops and flat bottoms • usually scattered across the sky
Cumulonimbus (Cb) (extends through all three levels)	• large heaped cloud that usually extends from low altitudes to the tropospause, where it normally forms an anvil top resembling cirriform cloud • usually produces heavy precipitation, often in the form of hail, combined with thunder and lightning

Middle-Level Clouds (2–4 km in polar regions, 2–7 km in temperate regions, and 2–8 km in tropical regions)

Altocumulus (Ac)	• a sheet of cloud broken up into individual clumps or rolls that are normally regularly arranged • the individual cloud elements are larger than for cirrocumulus clouds and smaller than for stratocumulus clouds
Altostratus (As)	• a featureless sheet of cloud • can make the sun appear to be shining through frosted glass • can occasionally produce light precipitation
Nimbostratus (Ns) (can extend through more than one level)	• a featureless sheet of cloud thick enough to block out the sun and moon • usually produces steady precipitation of moderate intensity that can cause the base of the cloud to appear diffuse

High-Level Clouds (3–8 km in polar regions, 5–14 km in temperate regions, and 6–18 km in tropical regions)

Cirrus (Ci)	• isolated fibrous clouds that appear as streaks scattered across the sky
Cirrocumulus (Cc)	• a sheet of cloud broken up into tiny individual clumps that are normally regularly arranged • the individual cloud elements are smaller than for altocumulus cloud
Cirrostratus (Cs)	• a featureless sheet of cloud resembling a veil over the sky, can produce a halo around the sun or moon • thin enough to clearly see the sun and moon

[a] For visual representations of these various cloud types, see the cloud chart that appears at the end of this text.

9.3.3 | Cirriform Clouds

All cirriform clouds occur in the upper troposphere. Due to the low water vapour content of the air at these heights, these clouds are thin and can have a fibrous appearance. Further, due to the coldness of the air at these heights, cirriform clouds are composed of ice crystals. It is these ice crystals that are responsible for the **haloes** associated with cirriform clouds (Figure 9.8). Ice-crystal clouds also tend to have ragged edges as compared to the more distinct edges of water-droplet clouds (Figure 9.9). This is because water droplets evaporate more easily than do ice crystals (Section 7.3); as a result, ice crystals can persist around the edges of a cloud, while water droplets cannot stray far from a cloud before evaporating. In *cirrus* clouds, ice crystals are often large enough to fall for quite a distance without evaporating. These falling ice crystals are called **fall streaks** (Figure 9.10).

HALO
A ring of light appearing around the sun or the moon due to the refraction of light by the ice crystals in a cloud.

FALL STREAKS
Ice crystals falling from a cirrus cloud.

Cirriform clouds are usually the first clouds to appear as a warm front approaches (Figure 13.14). Thus, their appearance in general, and the appearance of the haloes they cause in specific, can be a sign of cloudy, wet weather to come. An old weather saying expresses this nicely: "haloes around the moon or sun

FIGURE 9.9 | *Top:* The ragged edges of cirrus clouds compared to *Bottom:* the distinct edges of cumulus clouds.

means that rain will surely come." However, because wind can quickly blow cirrus clouds great distances, these clouds can often be seen when the weather system that caused them to form is a long way off.

9.3.4 | Cumuliform Clouds

Atmospheric stability (Section 8.4) is the major determinant of whether clouds will be cumuliform or stratiform. Cumuliform clouds generally form in unstable air as air parcels rise from a surface heated by the sun. They are also associated with cold fronts (Section 13.7) and can form as air is forced up the windward slopes of mountains. A species of cumuliform cloud can even form *downwind* of a mountain range, in the waves that develop when air flows over

FIGURE 9.8 | A cirrostratus cloud causing a halo to form around the sun.

mountaintops. As air blows through these waves, water vapour might condense as air rises toward the wave crest and then evaporate as air sinks toward the wave trough (Figure 9.11). Because the clouds that form in the wave crests are lens-shaped, they are called *lenticular* clouds. As these clouds normally form as middle-height clouds, the most common species is *altocumulus lenticularis*.

The nature of cumuliform clouds reflects the unstable nature of the air in which they form. Indeed, it is the turbulence associated with instability that gives these clouds their lumpy, cauliflower-like appearance. Cumuliform clouds do not cover the whole sky; instead, they produce individual *heaps* of cloud between which there is clear blue sky (Figure 9.12a). These clouds are produced by rising columns of air. In the clouds, the air continues to rise, while in the areas of blue sky, the air sinks (Figure 9.12b). Rising parcels of air are relatively small and so, too, are the widths of the clouds they produce. The smallest cumulus clouds can be anywhere from just under a kilometre to a few kilometres across, and they are often about as tall as they are wide. These small cumulus humilis clouds are created by updrafts of a few metres per second, while large *cumulonimbus* clouds are produced by much stronger updrafts of over 10 metres per second. Because of these strong updrafts, and the associated downdrafts, aircraft avoid flying through cumulonimbus clouds. Being formed and maintained by somewhat ephemeral bubbles of warm air rising from the surface, cumuliform clouds have relatively short life times, ranging from a few minutes to a few hours. When no longer fed by air rising from the surface, these clouds will dissipate as they mix with the surrounding drier air.

If they don't dissipate, these clouds may grow. Cumuliform clouds can often be seen to grow from cumulus humilis to larger *cumulus congestus* and, finally, to cumulonimbus. These changes are often observed to occur on warm summer afternoons. Recall that under these conditions, a mixed layer develops in the lowest kilometre or two of the atmosphere (Section 8.6). As Figure 9.13 shows, below the mixed layer is a shallow surface layer where the ELR is greater than the DALR; above the mixed layer is an

FIGURE 9.10 | Ice crystals falling from a cirrus cloud create fall streaks.

inversion. When the LCL lies well below the inversion, conditions are favourable for the formation of cumulus clouds. Rising air will reach its LCL and begin to condense; this point can be as little as 500 m above the surface. The LCL can be "seen" as the flat cloud bottoms so characteristic of cumulus clouds (Figure 9.14). The condensing cloudy air will continue to rise until it reaches the temperature of the surrounding air, somewhere in the inversion. The tops of the clouds will not be flat, however, because the momentum of the rising air will cause overshoot, and thus the cauliflower-like tops associated with cumulus clouds will appear. The cloud height ultimately depends on the difference between the height of the LCL and the height at which the air stops rising due to the inversion.

FIGURE 9.11 | a) Lens-shaped lenticular clouds. b) These clouds form in the crests of the waves that are created as air flows over mountaintops.

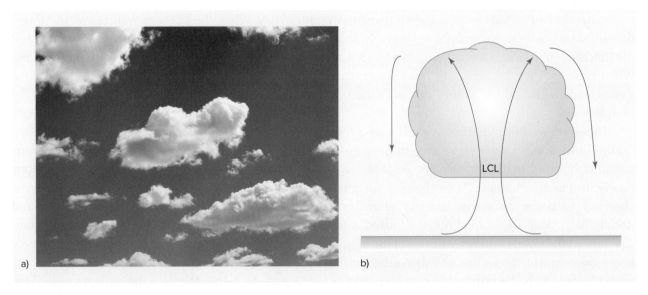

FIGURE 9.12 | a) Cumulus clouds generally appear as small heaps sitting in a blue sky. b) Air rises below and within cumulus clouds, while it sinks in the area of clear sky between these clouds.

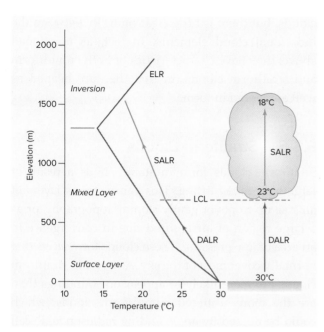

FIGURE 9.13 | The formation of a cumulus humilis cloud. The temperature of the surrounding air is shown by the ELR (blue line). Air rising from the ground will follow the DALR (red line) until it reaches its LCL, at which point it will begin to condense; the air will continue to rise and cool at the SALR (orange line), forming a cloud. The inversion creates a "cap" that prevents the air from rising past the point at which it reaches the same temperature as the surrounding air. How would the conditions shown on this diagram need to be different in order for a cumulus congestus or a cumulonimbus cloud to form?

Cumulus congestus clouds develop from growing cumulus humilis clouds; in turn, cumulonimbus clouds develop from growing cumulus congestus clouds (Figure 9.15). Whereas cumulus humilis clouds do not produce precipitation, cumulus congestus clouds can, and cumulonimbus clouds do. The rain from these clouds is typically very heavy and of short duration; it is quite different from the rain produced by stratiform clouds. Cumulonimbus clouds are unique in that they produce thunder and lightning, and often hail, as well as rain (Section 14.4). Cumulonimbus clouds are also distinguished from cumulus congestus and cumulus humilis clouds in that they contain ice crystals in addition to water droplets. The ice crystals change the appearance of the top of the cloud, making it more ragged and less distinct (Figure 9.15, bottom).

In order for cumulus congestus and cumulonimbus clouds to form, the air must be able to rise past the inversion. This could happen simply as a result of greater surface heating. Over the course of the day, surface heating can raise the temperature of the air enough that the air can rise past the inversion.

FIGURE 9.14 | A sky of cumulus humilis clouds. Note that the bases of these clouds are all at the same height, reflecting the uniform moisture conditions over this area.

FIGURE 9.15 | *Top:* Cumulus congestus clouds that have developed from cumulus humilis clouds. *Bottom:* A cumulonimbus cloud with an anvil top.

Alternatively, the rising air could be pushed through the inversion as a result of forced lifting. Either way, the cloud can continue to grow. If it grows to the point where its top reaches temperatures of about −10°C, ice crystals will form, making the cloud a cumulonimbus. These clouds are often able to grow until their tops reach the stable layer marking the tropopause, approximately 11 km above the surface. Here, the tops of cumulonimbus clouds spread out, forming anvils (Figure 9.15, bottom). Because they are composed of ice crystals, these **anvil tops** resemble cirrus clouds.

ANVIL TOP
The horizontally spreading top of a cumulonimbus cloud.

While the large cumulus clouds we've discussed so far tend to have moderate widths and can extend to many kilometres in height, stratocumulus clouds are usually wide and shallow. These clouds form when the capping inversion is only a few tens of metres above the LCL. Under such conditions, air will rise from the surface, reach the inversion very shortly after beginning to condense, and then stop rising. Thus, instead of growing in height, they flatten out like pancakes; as a result, they can block out enough sunlight to prevent further convection (Figure 9.16). When stratocumulus clouds form in this way, they look like cumulus clouds, but there is little clear blue sky between the individual cloud elements. In fact, as mentioned above, they have characteristics of both cumuliform and stratiform clouds, in that they are heaped as well as layered in form.

9.3.5 | Stratiform Clouds

Stratiform clouds form in stable air as a result of large-scale *forced* lifting. This can occur as layers of air rise up fronts or gently sloping topography, or as a large region of air is lifted due to convergence in an area of low pressure. These clouds form *layers* that normally cover most of the sky. As a result, stratiform clouds are more likely to appear on soundings than are the more scattered cumuliform clouds, which could be missed by an ascending radiosonde. Recall Figure 8.10, which shows an atmospheric sounding for Port Hardy, British Columbia, in which a cloud layer is evident where the air temperature and dewpoint temperature lines merge. It is quite likely that this layer cloud was a nimbostratus cloud, as it was raining at the time of the sounding. Often, cloud layers are quite featureless, sometimes doing nothing more than changing the colour of the sky from blue to white or grey (Figure 9.17). In stable conditions, vertical motions are suppressed, making these clouds somewhat flat and uninteresting.

Because stratiform clouds form as air rises slowly and continuously over large areas, they will not only cover large areas of hundreds to thousands of square kilometres, but they will also last much longer than cumulus clouds. This is because the conditions that produce the lift tend to stay in place longer than those that produce cumuliform clouds. Thus, while cumuliform clouds are fleeting, stratiform clouds tend to be persistent, at times lingering for days. In addition, typical updrafts in stratiform clouds are usually less than 10 *centimetres* per second; these are considerably weaker than the updrafts associated with cumuliform clouds. This means that, whereas air in a cumulonimbus cloud can rise from the surface to the tropopause in about an hour, air rising in a stratus cloud could take a day or more to reach the tropopause.

The three layered clouds—cirrostratus, altostratus, and stratus (Figure 9.18)—all have lumpier

FIGURE 9.16 | Stratocumulus clouds spread out over the sky, blocking sunlight.

FIGURE 9.17 | Layered stratus clouds cover the sky, giving it a white appearance.

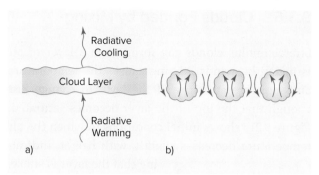

FIGURE 9.19 | The development of individual cloud elements in a layer of cloud. a) The bottom of the cloud layer is warmed by the absorption of infrared radiation from the ground. At the same time, the top of the cloud layer is cooled by the emission of infrared radiation to space. b) This leads to the development of small convection cells. In areas where the air is sinking, the air warms and the cloud evaporates, leaving blue sky.

counterparts, which result from the development of instability in the cloud layer. These clouds are cirrocumulus, altocumulus, and stratocumulus. Because they are of limited vertical extent and usually cover most of the sky, these clouds are still considered to be *layer* clouds. However, instead of being featureless, they are composed of individual elements between which blue sky can usually be seen. These cloud elements form as the top of the cloud layer is cooled by the emission of infrared radiation to space (Figure 9.19a). At the same time, the cloud base may be warmed by the absorption of infrared radiation from the ground. This makes the cloud layer unstable, so

that convection cells form in it. Where the air sinks, it warms slightly, and the sky clears; where the air rises, it remains cloudy (Figure 9.19b).

Altocumulus and cirrocumulus clouds can create beautiful skies, especially at sunrise and sunset, but they are not common (Figure 9.20). Stratocumulus clouds, on the other hand, are common and can be produced by processes other than the one described here. As explained in Section 9.3.4, above, stratocumulus clouds can be formed by the spreading out of cumulus clouds. They can also be formed by mixing, as can a few other types of clouds.

FIGURE 9.18 | Altostratus clouds are generally quite featureless. When the sun is visible through these clouds, it often looks as though it is shining through frosted glass.

FIGURE 9.20 | An uncommon and beautiful display of cirrocumulus clouds, produced in an unstable layer.

9.3.6 | Clouds Formed by Mixing

Stratocumulus clouds can form when a layer of air is mixed by winds. Mixing causes the temperature and moisture structure of the layer to change just enough that the top of the layer becomes saturated. Figure 9.21a shows initial conditions in which the air temperature decreases slightly with height, indicating that the layer is stable, and the dew-point temperature decreases with height, indicating that there is more moisture at the surface than there is higher up. In addition, notice that throughout this layer of air the dew-point temperature is close to the air temperature, indicating that the air is close to saturation.

MIXING CONDENSATION LEVEL
The height at which water vapour will condense as a result of the mixing of an atmospheric layer.

JET CONTRAIL
A long narrow cloud in the upper troposphere produced by the condensation of the water vapour in aircraft exhaust.

Figure 9.21b shows what happens once a wind has mixed the air layer from the surface up to a height of 1000 m. Recall that mixing can bring the temperature profile close to the dry adiabatic lapse rate (Section 8.1.1), so that air in the lower portion of the layer warms, while air in the upper portion cools. At the same time, mixing will make the water vapour content of the layer more uniform with height, so that there is less water vapour near the bottom of the layer and more water vapour near the top. The combination of cooling and moistening at the top of the air layer causes the layer to reach its dew-point temperature and begin to condense, forming a cloud. The base of this layer of stratocumulus cloud marks the **mixing condensation level.** Above the layer in which the mixing occurred, the air remains stable, creating the top of the cloud layer. Marine stratocumulus clouds can form by this process. These clouds are very common along west coasts, where currents cause the upwelling of cold water from below. The cold water cools the air, while the ocean moistens it. All it takes is a wind, and a layer of stratocumulus can form.

Mixing due to wind shear can produce clouds at any level of the atmosphere because of the vertical motions it creates. If there is enough moisture where this mixing occurs, clouds can form. An example of this occurs due to the turbulence created by strong wind shear in the jet streams (Section 12.2.5.2). Sometimes, a strip of *jet stream cirrus* clouds will form in the turbulent air along the poleward edge of a jet stream. When such clouds form, they make it easy to locate a jet stream from the ground.

Jet contrails, the long thin clouds left behind by jets, are formed when the warm, moist air from the

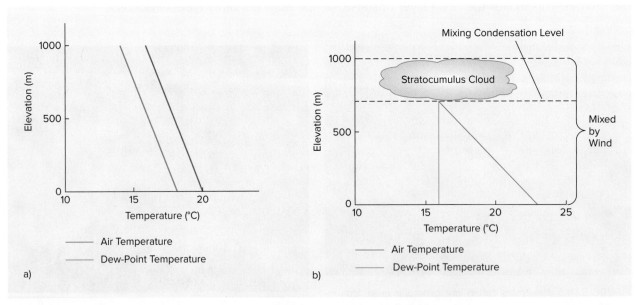

FIGURE 9.21 | a) Before the air layer is mixed, it is stable and close to saturation throughout. **b)** After the air is mixed, the top of the layer becomes saturated, as shown by the merging of the air temperature and dew-point temperature lines, forming stratocumulus clouds.

FIGURE 9.22 | Jet contrails from these Canadian Snowbird aircraft performing at an air show in Abbotsford, British Columbia, form when warm, moist air from the jets' exhaust mixes with colder air and condenses.

jet's exhaust is mixed with the cold, dry air of the upper troposphere (Figure 9.22). As the warm air mixes into the cold air, it will cool and, as a result, the large amount of water vapour it contains will likely condense. Jet contrails don't usually last long, as they quickly evaporate in the dry air characteristic of these heights.

9.4 | Fogs

As mentioned above, fogs are similar to clouds, except that they form near the surface rather than higher in the atmosphere. Likewise, the processes by which fogs form are similar to those that form clouds. Most fogs form by cooling. For example, **upslope fogs** can form as air rises and cools along slopes. Other fogs form by mixing or adding water vapour. Below, we will explore the most common ways in which fogs form.

9.4.1 | Radiation Fog

Radiation fogs form when a layer of moist air at the surface cools to its dew-point temperature by emitting more radiation than it is absorbing. This commonly occurs overnight, when the air is close to being saturated and conditions favour strong radiative cooling.

For strong radiative cooling, three conditions are needed. The first is clear skies. When skies are clear, much more longwave radiation will leave the surface than will return from the atmosphere. Thus, the net longwave radiation, L^*, will be a large negative number, indicating a large longwave loss (Section 6.6.1). It also helps if drier air overlies the moist surface air because, like clear skies, dry air absorbs and emits less longwave radiation than does moist air. The second condition that is needed for strong radiative cooling is light wind. Winds must be *light* because strong winds generate mechanical convection that will mix warm air downward from above, slowing the cooling near the surface. In addition, strong winds will mix in the drier air from above, increasing the dew-point depression. On the other hand, with no winds at all, the fog layer will be very shallow, or there may be no fog at all, only dew. This is because without some mixing, radiative cooling alone would take a very long time to produce a layer of cold air deep enough for fog to form in it. The final condition that is needed is long nights. When nights are longer, as they are in fall and winter, there is more *time* for the air to cool to its dew-point temperature (Example 9.3).

UPSLOPE FOG
A fog that forms as air rising up a slope cools adiabatically and condenses.

RADIATION FOG
A fog that forms when a layer of moist air at the surface cools radiatively to its dew-point temperature.

Example 9.3

At 6:00 PM, the air temperature is 7°C and the relative humidity is 75 per cent. How much does the air need to cool to form a radiation fog? If the air is cooling at an average rate of 0.4°C/hr, at what time will a radiation fog form?

Using Table 7.1, e_s is 1.001 kPa.

Use Equation 7.11.

$$0.75 = \frac{e}{1.001 \text{ kPa}}$$

$$e = 0.750 \text{ kPa}$$

Use e in Table 7.1 to determine that T_d is about 3°C.

This means that in order for a radiation fog to form, the air needs to cool 4°C. This will take 10 hours, so fog will begin to form at 4 AM.

Since long nights and moisture are both important to their formation, radiation fogs are most common in the fall. At this time, the air is still moist from the summer, and nights are beginning to get longer. During fall, an afternoon rain followed by clear skies often indicates that a radiation fog will form overnight. The depth of a radiation fog can vary from being very shallow and patchy to being about 300 m deep. Because the air in a radiation fog is cold, these fogs will drain downhill and sit in the lowest areas. At elevations above the fog, the clear skies responsible for the fog can be seen (Figure 9.23).

Even in the lowest spots, however, radiation fogs don't usually linger for long. As the day progresses, they usually dissipate. The process by which this occurs is commonly described as "burning off," but this phrase is a misleading description of what really happens. Fogs, like clouds, are good reflectors, and rather poor absorbers, of solar radiation. This means that first thing in the morning, when the sun is still low in the sky, solar radiation will be strongly reflected from the top of the fog. However, as the sun gets higher in the sky, more of its radiation will penetrate the fog and warm the surface. Warm air rising from the surface will then gradually begin to warm the foggy air. This will raise the saturation vapour pressure of the foggy air, and the water droplets of the fog will begin to evaporate. Because the fog warms from the bottom up, it will also dissipate from the bottom up. Thus, the fog appears to "lift." Lifting fogs can often form stratus cloud. In most cases, radiation fogs will have lifted by midday but, in some cases, they may linger for days. This can happen around the winter solstice, when the sun is very low in the sky; it can also occur if the fog is particularly thick. In both these cases, the sun is unable to penetrate the fog enough to sufficiently warm the ground.

9.4.2 | Advection Fog

Like radiation fogs, **advection fogs** form when air is cooled to its dew-point temperature. But, while radiation fogs form by *radiative* cooling, advection fogs form by *contact* cooling. In the formation of advection fogs, warm moist air is cooled as it is *advected* by wind over a cold surface. The wind moves the warm, moist air over the cold surface and also mixes it downward to cool by contact with the surface. Examples of cold surfaces involved in this process include cold water and ice- or snow-covered surfaces. Advection fogs tend to be deeper than radiation fogs; they can be up to 1 km deep. In addition, they can form at any time of day or night.

Both San Francisco and Newfoundland are famous for advection fogs. In San Francisco, these fogs form during the summer, when the water off the California coast is cold due to upwelling (Figure 9.24).

ADVECTION FOG
A fog that forms when warm, moist air cools to its dew-point temperature as it is blown over a cold surface.

FIGURE 9.23 | The cold air of a radiation fog drains downhill into the valley in Madonna del Sasso, Italy. Note the clear skies responsible for the fog at higher elevations.

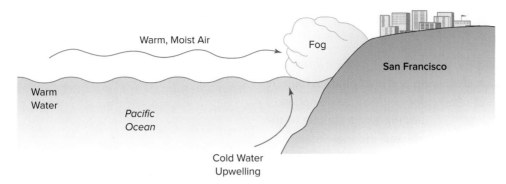

FIGURE 9.24 | Warm moist air blows toward the shore just off the coast of San Francisco where cold water is upwelling from below.

When warm moist air lying over the Pacific Ocean is blown eastward over this cold water, fogs form along the coast (Figure 9.25). Because these fogs tend to form mostly in summer, spring and fall temperatures can be higher than summer temperatures in this region.

Newfoundland's fogs, on the other hand, form year-round. In fact, some parts of Newfoundland are foggy more often than they are not. In these places fogs occur on average for more than 200 days of the year! Newfoundland's fogs are caused by the proximity of warm and cold ocean currents in the northwest Atlantic Ocean. Just off the coast, the cold Labrador Current moves southward. A little farther east, the warm Gulf Stream flows northward (Figure 9.26). When winds are right, warm moist air from above the Gulf Stream will blow over the cold waters of the Labrador current, and the air will cool to its dew-point temperature, forming fog.

Question for Thought

How is the role wind plays in forming radiation fogs different from the role it plays in forming advection fogs?

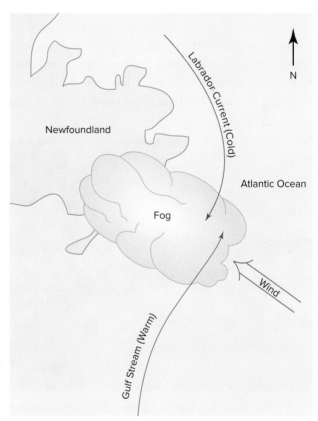

FIGURE 9.26 | The proximity of the cold southward-flowing Labrador Current and the warm northward-flowing Gulf Stream makes advection fogs very common in Newfoundland.

FIGURE 9.25 | Advection fog forms along the California coast as warm moist air is blown over the cold Pacific Ocean.

9.4.3 | Fogs Formed by Adding Vapour and Mixing

Unlike radiation and advection fogs that are formed by the cooling of air, **steam fogs** and **precipitation fogs** are formed by a combination of adding water vapour to the air and mixing. Steam fogs are so named because their thin and wispy appearance resembles steam rising from the surface. They look quite different from the generally deeper, and thicker, radiation and advection fogs. In fact, most steam fogs tend to look more like what we often call *mist* (Figure 9.27).

STEAM FOG
A fog that forms when water vapour evaporating from a warm, moist surface is mixed into colder air above that surface.

PRECIPITATION FOG
A fog that forms when water vapour resulting from the evaporation of raindrops causes saturation as it mixes into cold air.

Steam fogs form when the water vapour that has evaporated from a warm, moist surface rises and mixes into colder air above (Figure 9.28). The mixing process may bring the air to its saturation point, causing condensation and, thus, fog. It is common for steam fogs to form over lakes in the late summer or early fall, when the lake water is warm compared to the surrounding land. Such fogs will form when the cold air above the land moves over the lake. The

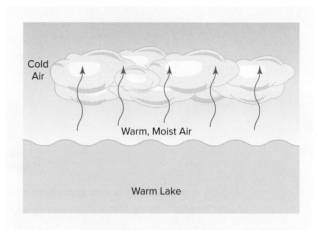

FIGURE 9.28 | The formation of steam fog. Warm, moist air lies directly over the lake. As colder air moves over the lake, the warm, moist air will rise, saturating the colder air.

steam-like appearance of these fogs results from the rising motion of the warm, moist air as it condenses. In addition to forming over lakes, these fogs also commonly form over well-watered surfaces such as golf courses and playing fields. A similar process is responsible for the "steam" that we might see rising from wet rooftops and puddles when the sun comes out after a rain, or the little clouds produced by our breath on a cold day.

Precipitation fog is the thin fog or mistiness that often accompanies rainfall (Figure 9.29). As raindrops fall, they are surrounded by water vapour. If these raindrops fall through cold air, as often occurs at a warm front (Section 13.5), the addition of water vapour to the cold air may saturate the air (Example 9.4). Because of its association with frontal precipitation, such fog is also often referred to as *frontal fog*.

Example 9.4

The air temperature is 4°C and the dew-point temperature is 3°C. How much water vapour must be added to the air by falling rain to form a fog?

Using Table 7.1, e_s is 0.813 kPa and e is 0.758 kPa.

Therefore, 0.055 kPa of water vapour must be added to the air to saturate it.

FIGURE 9.29 | Precipitation fog forms along this country road as raindrops fall through cold air, causing saturation.

9.5 | Chapter Summary

1. Water vapour can condense to form water droplets in two ways. The first is homogeneous nucleation, in which water droplets form by the chance collision of water vapour molecules. The second is heterogeneous nucleation, in which water droplets form when water vapour condenses on atmospheric aerosols. In the atmosphere, only those water droplets that form by heterogeneous nucleation can survive and grow into cloud droplets.

2. The curvature and solute effects help us understand how water droplets form and grow in the atmosphere. The curvature effect inhibits the formation of droplets because the relative humidity required for them to be in equilibrium with their environment increases as droplet size decreases. The solute effect promotes the formation of droplets because water vapour will condense on soluble aerosol at relative humidities below 100 per cent.

3. Atmospheric aerosols on which water vapour will condense are called cloud condensation nuclei (CCN). The most effective CCN are those aerosols that are wettable, soluble, and large or giant in size. CCN that are both wettable and soluble are hygroscopic.

4. Kohler curves provide the relationship between droplet size and equilibrium (saturation) vapour pressure for a given amount and type of solute. These curves show that, according to the solute effect, water vapour will begin condensing on soluble substances at relative humidities below 100 per cent. As a droplet grows, the solution becomes diluted, the curvature effect takes over, and the droplet will continue to grow only with relative humidities above 100 per cent.

5. Water droplets will not spontaneously freeze to form ice crystals until temperatures drop below about −40°C. Just as CCN help in the formation of water droplets in the atmosphere, ice nuclei assist in the formation of ice crystals in the atmosphere. Ice crystals can form when water vapour is deposited directly onto an ice nucleus. In addition, a super-cooled water droplet may freeze if it contains an ice nucleus and temperatures drop sufficiently, or if it collides with an ice nucleus. Clouds at temperatures between 0°C and −40°C contain both water droplets and ice crystals.

6. Clouds are classified based on height and form. This results in 10 major cloud types: stratus (St), strato-cumulus (Sc), cumulus (Cu), cumulonimbus (Cb), altocumulus (Ac), altostratus (As), nimbostratus (Ns), cirrus (Ci), cirrocumulus (Cc), and cirrostratus (Cs).

7. Cirriform clouds form high in the troposphere and are thin, wispy clouds composed of ice crystals. These clouds often cause haloes to appear around the sun or the moon.

8. Cumuliform clouds are heaped in form. They are considered low clouds because their bases are at the height of low clouds, but they can extend through the troposphere. They form in unstable air and can grow from the small cumulus humilis of a sunny afternoon to the towering cumulonimbus capable of producing heavy rain, hail, and thunder and lightning.

9. Stratiform clouds are layered in form and can occur at any height in the troposphere. They tend to form in stable air as a result of forced lifting. Featureless layered clouds such as stratus, altostratus, and cirrostratus can develop into somewhat more varied stratocumulus, altocumulus, and cirrocumulus as the cloud layer is destabilized.

10. Radiation and advection fogs both form when air is cooled to its dew-point temperature. Radiation fogs form overnight as a layer of moist surface air cools by emitting more radiation than it is absorbing. Advection fogs form when a layer of warm, moist air is cooled as it is blown across a cold surface.

11. Steam fogs and precipitation fogs form when the water vapour evaporated from a water surface is mixed into, and saturates, cold air. In the case of steam fogs, the water surface is often a lake; in the case of precipitation fogs, the water surface is a raindrop.

Review Questions

1. Why won't water droplets formed by homogeneous nucleation survive in the atmosphere?

2. How do the curvature effect and the solute effect influence the formation of water droplets in the atmosphere?

3. Under what conditions will a water droplet a) grow, b) be in equilibrium with its environment, and c) shrink?

4. How can a Kohler curve be used to describe the factors influencing the formation of a cloud droplet? What is the significance of the height of the peak in a Kohler curve?

5. What factors make good cloud condensation nuclei (CCN)? Why?

6. Why are ice nuclei important in the atmosphere?

7. What are the characteristics we use to classify clouds? Based on this, what are the ten major cloud types?

8. How do cumuliform clouds and stratiform clouds compare in terms of their characteristics and the conditions under which they are likely to form?

9. What are the differences between cumulus humilis, cumulus congestus, and cumulonimbus clouds?

10. How might a layer of stratiform cloud be destabilized?

11. How do radiation fogs and advection fogs form? How are their formations similar to one another?

12. How do steam fogs and precipitation fogs form? How are their formations similar to one another?

Suggestions for Research

1. Investigate why some airports have recurring problems with fog on runways, and note the ways in which these airports deal with fog.

2. Identify areas of the world where there is little to no precipitation but plenty of fog and low clouds. Describe the methods used by people and vegetation in these areas to extract moisture from the fog and low clouds.

References

Charlson, R.J., J.E. Lovelock, M.O. Andreae, and S.G. Warren. (1987). Oceanic phytoplankton, atmospheric sulphur, cloud albedo, and climate. *Nature 326*, 655–61.

Lovelock, J.E., (2006). *The revenge of Gaia: Why the Earth is fighting back—and how we can still save humanity.* London: Allen Lane.

Salby, Murry L. *Fundamentals of Atmospheric Physics.* Academic Press—Elsevier Science, San Diego, 1996.

Wallace, J.M., & P.V. Hobbs. (2006). *Atmospheric science: An introductory survey* (2nd ed.). Boston, MA: Elsevier Academic Press.

10
Precipitation

Learning Goals

After studying this chapter, you should be able to

1. *describe* the two methods by which cloud droplets can grow to produce precipitation,

2. *describe* the atmospheric conditions that lead to different types of precipitation,

3. *explain* how hailstones are formed,

4. *associate* the character of precipitation with the processes leading to the uplift of air, and

5. *account* for the spatial and temporal distributions of precipitation.

Precipitation occurs when water droplets or ice crystals in a cloud become large enough to overcome the updrafts within the cloud. The type of precipitation that reaches the ground depends largely on the temperature profile of the atmosphere below the cloud.

In Chapter 9, you learned that clouds are suspensions of water droplets, ice crystals, or both that form in updrafts of rising air. When the water droplets or ice crystals are large enough, they can overcome these updrafts to fall as **precipitation**. In this chapter, you will discover how tiny cloud droplets grow large enough to form precipitation. To most of us, the most common form of precipitation is rain, but there are, of course, other types of precipitation, including **drizzle**, **sleet**, **freezing rain**, and **snow** (Figure 10.1). As you will see, variations in the temperature profile of the atmosphere below the cloud determine the type of precipitation that reaches the ground. **Hail** is a little different from the other forms of precipitation, in that it will form only in the very strong updrafts of cumulonimbus clouds. Finally, you will discover how to apply the basic uplift mechanisms associated with the formation of clouds (Section 9.3) to account for the global distribution and character of precipitation.

PRECIPITATION
Any liquid or solid water particles that fall from the atmosphere and reach the ground.

DRIZZLE
A type of precipitation in the form of water droplets smaller than typical raindrops.

SLEET
A type of frozen precipitation that does not have a crystal structure.

FREEZING RAIN
A type of precipitation that will melt as it falls through the atmosphere and then refreeze upon contact with objects at the surface.

SNOW
A type of precipitation in the form of ice crystals.

HAIL
Frozen precipitation made up of concentric layers of alternating clear and opaque ice.

TERMINAL VELOCITY
Constant fall speed in still air reached when the force of gravity equals the opposing force created by the resistance of the air.

FIGURE 10.1 | In the winter, when temperatures are below freezing, many regions tend to experience precipitation as snowfall rather than rainfall. When heavy snowfall combines with strong winds, the result is often a blizzard, such as the one pictured here in Inuvik, Northwest Territories.

10.1 | Cloud Droplets versus Raindrops

Water particles, falling as precipitation, vary over a large range of sizes (Table 10.1). Tiny drizzle droplets are about 100 μm in radius, while typical raindrops have radii of about 1 mm. Hailstones, the largest form of precipitation, have radii of more than about 3 mm, and they have been known to reach sizes of greater than 5 cm in radius. In comparison, recall that cloud droplets are usually only about 10 μm in radius (Section 9.1.2). In fact, the only real difference between water particles that fall from clouds and those that don't is size—the ones that fall are bigger than the ones that don't. Because larger drops fall faster than smaller ones, larger drops are more likely to be able to overcome the updrafts and fall from the cloud as precipitation. Even when very tiny cloud droplets do begin to fall from clouds, their size, coupled with their slow fall speeds, cause them to evaporate very quickly. For precipitation to occur, therefore, the water particles in a cloud need to grow. Not all clouds form precipitation; in fact, most don't, simply because conditions must be right for water droplets, or ice crystals, to grow large enough to fall fast enough.

10.2 | Terminal Velocity

When objects are falling freely, as water drops can in the atmosphere, they eventually reach a constant speed known as **terminal velocity**. We can apply Newton's first law of motion to help show why falling objects reach terminal velocity. Recall that this law states that if an object is moving at a constant speed, the forces acting on it are balanced; conversely, if an object is accelerating, the forces acting on it are *not* balanced (Section 1.5).

For a water drop falling through the atmosphere, the force of gravity increases with the drop's mass, and the force of air resistance increases with the drop's velocity and surface area. (Air resistance also increases with the density of the air, we will ignore that here.) The drop is initially set in motion by the force of gravity. At first, gravity will be a much larger force than will be air resistance, and the drop will accelerate. As the drop falls faster and faster, the force of air resistance will gradually increase until it equals the force of gravity. At this point, the drop will no longer accelerate; it will have attained its terminal velocity and will continue to fall at a constant speed.

For larger drops, this balance will be achieved at a greater velocity than it will for smaller drops. This is because smaller drops have a larger surface area to mass ratio than do larger drops. In other words, for a given surface area, there is less mass in a small drop than there is in a larger one. It follows that since mass influences the force of gravity, while surface area influences the force of friction, a balance between forces will be achieved sooner for a smaller drop than for a larger one. The result is that a smaller drop will be falling more slowly than a larger one when it achieves balance. In reality, the situation is a little more complicated than this description suggests because airflow past drops bigger than about 50 µm in radius becomes turbulent, thereby offering increased resistance. Terminal velocity, therefore, initially increases very quickly, then more slowly, with size (Table 10.1).

A comparison of the terminal velocities in Table 10.1 to the typical updrafts in clouds shows why only the bigger drops can fall as precipitation. In Section 9.3, we saw that updrafts in stratiform clouds are typically less than 0.1 m/s, while updrafts in cumulonimbus clouds can be greater than 10 m/s. When the updrafts are greater than the terminal velocity of a drop, that drop will *rise* in the cloud. On the other hand, when the updrafts are less than the terminal velocity of a drop, that drop can *fall* in the cloud.

Table 10.1 shows that a drop greater than about 50 µm in radius could fall in a stratus cloud. It turns out, however, that drops need to be bigger than about 100 µm in radius to reach the ground, as drops smaller than this will evaporate almost as soon as they leave the cloud. Thus, 100 µm is considered to be a somewhat arbitrary boundary between cloud droplets and precipitation particles. Drops of this size can form *drizzle* but, for a cloud to produce drizzle, it can't be more than a few hundred metres above the surface. If the cloud is much higher, the tiny drops are likely to evaporate before reaching the ground. So, while low-lying stratus clouds can produce drizzle, most other sorts of clouds cannot. To fall from a cumulonimbus cloud, for example, drops must be much bigger—at least 2 mm in radius. The raindrops, or hailstones, that fall from a cumulonimbus cloud must be large in order to overcome the strong updrafts characteristic of these clouds.

In order to fall as precipitation, therefore, a water particle needs to be large enough to overcome the updrafts in a cloud and to survive the fall to Earth without evaporating. Because an average cloud droplet is only about 10 µm in radius, whereas an average raindrop is about 1 mm in radius, the *radius* of a cloud droplet needs to increase roughly one hundred times for the cloud droplet to become a raindrop (Figure 10.2). This is a *volume* increase of about one million times!

TABLE 10.1	Terminal velocities of water droplets by size.	
Type of Drop	**Droplet Radius**	**Terminal Velocity**
Typical Cloud Droplet	10 µm	0.01 m/s
Large Cloud Droplet	50 µm	0.3 m/s
Drizzle	100 µm	0.7 m/s
Typical Raindrop	1.0 mm	6.5 m/s
Large Raindrop	2.0 mm	10 m/s

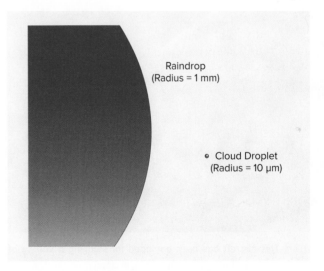

Raindrop
(Radius = 1 mm)

Cloud Droplet
(Radius = 10 µm)

FIGURE 10.2 | **The relative sizes of a typical raindrop and a typical cloud droplet in the atmosphere.**

10.3 | Growth of Cloud Droplets

Section 9.1 described the first two stages in the growth of a cloud droplet in the atmosphere. The first step is *heterogeneous nucleation*, in which water begins to condense on a cloud condensation nucleus (CCN). The second step is *diffusion*, in which water vapour continues to diffuse to the droplet and condense onto it. Recall, however, that there is a limit to how big drops can grow by condensation alone. This limit is usually about 10 μm in radius, although cloud droplets can range anywhere from about 2 μm to 50 μm in radius. To produce *precipitation*, these droplets must continue to grow.

It is not a simple matter to observe the processes at work producing precipitation in clouds. Direct observations from aircraft have been made of such things as cloud droplet size distributions, liquid water content, and updraft velocities (Figure 10.3). Cloud processes have also been studied in the lab and through computer simulations. Through such studies, we have concluded that there are two possible mechanisms by which droplets can grow in clouds: growth by collision and coalescence, which occurs in **warm clouds**, and growth by the Bergeron–Findeisen process, which occurs in **cold clouds**.

WARM CLOUD
A cloud in which the temperature is above 0°C throughout; such clouds contain water droplets only.

COLD CLOUD
A cloud in which the temperature is below 0°C in at least part of the cloud; such clouds contain a mixture of supercooled water droplets and ice crystals.

COLLISION EFFICIENCY
The probability that two droplets in a cloud will collide.

10.3.1 | Growth by Collision and Coalescence

In warm clouds, raindrops form by the process of *collision and coalescence* as water droplets move upward, downward, and sideways through a cloud. Because different-sized droplets will be moving at different speeds, the larger ones are likely to catch up with the smaller ones and collide with them. When they collide, it is possible that the droplets might join, or coalesce, to form a larger droplet. This larger droplet might then collide with another droplet and form an even larger droplet. By this simple process of collision and coalescence, droplets can grow very quickly, and rain can form. But droplets will not always collide upon catching up to each other and, even if they do collide, they will not always coalesce.

The probability that two droplets will collide is expressed as their **collision efficiency**. Collision efficiency is studied mathematically. The factors influencing it are complex, but it can be summarized as follows. Calculations show that as the size of the bigger droplet—known as the collector drop—increases, the likelihood of collision increases, but unless the collector drop is at least 20 μm in radius, the probability of collision is very low. This is because the terminal velocities of droplets smaller than 20 μm are so small that these tiny droplets tend to move together in the air rather than collide. Collision efficiency also depends on the *relative* sizes of the droplets. If there is a large difference in size between the drops, it is unlikely that they will collide because the smaller ones will tend to follow streamlines around the larger ones (Figure 10.4). As they get closer in size, the collision efficiency tends to increase.

Just as all droplets in a collector drop's path will not collide with it, not all colliding drops will coalesce. When air becomes trapped between the colliding drops, it can cause them to bounce apart. The

FIGURE 10.3 | This aircraft has been equipped for studying a variety of atmospheric processes, including those operating in clouds.

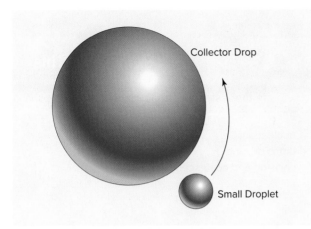

FIGURE 10.4 | A small droplet follows the streamlines around a larger droplet, thus avoiding collision.

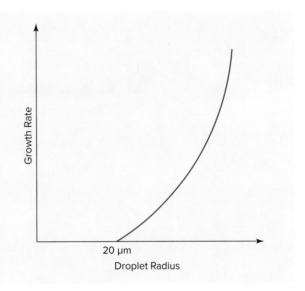

FIGURE 10.5 | The growth of droplets by collision and coalescence is an accelerating process, but droplets don't usually begin to grow by this process until they reach about 20 µm in radius.

likelihood that two colliding drops will coalesce is expressed as their **coalescence efficiency**. Coalescence efficiency is studied in the lab. Again, the relationship is complex, but it seems that colliding drops are most likely to coalesce when they are either very different or very similar in size. Experiments also show that electric fields, which are often present in clouds, increase the chances of coalescence.

From the above discussion, a requirement for the collision and coalescence process is that there be a variety of droplet sizes in the cloud, at least some of which have radii of 20 µm or larger. Indeed, models have shown that the process is unlikely to proceed without some droplets of at least this size. Because the collision and coalescence process favours larger droplets, it is an accelerating process (Figure 10.5). In addition, the process will be enhanced if there is an electric field. What's more, since droplets grow by moving through the cloud not only as they *fall* but also as they *rise*, the

process will be further enhanced in *deep* clouds with *strong updrafts*. These conditions will lengthen the potential path of a droplet moving through the cloud, thus increasing the droplet's chances of colliding, coalescing, and growing to raindrop size (Figure 10.6). It follows that, just as is observed, deep cumulonimbus clouds are capable of producing very large raindrops.

> **COALESCENCE EFFICIENCY**
> The probability that two colliding cloud droplets will coalesce.

Recall that marine clouds usually contain bigger drops than do continental clouds due to the more numerous CCN over continents (Section 9.1.2). As a result, marine clouds are more likely to produce precipitation by the collision and coalescence process than are continental clouds. In addition, they do not need to be as thick as continental clouds do in order to produce precipitation, and they can produce precipitation when they have smaller updrafts than those required in continental clouds.

In summary, cloud droplets begin to form by the process of nucleation; once they have formed in this way, they can continue to grow to about 10 µm in radius through the process of diffusion. At this point, the droplets are still small enough to be held in the cloud by updrafts, and they might begin to

Remember This

The collision and coalescence process requires that there be droplets of various sizes and that some droplets be bigger than 20 µm in radius; the process is enhanced when clouds are deep with strong updrafts, and when electric fields are present.

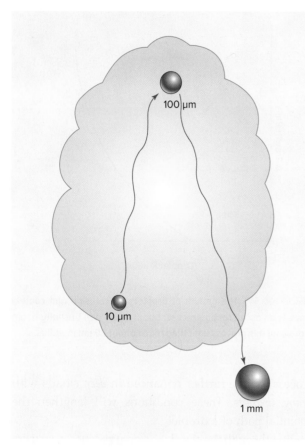

FIGURE 10.6 | A cloud droplet grows considerably as it is carried through a deep cloud with strong updrafts.

grow to produce raindrops by the process of collision and coalescence. Because the process of collision and coalescence is enhanced in deep clouds with strong updrafts, and because it occurs in warm clouds, it tends to be an important process in the large cumulus clouds that form in the tropics. However, precipitation is, of course, observed to fall from the much thinner stratiform clouds common in the mid-latitudes. The process responsible for precipitation from these *cold clouds* is outlined in the following section.

Question for Thought

How are the processes of growth by nucleation and diffusion different from the processes of growth by collision and coalescence?

Remember This

There are three steps in the formation of raindrops in warm clouds:

- nucleation,
- diffusion, and
- collision and coalescence.

10.3.2 | Growth by the Bergeron–Findeisen Process

In cold clouds, precipitation may form by the *Bergeron–Findeisen process*, or the *ice-crystal process*. Tor Bergeron and Walter Findeisen were the first scientists to describe this process, in the 1930s. Bergeron built on the work of Alfred Wegener, a German meteorologist who, in 1911, first determined that when ice crystals and water droplets exist together, the ice crystals will grow as water vapour is deposited on them rather than on the water droplets. Bergeron, a Norwegian meteorologist, applied this idea to explain the formation of precipitation in clouds. Findeisen, another German meteorologist, investigated the theory further through his work with cloud chambers.

As a result of the Bergeron–Findeisen process, ice crystals grow to the point where they are large enough to overcome updrafts. This means that all precipitation forming by this process begins as ice crystals, or what we know as snow. Since the Bergeron–Findeisen process is normally how precipitation forms in the mid-latitudes, it follows that mid-latitude precipitation usually begins as snow, even in the summer. Because it requires the presence of ice crystals, this process can *only* occur in cold clouds.

The Bergeron–Findeisen process is based on two facts first introduced in previous chapters. The first is that supercooled water droplets and ice crystals *coexist* in cold clouds at temperatures between about 0°C and −40°C (Section 9.2). The second is that the saturation vapour pressure over water is greater than the saturation vapour pressure over ice (Section 7.3). This means that when the air is saturated for water, it will be supersaturated for ice. This is shown in Example 10.1.

Remember that the supersaturation for liquid water that normally occurs in clouds is no more than about 1 per cent (Section 9.1.2). The basis of the Bergeron–Findeisen process, therefore, is that when water droplets and ice crystals coexist in a cloud, there will be a tendency for water vapour to diffuse from the water droplets to the ice crystals and deposit onto the ice crystals. As a result of this diffusion process, water droplets will evaporate and shrink, while ice crystals will grow.

Example 10.1

Show that when air is saturated for water at –10°C, it will be supersaturated for ice by 12 per cent.

Use Table 7.1.

At –10°C, the saturation vapour pressure over water is 0.29 kPa, while the saturation vapour pressure over ice is 0.26 kPa. If the air is saturated for water, then the vapour pressure is 0.29 kPa. Using Equation 7.11, the relative humidity for ice is

$$RH = \frac{0.29 \text{ kPa}}{0.26 \text{ kPa}} \times 100\% = 112\%$$

In detail, the Bergeron–Findeisen process occurs as follows. A cloud begins to form when the air becomes slightly supersaturated for liquid water. At this point, the relative humidity is just above 100 per cent, so water droplets form and grow. As the air continues to rise and cool, some ice crystals form. Now ice crystals and water droplets are both growing in the cloud, but the supersaturation is greater for the ice crystals than it is for the water droplets, so the ice crystals grow a little faster. As the droplets and crystals grow, the water vapour in the cloud is reduced and, as a result, the relative humidity decreases. Eventually, the relative humidity drops below 100 per cent for water; at this point, the air is subsaturated for water but still supersaturated for ice. As a result, the water droplets begin to evaporate. This evaporation adds water vapour to the air, maintaining the supersaturation for ice and allowing the ice crystals to continue growing. The ice crystals grow as the water droplets shrink; in other words, the ice crystals grow at the expense of the water droplets. This growth is most rapid at cloud temperatures of between about –8°C and –16°C because it is within this temperature range that the difference between the saturation vapour pressure over water and that over ice is greatest (Figure 7.5).

Once ice crystals reach a certain size through deposition, they are likely to continue to grow by two other processes: **accretion** and **aggregation** (Figure 10.7). Either way, the ice crystals can continue to grow until they can overcome the updrafts within the cloud and fall from the cloud as precipitation. As the crystals fall, they usually melt, reaching the ground as rain. Radar observations that can "see" the different signals from snow, melting snow, and rain confirm that much rain begins as snow.

ACCRETION
A process by which ice crystals grow by colliding with supercooled water droplets that then freeze onto them.

AGGREGATION
A process by which ice crystals grow by colliding with other ice crystals that then stick together.

Question for Thought

Ever since we determined how clouds produce precipitation *naturally*, we have attempted to modify existing clouds to produce precipitation *artificially*. What could be done to make a *warm* cloud produce precipitation? What could be done to make a *cold* cloud produce precipitation?

FIGURE 10.7 | Accretion and aggregation are the processes by which snowflakes, such as the one shown here, form.

10.4 | Precipitation Types

Whether or not ice crystals melt before they reach the ground depends on the temperature of the air below the cloud. One way to forecast what type of precipitation will fall is to apply the following rules of thumb using atmospheric soundings. If there are no layers greater than about 200 m deep with temperatures above freezing, ice crystals will not melt as they fall and will reach the ground as *snow* (Figure 10.8a). It follows that it can be snowing even when surface temperatures are above freezing. If there is a layer of air at the surface that is deeper than about 400 m with temperatures above freezing, ice crystals will melt, and the precipitation will reach the surface as *rain* (figures 10.8b and 10.9). If the above-freezing air layer at the surface is somewhere between 400 and 200 m deep, the precipitation will partially melt, producing a mixture of snow and rain at the surface (Figure 10.8c).

In addition to soundings, meteorologists can use upper-air charts that show the thickness of the 1000 to 500 hPa layer for forecasting precipitation type (Section 3.7). The rule of thumb in this case is that locations north of the 5400 m thickness line will receive snow, while locations south of it will receive rain. When thickness is less than 5400 m, the air layer is cold enough that the snow will not melt as it falls.

It may sound easy to predict precipitation type using the above rules, but this is not always the case, because falling precipitation can influence the temperature of the air through which it falls (Vasquez, 2011, p. 123). If precipitation falls through a layer of dry air, some evaporation will occur. Because evaporation can cool the air considerably, what is forecast as rain might instead reach the surface as snow. For example, precipitation could be falling through dry air that is at a temperature a few degrees *above* zero. Because the air is dry, the wet-bulb temperature could be *below* zero (Section 7.6). As the falling precipitation evaporates, the air temperature will drop to roughly the level of the wet-bulb temperature. As a result, the falling precipitation will not melt on the way down; instead, it will reach the surface as snow. When the air temperature is close to freezing, therefore, meteorologists must consider the dryness of the air when they forecast precipitation type.

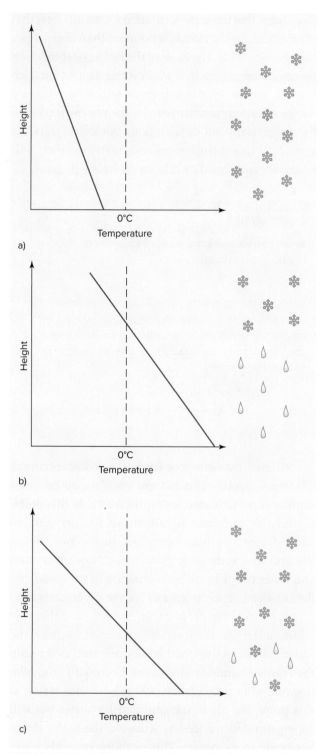

FIGURE 10.8 | a) If temperatures below the cloud are cold (i.e., below 0°C), or any warm layers are no more than about 200 m deep, precipitation will reach the ground as snow. **b)** If a warm air layer (i.e., air with temperatures above 0°C) at the surface is at least 400 m deep, ice crystals are likely to completely melt and reach the ground as rain. **c)** If warm air near the surface is between about 400 and 200 m deep, precipitation is likely to be a mix of rain and snow.

Dry air layers can affect precipitation in other ways as well. If precipitation falls into very dry air below the cloud, it can completely evaporate before reaching the ground. This will produce **virga** that, strictly speaking, is not precipitation *because* it doesn't reach the ground (Figure 10.10). Sometimes rain will evaporate only partially as it falls, so that it reaches the surface as *drizzle*. But evaporation does not have to occur in order for drizzle to be produced; recall from Section 10.2 that very low stratus clouds with weak updrafts can also produce drizzle-sized drops.

In addition to rain and snow, precipitation might also reach the ground in the form of *sleet* or *freezing rain*. For sleet to occur, the falling snow first melts into rain as it falls through a warm layer of air. It then refreezes in the form of small ice pellets as it falls through a surface layer of subfreezing temperatures. In order for sleet to form, the freezing layer near the surface must be at least 250 m deep (Figure 10.11a). Freezing rain forms in much the same way except that the subfreezing layer at the surface is shallower (Figure 10.11b). Because of the shallowness of the subfreezing layer, the precipitation does not refreeze until it comes into contact with the cold surface. Freezing rain can be a very destructive form of precipitation because the weight of the ice that forms can easily break tree branches and electrical

FIGURE 10.9 | The snow line on Vancouver's North Shore Mountains. Because Vancouver lies at sea level while the mountains rise to altitudes of 1000 m and higher, the tops of these mountains can receive snow while the city receives rain.

wires and it can make roads dangerously slippery (Figure 10.12). To forecast these types of precipitation, meteorologists must find and measure the depth of warm layers of air (those with temperatures above 0°C) and cold layers of air (those with temperatures below 0°C); they can obtain this information by carefully analyzing soundings.

VIRGA
Water particles that fall from a cloud but do not reach the ground.

Remember This

The following rules of thumb are useful for forecasting the type of precipitation that will reach the ground:

- snow will fall when there are no warm (T > 0°C) layers of air deeper than 200 m;
- a mixture of snow and rain will fall when there is a warm (T > 0°C) layer of air at the surface between 200 and 400 m deep;
- rain will fall when there is a warm (T > 0°C) layer of air at the surface deeper than 400 m;
- sleet will fall when a warm (T > 0°C) layer of air at least 200 m deep overlies a cold (T < 0°C) surface layer of air at least 250 m deep; and
- freezing rain will develop when conditions are the same as for sleet, except that the cold (T < 0°C) surface layer of air is less than 250 m deep.

FIGURE 10.10 | Precipitation falling into dry air produces virga, which appears as dark streaks below the cloud.

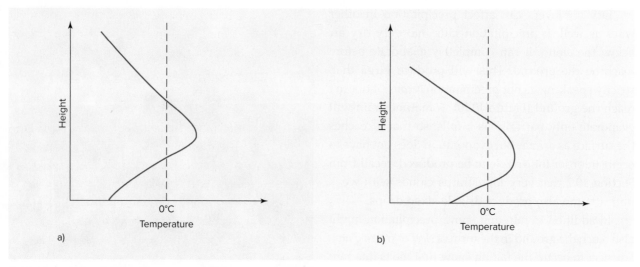

FIGURE 10.11 | a) Sleet is likely to form when a warm layer of air (i.e., a layer with temperatures greater than 0°C) that is at least 200 m deep overlies a cold layer of air (i.e., a layer with temperatures less than 0°C) at the surface that is at least 250 m deep. b) Freezing rain is likely to occur when the cold air at the surface is less than 250 m deep.

10.5 | Hail

Hail is a type of frozen precipitation that results from extreme growth by accretion; this can occur only under the unique conditions present in a cumulonimbus cloud, particularly during a thunderstorm (Section 14.4). Hailstones are generally larger than 5 mm in

GRAUPEL
Lumps of ice that begin as ice crystals but become rounded as supercooled water freezes onto them.

FIGURE 10.12 | Freezing rain can cause great destruction when it freezes on contact with surface objects, as shown in this photograph taken after a storm in Summerside, PEI, in 2008.

diameter (Figure 10.13); for hail to be considered severe, however, the hailstones must be larger than 2 cm in diameter. As hailstones get larger, they fall faster and become increasingly more destructive. Some very large hailstones can fall at rates of over 150 km/h.

Hailstones form in cumulonimbus clouds as ice accumulates on ice particles, which are either frozen raindrops or **graupel**. Recall that cumulonimbus clouds are characterized by strong updrafts and, therefore, turbulence. It is this turbulence that keeps the ice particles in the cloud and moves them rapidly from one part of the cloud to another. As the ice particles are carried up above the freezing level, they collide with supercooled water droplets that freeze onto them, allowing them to continue to grow to hailstone size by the process of accretion (Figure 10.14).

In parts of the cloud where the liquid water content is low, the hailstones will accumulate opaque ice; in parts of the cloud where the liquid water content is high, the hailstones will accumulate clear ice. This is because the amount of latent heat that is released as the water freezes onto the forming hailstones varies between different regions of the cloud. Where there is less water accumulating on the hailstone, less latent heat is released; as a result, a hailstone will be colder than it would be where there is more liquid water and more latent heat. Water will freeze more quickly onto the hailstones where they are colder, and

FIGURE 10.13 | Hailstones, such as the one shown here, are generally larger than 5 mm in diameter; they are formed by the process of accretion.

the ice that forms will be opaque because it contains air bubbles. Where the hailstones are warmer, they will accumulate a layer of water that will freeze more slowly, allowing the air to escape. In these regions of the cloud, the hailstones accumulate clear ice. As a result, individual hailstones are made up of alternating concentric layers of clear and opaque ice. Eventually, a

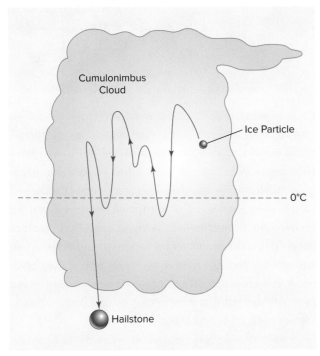

FIGURE 10.14 | As ice particles are carried up and down in the cloud, they grow by accumulating layers of ice in the cloud's subfreezing portion. Eventually, these ice particles will grow large enough to fall from the cloud as hailstones.

hailstone will be big enough to overcome the updrafts and will fall from the cloud. Clearly, the more vigorous the updrafts, the bigger will be the hailstones.

10.6 | Snow-Making

Once we understand the workings of a natural process, it can be tempting to try to imitate it, especially when there is potential for economic gain. Ever since downhill skiing became popular in the middle of the last century, ski resorts have become big moneymakers. But we can't always count on the snow to fall at certain times of the year just because it has in the past. For this reason, many ski resorts are now equipped with snow-making machines for those times when the snow does not come.

The ice-crystal process for precipitation formation, described in Section 10.3.2, has certain requirements, the most obvious of which are water and cold temperatures. But these alone will not produce snow. Often, a cold cloud will not produce precipitation because, although it contains supercooled water droplets, it does not contain ice crystals. This can happen because, in nature, ice nuclei tend to be far less abundant than condensation nuclei. So, in addition to moisture and cold, ice nuclei are needed as well.

Most snow-making machines are like big guns: the ingredients go in one end and freshly made snow is sprayed out the other. Water and compressed air are the two main ingredients. The compressed air has three functions. First, it helps to break the stream of water into tiny droplets. Second, it helps to blow these water droplets out of the snow gun. Third, it helps to cool the air—as the compressed air leaves the gun, it will expand and, therefore, cool adiabatically (see Section 4.3). Ice nuclei are often another ingredient that goes into snow guns. Although air usually contains some ice nuclei, adding more ensures that they will be in good supply and that the water droplets will become ice crystals. Since ice nuclei tend to become active at a variety of different temperatures, it is most effective to add nuclei that are active at high temperatures. It is not possible, however, just to put water, air, and ice nuclei in one end of the snow gun and expect snow to come out the other. Weather conditions must also be right.

It certainly needs to be cold to make snow, but humidity is important as well. Therefore, measurements of both air temperature and wet-bulb temperature are taken to determine whether conditions are right for making snow. When the air is dry, evaporation is more likely to occur, and evaporation will cool the air. It follows that if the air is very dry, it may be possible to make snow even when temperatures are *above* freezing. The dryness means that evaporation could chill the air enough that snow will form. On the other hand, if the air contains a lot of water vapour, the temperature needs to be several degrees *below* freezing for successful snow-making.

10.7 | The Distribution and Character of Precipitation

In some locations, most rain falls when it is warm; the wettest time of day might be the afternoon, and the wettest time of year might be the summer. In other places, rain is equally probable at any time. Sometimes rainfall can be intense, but of short duration; at other times, rain can fall steadily for hours. Where there are mountains, the *spatial* pattern of precipitation will be strongly affected. Recall that very wet regions are often found on the windward sides of mountains, while deserts can develop on the leeward sides of mountains. Much of this variability can be accounted for by the uplift mechanisms responsible for forming the clouds that produce the precipitation (Section 9.3).

Remember This

Recall the five basic mechanisms by which air can rise to form clouds:

- convection,
- orographic lift,
- convergence,
- frontal lift, and
- divergence aloft.

In the Field

Studying Precipitation Patterns on Regional and Global Scales
Ronald Stewart, University of Manitoba

Precipitation is one of the most important aspects of the global water cycle, and it is also prone to extremes. Every day we hear of hardship being experienced somewhere in the world as a result of too much or too little precipitation. However, weather and climate models have a great deal of difficulty dealing with precipitation and, given the importance of the future occurrence of precipitation, much needs to be done to address this issue.

I examine various scientific issues related to precipitation. My background is in examining precipitation on small scales, down to individual precipitation particles, so I base my research on that small-scale perspective and apply it as appropriate, even on global scales.

For my continuing research on individual particles, I focus on winter precipitation. This sort of precipitation has not received the global attention it deserves, mainly because most regions of the world are more concerned with rain. As a result, many fundamental issues remain under-examined; as a consequence, we don't even have appropriate definitions of some types of particles, especially those that are made up of a combination of liquid and solid matter. Even some of the formal definitions that we do have are incomplete—for example, our definition of ice pellets does not account for the fact that many of these pellets actually contain liquid water inside or that they often occur as a collection of individual particles. Imagine, in the twenty-first century, that something so seemingly simple is still not adequately addressed!

I have made many detailed measurements of winter precipitation over Canada, across the Arctic and from Atlantic Canada to the West Coast. I have collected samples of precipitation particles, I have photographed them, and I have documented the surrounding atmospheric conditions that led to their formation. I have also invited many of my undergraduate and graduate students to join in on this research. The projects are relatively small in scope, so students can do them on their own, and they have uncovered many fascinating insights into the features of the precipitation. They have found, for example, that ice pellets sometimes look like partial shells of ice, that rain drops just above freezing actually contain some embedded ice, that ice pellets can be so rough they are like balls of Velcro, and that some falling snow is reddish in colour. My students have gone on to various careers across Canada and around the globe, but, even now when I meet them, we often end up talking about the enjoyment of carrying out this work.

I also study how storms produce the larger environment for winter precipitation. To do this, I have flown over 600 hours in research aircraft through storms over North America. Some of the onboard instruments are capable of producing images of individual particles aloft as well as recording the temperature, moisture, and wind conditions. Such research has also led to many harrowing experiences, as the aircraft are often tossed around in violent storms.

I also lead efforts globally to better understand how and why the climate produces excesses and deficits of precipitation. I am currently studying what is similar and what is different between droughts, regardless of where they occur on Earth. There are numerous fundamental challenges that have not yet been adequately met, such as unravelling the different pathways through which the lack of precipitation is sustained over long periods. For example, anomalously low precipitation can be produced within descending air aloft that leads to clear skies, through the horizontal movement of dry air into a region, through the elimination of falling precipitation aloft by evaporation or sublimation before reaching the surface, or through reduced evapotranspiration or evaporation from dry surface conditions. We know that one or more of these different factors occur over the course of a drought, but we don't yet know how or why they occur at a given time or in a given order. If we could better understand the evolving driving mechanisms for producing and prolonging drought, we would be in an improved position to predict when droughts might occur. Such predictions would have huge implications for societies across the globe. I work with many scientists around the world who utilize, for example, global climate models and satellite information in order to address the whole issue. Our work is a good example of the ways in which a strong international team can work together to address a big scientific problem.

Overall, there are many scientific issues related to precipitation. Although any one person can address only a few of these issues, one can apply a basic expertise and perspective to a range of issues and make a difference.

RONALD STEWART is professor and head of the Department of Environment and Geography at the University of Manitoba. He is also a fellow of the Royal Society of Canada. His research focuses on regional climate, storms, and precipitation.

When the surface is strongly heated, as it might be on a hot summer afternoon, pockets of warm air can rise and form cumulus clouds. The larger cumuliform clouds—cumulus congestus and cumulonimbus—bring heavy rain, which often falls in large drops, or hail. As cumulus clouds are limited in horizontal extent, the precipitation they produce is usually localized. Also typical of these showers is their tendency to stop as quickly as they began; their duration is commonly measured in minutes. But often, just as the skies have cleared, new clouds quickly build up and the rain begins again. Such intense, intermittent rain storms may occur several times in the late afternoon of a hot day before the surface begins to cool in the evening.

Precipitation formed in this way is also often accompanied by thunderstorms (Section 14.4) and is referred to as **convectional precipitation**. In many midlatitude locations, summer precipitation is convectional and tends to fall mostly in the afternoons (Figure 10.15). This often makes summer the wet season. Near the equator, convectional precipitation is common year-round, and heavy rain can be expected almost every afternoon (Figure 10.16).

CONVECTIONAL PRECIPITATION
Precipitation that falls from clouds produced by surface heating.

OROGRAPHIC PRECIPITATION
Precipitation that falls from clouds produced by air rising along a topographic barrier.

RAIN-SHADOW DESERTS
Deserts that form on the leeward sides of mountains.

ISOHYETS
Lines on a map connecting points of equal precipitation.

The precipitation that results when air is forced up the windward side of a mountain is **orographic precipitation**. Wherever mountain ranges lie perpendicular to the prevailing wind, their windward sides will be wet and their leeward sides will be dry. The dry areas that form on the leeward sides of mountains are known as **rain-shadow deserts**. Most mid-latitude deserts are formed in this way; some of them are among the driest locations on Earth.

In contrast, the windward sides of mountains are some of the wettest locations on Earth, particularly where the mountains lie along the coast, as they do on the west coast of North America. Figure 10.17 shows the global distribution of precipitation using **isohyets** and shading. Notice that the northwest coast of North America is very wet. Westerly winds also blow off the Atlantic Ocean onto the west coast of Europe, but this region is not as wet as the west coast of North America because there are no north–south trending mountains. Notice also that the interior of British Columbia is dry, as this area lies in the lee of the mountains. In fact, the spatial variability of precipitation in the province of British Columbia is extremely complex. As air moves

eastward across the province, it encounters mountain ranges that trend north-south. This topography leads to alternating wet and dry zones across the province, with the wet zones getting steadily drier with distance from the coast (Figure 10.18).

The city of Vancouver, with the Coast Mountains to the north, is a good example of the spatial variability in precipitation that can result from geographical influences, even at a small scale (Figure 10.19). Vancouver airport receives just under 1000 mm of precipitation a year. This relatively small amount can be attributed to the airport being influenced by the rain shadow of the Vancouver Island Mountains immediately to the west. Only about 25 km to the northeast, annual precipitation totals are more than two-and-a-half times the amount recorded at the airport due to the mountains along Vancouver's north shore.

The importance of the prevailing winds, combined with mountain ranges, in creating precipitation patterns is nicely illustrated along the west coast of South America. The Andes mountains run parallel to this coast. In the tropics, prevailing winds are from the east (Section 12.2). These easterly winds result in a rain-shadow desert along the coast of southern Peru and northern Chile (Figure 10.17). This is the Atacama

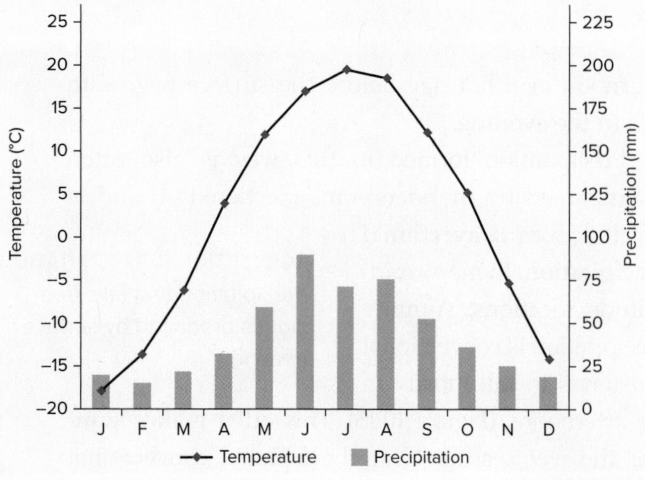

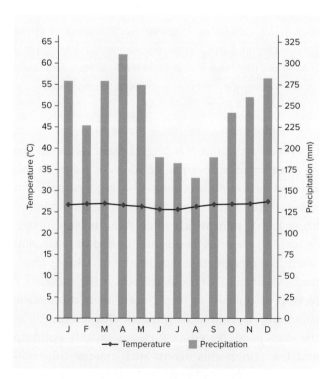

FIGURE 10.15 | Winnipeg, Manitoba, is a mid-latitude location that experiences convectional precipitation in the summer.

FIGURE 10.16 | Iquitos, Peru, is an equatorial location that experiences high amounts of convectional precipitation year-round.

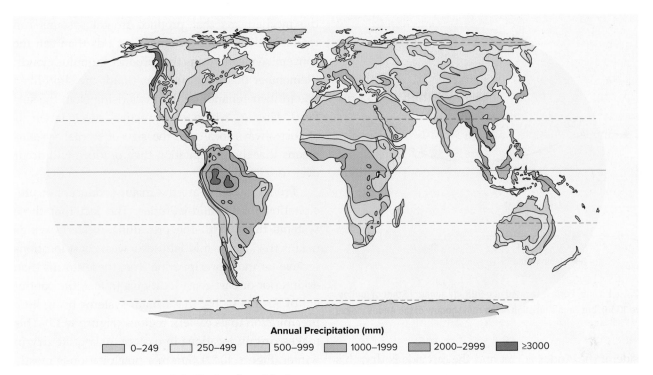

Annual Precipitation (mm)

0–249   250–499   500–999   1000–1999   2000–2999   ≥3000

FIGURE 10.17 | Map of the global distribution of precipitation.

Source: Adapted from Q.H. Stanford. (2008). *Canadian Oxford world atlas* (6th. ed.). Don Mills, ON: Oxford University Press Canada, p. 125.

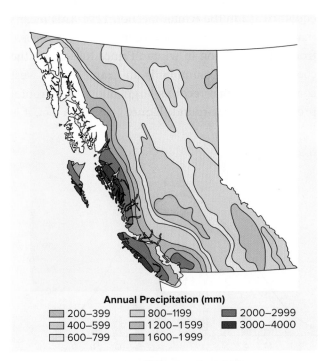

Annual Precipitation (mm)

200–399   800–1199   2000–2999
400–599   1200–1599   3000–4000
600–799   1600–1999

FIGURE 10.18 | Precipitation in British Columbia is strongly controlled by both distance from the coast and topography, as shown by this map of average annual precipitation totals. The interior of the province is dry, as this area lies to the leeward of the Coast Mountains. Just eastward, precipitation totals begin to rise again on the windward side of the Rocky Mountains.

Source: Adapted from Robin W. Pigott and Bill Hume (2009). *Weather of British Columbia.* Lone Pine Publishing.

Desert, one of the driest deserts on Earth. The climate of Lima, Peru (Figure 10.20), on the west side of the Andes, provides a sharp contrast to that of Iquitos, Peru (Figure 10.16), on the east side of the Andes. Further south, in the mid-latitudes, the prevailing winds are from the west (Section 12.2). Here, the *west*

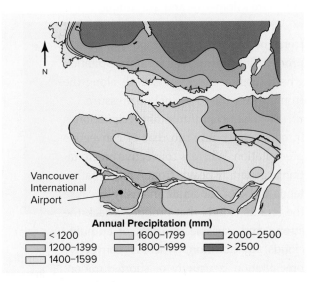

Annual Precipitation (mm)

< 1200   1600–1799   2000–2500
1200–1399   1800–1999   > 2500
1400–1599

FIGURE 10.19 | The distribution of mean annual precipitation in Vancouver, British Columbia.

Source: Adapted from Tim Oke and John Hay (1994). *The Climate of Vancouver.* B.C. Geographical Series, Number 50.

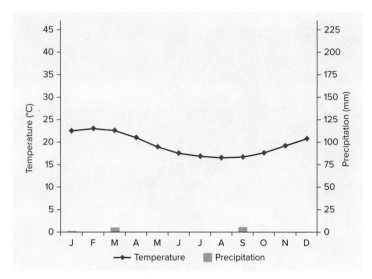

FIGURE 10.20 | Lima, Peru, is located at a latitude similar to that of Iquitos, Peru (Figure 10.16), but Lima's location in the rain shadow of the Andes makes it a desert.

side of the Andes is wet and the *east* side is dry; this rain-shadow desert is the Patagonia Desert that lies mostly in Argentina (Figure 10.17).

Precipitation produced in the frontal systems of the mid-latitudes is known as **frontal precipitation**. Frontal systems, also known as *extratropical cyclones* or *mid-latitude cyclones*, are large areas of low pressure that form along fronts (chapters 13 and 14). These storms usually cover thousands of square kilometres and are unique to the mid-latitudes because it is only there that fronts occur. Air rises in frontal systems to produce stratiform clouds and precipitation where there is *convergence*, both along the fronts and in the centre of low pressure.

The character of frontal precipitation is in complete contrast to that of convective precipitation. Whereas convective precipitation is generally heavy, frontal precipitation tends to be of moderate intensity because the stratus clouds that produce it lack the strong and variable updrafts of cumulus clouds. In addition, whereas convective precipitation is usually of short duration, frontal precipitation can continue for hours, even days, with little break. This is because

FRONTAL PRECIPITATION
Precipitation that falls from clouds produced in frontal systems.

the mechanisms that produce frontal systems can maintain themselves for longer periods than can the convective mechanisms that produce cumulus clouds. (Remember that most cumulus clouds are short-lived due to their reliance on thermals rising from the surface.) Finally, whereas convective precipitation is almost always localized, the size of frontal systems means that the precipitation they produce will occur over very large areas.

Frontal systems are the major producers of precipitation in the mid-latitudes. The fact that these systems travel with the prevailing westerly winds means that, in the mid-latitudes, west coast locations receive far more precipitation from these storms than do interior or east coast locations. In fact, the continent of Asia is so big that frontal systems bring little precipitation to its eastern regions (Figure 10.17). This makes certain locations in eastern Asia quite dry in winter (Figure 10.21). Summer precipitation is mostly convectional in these locations.

The path of frontal systems tends to shift with the seasons, moving poleward in the summer and equatorward in the winter (Section 12.2). This means that any mid-latitude location can potentially receive frontal precipitation in winter. Prince Rupert, on the coast of British Columbia, is an example of a west coast location that receives large amounts of frontal precipitation year-round (Figure 10.22b). In fact, it is

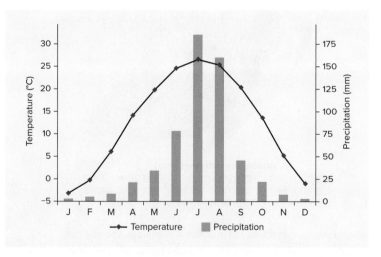

FIGURE 10.21 | Bejing, China, does not receive significant frontal precipitation in the winter because it is a mid-latitude location situated on the east side of a very large continent. The relatively high amount of precipitation that falls in the summer is due mostly to convection.

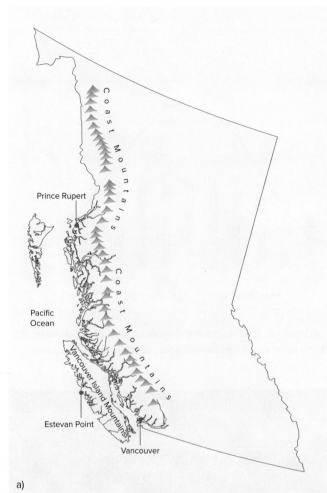

a)

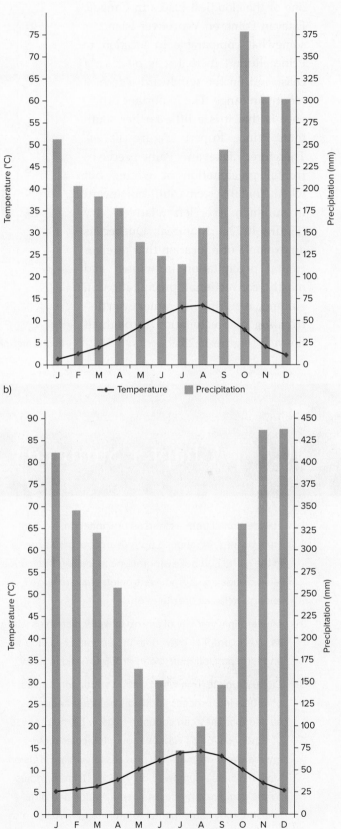

b)

c)

FIGURE 10.22 | a) A map of British Columbia showing the location of Prince Rupert and Estevan Point, which are both mid-latitude locations. b) Prince Rupert receives large amounts of precipitation year-round. Not only does it always lie in the path of frontal systems, it is also on the windward side of British Columbia's coastal mountain range. c) Estevan Point experiences considerably more precipitation in the winter than it does in the summer. This is because the average path of frontal systems shifts northward in the summer.

one of the cloudiest places in Canada. Estevan Point, on Vancouver Island, is somewhat comparable in location to Prince Rupert in that it is on a west coast and on the windward side of a mountain range. The important difference is that it is a little farther south than Prince Rupert (Figure 10.22a). Therefore, Estevan Point receives frontal precipitation in winter, but when frontal systems shift northward in summer, it is left relatively dry (Figure 10.22c). Montreal, Quebec, is an example of a location that receives frontal precipitation in winter, and mostly convectional precipitation in summer, resulting in a rather even distribution of precipitation throughout the year (Figure 10.23).

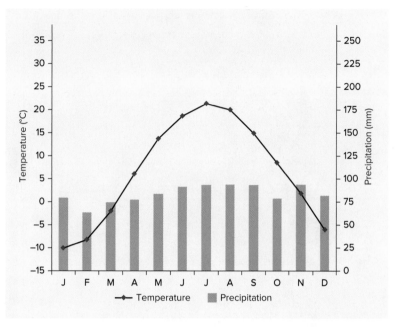

FIGURE 10.23 | Montreal, Quebec, is a mid-latitude location that receives mostly frontal precipitation in the winter and convectional precipitation in the summer.

10.8 | Chapter Summary

1. In order to fall from a cloud as precipitation, a drop must be large enough to both overcome the updrafts in the cloud and not evaporate once leaving the cloud. Most clouds don't produce droplets that are large enough to fall as precipitation.

2. The terminal velocity of a drop of water increases with its size. In order to overcome the updrafts in a cloud and form precipitation, cloud droplets need to grow.

3. Cloud droplets form when water vapour condenses onto CCN in a process known as *heterogeneous nucleation*. They can continue to grow by the process of condensation as water vapour molecules move toward water droplets in a process known as *diffusion*. Most droplets will grow to only about 10 μm in radius by the combined processes of nucleation and diffusion, but they can continue to grow to raindrop size by either the collision and coalescence process, or by the Bergeron–Findeisen process.

4. The collision and coalescence process forms precipitation in warm clouds. As droplets move through the cloud, those that are moving faster should catch up to, collide with, and coalesce with smaller droplets. Eventually, some drops will be big enough to overcome the updrafts in the cloud. This process can work only if the cloud contains droplets in a variety of sizes, some of which must be larger than about 20 μm in radius, and the process is enhanced when clouds are deep with strong updrafts, and when electric fields are present.

5. The Bergeron–Findeisen process forms precipitation in cold clouds. These clouds, in which temperatures are between 0°C and –40°C, contain supercooled water droplets as well as ice crystals. Because the saturation vapour pressure for the water droplets is greater than that for the ice crystals, the water droplets will tend to evaporate, and the ice crystals will tend to grow as water vapour is deposited onto them. These ice crystals may continue to grow by accretion and aggregation until they are big enough to overcome the updrafts in the cloud.

6. Any precipitation formed by the ice-crystal process will begin as snow. The temperature profile in the air below the cloud will then determine the precipitation type that reaches the ground. If temperatures remain below freezing, it will be snow. If there is a deep enough layer of above-freezing temperatures at the surface, it will be rain. When the snow melts then refreezes, precipitation will reach the ground as either sleet or freezing rain.

7. Hail is a type of frozen precipitation made up of concentric layers of ice. Hailstones form in cumulonimbus clouds as the turbulence in the cloud moves ice particles around, allowing them to grow by accretion of supercooled water. The stronger the updrafts in these clouds, the bigger will be the hailstones.

8. One practical application of understanding precipitation processes is snow-making. An understanding of the Bergeron–Findeisen process and nucleation below freezing has allowed us to successfully make snow at ski resorts.

9. Depending on how the air is lifted, precipitation can be of three types: convective, orographic, or frontal. Convective precipitation tends to be intense, localized, and of short duration. It most commonly occurs on warm afternoons. Orographic precipitation falls on the windward sides of mountains, making these locations some of the wettest places on Earth. In contrast, the leeward sides of mountains can become rain-shadow deserts and are some of the driest places on Earth. Frontal precipitation is confined to the mid-latitudes and tends to fall steadily for hours over large areas. Because the path of frontal systems shifts with the seasons, it can influence the seasonal distribution of precipitation in mid-latitude locations.

Review Questions

1. How and why are cloud droplets and rain drops different from each other?

2. What is a warm cloud? How does precipitation form in warm clouds?

3. What is a cold cloud? How does precipitation form in cold clouds?

4. Why is a *cold* stratus cloud likely to produce precipitation, while a *warm* stratus cloud is unlikely to do so?

5. How does the temperature stratification below a cloud influence what type of precipitation will reach the ground?

6. Snow, sleet, freezing rain, and hail are all forms of frozen precipitation. How are they different from each other?

7. In what two ways can drizzle occur?

8. Why might precipitation falling through dry air cause cooling?

9. How and why does hail form in cumulonimbus clouds?

10. How do convectional precipitation and frontal precipitation differ?

11. What factors influence the spatial and temporal distribution of precipitation?

Suggestions for Research

1. Explore the various methods that people have used to produce rain by cloud seeding. Comment on the success of such endeavours, and suggest additional studies that could be conducted to improve cloud-seeding techniques.

2. Visit the website of Environment Canada's Weather Office and obtain precipitation data for your city or region. Look at historical precipitation trends, and suggest practical uses for this information.

References

Vasquez, T. (2011). *Weather analysis and forecasting handbook*. Garland, TX: Weather Graphics Technologies.

11
Winds

Learning Goals

After studying this chapter, you should be able to

1. *describe* the forces that act on the air to create winds;

2. *show* how these forces interact to produce the winds we observe in the upper air and at the surface;

3. *describe* how horizontal and vertical motions are related in high- and low-pressure systems, and *describe* how the vertical motions are associated with weather;

4. *explain* why there is a relationship between the horizontal temperature gradient of an atmospheric layer and the change in winds with height through the layer;

5. *appreciate* the value of the thermal wind concept; and

6. *explain* how differences in heating and cooling at the local scale can cause winds.

Winds can have a powerful impact on conditions at the surface. The storm waves shown here, crashing against the shores of Cape Breton Highlands National Park, are being generated as strong winds blow across the ocean's surface.

In Chapter 3, you were introduced to weather maps. An important feature of these maps, whether they are for the surface or the upper air, is that they show the *horizontal* variation of pressure. It is this variation in pressure, shown using *isobars* on surface maps and *height contours* on upper-air maps, that is the driving force behind **wind**. This motion, as with any motion, results from the action of *forces* and can be explained using Newton's laws (Section 1.5).

This chapter will begin with a discussion of the forces that act on the air. These forces are gravity, the pressure gradient force, the **Coriolis force**, the **centripetal force**, and **friction**. Considered along with Newton's laws of motion, these forces can be used to explain the relationship between the winds we observe and the patterns of isobars or height contours on weather maps. In general, upper-air winds flow parallel to the height contours on upper-air maps, while surface winds tend to flow at an angle across the isobars on surface weather maps (Figure 11.1).

WIND
The mostly horizontal movement of air relative to Earth's surface.

CORIOLIS FORCE
An apparent force, resulting from the rotation of Earth, that causes objects moving freely above Earth's surface to appear to be deflected from a straight-line path, relative to Earth's surface.

CENTRIPETAL FORCE
The net force, directed toward the centre of rotation, that acts on a rotating object.

FRICTION
The force that resists motion whenever objects move, or try to move, relative to each other.

Remember This

Remember that Newton's first law states that as long as there is no *unbalanced* force acting on an object, a stationary object will remain stationary, and a moving object will continue to move at a constant velocity in a constant direction. Newton's second law states that the acceleration of an object is proportional to the unbalanced force acting on the object and inversely proportional to the object's mass; the direction of the acceleration is the same as the direction of the unbalanced force. Newton's second law can be expressed as a = F/m.

At the end of this chapter, you will discover how local wind systems operate. These systems occur over distances of a few kilometres to several hundred kilometres, and their winds are active over periods of hours. In Chapter 12, you will learn about wind systems at the *planetary scale*. At this much larger scale, distances are measured in thousands of kilometres, and the winds are considered to be more or less permanent. At both scales, the wind systems are caused by pressure gradients that have resulted from temperature gradients (Section 3.6).

11.1 | Forces That Influence Atmospheric Motion

11.1.1 | Pressure Gradient Force

Of the forces acting on the air, the *pressure gradient force* is the one that *drives* the wind. The others will act to slow wind down, or change its direction, but they cannot create winds out of still conditions. Recall that the pressure gradient force results from differences in pressure. A pressure gradient (PG) is simply the change in pressure, ΔP, with distance, Δx, as shown by Equation 11.1.

$$PG = \frac{\Delta P}{\Delta x}$$

(11.1)

We can use weather maps to determine pressure gradients, as shown in Figure 11.2.

The pressure gradient *force* (PGF) is proportional to the pressure gradient (Equation 11.2), and it causes air to move from regions of high pressure to regions of low pressure

$$PGF = -\frac{1}{\rho}\left(\frac{\Delta P}{\Delta x}\right)$$

(11.2)

There is a negative sign in Equation 11.2 because the pressure *decreases* as the distance *increases* in the direction of the pressure gradient force (Figure 11.3). (This negative sign can be ignored in most calculations.) Equation 11.2 indicates that the units of the pressure gradient force are m/s^2. As shown by Newton's second law, these are the units of acceleration, or force per unit mass.

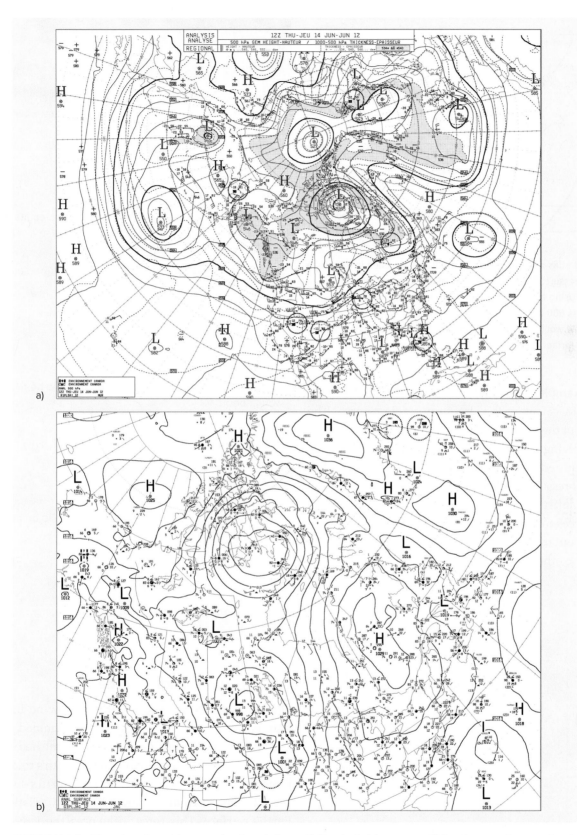

FIGURE 11.1 | a) An upper-air weather map showing winds paralleling the height contours. b) A surface weather map showing winds crossing the isobars at an angle. Winds are represented by small, flagged arrows on both maps. These wind arrows also show that winds are much faster on the upper-air map than they are on the surface map. Why do you think this is?

Source: Environment Canada, Weather Office, www.weatheroffice.gc.ca/analysis/index_e.html, 14 June 2012.

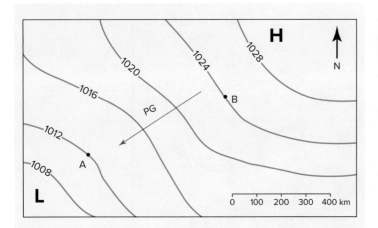

FIGURE 11.2 | On this map, the pressure gradient's direction is from high pressure to low pressure. The isobars show that the change in pressure from A to B is 12 hPa. Using the map scale, we can determine that the distance between A and B is 600 km. The pressure gradient from A to B is, therefore, 2 hPa/100 km. How would the pressure gradient change if the isobars were closer together or farther apart?

Example 11.1 shows how Equation 11.2 can be used to calculate the pressure gradient force. The speed of the resulting wind can then be calculated by multiplying this force by the time over which it acts. As required by Newton's laws, the longer an *unbalanced* pressure gradient force acts, the faster will be the resultant wind. Of course, in reality, winds do not accelerate indefinitely. Nor do they blow themselves out by eliminating pressure differences. This is because other forces are involved.

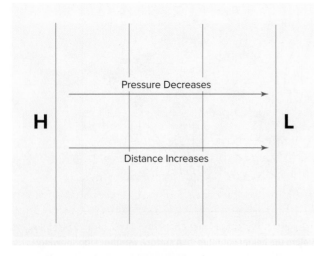

FIGURE 11.3 | As distance increases in the direction of the pressure gradient force, pressure decreases.

Question for Thought

How could you use Newton's second law to show that Equation 11.2 expresses the pressure gradient force as a force per unit mass?

Example 11.1

a) Calculate the pressure gradient force that results from a pressure gradient of 2 hPa/100 km. Use a value of 1.25 kg/m^3 for the density of air.

Use Equation 11.2, and convert the pressure gradient to units of pascals per metre.

$$\text{PGF} = \left(\frac{1}{1.25 \text{ kg/m}^3}\right)\left(\frac{200 \text{ kg} \cdot \text{m}^{-1} \cdot \text{s}^{-2}}{100,000 \text{ m}}\right)$$

$$= 0.0016 \text{ m/s}^2$$

b) If this force acted for one hour, what would be the wind speed, V?

$$V = a\,t$$

$$= 0.0016 \text{ m/s}^2 \times 3600 \text{ s}$$

$$= 5.8 \text{ m/s}$$

c) If this force acted for two hours, what would be the wind speed, V?

$$V = 0.0016 \text{ m/s}^2 \times 7200 \text{ s}$$

$$= 11.5 \text{ m/s}$$

The longer the pressure gradient force acts, the faster will be the winds.

11.1.2 | The Coriolis Force

One of the other forces that influences wind behaviour is the *Coriolis force*. This force is named after Gustave-Gaspard de Coriolis, who identified it in 1835. The Coriolis force occurs due to Earth's rotation; it makes objects moving freely above Earth's surface seem to deflect from a straight-line path, relative to Earth's surface. This force influences the *large-scale* movement of air and water. It can also influence the way we perceive the paths of long-range military projectiles like bullets, artillery shells, and missiles.

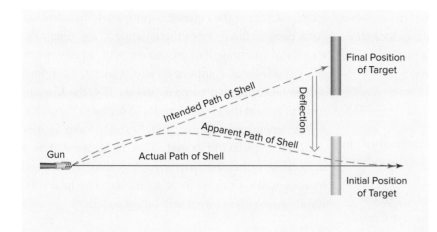

FIGURE 11.4 | From an aerial perspective, we can see why an artillery shell will miss its target if Earth's rotation is not taken into account. The shell is aimed to reach the target's initial position but, as the shell travels toward the target, the target moves along with Earth's rotation. Thus, the *actual path* of the shell ends up being to the right of the *intended path*. If we were to view the same process from Earth's surface, it would seem as though the shell has taken the *apparent path*, because we would not be aware that the target has moved; thus, it would seem as though the shell has been deflected. (The gun is, of course, moving with the rotating Earth as well, so the movement of the target discussed here is actually its movement *relative* to the gun.)

Such things are somewhat more tangible than air and water, so they can be used to help understand the Coriolis force.

As a military projectile heads toward its target, it will move freely above the surface in a straight line. Meanwhile, the target, fixed to Earth's surface, will move as Earth rotates, so that the projectile is heading toward a *moving* target. If this movement is not accounted for, the projectile will miss the target. From an aerial perspective, we would be able to see why (Figure 11.4). The projectile travels in a straight line, but the target will have moved by the time the projectile reaches the target's initial position. When we are on the ground, we are not aware that the target is moving because we are *also* moving. Thus, it *appears* as though the projectile is deflected because, as we see it, the projectile starts off heading toward the target, but then ends up to the right of the target. In other words, it appears that it is acted upon by a force. (Recall that, according to Newton's first law, a change in direction is the result of the action of an unbalanced force.) This apparent force is the *Coriolis force*.

In contrast to the pressure gradient force, the Coriolis force is not a *driving* force; it is a *deflecting* force. It acts only once the air is in motion, and it acts at right angles to the path of motion (Figure 11.5). In the northern hemisphere, the Coriolis deflection is to the *right* of the path of motion; in the southern hemisphere, it is to the *left*. This difference between the hemispheres occurs because the direction of Earth's rotation differs based on perspective. Looking down from above the North Pole, we would see Earth

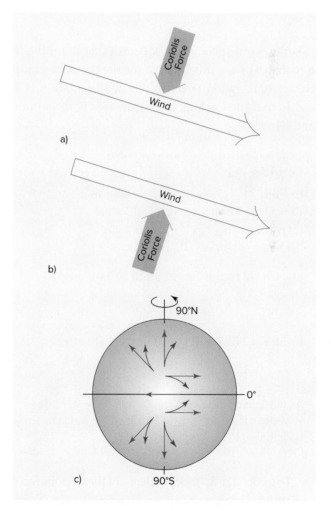

FIGURE 11.5 | The Coriolis force acts at right angles to the path of motion. a) In the northern hemisphere, it produces deflection to the *right* of the path of motion. b) In the southern hemisphere, the deflection is to the *left* of the path of motion. c) The Coriolis force acts everywhere on Earth except at the equator. Red lines show the intended path, while blue lines show the actual path.

rotating counterclockwise; looking down from above the South Pole, we would see Earth rotating clockwise.

To define the Coriolis force mathematically, we begin by defining the Coriolis parameter, f_c.

$$f_c = 2\,\Omega \sin \varphi \tag{11.3}$$

In Equation 11.3, φ is latitude, and Ω is Earth's rotation rate of 7.27×10^{-5} rad/s, or 360° per day. (The radians [rad] in the units can be ignored here, leaving s^{-1} as the units for the Coriolis parameter.) Making use of the Coriolis parameter, we can express the Coriolis force (CF) as shown by Equation 11.4.

$$CF = f_c\,V \tag{11.4}$$

V is the wind speed in m/s. Notice that the units in Equation 11.4 are m/s^2, just as for the pressure gradient force. Equation 11.4, together with Equation 11.3, shows that the Coriolis force increases with latitude and wind speed.

Example 11.2

Determine the Coriolis force for a wind speed of 10 m/s at a) 30° latitude and b) 60° latitude.

Use Equation 11.4.

a) $CF = 2\,(7.27 \times 10^{-5}\ s^{-1})\,(\sin 30°)\,(10\ m/s)$

$= 7.3 \times 10^{-4}\ m/s^2$

b) $CF = 2\,(7.27 \times 10^{-5}\ s^{-1})\,(\sin 60°)\,(10\ m/s)$

$= 1.3 \times 10^{-3}\ m/s^2$

As the latitude increases, the Coriolis force increases. How does the Coriolis force change if the wind speed is increased to 20 m/s in each case?

To understand the dependence of the Coriolis force on latitude, imagine a person standing on the North Pole for a day. The person will rotate 360°. Now imagine a person standing on the equator for a day. This person will travel a distance equivalent to the circumference of Earth, about 40,000 km, but will not rotate. The movement experienced at the North Pole is completely rotational; that at the equator is completely *translational*. As a person moves from the equator to the poles, the rotational component of movement will *increase*, while the translational component will *decrease*. Therefore, as can be confirmed using Equation 11.4, the Coriolis force is zero at the equator and maximum at the poles.

As mentioned above, the Coriolis force is also dependent on wind speed. This is because it is a deflecting, rather than a driving, force. As wind speed increases, the Coriolis force increases. The faster the wind moves, the more it will be deflected.

Remember This

The Coriolis force

- affects wind *direction*, but not wind *speed*;
- causes deflection to the *right* of the path of motion in the northern hemisphere and to the *left* of the path of motion in the southern hemisphere;
- increases with latitude; and
- increases with wind speed.

11.1.3 | Centripetal Force (or Acceleration)

When winds follow a curved path, the *centripetal force* becomes important. This force acts toward the centre of rotation, continually pulling in toward that centre. It is the force necessary to keep an object moving along a curved path. When an object moves in a curved path, it is always changing direction; therefore, it is accelerating. Recall that Newton's first law tells us that when something is accelerating, the forces on it are unbalanced. This imbalance, or *net* force, is the centripetal force—more accurately, it is the centripetal *acceleration*. This seeming confusion arises because rotation is not adequately covered in Newton's laws.

The centripetal force is easily visualized. When a rock is tied to a string and spun around, the string produces the centripetal force. In this case, it is a *mechanical* force. If the string were to break, the rock would fly off in a straight line that is tangential to the circle (Figure 11.6). Another example comes from the revolution of Earth around the sun. In this case, the centripetal force is the *gravitational* force between Earth and the sun.

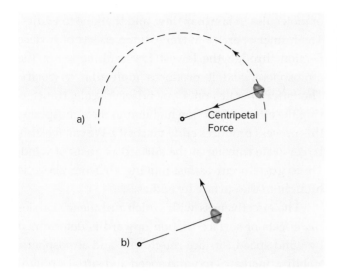

FIGURE 11.6 | a) When a rock rotates on the end of a string, the string provides the centripetal force that continually pulls the rock in toward the centre. b) When the string breaks, the rock flies off in a straight line that is tangential to the circle.

The centripetal force, C_LF, can be represented as a force per unit mass.

$$C_LF = \frac{V^2}{R}$$

(11.5)

In Equation 11.5, V, once again, is the wind speed, and R is the radius of curvature. For winds, this is the curvature of the isobars on a weather map. As the isobars get straighter, the radius approaches infinity and the centripetal force approaches zero. Like the Coriolis force, the centripetal force is not a driving force; it acts only once the air is moving, making it dependent on wind speed, as shown in Equation 11.5.

Example 11.3

Calculate the centripetal force for winds with a speed of 5 m/s at 500 km from the centre of a low-pressure system.

Use Equation 11.5.

$$C_LF = \frac{(5\text{ m/s})^2}{500,000\text{ m}}$$

$$= 0.00005\text{ m/s}^2$$

Question for Thought

How do equations 11.4 and 11.5 show that the Coriolis force and the centripetal force are not *driving* forces?

Equation 11.5 shows that the force required to keep an object moving along a curved path is proportional to the object's velocity and inversely proportional to the radius of the curve. The faster an object travels, and the smaller the radius of the curve, the greater the force required to keep the object in the curve. That's why drivers should slow down when entering a curved portion of the road, especially when the curve is tight. In the case of a car rounding a curve, the centripetal force comes from friction. This means that if a car is travelling fast on a tight turn, friction needs to be high. So, if the road is wet, the car needs to slow down even more.

Remember This

Forces have both a direction and a magnitude. For the *pressure gradient force*, the direction is from high pressure to low pressure, and the magnitude is represented by Equation 11.2. For the *Coriolis force*, the direction is perpendicular to the direction of the motion, and the magnitude is represented by Equation 11.4. For the *centripetal force*, the direction is toward the centre of rotation, and the magnitude is represented by Equation 11.5.

11.1.4 | Friction

Whenever objects move, or try to move, relative to each other, the force of *friction* resists that motion. It follows that friction acts in the direction opposite to that of the motion, thus slowing the motion. In addition, friction increases with the speed of the motion.

In the atmosphere, friction arises as air moves relative to Earth's surface. As a result, wind speeds increase away from Earth's surface (Figure 11.7). Recall that a change in wind speed or direction with height

VISCOSITY
Resistance to flow.

MOMENTUM
The product of mass and velocity.

LAMINAR FLOW
A type of flow in which layers of the fluid flow over each other with no disruption between the layers.

EDDIES
Irregular whirling motions that characterize turbulent flow.

MOLECULAR VISCOSITY
Small-scale resistance to flow due to the random motions of molecules in laminar flow.

EDDY VISCOSITY
Large-scale resistance to flow due to the random, irregular motions associated with turbulence.

is known as *wind shear*. However, because turbulence carries the effects of friction up into the atmosphere, this increase in wind speed with height occurs gradually, rather than suddenly. This is possible because friction also occurs *within* the air, as air molecules move relative to each other. Such *internal* friction, or **viscosity**, arises due to **momentum** exchange—the continual process by which slower-moving air near the surface is exchanged with faster-moving air above.

There are two types of momentum exchange: that due to the random motions of molecules in **laminar flow**, and that due to **eddies**. The former operates to transfer the effects of surface friction through the lowest few millimetres of the atmosphere, and it produces **molecular viscosity**. The latter operates on a much larger scale, transferring the effects of friction higher into the atmosphere; this process produces **eddy viscosity**. We can feel this larger-scale transfer at the surface as gusts of wind. These gusts occur as fast-moving air from above is brought to the surface by eddies.

The effectiveness with which turbulence carries the effects of surface friction upward is determined by wind speed, surface roughness, and atmospheric stability. Increases in wind speed and surface roughness will increase *mechanical* turbulence, while surface heating, and the instability it creates, will increase *thermal* turbulence (Section 4.4.2). Thus, turbulence is maximized when fast wind speeds and an unstable atmosphere occur over a rough surface such as a forest or an urban area. In unstable air, the vertical motions associated with mechanically created eddies will be enhanced, increasing turbulence.

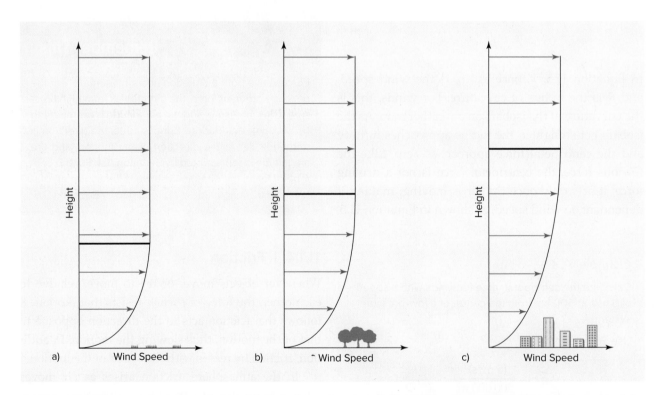

FIGURE 11.7 | The change in wind speed with height for surfaces of varying roughness. As surface roughness increases from a) to c), the wind shear decreases, and the height of the planetary boundary layer (shown by the black lines) increases. For a given surface roughness, this effect is greater in unstable air and smaller in stable air.

On the other hand, in stable air, the vertical motions of mechanically created eddies will tend to be suppressed (Figure 11.8).

As shown in Figure 11.7, the depth of the **planetary boundary layer** increases with surface roughness, because the increased mechanical turbulence associated with this roughness carries surface effects higher. Greater thermal turbulence also increases the depth of the planetary boundary layer, so that it tends to be greater with a warmer surface and smaller with a cooler surface. With the very strong surface heating that leads to thunderstorms (Section 14.4), surface effects can be carried to the tropopause. On average, however, the planetary boundary layer is about 1 to 2 km in depth on a clear day.

In addition, turbulence is linked in rather complex ways with the vertical wind shear, or the *rate* at which wind speed increases with height. Figure 11.7 shows that wind shear decreases with increasing surface roughness. Wind shear is also less in an unstable atmosphere. Turbulence helps reduce wind shear because as turbulence increases, the vertical transport of momentum increases; the result is that wind speeds tend to become more uniform with height. However, this relationship is complex because wind shear, in turn, generates turbulence.

> **PLANETARY BOUNDARY LAYER**
> The layer of air closest to the surface that is influenced by friction from the surface.

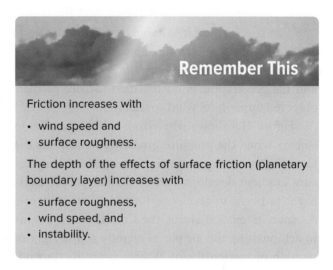

Remember This

Friction increases with

- wind speed and
- surface roughness.

The depth of the effects of surface friction (planetary boundary layer) increases with

- surface roughness,
- wind speed, and
- instability.

Here we have seen how friction associated with air moving over the surface affects the vertical wind *profile* near the surface. In Section 11.2.3, we will examine how friction affects the direction and speed of winds in the planetary boundary layer. First, however, we will consider the much simpler case of upper-air winds, which are unaffected by surface friction.

11.2 | The Winds

As the forces described above act on the air, they produce the winds that we observe. The pressure gradient force is the force that starts the air moving. As the wind accelerates, the effect of the other forces will gradually increase because these other forces depend on wind speed. Feedback processes operate among the forces until they become balanced. To explain the winds we observe, and to develop equations for predicting wind speeds, we will apply Newton's laws of motion.

FIGURE 11.8 | When air flows over an obstacle (red arrow), the obstacle generates mechanical eddies (blue arrow). a) When the atmosphere is stable, the eddies will remain small. b) When the atmosphere is unstable, the eddies can grow as the vertical motions are enhanced by thermal turbulence. How would conditions at the ground be different in these two situations?

11.2.1 | Geostrophic Winds

Geostrophic winds are idealized winds that develop when the pressure gradient force and the Coriolis force are balanced. When plotted as arrows on a weather map, these winds are observed to *parallel* the isobars. Geostrophic winds occur under conditions of straight flow above the planetary boundary layer, so there is no centripetal force, and the effects of friction are negligible. Although geostrophic winds are an approximation, upper-level winds are usually observed to blow more or less parallel to isobars, and the geostrophic wind equation usually predicts observed upper-level wind speeds well.

> **GEOSTROPHIC WIND**
>
> A wind, depicted on a weather map as blowing parallel to straight isobars or height contours, that develops when the pressure gradient force is balanced by the Coriolis force.

Figure 11.9 shows why winds will be parallel to isobars when the pressure gradient force equals the Coriolis force. Imagine an air parcel at Point 1, a pressure gradient develops around this air parcel, causing it to begin to accelerate toward lower pressure. As soon as motion starts, the Coriolis force begins to act, pushing the air parcel slightly to the right of its path of motion (Point 2). Because of its dependence on wind speed, the Coriolis force will be very small at this point. However, with the pressure gradient force greater than the Coriolis force, the wind will continue to accelerate. As the wind gets faster, the Coriolis force will gradually increase, so that the wind is turned more and more to the right of its path (Point 3). Notice that the Coriolis force is changing direction as it increases, because it always acts at right angles to the wind. Eventually, a wind speed is reached at which the Coriolis force will have increased enough to be equal to the pressure gradient force. In addition, the Coriolis force will have turned to the point that it is acting in the opposite direction to the pressure gradient force (Point 4). At this point, therefore, the Coriolis force is equal in magnitude and opposite in direction to the pressure gradient force. The result is that winds blow, at a steady speed, parallel to the isobars. In the northern hemisphere, low pressure will lie to the left of the path of motion, and high pressure will lie to the right. In the southern hemisphere, the wind will flow in the opposite direction.

Because a feedback operates between the two forces, a balance can be maintained. For example, if the wind were to speed up, a corresponding increase in the Coriolis force would turn the flow toward higher pressure. Going against the pressure gradient, the winds would slow. As a result, the Coriolis force would decrease, and the forces would again be balanced.

In reality, the development of winds is not as straightforward as Figure 11.9 suggests, because the air will always be moving and adjusting to changes in the pressure gradient. Figure 11.9 simply provides an explanation of how the interplay of forces creates the winds we observe. This explanation allows us to produce an equation that can be used for calculating the velocity of geostrophic winds.

To do this, we begin by setting the pressure gradient force as equal to the Coriolis force, as shown in Equation 11.6a. We then solve this equation for the geostrophic wind speed, V_g.

$$-\frac{1}{\rho}\left(\frac{\Delta P}{\Delta x}\right) = f_c\, V_g \qquad (11.6a)$$

$$V_g = -\left(\frac{1}{\rho\, f_c}\right)\left(\frac{\Delta P}{\Delta x}\right) \qquad (11.6b)$$

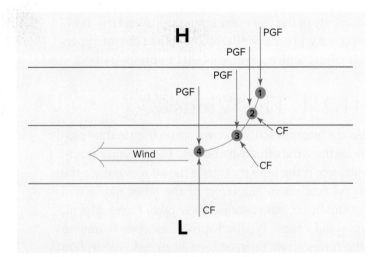

FIGURE 11.9 | The pressure gradient force and the Coriolis force come into balance to create geostrophic winds. Note that this example represents the direction the wind would take in the northern hemisphere.

Because geostrophic winds occur in the upper air, and because upper-air charts show pressure variation using height contours instead of isobars, it is more practical to use *height* gradients rather than *pressure* gradients in Equation 11.6. Keep in mind that height gradients are equivalent to pressure gradients, because high heights correspond with high pressure, and low heights correspond with low pressure (Section 3.7). Using the hydrostatic equation (Equation 3.9), we can substitute $\rho\, g\, \Delta z$ for ΔP in Equation 11.6b and obtain Equation 11.7.

$$V_g = -\left(\frac{g}{f_c}\right)\left(\frac{\Delta z}{\Delta x}\right) \tag{11.7}$$

Example 11.4

a) At latitude 50° and a height of about 5 km in the atmosphere, the pressure changes 2 kPa over a distance of 700 km. The standard atmosphere shows that at this height, air density is about 0.7 kg/m³. What is the geostrophic wind speed?

Use Equation 11.3 to calculate the Coriolis parameter.

$$f_c = 2\ (7.27 \times 10^{-5}\ \text{s}^{-1})\ (\sin 50°)$$

$$= 0.00011\ \text{s}^{-1}$$

Use Equation 11.6b to calculate the geostrophic wind speed.

$$V_g = \left[\frac{1}{(0.7\ \text{kg/m}^3)\ (0.00011\ \text{s}^{-1})}\right]\left(\frac{2\ \text{kPa}}{700\ \text{km}}\right)$$

$$= 37.1\ \text{m/s}$$

b) At the same latitude, the height of the 500 hPa surface increases 20 m over a distance of 150 km. Calculate the geostrophic wind speed in this case.

Since the latitude is the same as in a), we can use the same Coriolis parameter, but this time we use it in Equation 11.7.

$$V_g = \left(\frac{9.8\ \text{m/s}^2}{0.00011\ \text{s}^{-1}}\right)\left(\frac{20\ \text{m}}{150{,}000\ \text{m}}\right)$$

$$= 11.9\ \text{m/s}$$

Equations 11.6 and 11.7 both show that the geostrophic wind speed increases with the pressure gradient, as we would expect. This means that wind speeds will be greatest where isobars, or height contours, are closest together. The equations also show that, due to its dependence on the Coriolis parameter, geostrophic wind speed is dependent on latitude. Figure 11.9 helps to show why this is. Recall that the Coriolis force is greater at higher latitudes than it is at lower latitudes, for a given wind speed. With a greater Coriolis force, the pressure gradient force and the Coriolis force will come into balance at a lower wind speed than they would with a smaller Coriolis force. In other words, the Coriolis force can catch up to the pressure gradient force more quickly. This means that, for the same pressure gradient force, geostrophic winds are slower near the poles than they are near the equator. In addition, because the Coriolis parameter is zero at the equator, geostrophic winds are undefined at the equator. Very close to the equator, pressure gradients will tend to blow themselves out. The dependence of geostrophic winds on latitude is shown in Example 11.5.

Example 11.5

In Example 11.4, a pressure gradient of 2 kPa over 700 km was shown to produce a geostrophic wind speed of 37.1 m/s at latitude 50°. What would the wind speed be at latitude 20° for the same pressure gradient?

Use Equation 11.3 to calculate the Coriolis parameter.

$$f_c = 2\ (7.27 \times 10^{-5}\ \text{s}^{-1})\ (\sin 20°)$$

$$= 0.00005\ \text{s}^{-1}$$

Use Equation 11.6b to calculate the geostrophic wind speed.

$$V_g = \left[\frac{1}{(0.7\ \text{kg/m}^3)\ (0.00005\ \text{s}^{-1})}\right]\left(\frac{2\ \text{kPa}}{700\ \text{km}}\right)$$

$$= 81.6\ \text{m/s}$$

Geostrophic wind speeds are greater at lower latitudes than they are at higher latitudes, for the same pressure gradient.

11.2.2 | Gradient Winds

Like geostrophic winds, **gradient winds** occur above the planetary boundary layer and, when plotted on maps, flow parallel to isobars or height contours. Unlike geostrophic winds, however, gradient winds apply when these lines are *curved*. On a weather map, therefore, gradient winds will *look* just like geostrophic winds; but, because of the curvature of the isobars, the forces involved and, therefore, the speed of the winds will be different. In curved flow, wind is constantly changing direction and, according to Newton's first law, this means that the forces on

GRADIENT WIND
A wind, depicted on a weather map as flowing parallel to curved isobars or height contours, that develops when the pressure gradient force and the Coriolis force are unbalanced.

CYCLONE
An area of low pressure around which winds blow counterclockwise in the northern hemisphere and clockwise in the southern hemisphere.

ANTICYCLONE
An area of high pressure around which winds blow clockwise in the northern hemisphere and counterclockwise in the southern hemisphere.

the wind must be unbalanced. As noted above, the imbalance of forces, or the *net* force, is equal to the centripetal force, or centripetal acceleration. Gradient winds can be best understood by considering how the imbalance of forces develops and why this imbalance is needed to keep the wind moving in a curved path.

Consider first an air parcel in simple geostrophic flow around a low-pressure system in the northern hemisphere (Figure 11.10a). At first, the air parcel begins to move in a straight line and parallel to the isobars, according to the balance between the pressure gradient force directed inward toward the centre of low pressure and the Coriolis force directed outward (Point 1). If the air parcel continues to move in a straight line, it will begin to slow because the pressure gradient force will begin to act *against* it (Point 2). As the air slows, the Coriolis force will decrease, making it smaller than the pressure gradient force, so that the wind follows a curved path (Figure 11.10b). The *difference* between the pressure gradient force and the Coriolis force is the *net* force that we call *the centripetal force*. As winds flow around an area of low pressure, therefore, they need to be slightly *slower*

than geostrophic winds in order to continue to follow a curved path. Notice that these winds will travel around the low in a *counterclockwise* direction in the northern hemisphere; in the southern hemisphere, winds blow *clockwise* around lows. Such circular areas of low pressure are known as **cyclones**.

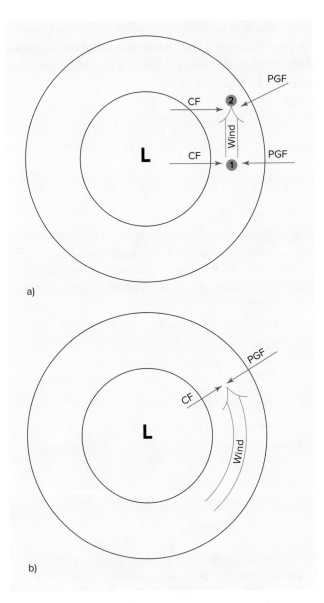

a)

b)

FIGURE 11.10 | Gradient wind flow around an area of low pressure in the northern hemisphere. a) At Point 1, an air parcel is moving under the influence of the balance between the pressure gradient force and the Coriolis force, just as it would for geostrophic flow. If the air parcel continues to move in a straight line, the forces on it will no longer be balanced by the time it reaches Point 2. This will slow the air parcel, thus reducing the Coriolis force. b) With the Coriolis force less than the pressure gradient force, the wind follows a curved path around the low.

On the other hand, circular areas of high pressure are known as **anticyclones**. Air moves clockwise around anticyclones in the northern hemisphere and counterclockwise around them in the southern hemisphere. Winds around anticyclones must be slightly *faster* than geostrophic winds because, in this case, the Coriolis force must be greater than the pressure gradient force to maintain curving flow (Figure 11.11).

The above is a simplified explanation of why, for the same pressure gradient force, winds are slightly faster around anticyclones than they are around cyclones. It is also possible to provide a mathematical explanation. For both cyclones and anticyclones, the difference between the pressure gradient force and the Coriolis force is the centripetal force, as shown in Equation 11.8.

$$PGF - CF = C_L F \qquad (11.8)$$

Since gradient wind speed is different depending on whether the wind is around a cyclone or around an anticyclone, we need an equation for each case. As a convention, the radius of curvature, R, is positive for counterclockwise flow and negative for clockwise flow. Applying this convention, using V_{gr} to represent gradient wind speed, and substituting the

equations for the forces into Equation 11.8, we obtain Equation 11.9a for a cyclone and Equation 11.9b for an anticyclone.

$$-\frac{1}{\rho}\left(\frac{\Delta P}{\Delta x}\right) - f_c V_{gr} = \frac{V_{gr}^2}{R} \qquad (11.9a)$$

$$-\frac{1}{\rho}\left(\frac{\Delta P}{\Delta x}\right) - f_c V_{gr} = -\frac{V_{gr}^2}{R} \qquad (11.9b)$$

These two equations are the gradient wind equations. Notice that if the flow is straight, the radius of curvature becomes infinitely large, and the term on the right approaches zero. Under these conditions, what's left is the geostrophic wind equation. It follows that the geostrophic wind equation is just a special case of the gradient wind equation that applies for straight flow.

To show mathematically that gradient wind speeds vary depending on whether the flow is cyclonic or anticyclonic, we can substitute Equation 11.6a for $-1/\rho\ (\Delta P/\Delta x)$ in equations 11.9a and 11.9b. With some rearranging, we obtain Equation 11.10a for a cyclone and Equation 11.10b for an anticyclone.

$$V_{gr} = V_g - \left(\frac{1}{f_c}\right)\left(\frac{V_{gr}^2}{R}\right) \qquad (11.10a)$$

$$V_{gr} = V_g + \left(\frac{1}{f_c}\right)\left(\frac{V_{gr}^2}{R}\right) \qquad (11.10b)$$

For winds around a cyclone (Equation 11.10a), we *subtract* the second term on the right from the geostrophic wind speed, so that winds around cyclones are *subgeostrophic*, or *slower* than geostrophic winds. For winds around an anticyclone (Equation 11.10b), we *add* the second term on the right to the geostrophic wind speed, so that winds around anticyclones are *supergeostrophic*, or *faster* than geostrophic winds. This means that, *for the same pressure gradient*, winds curving around an anticyclone will be faster than they would be for straight flow, and winds curving around a cyclone will be slower than they would be for straight flow. This is shown in Example 11.6.

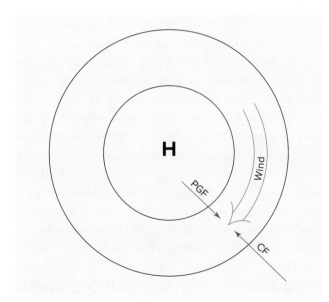

FIGURE 11.11 | Gradient wind flow around an area of high pressure in the northern hemisphere. With the pressure gradient force less than the Coriolis force, the wind follows a curved path around the high.

Example 11.6

For a geostrophic wind speed of 5 m/s at a latitude of 50°, calculate the gradient wind speed around a) a cyclone and b) an anticyclone. Use a radius of curvature of 250 km.

First, solve Equation 11.10 for V_{gr} using the quadratic formula.

$$V_{gr} = \frac{2 V_g}{1 \pm \sqrt{\left(1 + \dfrac{4 V_g}{f_c R}\right)}}$$

We will use the solution with the plus sign in front of the radical because then when R gets very large, as it will for straight flow, V_{gr} will equal V_g.

For the Coriolis parameter, f_c, we can use the value for latitude 50° that we found in Example 11.4: 0.0001 s^{-1}.

a) Use a positive radius of curvature for cyclonic flow.

$$V_{gr} = \frac{2 \, (5 \text{ m/s})}{1 + \sqrt{1 + \dfrac{4 \, (5 \text{ m/s})}{(0.0001 \text{ s}^{-1}) \, (250,000 \text{ m})}}}$$

$$= 4.3 \text{ m/s}$$

b) Use a negative radius of curvature for anticyclonic flow.

$$V_{gr} = \frac{2 \, (5 \text{ m/s})}{1 + \sqrt{1 - \dfrac{4 \, (5 \text{ m/s})}{(0.0001 \text{ s}^{-1}) \, (250,000 \text{ m})}}}$$

$$= 6.9 \text{ m/s}$$

For a geostrophic wind speed of 5 m/s, wind speed around a cyclone will be slightly less at 4.3 m/s, and wind speed around an anticyclone will be slightly higher at 6.9 m/s.

We can use the solved form of Equation 11.10 (see Example 11.6.) to show that the speed of flow around an anticyclone is limited.

$$V_{gr} = \frac{2 V_g}{1 \pm \sqrt{\left(1 + \dfrac{4 V_g}{f_c R}\right)}} \tag{11.10}$$

RIDGE
An elongated area of high pressure.

TROUGH
An elongated area of low pressure.

If the wind speed is greater than $(f_c R)/4$, then the expression under the square root sign is negative, and the result is undefined. (Remember that R is negative for anticyclonic flow.) In other words, there is a physical limit to how large the pressure gradient around an anticyclone can get. There is no such limit for wind flow around cyclones. Therefore, while wind speeds around cyclones can get very fast, wind speeds cannot get very fast around anticyclones.

What is most essential in the above discussion of gradient winds, however, is that for the *same pressure gradient force*, wind speeds are faster around highs than they are around lows. Although this result may seem trivial, it has significant implications, because changes in wind speed in the upper air can influence surface pressure. This can be shown using an upper-air chart (Figure 11.12).

Notice the wavy pattern made by the height contours in Figure 11.12. A wave that meanders northward denotes a **ridge** of high pressure; moving outward from the *axis* of the ridge, heights, or pressures, get lower. Similarly, a wave meandering southward denotes a **trough** of low pressure; moving outward from the *axis* of the trough, heights, or pressures, get higher. Remember that on an upper-air chart, high height contours correspond to high pressures and warm temperatures, and low height contours correspond to low pressures and cold temperatures (Section 3.7). This association reveals that a ridge is a region of warmer temperatures extending north, while a trough is a region of colder temperatures extending south.

As winds flow around a ridge, they will be turning clockwise, or anticyclonically, and travelling *faster* than geostrophic winds (Figure 11.12). On the other hand, as winds flow around a trough, they will be turning counterclockwise, or cyclonically, and travelling *slower* than geostrophic winds. Thus, *speed divergence* occurs when wind speeds up as it flows from a trough to a ridge, while *speed convergence* occurs when wind slows down as it flows from a ridge to a trough (Figure 11.13). In the former case, there is a net outflow of air from the region; in the latter case, there is a net inflow of air into the region. When divergence occurs in the upper part of the troposphere, the net outflow can remove mass from the air column, decreasing the surface pressure. Conversely, convergence in the upper troposphere can lead to an increase of surface pressure. Therefore, convergence and divergence in the upper air are very important to the formation and development of the

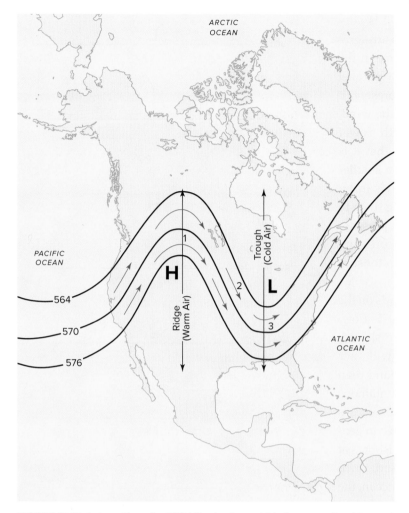

FIGURE 11.12 | A portion of a 500 hPa chart on which the axes of a ridge and a trough have been labelled. The curved lines are height contours in decameters (1 decametre = 10 metres), and the arrows are winds. At Point 1, the winds are turning anticyclonically and are faster than the winds at Point 2, where the flow is straight. At Point 3, the winds are turning cyclonically and are slower than the winds at Point 2.

large high- and low-pressure systems that influence mid-latitude weather.

Surface highs are likely to form just *upstream* of an upper-air trough, and surface lows are likely to form just *downstream* of an upper-air trough (Figure 11.13). These systems are deep in that they extend through the troposphere. In addition, they *tilt* with height. In the northern hemisphere, high-pressure systems tilt upward toward the southwest, and low-pressure systems tilt upward toward the northwest (Section 14.2). Our look at surface winds in the next section further shows how the activity of surface air is linked to the activity of upper air, this time through vertical motions.

11.2.3 | Surface Winds

Observations show that winds within the planetary boundary layer, where friction is significant, do not flow parallel to isobars when plotted on a map but, instead, cross these lines at an angle toward lower pressure. To explain why, we will begin by considering geostrophic flow and visualizing what happens if the force of friction is added to the effects of the pressure gradient force and the Coriolis force. (We can ignore

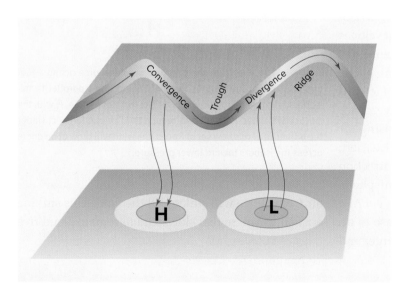

FIGURE 11.13 | Directly below areas of convergence in the upper airflow, the buildup of mass in the air column will lead to the formation of an area of high pressure at the surface. This high will extend through the troposphere, tilting upward to the southwest, where it forms a ridge in the upper air. Directly below areas of divergence in the upper airflow, the loss of mass in the air column will lead to the formation of an area of low pressure at the surface. This low will extend through the troposphere, tilting upward to the northwest, where it forms a trough in the upper air.

the centripetal force at the surface because it is very small compared to friction.) When the pressure gradient force is balanced with the Coriolis force in the absence of friction, the resulting winds blow parallel to isobars (Figure 11.14a). When friction is present, it will act in the direction opposite to that of the wind, slowing the wind down (Figure 11.14b). With slower winds, the Coriolis force will decrease, so that it is less than the pressure gradient force. With a smaller Coriolis force, the winds will be more strongly influenced by the pressure gradient force, causing them to cross the isobars at an angle toward lower pressure (Figure 11.14c). For surface flow, then, the pressure gradient force is equal and opposite to the vector sum of the Coriolis force and the friction force.

The magnitude of the friction force depends on the roughness of the surface over which the air is flowing. As a result, the amount that winds are slowed, and the angle at which they cross isobars, will vary with the terrain. Roughness is greater over land than over water, and it increases even more in mountainous areas. As air flows over rugged terrain, friction causes it to slow to about half the speed it would have in geostrophic flow; in such cases, wind will cross isobars at an angle of about 45 degrees. On the other hand, when air flows over smoother surfaces, such as the ocean, the wind speed will not be reduced by quite as much, and the winds usually cross the isobars at much smaller angles—typically angles of about 5 or 10 degrees. As friction decreases upward through the planetary boundary layer, winds gradually become faster and shift directions until, once friction is no longer important, they flow roughly parallel to the isobars.

As you might expect, surface winds also cross isobars at angles in relation to cyclones and anticyclones (Figure 11.15). As air flows in a counterclockwise direction around a cyclone in the northern hemisphere, it will cross isobars at an angle flowing *inward* toward the centre of the low. Such flow causes convergence at the surface and forces air to rise. This rising air will cool and likely form clouds (Figure 11.16, left; Section 9.3). As air flows in a clockwise direction around an anticyclone in the northern hemisphere, it will cross isobars at an angle and flow *outward* away from the centre of the high. In the case of an area of high pressure, surface air will diverge, and air from above will sink to fill the void. This sinking air will warm, so that skies tend to remain cloud free (Figure 11.16, right).

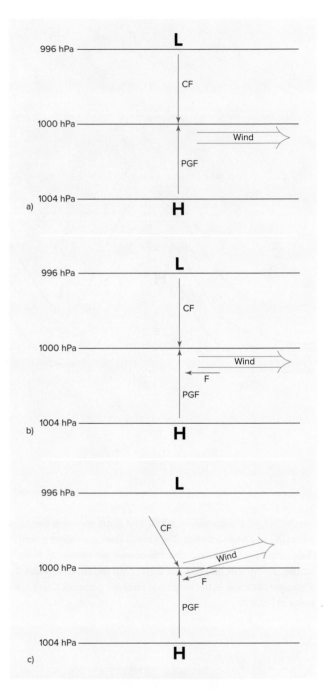

FIGURE 11.14 | a) With the pressure gradient force acting equal and opposite to the Coriolis force, the wind flows parallel to the isobars, as for geostrophic winds. **b)** The force of friction, *F*, with the surface will slow the wind and reduce the Coriolis force. **c)** Under the influence of this new set of forces, the wind will blow at an angle across the isobars toward lower pressure.

The above discussion has shown one of the ways in which horizontal motions in the atmosphere are linked to the vertical motions that are so important for creating weather. Despite their importance, vertical motions are generally much slower than horizontal motions. As air rises in a low-pressure system, it may travel at a rate of only one or two *centimetres* per second. In contrast, horizontal wind speeds are typically several *metres* per second.

Remember from Section 11.2.2 that air diverges above surface lows and converges above surface highs. As long as the divergence aloft is greater than the convergence at the surface, low-pressure areas will intensify because there will be a net outflow of air from the column. Likewise, as long as convergence aloft is greater than divergence at the surface, high-pressure areas will intensify because there will be a net inflow of air into the column. This convergence and divergence aloft also helps enhance the vertical motions associated with areas of high and low pressure. Because the inversion in the stratosphere effectively prevents tropospheric air from rising into the stratosphere, convergence in the upper troposphere will lead to sinking, and divergence in the upper troposphere will help to pull air up from below. Figure 11.17 shows how horizontal convergence and divergence, created by winds at the surface and aloft, is linked to the vertical motions in high- and low-pressure systems. This diagram highlights the three-dimensional structure of the deep high- and low-pressure systems associated with mid-latitude weather (Section 14.2). Also important in the interactions between the various layers of the atmosphere is the **thermal wind** concept.

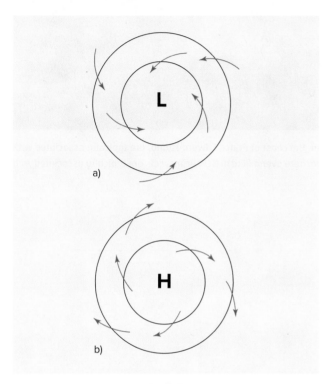

FIGURE 11.15 | a) Wind flow around a cyclone in the northern hemisphere. As air flows in toward the centre of the low, it will be forced to rise. **b)** Wind flow around an anticyclone in the northern hemisphere. As air flows out of the centre of the high, it will be replaced by air sinking from above.

THERMAL WIND
The change in geostrophic wind with height due to the horizontal variation of temperature.

Remember This

- Low-pressure areas are associated with converging air at the surface and diverging air aloft, which leads to rising air and the formation of clouds. As long as the upper-level divergence is greater than the surface convergence, the low will intensify.
- High-pressure areas are associated with diverging air at the surface and converging air aloft, which leads to sinking air and clear skies. As long as the upper-level convergence is greater than the surface divergence, the high will intensify.

11.3 | Variation of Winds with Height: The Thermal Wind

Recall from Chapter 3 that temperature differences in the atmosphere create pressure gradients above Earth's surface; also recall that, at a given height, the direction of these pressure gradients is from areas of warm air to areas of cold air. Figure 11.18 depicts the sloping pressure surfaces that create pressure gradients. Note that the slope of these lines—and therefore the pressure gradient—increases with height. The pressure surfaces slope because of the horizontal

FIGURE 11.16 | *Left:* Overcast conditions, such as those shown here on the coast of Prince Edward Island, are typically associated with an area of low pressure. *Right:* Clear-sky conditions, such as those shown here over a field in New Brunswick, are typically associated with an area of high pressure.

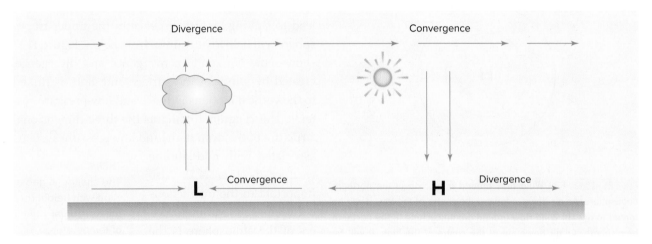

FIGURE 11.17 | The relationship between horizontal and vertical motions in low-pressure systems (on the left) and high-pressure systems (on the right). The divergence and convergence in the upper air result from the changes in wind speed depicted in Figure **11.12**.

temperature gradient and the fact that pressure decreases more slowly with height in warm air than it does in cold air. Because pressure gradients increase with height, the geostrophic wind speed will also increase with height.

If the temperature gradient were to increase from that shown in Figure 11.18, the pressure surfaces would slope more steeply, increasing the pressure gradient at a given height. In addition, the pressure gradients would increase more rapidly with height. An increase in the horizontal temperature gradient, therefore, will lead to a greater *change* in the

geostrophic wind with height. Thus, the change in geostrophic wind with height is proportional to the horizontal temperature gradient.

The *change* in the geostrophic wind with height that occurs due to the horizontal variation of

Question for Thought

Why do wind speeds continue to increase above the planetary boundary layer?

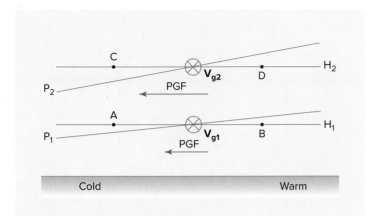

FIGURE 11.18 | An atmospheric condition in the northern hemisphere in which there is a horizontal temperature gradient but no surface pressure gradient. The temperature gradient causes pressure surfaces (blue lines) to slope upward from an area of cold air (left) to an area of warmer air (right). This creates a pressure gradient from right to left, above the surface, that increases with height. (Note that the pressure difference between points C and D is greater than that between points A and B.) This pressure gradient produces geostrophic winds, **V**$_{g1}$ and **V**$_{g2}$, that blow with low pressure on the left of the motion. (In this diagram these winds are blowing into the page.) The speed of these winds increases with height.

temperature is the thermal wind. The thermal wind relationship states that the *difference* between the geostrophic wind at the top of an atmospheric layer and that at the bottom of an atmospheric layer is proportional to the mean horizontal temperature gradient of the layer. Despite its name, the thermal wind it is not an *actual* wind; instead, it is a *measure* of the wind shear, or the change in wind with height. The name *thermal wind* arises from the fact that the concept *quantifies* the relationship between winds and temperature; it does so by using **vectors** and the thermal wind equation.

The thermal wind vector, V_T, for an atmospheric layer is the difference between the vector for the geostrophic wind at the top of the layer, V_{g2}, and the vector for the geostrophic wind at the bottom of the layer, V_{g1}.

$$\mathbf{V_T} = \mathbf{V_{g2}} - \mathbf{V_{g1}} \qquad (11.11)$$

Thermal winds can, therefore, be calculated using vector math when the upper and lower winds are known. For the simple case shown in Figure 11.18, the isotherms, at any height, will parallel the isobars. Under these conditions, the atmosphere is said to be **barotropic**, and the geostrophic winds will get faster with height, but they will not change direction with height. This is shown in Figure 11.19, along with the corresponding vector diagram for the thermal wind.

The upper-air pressure gradients depicted in figures 11.18 and 11.19 were shown to develop with no pressure gradient at the surface. When there is a pressure gradient at the surface, upper-level pressure gradients develop as a result of a combination of the horizontal temperature gradient *plus* the surface pressure gradient (Figure 11.20). (If there is a surface pressure gradient but no temperature gradient, then the pressure gradient will be constant with height, and so will be the geostrophic winds.) Rearrangement of Equation 11.11 to produce Equation 11.12, as shown below, provides another way to see this because now the upper-air wind, V_{g2}, is shown to result from the surface pressure gradient, as represented by V_{g1}, *plus* the temperature gradient of the layer, as represented by V_T. Under these conditions, isotherms may not parallel isobars, and winds will change direction, as well as speed, with height (Figure 11.21). When isotherms are not parallel to the isobars, the atmosphere is said to be **baroclinic**.

$$\mathbf{V_{g2}} = \mathbf{V_{g1}} + \mathbf{V_T} \qquad (11.12)$$

We can obtain the thermal wind equation by substituting the geostrophic wind equation (Equation 11.7) for V_{g1} and for V_{g2} in Equation 11.12 and simplifying.

$$\mathbf{V_T} = -\left(\frac{g}{f_c}\right)\left(\frac{\Delta h}{\Delta x}\right) \qquad (11.13)$$

VECTOR
A concept used to represent phenomena that have both speed and direction.

BAROTROPIC
The atmospheric condition in which isotherms parallel isobars.

BAROCLINIC
The atmospheric condition in which isotherms cross isobars.

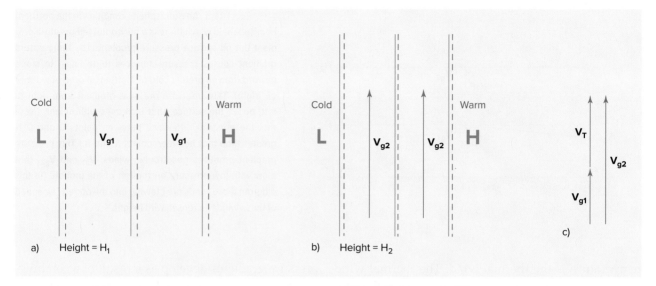

FIGURE 11.19 | Map views for the atmospheric condition depicted in Figure 11.18. The solid blue lines are isobars with low pressure on the left and high pressure on the right. The dashed red lines are isotherms with cold temperatures to the left and warm temperatures to the right. a) At a height of H_1, the geostrophic wind, V_{g1}, flows parallel to the isobars, with low pressure to the left of the motion. b) At a height of H_2, the direction of the pressure gradient has not changed, but it has increased in magnitude. V_{g2} is faster than V_{g1}, but it flows in the same direction. c) A vector diagram shows that the thermal wind, V_T, is the difference between the winds at H_2 and those at H_1. Notice that the thermal wind flows parallel to the isotherms, with colder temperatures to the left of motion.

Equation 11.13 gives the thermal wind as a function of the thickness gradient, $\Delta h/\Delta x$, of the atmospheric layer. Recall from the hypsometric equation that thickness is a function of temperature, so the thickness gradient represents the average temperature gradient of the atmospheric layer (Section 3.5). Equation 11.13 shows

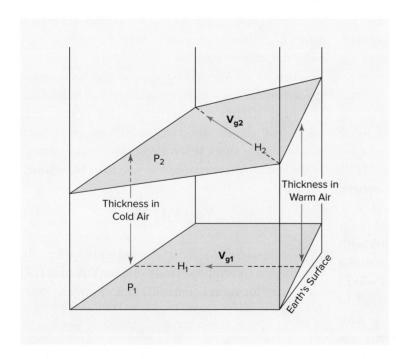

FIGURE 11.20 | An atmospheric condition in the northern hemisphere in which there is a temperature gradient, from warm air on the right to cold air on the left, *and* a surface pressure gradient, from low pressure at the front of the diagram to high pressure at the back. The surface pressure gradient is depicted by the sloping pressure surface, P_1. The geostrophic wind, V_{g1}, blows along P_1 at a constant height, H_1, with low pressure to the left of the motion. A higher pressure surface, P_2, slopes more steeply and in a different direction than P_1. P_2 is highest where the surface pressure is highest, and the thickness is greatest in the warm air; P_2 is lowest where the surface pressure is lowest, and the thickness is least in the cold air. Thus the slope of P_2 arises from the combination of the temperature gradient *and* the surface pressure gradient. The geostrophic wind, V_{g2}, blows along P_2 at a constant height, H_2, with low pressure to the left of the motion. It is faster than V_{g1}.

Source: Adapted from R.B. Stull. (2000). *Meteorology for scientists and engineers* (2nd ed.). Pacific Grove, CA: Brooks/Cole.

An upper-air pressure gradient results from the combination of a surface pressure gradient and a horizontal temperature gradient.

- With a surface pressure gradient only, the upper-air pressure gradients will be the same as the surface pressure gradient. Winds will be constant with height.
- With a horizontal temperature gradient, the upper-air pressure gradients will vary with height. Winds will vary with height. The relationship between the changes in winds with height and the horizontal temperature gradient is the thermal wind relationship.

that the thermal wind increases with the thickness gradient. Where there is a large thickness gradient, thermal winds will be large. Where thermal winds are large, the change in wind with height, or wind shear, is large.

Notice the similarity between the thermal wind equation and the geostrophic wind equation.

Geostrophic winds are proportional to the *height* gradient, while thermal winds are proportional to the *thickness* gradient. In addition, geostrophic winds blow parallel to the *height* gradient, with low heights to the left of the flow, while thermal winds blow parallel to the *thickness* gradient, with smaller

Example 11.7

If the thickness of the 1000–500 hPa layer decreases 300 m toward the north over a distance of 1000 km, what is the magnitude of the thermal wind? What is the direction of the thermal wind? Use a Coriolis parameter of 1×10^{-4} s^{-1}.

Use Equation 11.13.

$$\mathbf{V_T} = \left(\frac{9.8 \text{ m/s}^2}{1 \times 10^{-4} \text{ s}^{-1}} \right) \left(\frac{300 \text{ m}}{1{,}000{,}000 \text{ m}} \right)$$

$$= 29.4 \text{ m/s}$$

The direction of the thermal wind is westerly because cold air is to the north. (Winds are named for the direction they are coming *from*.)

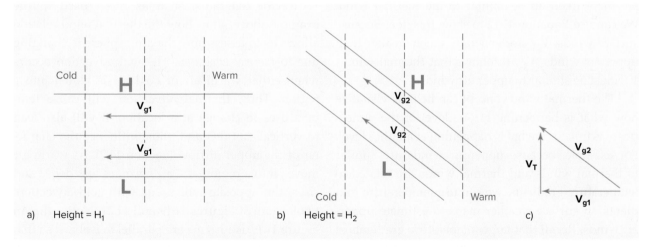

FIGURE 11.21 | Map views for the atmospheric condition depicted in Figure 11.20. The solid blue lines are isobars and the dashed red lines are isotherms. a) At a height of H$_1$, the geostrophic wind, **V$_{g1}$**, flows parallel to the isobars. b) At H$_2$, the direction and magnitude of the pressure gradient have changed. **V$_{g2}$** flows faster and in a different direction than **V$_{g1}$**. The isotherms show that the temperature gradient has not changed. c) A vector diagram shows that the thermal wind, **V$_T$**, is the difference between the winds at H$_2$ and those at H$_1$. Notice that the thermal wind flows in a different direction than either the upper or lower wind. Notice also that the thermal wind flows parallel to the isotherms, with colder temperatures to the left of the motion, just as it did in Figure 11.19.

thicknesses, or colder air, to the left of the flow. The relationship between the thermal wind and the temperature gradient can be seen in both Figure 11.19 and Figure 11.21.

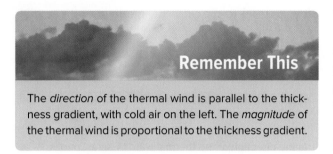

Remember This

The *direction* of the thermal wind is parallel to the thickness gradient, with cold air on the left. The *magnitude* of the thermal wind is proportional to the thickness gradient.

In Figure 11.21, isotherms cross the isobars, as is usually the case in the lower half of the troposphere (Figure 11.22a). In Figure 11.19, isotherms run parallel to the isobars, as is usually true in the upper half of the troposphere (Figure 11.22b). These conditions arise because there is not necessarily a relationship between temperature and pressure at the surface, whereas temperature gradients are a more important control of the slope of isobaric surfaces and, therefore, the pressure gradients farther above the surface. As a result, actual winds near the surface may be quite different from the thermal wind, while upper-air winds are more likely to be similar to the thermal wind. We can use Equation 11.12 to show this, too. Because surface winds, V_{g1}, are generally much weaker than upper-air winds, V_{g2}, it follows that thermal winds, V_T, must be similar to upper-air winds.

The thermal wind concept can help us visualize how what is happening at one level of the atmosphere is linked to what is happening at another level. For example, because upper-air winds are similar to thermal winds and thermal winds are related to temperature gradients, we can use temperature gradients on surface weather maps to estimate upper-air winds. Recall that large temperature gradients at the surface are indicated as *fronts* (Section 3.7). On the surface weather map shown in Figure 11.23a, there is a large cold front, and thus a large temperature gradient, in the vicinity of Canada's east coast. Notice that this cold front is associated with a strong pressure gradient, and fast winds from the southeast, at the

500 hPa level; notice also that the thickness contours and the height contours on the 500 hPa chart roughly parallel the surface cold front (Figure 11.23b). This shows that the upper wind and the thermal wind are very similar and that they can be predicted using the temperature pattern, or front, at the surface.

Just as a surface chart allows us to approximate upper-air winds, an upper-air chart allows us to estimate surface winds. An example of this technique is shown on Figure 11.23b. In an area over the Pacific Ocean where the thickness contours cross the height contours on the 500 hPa chart, two vector diagrams have been drawn. (These vector diagrams are similar to the one in Figure 11.21c.) The upper-air wind vectors, V_{g2}, and the thermal wind vectors, V_T, were drawn using the height contours and the thickness contours, respectively. Using these wind vectors, the surface wind vectors, V_{g1}, could then be estimated. The surface weather map for the same time (Figure 11.23a) shows that this technique predicts the surface wind direction quite well; as would be expected, these surface winds rotate counterclockwise around the large low-pressure system over the ocean. As these examples show, once you are familiar with the thermal wind concept, you will be able to look at a map of one layer of the atmosphere and imagine what is going on in another layer.

Vector diagrams, such as those used in the example above, show how the thermal wind relation allows us to assess how the atmosphere is changing due to *thermal advection*. Thermal advection occurs when either warm air or cold air is blown into a region. Thus, thermal advection will cause temperatures to change at a location; it will also lead to vertical motions and, thus, influence the character of the upper airflow (Section 14.2). As warm air moves into a region, the air becomes less dense and rises; the opposite will occur with cold advection. Look again at Figures 11.19 and 11.21. Notice that in Figure 11.19, isotherms are parallel to isobars, so that winds blow parallel to the temperature gradient. Under these conditions, there will be no temperature advection. On the other hand, in Figure 11.21, the winds blow across the isotherms. In this case, the winds are producing warm advection (notice that they are bringing in warm air). It follows that

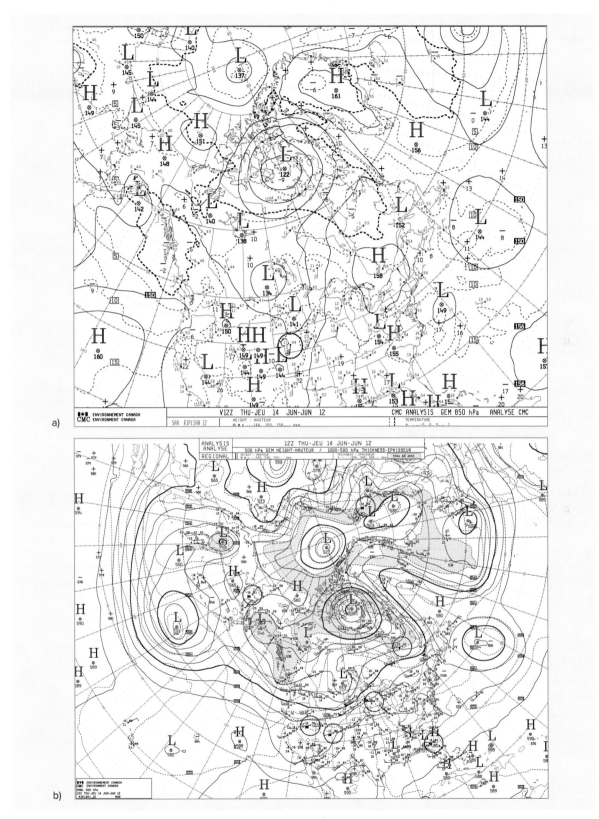

FIGURE 11.22 | a) An 850 hPa chart showing that isotherms (dashed lines) tend to cross the height contours (solid lines). b) A 500 hPa chart showing that thickness contours (dashed lines) roughly parallel the height contours (solid lines) over much of the map.

Source: Environment Canada, Weather Office, www.weatheroffice.gc.ca/analysis/index_e.html, 14 June 2012.

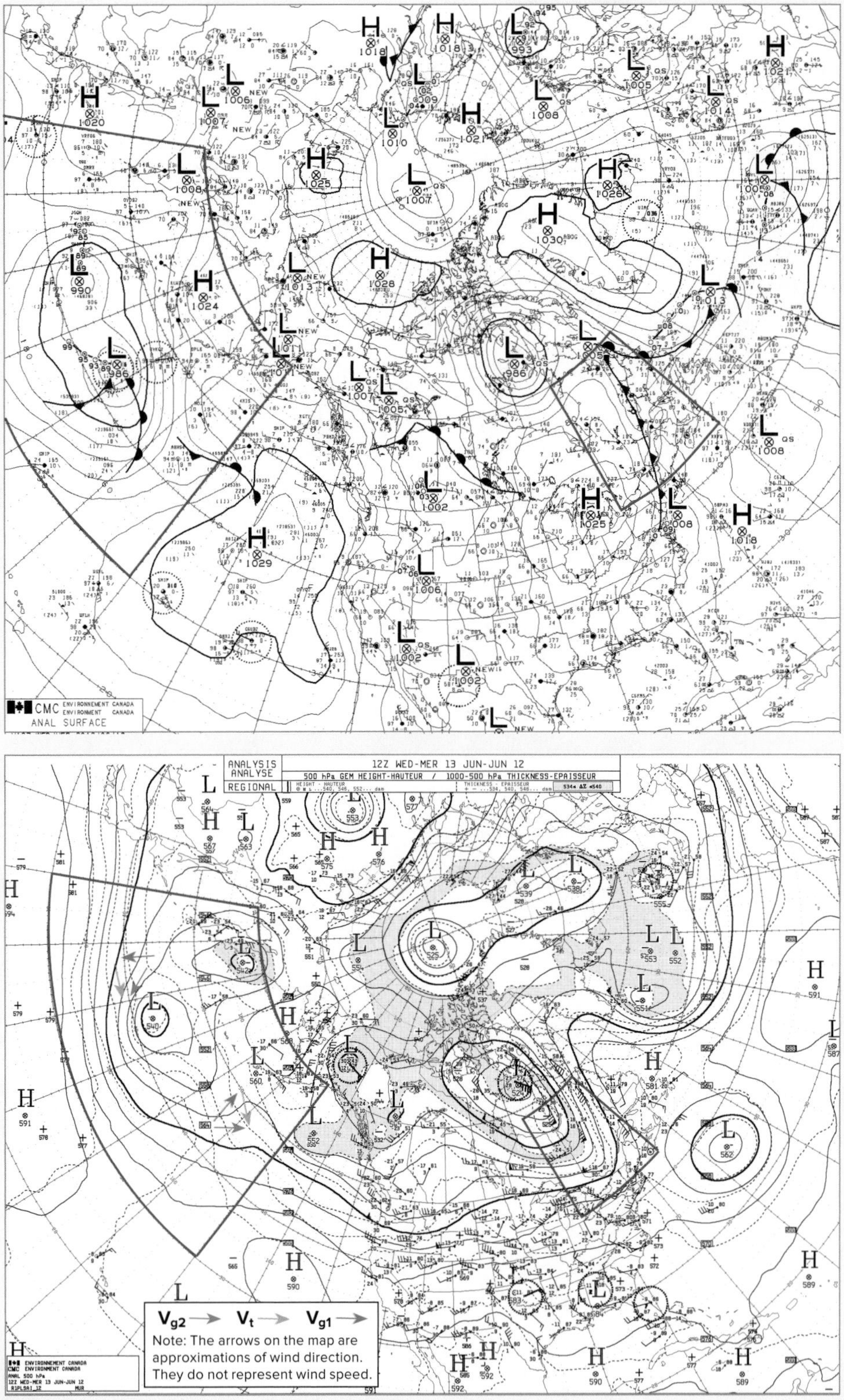

FIGURE 11.23 | a) A surface weather map and **b)** a 500 hPa chart, for the same time. The area highlighted on the right of both maps shows that a cold front at the surface corresponds to fast winds at the 500 hPa level. The area highlighted on the left of the charts shows how we can use vectors diagrams drawn on an upper-air chart to estimate wind-direction at the surface.

Source: Environment Canada, Weather Office, www.weatheroffice.gc.ca/analysis/index_e.html, 13 June 2012.

Look back at the vector diagrams over the Pacific Ocean in Figure 11.23b. Is warm or cold advection occurring in each case?

temperature advection occurs under baroclinic conditions but not under barotropic conditions.

When we know the change in wind direction with height, the thermal wind relation allows us to determine whether warm or cold advection is occurring. We can do this by using vector diagrams, as shown in figures 11.21c and 11.23b. When winds are **veering**, warm advection is occurring (Figure 11.24a). On the other hand, when winds are **backing**, cold advection is occurring (Figure 11.24b). Since the thermal wind relation always holds, backing winds always produce cold advection, and veering winds always produce warm advection. Example 11.8 shows how we might use a thickness chart to estimate thermal advection.

VEERING
Rotating clockwise with height.

BACKING
Rotating counterclockwise with height.

Example 11.8

The diagram below shows thickness contours, in metres, on a portion of a 1000 to 500 hPa thickness chart. A vector, **V**$_{g2}$, is drawn on this map to represent the upper-air wind. The speed of this wind is 25 m/s; the length of the arrow is drawn proportional to the wind speed. Assuming that the speed of the thermal wind is 20 m/s, draw a vector diagram to scale and use it to determine whether cold or warm advection is occurring. Use your diagram to estimate the speed of the surface wind.

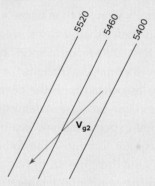

First, draw the thermal wind vector, **V**$_T$, on the diagram, so that it flows with cold air (lower thickness) to the left of the motion and has a length about 4/5 that of **V**$_{g2}$. Next, using the upper-air wind vector and the thermal wind vector, draw in the surface wind vector, **V**$_{g1}$. The winds are backing with height; thus, cold advection is occurring. The length of the surface wind vector indicates that its speed is about 10 m/s.

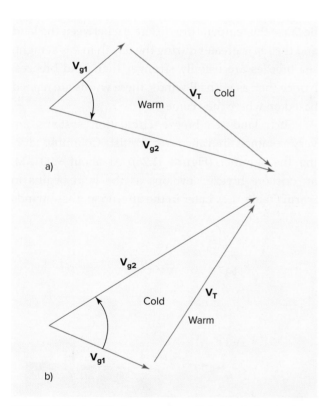

a)

b)

FIGURE 11.24 | a) Winds are veering with height, indicating warm advection. b) Winds are backing with height, indicating cold advection. The direction of the thermal wind arrow tells us the direction of the temperature gradient. Notice that the direction has changed between a) and b).

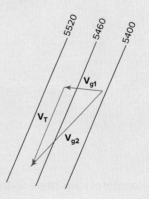

11.4 | Local Winds

Like all winds, winds occurring at the local scale, or the **mesoscale**, are driven by pressure differences. In the case of local winds, these pressure differences are linked to temperature differences that usually occur due to *variations in surface properties* or *variations in terrain*.

Land breezes and **sea breezes** are examples of local winds that develop due to temperature differences resulting from variations in surface properties.

MESOSCALE
A scale of a few kilometres to a few hundred kilometres.

LAND BREEZE
A breeze that flows from the land to the ocean at night.

SEA BREEZE
A breeze that flows from the ocean to the land during the day.

Recall from Section 6.6.3 that water surfaces heat and cool more slowly than land surfaces. As a result, land surfaces are usually warmer than water surfaces during the day and colder than water surfaces at night. These surface temperature differences promote the development of thermal circulations, or convection cells, which are associated with the development of thermal high- and low-pressure systems (Section 3.6).

As you might expect, land and sea breezes tend to develop in coastal locations. The following illustration reveals how. Imagine an initial condition in which there are no horizontal temperature or pressure gradients. This means that pressure surfaces are flat and there is no airflow. When the sun rises, the land will warm more than the water. In the warm air over the land, pressure will decrease more slowly with height than it will in the colder air over the water (Figure 11.25a). This difference will create a pressure gradient above the surface directed from the land to the water. As air flows along this upper pressure gradient, the mass of air over the land will decrease, while the mass of air over the water will increase. With more mass over the water, the surface pressure will increase over the water, and with less mass over the land, the surface pressure will decrease over the land. These changes in surface pressure occur due to the heating differences between the water and the land; thus, they are *thermal* pressure systems.

This pressure gradient at the surface results in an onshore flow, or *sea breeze*, during the day. These winds generally travel anywhere from about 50 to 100 km inland. They are usually strongest in the afternoon, as it is then that the greatest temperature contrasts between the land and the water occur. At roughly 1 km above the surface, the pressure pattern reverses itself, creating a seaward return flow. At night, the land will be colder than the ocean, and a surface offshore flow, or *land breeze*, will develop (Figure 11.25b). Because the temperature difference between the land and the sea is greater during the day than it is at night, sea breezes are usually stronger than land breezes. Notice that, as with all winds, these winds are named based on where they come *from*.

Such land–sea breeze circulation systems are very common in Vancouver, British Columbia, during the summer (Figure 11.26). At about 9:00 AM, an onshore breeze develops as the land begins to warm (Table 11.1). Later in the afternoon, these winds

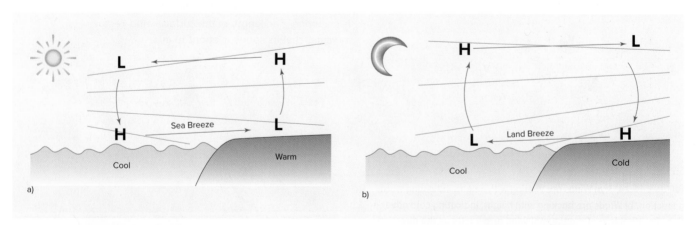

FIGURE 11.25 | a) The development of a sea breeze circulation system during the day, when a thermal low-pressure system develops over the warmer land and a thermal high-pressure system develops over the colder water. **b)** The development of a land breeze circulation system at night under the opposite conditions.

FIGURE 11.26 | Sunrise in Vancouver. As the sun rises and the land becomes warmer than the water, a sea breeze will develop.

reverse, and a weaker offshore breeze develops. One of the consequences of this airflow in the Vancouver area is that air pollutants from the city are blown up the Fraser Valley during the day. As a result, air pollution concentrations can become higher in the valley than they are in the city (Section 17.1).

In contrast to land and sea breezes, **mountain** and **valley winds** are examples of local winds that develop due to temperature differences that result from variations in terrain (Figure 11.28). As with land–sea breeze systems, mountain–valley wind systems occur on

MOUNTAIN WIND
Air blowing downslope.

VALLEY WIND
Air blowing upslope.

TABLE 11.1 | Hourly weather data for a clear summer day at Vancouver International Airport. The observed wind directions show the shift, at about 08:00 or 09:00, from an easterly land breeze to a westerly sea breeze. In the evening, the wind direction shifts back to easterly. Both wind speed and wind direction are measured using an anemometer (Figure 11.27).

Time	Temperature (°C)	Dew-Point Temperature (°C)	Relative Humidity (%)	Wind Direction (10's deg.)[a]	Wind Speed (km/h)	Visibility (km)	Pressure (kPa)	Weather
00:00	14.6	10.3	75	10	9	48.3	101.92	Clear
01:00	14.9	10.5	75	11	6	48.3	101.94	Clear
02:00	15.3	10.0	71	13	9	48.3	101.96	Clear
03:00	14.7	9.9	73	10	4	48.3	101.96	Clear
04:00	14.0	9.6	75	14	4	48.3	101.97	Clear
05:00	12.9	9.0	77	10	9	48.3	101.98	Clear
06:00	14.9	10.2	73	9	9	48.3	102.01	Mainly Clear
07:00	15.6	10.5	72	14	4	48.3	102.03	Mainly Clear
08:00	17.3	10.9	66	—	0	48.3	102.04	Mainly Clear
09:00	17.8	11.7	67	23	11	48.3	102.05	Mainly Clear
10:00	18.6	12.0	65	26	15	48.3	102.02	Mainly Clear
11:00	18.6	12.3	67	25	9	48.3	102.01	Mainly Clear
12:00	18.9	12.3	66	25	11	48.3	101.99	Mainly Clear
13:00	19.5	12.3	63	26	15	48.3	101.96	Mainly Clear
14:00	20.3	11.8	58	24	13	48.3	101.93	Mainly Clear
15:00	20.5	11.9	58	22	13	48.3	101.90	Mainly Clear
16:00	20.7	11.1	54	23	13	48.3	101.85	Mainly Clear
17:00	20.9	11.6	55	23	13	48.3	101.80	Mainly Clear
18:00	21.2	11.3	53	21	11	48.3	101.74	Mainly Clear
19:00	20.5	11.0	54	20	13	48.3	101.73	Mainly Clear
20:00	19.6	10.8	57	20	6	48.3	101.73	Clear
21:00	17.0	11.4	70	15	11	48.3	101.77	Clear
22:00	16.3	11.3	72	10	6	32.2	101.80	Clear
23:00	15.1	10.7	75	10	11	32.2	101.79	Clear

[a] Wind directions are given as azimuths in tens of degrees. For example, the number 9 denotes an azimuth of 90°, or an east wind.
Source: Vancouver International Airport.

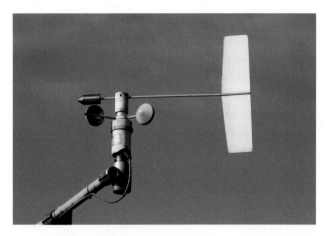

FIGURE 11.27 | Anemometers, such as the one shown here, are used to measure wind speed and direction.

a daily cycle, flowing one way during the day and the opposite way at night. Unlike land–sea breeze systems, however, they do not generally form closed cells. During the day, air along the mountain slopes will warm more than air at the same height over the valley, making pressure slightly lower along the slopes. The warm, less dense air will rise up the mountainside as a *valley wind* (Figure 11.29a). It is the pressure gradient that causes the air to rise *up the slope* rather than straight up. As air rises up the mountainside, air flows up the valley to replace the rising air. The rising air can often form clouds near, or just above, the mountaintops. In fact, during the summer, these valley winds make afternoon thundershowers quite common in some mountainous areas. At night, the air along the mountain slopes cools off more quickly than the air at the same height over the valley. This cooling reverses the pressure gradient, causing cold, dense air to drain down the slope and out of the valley as a *mountain wind* (Figure 11.29b). If this air is forced to flow through topographic constrictions as it descends, the winds can become very fast.

The exact timing of these winds can vary. When a slope is oriented toward the east, valley winds will begin much earlier in the morning than they will on mountain slopes that face west. In addition, valley winds tend to be more common in summer, while mountain winds tend to be more common in winter.

FIGURE 11.28 | The rough terrain of the Rocky Mountains causes mountain–valley wind systems to develop.

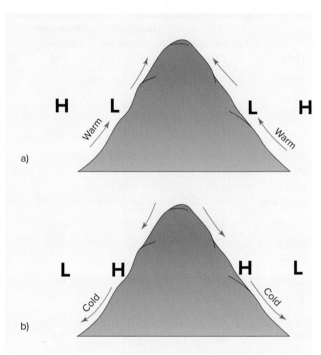

FIGURE 11.29 | a) During the day, warm air rises up the slopes of the mountain, creating a valley wind. The pressure adjacent to the slope is slightly lower than the pressure at the same height over the valley. b) At night, cold air flows down the slopes of the mountain, creating a mountain wind. The pressure adjacent to the slope is slightly higher than the pressure at the same height over the valley.

11.5 | Chapter Summary

1. The pressure gradient force drives the wind. This force has a magnitude proportional to the pressure gradient and is directed from high pressure to low pressure.

2. The Coriolis force causes winds to be deflected to the right of their path in the northern hemisphere and to the left of their path in the southern hemisphere. The Coriolis force increases with latitude and wind speed.

3. The centripetal force, or acceleration, keeps the winds moving in a curved path by pulling toward the centre of rotation. This force has a magnitude proportional to the wind speed and inversely proportional to the radius of curvature.

4. The force of friction slows winds near Earth's surface, so that wind speed increases with height. The effects of surface friction are carried upward by turbulence. Turbulence increases with rougher surfaces, faster winds, and greater instability. As turbulence increases, wind shear decreases, and the effects of surface friction are carried higher above the surface. The depth of the atmosphere to which the effects of surface friction are transferred is referred to as the *planetary boundary layer*.

5. The geostrophic wind equation approximates the wind speed for straight flow above the planetary boundary layer. Without friction or curvature, the pressure gradient force is balanced by the Coriolis force, and the winds flow parallel to the isobars or contours.

6. The gradient wind equation approximates the wind speed for curved flow above the planetary boundary layer. The imbalance between the pressure gradient force and the Coriolis force keeps the air moving along a curved path by creating a centripetal force that pulls the air in toward the centre of rotation. For the same pressure gradient, gradient winds are faster around highs than they are around lows.

7. In areas of low pressure, air converges at the surface, rises, and diverges aloft. The rising air leads to the formation of clouds. If upper-level divergence is greater than surface convergence, the low will intensify. In areas of high pressure, air diverges at the surface, sinks from above, and converges aloft. The sinking air leads to clear skies. If upper-level convergence is greater than surface divergence, the high will intensify.

8. Horizontal temperature gradients create upper-air pressure gradients that change with height. As a result of horizontal temperature gradients, therefore, winds change with height. We can use the thermal wind concept to study the relationship between the change in winds with height in an atmospheric layer and the average temperature gradient of that layer. This concept helps provide us with an appreciation that the layers of the atmosphere work together as a system.

9. Differences in heating between land and water can cause land–sea breeze circulation systems. During the day, a thermal high forms over the cooler water, while a thermal low forms over the warmer land; this creates a sea breeze. At night, the pressure gradient is reversed, creating a land breeze.

10. Warming causes air to rise up slopes as a valley wind. Cooling causes air to sink down slopes as a mountain wind.

Review Questions

1. What determines the magnitude and direction of the pressure gradient force?

2. What is the relationship between the Coriolis force and wind speed? What is the relationship between the Coriolis force and latitude? In what direction does the Coriolis force act?

3. In what direction does the centripetal force act? What determines the magnitude of this force?

4. What determines the magnitude of the friction force?

5. Why does the effect of surface friction extend upward into the atmosphere? What factors determine the depth of the atmosphere that will be influenced by surface friction?

6. How could you apply Newton's first law to explain why geostrophic winds parallel isobars?

7. Why are gradient winds faster around high-pressure areas than they are around low-pressure areas?

8. Why do surface winds cross isobars at an angle?

9. How do convergence and divergence lead to vertical motions? Describe the patterns of convergence, divergence, and vertical motions in high- and low-pressure systems.

10. Why are low-pressure systems associated with cloudy weather, while high-pressure systems are associated with clear weather?

11. How do temperature gradients produce upper-level pressure gradients? Why do such pressure gradients increase with height?

12. What is the relationship between changes in wind direction with height and thermal advection?

13. How do land–sea breeze circulation systems develop? How do valley and mountain winds develop?

Problems

1. Calculate the Coriolis parameter, f_c, for
   a) 10° latitude and
   b) 80° latitude.

2. For each latitude in Problem 1, calculate the Coriolis force if the wind speed is 15 m/s. If the wind is blowing from west to east in the northern hemisphere, what is the direction of the Coriolis force?

3. a) Calculate the geostrophic wind for the height contours shown in the diagram, below. The heights are given in decametres, the distance between the contours is 500 km, and the Coriolis parameter is 1×10^{-4} s^{-1}.
   b) At which point—1, 2, or 3—does the wind speed calculated in a) best apply?

4. Calculate the geostrophic wind velocity in each case.
   a) The latitude is 50°, the air density is 0.8 kg/m³, and the pressure gradient is 1.7 kPa/100 km.
   b) The Coriolis parameter is 1×10^{-4} s^{-1} and the height gradient on the 500 hPa surface is 0.8 m/km.

5. For a geostrophic wind speed of 12 m/s, a radius of curvature of 1000 km, and f_c of 7×10^{-5} s^{-1}, calculate
   a) the gradient wind speed for a cyclone and
   b) the gradient wind speed for an anticyclone.

6. Use a geostrophic wind velocity of 35 m/s to do the following.
   a) Calculate the speed of this wind around a cyclone at a radius of curvature of 500 km.
   b) Calculate the speed of this wind around an anticyclone at a radius of curvature of 500 km.
   c) Account for the difference in your answers to a) and b).

7. Calculate the magnitude of the thermal wind for a change in thickness of 380 m over 1000 km. The Coriolis parameter is 1×10^{-4} s^{-1}.

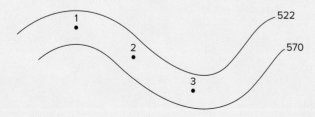

8. The diagram on the right shows thickness contours, in metres, on the 500 hPa chart. The distance between the three contour lines is 350 km. Also drawn on the diagram is the 500 hPa wind vector; its speed is 36 m/s.

a) Calculate the magnitude of the thermal wind, given that the Coriolis parameter is 1×10^{-4} s^{-1}.

b) Draw the thermal wind on the diagram. Use a scale based on the length of the 500 hPa wind vector.

c) Draw the surface wind on the diagram. Estimate the speed of the surface wind.

d) Is either warm or cold advection occurring? If so, which one?

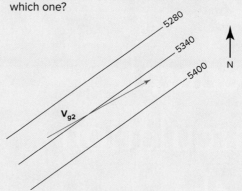

Suggestions for Research

1. Investigate the methods that people around the world use to protect their crops and garden plants from extreme wind, sunlight, heat, frost, and other weather elements. Explain which methods would be most effective in your region.

2. Read some recent news and journal articles that describe how wind is being used as a renewable energy source. List the advantages and disadvantages of relying on wind energy rather than energy produced by burning fossil fuels.

12

General Circulation of the Atmosphere

Learning Goals

After studying this chapter, you should be able to

1. *give reasons* for the need to transfer both energy and momentum from tropical latitudes toward polar latitudes;

2. *describe* and *account* for the pressure and wind patterns of the general circulation;

3. *explain* the existence of the subtropical jet streams and the polar front jet streams;

4. *appreciate* how the polar front jet streams influence mid-latitude weather;

5. *describe* how models are used to help us understand the general circulation of our planet;

6. *explain* how the general circulation accomplishes the necessary transfer of heat and momentum;

7. *use* the complete general circulation model to account for the steadiness of tropical weather compared to the variability of mid-latitude weather;

8. *appreciate* how vorticity impacts the flow of the atmosphere; and

9. *apply* the conservation of vorticity to explain why the mid-latitude upper airflow forms waves.

The subtropical jet stream shown here as a band of cloud, above Enchanted Rock, in Texas, is part of the planetary circulation system. This system is partly responsible for variations in weather and climate across the globe.

In Chapter 6, you saw that, although there is a balance between incoming solar radiation and outgoing longwave radiation for the Earth-atmosphere system as a whole, incoming and outgoing radiation are not balanced at individual latitudes. Recall that it is this imbalance that drives the global-scale system of pressure and winds known as the *general circulation* of the atmosphere. You can think of this system as a much larger and far more complex version of the land–sea breeze circulation system described in Section 11.4.

In this chapter, you will learn more about the general circulation of the atmosphere and the complex interaction of factors that influences it; you will also see how this circulation begins to provide an explanation for the global-scale variations in climate. As you will discover, this circulation system operates quite differently in the tropics than it does outside the tropics. This difference accounts for the relatively steady weather of the tropics compared to the much more variable **extratropical** weather.

EXTRATROPICAL
Relating to regions lying poleward of the tropics.

12.1 | Transport of Energy and Momentum

Recall from Section 6.5 that there is a *surplus* of radiation between roughly 40° N and 40° S, while there is a *deficit* of radiation poleward of these latitudes. As a result, there must be a poleward transfer of energy to prevent the polar regions from getting colder while the tropical regions get warmer (Figure 12.1, left side). This heat transport is the driving force behind the general circulation of the planet.

Also necessary, but less evident, is the transfer of *momentum*. Momentum is the product of mass and velocity. For a constant mass, an increase in velocity will lead to an increase in momentum. Because Earth's surface and the winds flowing over it make up a rotating system, we consider **angular momentum**, which is the product of mass, velocity, and the distance from the centre of rotation, *r*.

ANGULAR MOMENTUM
The momentum associated with motion along a circular path.

$$\text{Angular Momentum} = m \times v \times r \qquad (12.1)$$

In a rotating system, angular momentum is conserved. If any one of the variables decreases, another must increase to compensate, and vice versa. An example is that of a spinning skater. As the skater pulls in his or her arms, the distance from the centre of rotation to the point on the skater's body farthest from this centre decreases; as a result, the speed of the spin will *increase* to conserve momentum.

Conservation of momentum can often be facilitated by a *transfer* of momentum. Recall from Section 11.1.4 that momentum is transferred between Earth and the atmosphere by friction. Because Earth rotates from west to east, easterly winds transfer momentum from Earth to the atmosphere. In the case of westerly winds, momentum is transferred from the atmosphere to Earth. Considered another way, easterly winds could slow the rotation of Earth if there were no westerly winds, and westerly winds could speed up the rotation of Earth if there were no easterly winds. It follows that there must be both easterly and westerly winds on Earth, and that momentum in the atmosphere must be transferred from the easterly winds to the westerly winds, in order to maintain the current momentum of Earth and its atmosphere. As we will see, surface winds in the tropics are easterly, while surface winds in the mid-latitudes are westerly. Consequently, momentum must be transferred poleward from the easterlies of the tropics to the westerlies of the mid-latitudes (Figure 12.1, right side).

The general circulation, therefore, must transport both energy and momentum toward the poles. In the tropics, this is accomplished by a simple convection cell circulation; in the mid-latitudes, it is accomplished by weather systems in the form of eddies and waves.

12.2 | Descriptions of the General Circulation

As suggested above, and in Chapter 6, the general circulation is a very complex system. Therefore, to explain it, we'll begin by examining a *simple* model of this system, which is based on assumptions that simplify the real-word processes involved; then, we will consider the *real* situation.

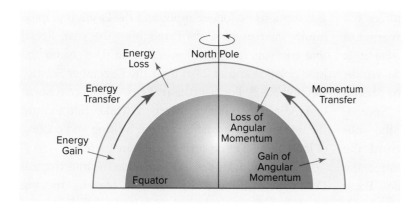

FIGURE 12.1 | The left side of the diagram shows that there is a net gain of energy in the tropical latitudes and a net loss in the polar latitudes. The right side of the diagram shows that easterly winds in the tropical latitudes gain momentum from Earth, while the westerly winds of the mid-latitudes lose momentum to Earth.

Source: Adapted from J. Marshall & R. A. Plumb. (2007). *Atmosphere, ocean, and climate dynamics: An introductory text.* Boston, MA: Elsevier Academic Press, p. 140.

12.2.1 | The Simple Case

Figure 12.2 provides a simplified description of the annually averaged pressure and wind patterns of Earth for both the surface and the upper air. For this simple case, two assumptions are made. First, we assume that the sun is always overhead at the equator, so that the equator receives more solar radiation than does anywhere else and there are no seasons. Second, we assume that Earth's surface is homogeneous—composed of one surface type.

Figure 12.2a shows an alternating pressure pattern at the surface, with low pressure straddling the equator and ringing the mid-latitudes, and high pressure in the subtropics and at the poles. These pressure systems, from equator to pole, are the **equatorial low**, the **subtropical highs**, the **subpolar lows**,

and the **polar highs**. Surface winds form in response to these pressure patterns. First, air flowing from the subtropical highs toward the equatorial low is deflected slightly by the Coriolis force, creating northeast winds in the northern hemisphere and southeast winds in the southern hemisphere. (Recall that winds are named for the direction they come from.) These are the **trade winds**. The equatorial low

EQUATORIAL LOW
A region of low pressure that develops at, or near, the equator.

SUBTROPICAL HIGH
A region of high pressure occurring in the subtropics.

SUBPOLAR LOW
A region of average low pressure occurring in the mid-latitudes.

POLAR HIGH
A region of high pressure occurring at the poles.

TRADE WINDS
The steady winds of the tropical region.

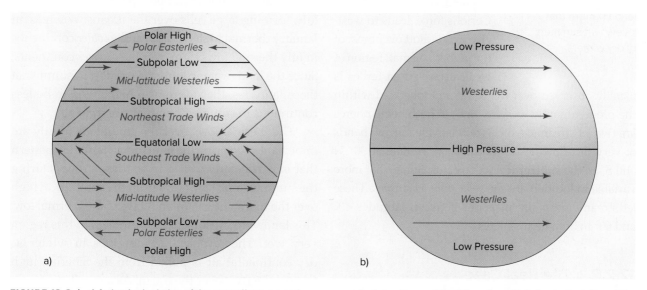

FIGURE 12.2 | a) A simple depiction of the annually averaged pressure and wind patterns at Earth's surface. b) A simple depiction of the annually averaged pressure and wind patterns aloft.

INTERTROPICAL CONVERGENCE ZONE (ITCZ)
The location in the tropics where the northeasterly trade winds converge with the southeasterly trade winds.

SUBTROPICAL JET STREAM
A narrow band of very fast westerly wind that occurs near the top of the troposphere at about 30° N or S.

ALEUTIAN LOW
The subpolar low over the northern Pacific Ocean.

ICELANDIC LOW
The subpolar low over the northern Atlantic Ocean.

CANADIAN HIGH
A thermal high that forms over Canada in the winter.

SIBERIAN HIGH
A thermal high that forms over Siberia in the winter.

PACIFIC HIGH
The subtropical high over the northern Pacific Ocean.

BERMUDA HIGH
The subtropical high over the northern Atlantic Ocean.

MONSOON
A circulation pattern that leads to very wet summers and very dry winters.

is also referred to as the **Intertropical Convergence Zone (ITCZ)** because it is here that the trade winds, from Earth's two hemispheres, converge. Second, air flowing from the subtropical highs toward the subpolar lows creates the *mid-latitude westerlies*. These winds experience much greater deflection than do the trade winds because of the poleward increase in the strength of the Coriolis force. Finally, air flowing from the polar highs to the subpolar lows creates winds known as the *polar easterlies*.

The far simpler pattern shown in Figure 12.2b depicts pressure and wind patterns aloft, or in what is roughly the upper half of the troposphere. On average, the pressure above the equator is high, and the pressure above the poles is low. The resulting pressure gradient, combined with the Coriolis force, leads to westerly winds aloft over most of the planet. (A small region of weak upper-air easterlies is usually observed over the equator.) Embedded within the predominantly westerly flow in both hemispheres are two jet streams, which are relatively narrow bands of very fast-flowing air. Located on average at 30° N and S are the **subtropical jet streams**. Somewhat more variable in location are the *polar front jet streams*. These jet streams normally meander between latitudes 40° and 60° in both hemispheres.

12.2.2 | The Real Case

Figure 12.3 shows the *observed* pressure and wind patterns for January and July. For the real case, the assumptions no longer apply, and the location of maximum solar input shifts throughout the year. Recall that the sun is directly overhead at 23½° N on the June Solstice and at 23½° S on the December Solstice (Section 5.8.2). Further, the real case does not assume that the surface is homogeneous, so it takes into account the influence of the differences in heating and cooling of land and water on surface pressure patterns.

The maps in Figure 12.3 show that the Intertropical Convergence Zone shifts with the seasons, moving south in January and north in July. This shift is more pronounced over land than it is over the oceans. Although it is less obvious, the other pressure systems shift seasonally as well. In January, the subpolar lows are dominant as *cells* over the oceans in the northern hemisphere. The cell in the Pacific Ocean is referred to as the **Aleutian low**, while the cell in the Atlantic Ocean is called the **Icelandic low**. The cellular pattern develops because the relative cold of the continents causes pressure to be higher over land than over water (Section 3.6). The thermally induced high-pressure cells above the continents are the **Canadian high** and the **Siberian high**. The much larger size of the Eurasian continent, compared to the North American continent, makes the Siberian high a much stronger high than is the Canadian high. The subtropical highs—the **Pacific high** and the **Bermuda high**—are relatively insignificant in the northern hemisphere in January.

On the other hand, these two subtropical highs dominate the northern-hemisphere circulation in July, forming large cells over the oceans. Whereas, in January thermal highs form over the *colder* continents, in July thermal lows form over the *warmer* continents. Since the land–water contrast is greatest in summer at these latitudes, the subtropical highs tend to be less continuous in summer than they are in winter.

The thermal lows over the continents in July are important in creating the **monsoon** circulation pattern that is particularly strong in southeast Asia. During the summer, moist air flows from the subtropical high over the Indian Ocean inland toward the thermal low. This landward airflow makes summers in this region very wet. The reverse flow develops in winter as dry continental air is pulled from the Siberian high toward the ITCZ over the ocean. As a result, monsoon climates are characterized by very dry winters. In fact, monsoon climates exhibit a greater annual *range* in precipitation amounts than do any other climate type (Figure 12.4).

12.2.3 | The Tropical Circulation

The circulation of the tropics is in the form of a cell known as the **Hadley cell** (Figure 12.5). A Hadley cell is a thermally direct cell with warm air rising near the equator and colder air sinking in the subtropics. The strong surface heating in equatorial regions leads to the development of the equatorial low,

HADLEY CELL
A thermally driven convection cell located between latitude 30° N and the equator, or between latitude 30° S and the equator.

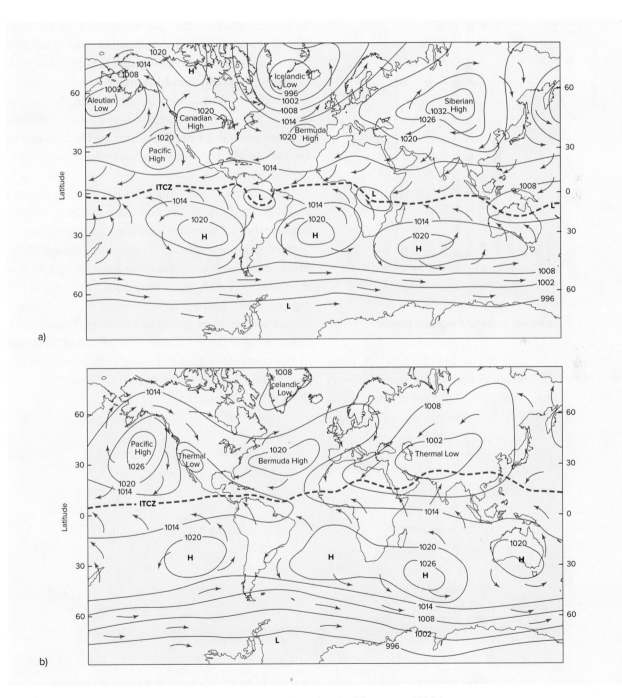

FIGURE 12.3 | The observed pressure and wind patterns at the surface in a) January and b) July.

Source: Adapted from Ahrens, C. Donald. *Meteorology today: An introduction to weather, climate, and the environment*, (9th ed.). California: Brooks/Cole, Cengage Learning, 2009, p. 263.

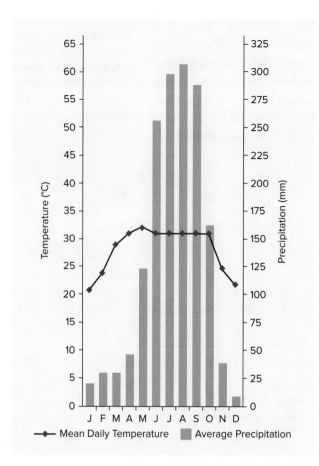

FIGURE 12.4 | A climate graph for a monsoon climate.

or the ITCZ, which is a *thermal* low (Section 3.6). Here, the trade winds converge, and warm, moist air rises from the surface, creating cumulonimbus clouds and

copious amounts of rainfall (Figure 12.6; Figure 10.17). In addition, the ITCZ is also known for being a location of very weak and variable winds due to a lack of strong pressure gradients. In the days of sailing ships, this zone was named *the doldrums* because sailors could often be stalled for days by the slack winds.

Upon reaching the tropopause, the air rising in the ITCZ diverges and heads toward the poles of both hemispheres. This air sinks back to the surface at roughly 30° N and S, creating the subtropical highs. In the centre of the subtropical highs is another region of light and variable surface winds similar to the doldrums. This area was also named during the days of sailing ships: it was called *the horse latitudes* because when ships had been drifting for days due to the lack of wind, the starving horses were eventually thrown overboard.

In contrast to the equatorial low, the subtropical highs have dynamic, rather than thermal, origins because they form due to sinking air. As a consequence of the sinking air, *subsidence inversions* develop; these inversions effectively prevent air from rising far from the surface (Section 8.5.2). As a result, in many of the regions dominated by the subtropical highs, skies are almost always clear and little rain falls (Figure 12.7). Thus, many of the great deserts of the world form due to the subtropical highs (Figure 12.8). Examples are the Sahara Desert of northern Africa (Figure 12.9), and the deserts of Australia.

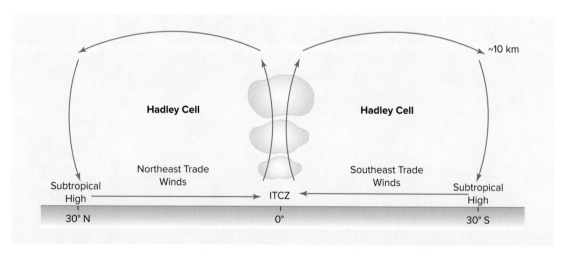

FIGURE 12.5 | The Hadley cells of the tropics, showing the location of the ITCZ, the subtropical highs, and the trade winds. What forces are acting to create the easterly trade winds?

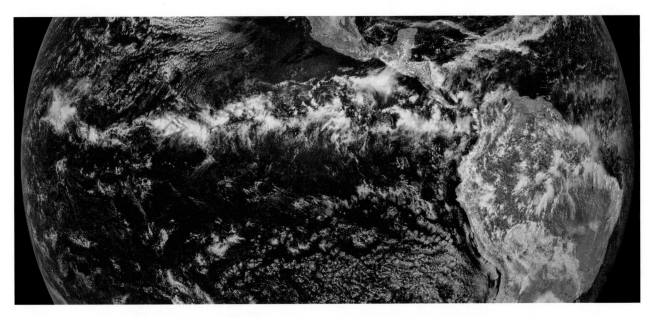

FIGURE 12.6 | The ITCZ can be seen on this satellite image as a band of cloud circling the planet near the equator.

Notice that Figure 12.8 shows that the subtropical deserts tend to occur on the western sides of continents. This is because the air is drier and the subsidence is stronger on the eastern sides of the subtropical highs than it is on their western sides. As air flows equatorward on the eastern sides of the highs, it flows over cold water and, as a result, picks up little water vapour. In addition, the subsidence inversions on the eastern sides of the highs are often only about 500 m above the surface, making it almost impossible for air to rise high enough to form clouds. In contrast, air on the western sides of the highs has travelled over the warm tropical oceans, picking up water vapour on the way. This additional moisture, coupled with the weaker subsidence, means that clouds and precipitation are more likely to develop. Therefore, east coasts in subtropical latitudes are much wetter than west coasts are. For example, on the eastern side of the Pacific high (Figure 12.3) lie the deserts of southern California and northern Mexico, while on the western side of the Pacific high lie the much wetter regions of southeast Asia. Similar patterns occur in relation to the other subtropical highs, in both the northern and southern hemispheres.

To complete the Hadley-cell circulation, the trade winds flow from the subtropical highs back to the ITCZ. These are very steady, reliable winds that blow over half the surface of Earth. It is the steadiness of these winds that has earned them their name, which comes from the historical use of the word *trade* to mean "on a steady course." As these winds blow across the water toward the west, they bring clouds and rain to east coasts in tropical latitudes. For example, the Hawaiian Islands, though small, receive considerably more rain on their east coasts than on their west coasts (Figure 12.10).

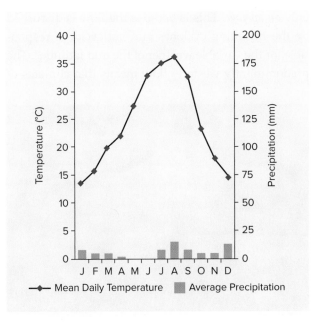

FIGURE 12.7 | A climate graph for a subtropical desert.

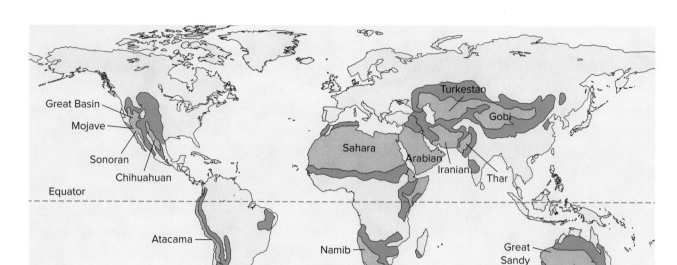

FIGURE 12.8 | The deserts of the world. The deserts of the subtropics form due to the subtropical highs. Most of the mid-latitude deserts are rain-shadow deserts.

12.2.4 | The Extratropical Circulation

While surface air diverging from the subtropical highs flows toward the ITCZ, forming the trade winds, it also flows poleward, creating the mid-latitude westerlies. These winds are not nearly as steady as the trade winds; in fact, they are westerly only *on average*. This is because the flow is disturbed by the travelling cyclones and anticyclones responsible for the variable weather of the mid-latitudes. The predominantly westerly flow means that climates of

west-coast locations in the mid-latitudes are more strongly influenced by the moderating effects of the ocean than are climates of east-coast locations. The annual temperature range on west coasts, therefore, is usually much smaller than it is on east coasts. In particular, east-coast winters are colder than west-coast winters (Figure 12.11). For example, for the same latitude, winter temperatures are much higher on Canada's west coast than they are on Canada's east coast (Figure 12.12).

FIGURE 12.9 | The Sahara Desert. What accounts for the extreme diurnal temperature ranges of subtropical deserts? (See Section 6.6.3.)

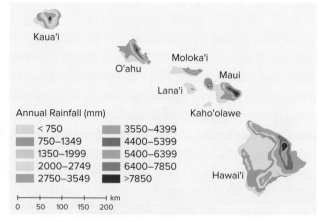

FIGURE 12.10 | A map of precipitation for the Hawaiian Islands.

Source: Based on the Online Rainfall Atlas of Hawai'i at http://rainfall.geography.hawaii.edu/

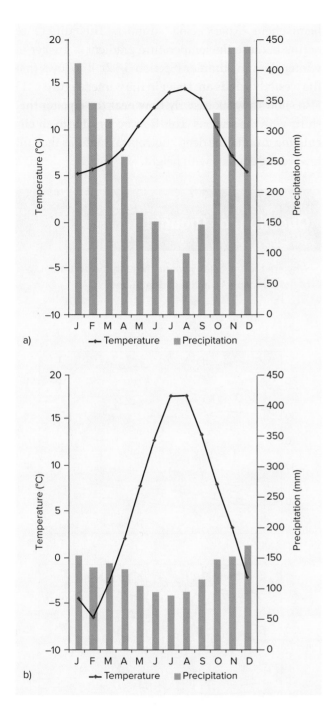

a)

b)

FIGURE 12.11 | a) Estevan Point, British Columbia, is located on the west coast of Canada. b) Sydney, Nova Scotia, is located on the east coast. The annual temperature range in Estevan Point is smaller than that in Sydney. Why do oceans tend to have a moderating influencing on temperature? Why does Estevan Point receive more precipitation than Sydney receives?

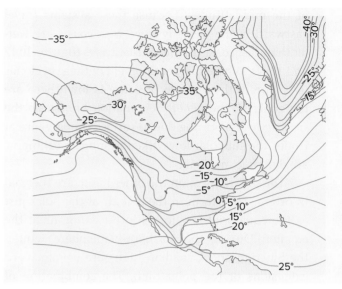

FIGURE 12.12 | Map of January temperatures in North America. Notice how the isotherms bend southward over the continent, especially toward the east.

Source: Adapted from Strahler and Archibold, 2011, p. 84.

the southern hemisphere. In fact, without land masses to slow the flow, the westerly winds can reach high speeds as they blow over the southern oceans and circle the continent of Antarctica. Like the doldrums and the horse latitudes, these winds were named by sailors, who referred to them as the roaring forties, the furious fifties, and the screaming sixties, after the latitudes at which they are found.

Just as warm air flows toward the poles from the *subtropical* highs, cold air flows toward the equator from the *polar* highs. This flow is deflected by the Coriolis force, producing the polar easterlies. Because these winds cover only a small portion of Earth's surface, they are a relatively insignificant part of the general circulation. However, they do play an important role in forming fronts, which develop where the cold polar easterlies meet the warmer westerlies, usually somewhere between 40° and 60° latitude. These fronts are sometimes conceptualized as one continuous front called the **polar front**; in reality, however, they are not continuous. At fronts, warm air rises over cold air, creating troughs of low pressure that can develop into low-pressure systems (Section 14.2). These systems are known as *mid-latitude cyclones*, or simply *frontal systems*, and they are responsible for most of the precipitation

POLAR FRONT

An idealized front that represents the meeting of cold polar air with warm tropical air.

Notice that Figure 12.3 shows a much more regular westerly flow in the southern hemisphere than in the northern hemisphere. This regularity is due to there being almost no land masses in the mid-latitudes of of the mid-latitudes. Because these systems travel eastward in the westerly flow, west coasts are wetter than east coasts in the mid-latitudes (figures 10.17 and 12.11). Where the development of these cyclones is favoured, the average pressure is low; these are the zones of the general circulation known as the *subpolar lows*. Forming due to surface convergence at fronts, the subpolar lows, like the subtropical highs, are dynamic in origin.

The polar highs, on the other hand, are thermal in origin. They form in the *cold* air at the poles just as the equatorial low forms in the *warm* air at the equator. During the northern-hemisphere winter, the Siberian and Canadian highs are equatorward extensions of the polar high. The combination of high pressure and cold, dry air makes the poles the least stormy areas on Earth. The storms that do occur in these regions are normally frontal systems that have strayed from the mid-latitudes.

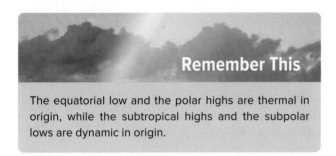

Remember This

The equatorial low and the polar highs are thermal in origin, while the subtropical highs and the subpolar lows are dynamic in origin.

12.2.5 | The Circulation Aloft

Because both the equatorial low and the polar high are thermal in origin, they are *shallow* pressure systems (Section 3.6). The equatorial low gradually reverses to a high with height, while the polar high gradually reverses to a low with height (Figure 12.13). Figure 12.14 shows how the poleward decrease in temperature creates both this *reversal* of the pressure gradient with height as well as the *increase* in the pressure gradient with height. The upper-air westerlies, of both hemispheres, are formed as air flows along this pressure gradient and is deflected by the Coriolis force. These winds blow with cold air on the left, as is predicted by the thermal wind relation (Section 11.3). Figure 12.13 shows that the upper-air pressure gradient is greater in the hemisphere experiencing winter than it is in the

hemisphere experiencing summer. This difference occurs because the temperature gradient is greater in winter than in summer (Section 5.8.2). It follows that the westerly winds are faster in the winter. Figure 12.13 also shows a weak easterly flow near the equator that shifts with the seasons. This flow occurs due to air circulating around the deep subtropical highs as they tilt toward the equator with height.

Question for Thought

Why are there westerly winds through the entire depth of the troposphere in the mid-latitudes?

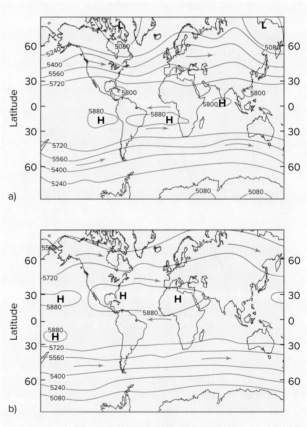

FIGURE 12.13 | a) The average January 500 hPa surface. **b)** The average July 500 hPa surface. What is the cause of the upper-air pressure gradient shown on these maps?

Source: Adapted from Ahrens, C. Donald. *Meteorology today: An introduction to weather, climate, and the environment*, 9th ed. California: Brooks/Cole, Cengage Learning, 2009, p. 266.

FIGURE 12.14 | The equator-to-pole temperature gradient leads to a surface pressure gradient directed from the poles to the equator. This pressure gradient reverses, and strengthens, with height.

Embedded within the upper airflow of each hemisphere are two major jet streams: the subtropical jet stream and the polar front jet stream (Figure 12.15). Both jet streams are high-altitude, fast-moving westerly winds that have average speeds of about 150 km/hr and exhibit their maximum strengths somewhere between the 300 and 200 hPa pressure surfaces. Apart from these similarities, these jet streams are quite different in terms of both their origins and their character.

12.2.5.1 | The Subtropical Jet Streams

The subtropical jet streams are located on average at about 30° N and S. These latitudes correspond with the location of sinking air in the Hadley cells and the development of the subtropical highs at the surface. The subtropical jet streams result partially from the need to conserve angular momentum. As air flows poleward in the upper branch of a Hadley cell, it becomes increasingly westerly due to the increase in the Coriolis force. At the same time, its distance from Earth's axis of rotation decreases (Figure 12.16). According to the conservation of angular momentum (Equation 12.1), if distance from the centre of rotation decreases, velocity must increase. This combination of circumstances creates the fast, westerly flow of the

subtropical jet streams, which have wind speeds of about 150 km/h. However, these winds are not nearly as fast as would be predicted from the conservation of angular momentum alone because turbulence will

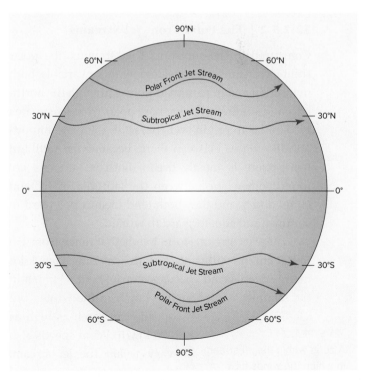

FIGURE 12.15 | The approximate location of the jet streams.

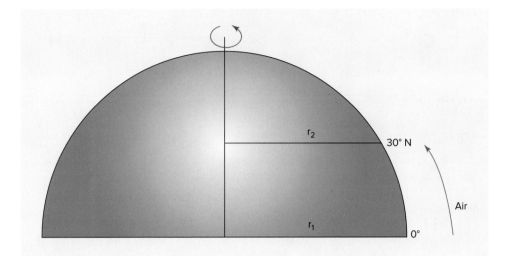

FIGURE 12.16 | As air travels northward, its distance from Earth's axis of rotation decreases. The radius at the equator, r_1, is greater than the radius at 30° N, r_2.

slow the air (Example 12.1). Therefore, the conservation of angular momentum does not provide a complete explanation for the behaviour of the subtropical jet streams. Poleward of these jet streams, the simple Hadley cells break down because conservation of angular momentum would require increasingly faster winds.

12.2.5.2 | The Polar Front Jet Streams

Compared to the subtropical jet streams, the polar front jet streams are considerably more variable, both in location and in pattern of flow. Although the northern- and southern-hemisphere polar front jet streams are similar in origin and character, there are important differences between them, because the southern hemisphere is mostly ocean. Both jet streams are found somewhere between the latitudes of 40° and 60° N and S, and they both tend to shift poleward in summer and equatorward in winter.

These jet streams form bands that range from 150 to 500 km in width, and they are generally a few kilometres deep. Wind speeds average about 80 km/h in the summer and about 160 km/h in the winter, but speeds can reach as high as 400 km/h. Wind speeds also vary *within* the jet stream; they decrease from central **jet streaks** to the edges of

JET STREAK
A zone within the jet stream in which the winds flow the fastest.

the band. The wind shear created by this variation in wind speeds produces turbulence within the jet stream. This turbulence, in turn, can cause clouds to

---

Example 12.1

At the equator, Earth spins at a rate of about 1670 km/h. At 30° latitude, where the distance to the centre of rotation has decreased to 0.866 (0.866 = cos 30°) of what it was at the equator, Earth spins at a rate of about 1446 km/h. Estimate the speed of the subtropical jet streams, based on the conservation of momentum alone.

Assume that conditions are calm at the equator, so that an air parcel of unit mass moving above the equator will be travelling at 1670 km/h. Now, use Equation 12.1, and set the angular momentum at the equator equal to the angular momentum at 30° latitude.

$$1670 \text{ km/h} \times 1 = V \times 0.866$$

$$V = 1928 \text{ km/h}$$

Next, subtract the speed of rotation at 30° latitude from 1928 km/h to obtain the speed of the air at 30° latitude.

$$1928 \text{ km/h} - 1446 \text{ km/h} = 482 \text{ km/h}$$

This is much faster than the actual speed of the subtropical jet streams, which is about 150 km/h.

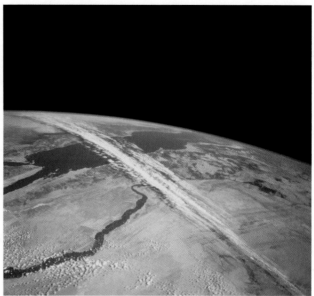

FIGURE 12.17 | *Left:* Jet stream cirrus seen from the ground. *Right:* Jet stream cirrus seen from space.

form, making a jet stream's position identifiable from the ground and from space (Section 9.3.6; Figure 12.17).

At times, the jet-stream flow is discontinuous; at other times, it splits into two, rejoining once again farther downstream. As the jet stream flows, it can remain more or less parallel to the latitude lines—producing **zonal** flow—but, more commonly, it will meander, taking on a **meridional** flow (Figure 12.18). This meandering is highly variable; the waves associated with it are called **Rossby waves**, or *long waves* (Section 12.5). There are usually anywhere from three to seven of these waves circling the planet as part

of the upper airflow, and their wavelengths are normally about 3000 to 4000 km (Figure 12.19).

Each polar front jet stream forms above a polar front. The large temperature gradients associated with these fronts interrupt the generally gradual temperature gradients that exist from the equator to

ZONAL
Relating to or varying in the east–west direction.

MERIDIONAL
Relating to or varying along a meridian, or in the north–south direction.

ROSSBY WAVES
Waves in the upper-air westerly flow of the mid-latitudes, with wavelengths of several thousand kilometres.

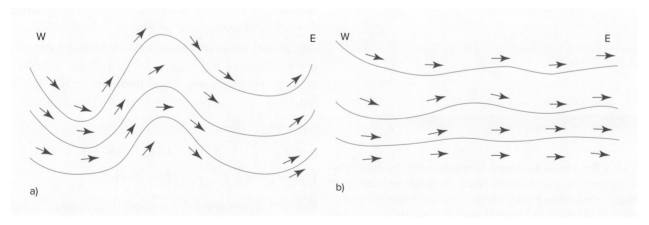

FIGURE 12.18 | a) Meridional flow. b) Zonal flow.

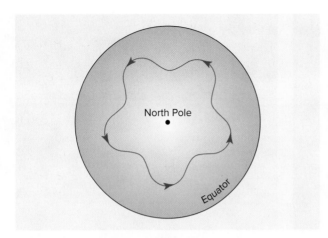

FIGURE 12.19 | Viewed from far above the North Pole, the Rossby waves of the upper airflow follow a meandering path.

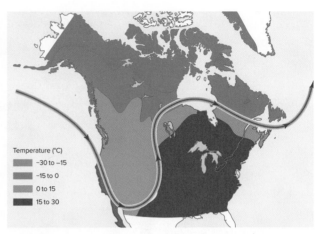

FIGURE 12.21 | The jet stream's path (shown in blue) during March 2012 caused temperatures to be unseasonably cold in western Canada and unseasonably warm in eastern Canada.

the poles (Figure 12.20), creating strong pressure gradients that result in fast winds. These pressure gradients increase with height, so the jet stream occurs near the top of the troposphere. Because the temperature contrast across fronts is greater in winter than in summer (Section 5.8.2), the jet streams are fastest in winter. In addition, just as the pressure systems of the general circulation shift with the seasons, so too do the polar front jet streams.

Because mid-latitude cyclones form in association with the polar front jet streams (Section 14.2), they are

carried from west to east, and places located along the jet streams' paths generally experience stormier weather than do places north or south of the jet streams. In particular, the seasonal shift in the *average* position of the jet streams can impact mid-latitude climates. Such an impact was shown in Section 10.7. Estevan Point, British Columbia, experiences a drier summer than does Prince Rupert, British Columbia, because of the northward shift of the jet stream in summer (Figure 10.22). As the jet stream shifts north, it effectively carries the storms with it, leaving Estevan Point *relatively* dry. Farther south in California, summers are extremely dry, as the summer jet stream rarely moves that far south.

The amplitude and position of the Rossby waves can also have a strong influence on mid-latitude weather. When these waves form the great, looping meanders characteristic of strong meridional flow, unseasonal temperatures can occur. Locations to the north of one of these great loops will have below-average temperatures, while locations to the south will have above-average temperatures (Figure 12.21).

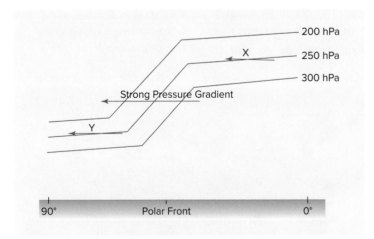

FIGURE 12.20 | The gradual decrease in temperature from the equator to the poles is interrupted by a much larger temperature drop at the polar front. This large drop in temperature creates a large pressure gradient that produces the polar front jet stream. The pressure gradients at points X and Y are much weaker than the one above the polar front. (Compare the upper part of this diagram to the upper part of Figure 12.14 in which a constant surface temperature gradient was assumed.)

12.3 | Explanations of the General Circulation

The underlying *cause* of the general circulation is the need to distribute energy and momentum, but simply identifying this cause is not enough to *explain* all the complexities of the observed circulation. In order to explain our observations of the general circulation, we use models.

12.3.1 | Early Conceptual Models

In 1735, George Hadley proposed a single-cell model to explain the trade winds (Figure 12.22). This model represents the general circulation in each hemisphere as a simple convection cell that is driven by the difference in temperature between the equator and the poles. (This cell is similar to the one associated with the land–sea breeze circulation system described in Section 11.4, but it occurs at a much larger scale.) Hadley's model recognizes that heating at the equator leads to low surface pressure, while cooling at the poles leads to high surface pressure. In this model, warm air rises at the equator and then flows toward the poles aloft, while cold air sinks at the poles and then flows toward the equator at the surface; this flow forms a thermally direct cell. When we include the effects of Earth's rotation, the surface flow becomes easterly, like the trade winds, and the upper flow becomes westerly.

Observation shows that Hadley's model does not fully account for the behaviour of the general circulation. Although upper-air winds *are* mostly westerly, surface winds are not all easterly. In addition, the model does not account for the subtropical highs and the subpolar lows. Further, such a cell as the one

Hadley proposed *could not* exist—the conservation of angular momentum would require air heading toward the poles in the upper flow to get so fast that the flow would become unstable and break down. In addition, the predicted easterly winds at the surface might slow the rotation of Earth if they were not balanced by westerly winds. Finally, if the winds were entirely zonal, as this model supposes, this single cell could not accomplish the necessary transfer of heat.

To improve upon this simple model, William Ferrel proposed a three-cell model in 1856 (Figure 12.23). In his model, Ferrel reduced the extent of the thermally direct cell of Hadley's model to the tropics; thus, he described the Hadley cell that we continue to recognize as a major part of Earth's general circulation. To the Hadley cell, Ferrel added a mid-latitude cell, which has been called the Ferrel cell, and a polar cell. The Ferrel cell is thermally *indirect*—in it, the *warmer air sinks* and the *colder air rises*—and the polar cell, like the Hadley cell, is thermally direct. Thus, Ferrel's model accounts for the existence of the subtropical highs and subpolar lows; it also accounts for the surface winds: the trade winds, the mid-latitude westerlies, and the polar easterlies. However, it predicts upper-air easterly winds in the mid-latitudes. In reality, winds are westerly throughout the troposphere in the mid-latitudes. Clearly, this model does not provide a complete explanation of our observations.

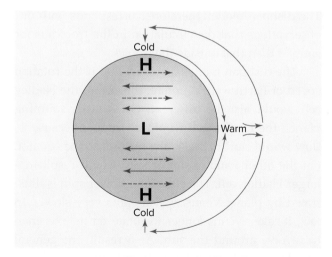

FIGURE 12.22 | A single-cell model of Earth's general circulation. Warm air rises at the equator and cold air sinks at the poles. Surface winds are easterly (solid arrows) and upper winds (dashed arrows) are westerly. (The red arrows on the right of the diagram represent convection cells.)

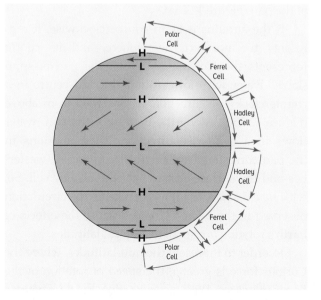

FIGURE 12.23 | The three-cell model of Earth's general circulation.

12.3.2 | Dishpan Models

In addition to the *conceptual* models described above, *physical* models can be used to account for the observed general circulation of Earth. One such model is known as the *dishpan model*. The "dishpan" is a doughnut-shaped pan, or annulus (Figure 12.24). Water is placed in the pan to represent the atmosphere. A dye, or metal filings, is placed in the water so that flow patterns can be observed. The centre is cooled to represent the poles, and the outside edge is warmed to represent the equator. The pan can be spun to represent Earth's rotation. Recall that these two factors—*differential heating* and *rotation*—are two important controls on the general circulation. In the dishpan model, both can be varied.

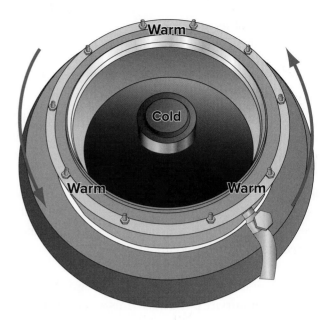

FIGURE 12.24 | The rotating annulus, or "dishpan," used for a dishpan model. The cold centre represents Earth's poles, and the warmed edges represent the equator.

Question for Thought

In what ways do the conditions of the dishpan model differ from real conditions?

Without rotation, the difference in heating between the edge and the centre of the pan creates a simple thermal circulation, similar in nature to the Hadley cell. Warm water rises on the outside rim of the pan and travels inward to the centre of the pan, where it cools, sinks, and travels back out to the edge of the pan (Figure 12.25a).

If the annulus is spun counterclockwise, it represents the northern hemisphere, and its centre represents the North Pole. When the pan is spun slowly, water will flow around the cold centre in a counterclockwise direction, as observed from above (Figure 12.25b). Along the bottom of the pan, water flows in a clockwise direction. These motions in the pan correspond well with the upper-air westerlies and the surface easterlies of the Hadley cell. So, with *slow rotation*, the flow represents the circulation observed in the *tropics*. This is because the effects of Earth's rotation are small in tropical latitudes.

In order to represent the mid-latitudes, where the Coriolis force is greater, the speed of rotation of the pan is increased. With faster rotation, the flow pattern

changes to a series of waves (Figure 12.25c). This pattern occurs because the very fast flow, which is required to conserve angular momentum in the fluid as it moves toward the centre, causes the simple flow to break down. (As the rotation speed increases, the fluid flow needs to get even faster to conserve angular momentum.) The waves in this pattern represent the Rossby waves. So, with *fast rotation*, the flow represents the circulation observed in the *mid-latitudes*. The dishpan model, therefore, corresponds with our observations that the circulation in the tropics is far simpler than that outside the tropics.

The dishpan model shows us that if the rotation speed of Earth were different, the size of the Hadley cell would also be different. On a faster-spinning planet, the Hadley cell would be smaller because its flow would start to break down closer to the equator. On the other hand, a slower spin would result in a larger Hadley cell. The effect of a slower spin is illustrated by planet Venus, which spins very slowly. In fact, it takes Venus longer to rotate on its axis than to revolve around the sun. As a result, the general circulation on Venus is far simpler than it is on Earth, and heat is transferred poleward much more efficiently. Despite Earth's faster rotation speed, however,

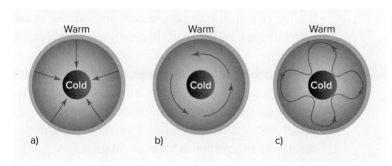

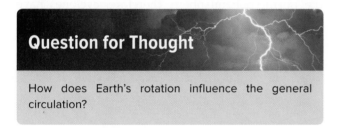

FIGURE 12.25 | Variations in the dishpan model. a) With no rotation, water at the surface flows in toward the centre, where it is cooled and sinks. b) With slow counterclockwise rotation, water at the surface begins to flow in a counterclockwise direction around the cold centre. c) With fast counterclockwise rotation, water at the surface continues to flow in a counterclockwise direction, but a wave pattern develops in the flow.

heat and momentum are still transferred poleward as required.

Question for Thought

How does Earth's rotation influence the general circulation?

12.3.3 | A Complete Conceptual Model

We can now move toward developing a complete conceptual model of Earth's general circulation, to account for the necessary transfer of heat and momentum described at the beginning of this chapter. First, we can recognize that the tropical circulation transfers heat toward the poles because, with the relatively small Coriolis force near the equator, winds in tropical regions maintain a relatively strong meridional component. For example, the trade winds are not direct east-to-west winds, but are instead *north*easterly and *south*easterly. Thus, they bring colder air into tropical regions. Likewise, the winds aloft in the tropics are not quite westerly, so they transfer warm winds poleward.

Second, we can observe that although the extratropical circulation is more complex, it still accomplishes heat transfer. In the upper airflow, a trough of low pressure is a wave of cold air moving toward the equator, while a ridge of high pressure is a wave

of warm air moving toward the poles (Figure 12.26a). Thus the Rossby waves transfer heat. We often experience this heat transfer as unseasonal temperatures, as was shown in Figure 12.21. Closer to the surface, in the mid-latitudes, the large-scale flow is in the form of eddies rather than waves. These eddies are the cyclones and anticyclones responsible for mid-latitude weather. They transfer heat because they, too, add a strong meridional component to the flow: in the northern hemisphere, air flows southward on the west side of cyclones and northward on their eastern sides, while the opposite occurs with anticyclones (Figure 12.26b).

Finally, we can see that the general circulation transfers *momentum* as well as *heat* poleward. Remember that in order to maintain the westerly winds of the mid-latitudes, momentum must be transferred from the tropics. This momentum transfer is accomplished because the waves of the upper flow tend to move from the southwest to the northeast. Figure 12.27 shows that such movement carries westerly momentum toward the poles.

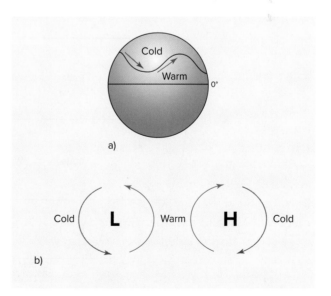

FIGURE 12.26 | a) Rossby waves transfer warm air northward and cold air southward in the northern hemisphere. b) The cyclones and anticyclones of the mid-latitudes also transfer warm air northward and cold air southward in the northern hemisphere.

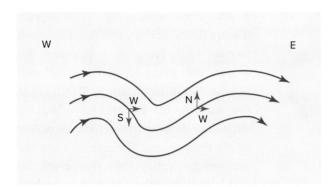

FIGURE 12.27 | Transfer of westerly momentum by the waves of the upper airflow.

VORTICITY
A measure of spin in a fluid.

Considering all of the above, we can construct a complete conceptual model of Earth's general circulation that comprises a large thermally direct cell (the

Hadley cell) in the tropics, waves and eddies in the mid-latitudes, and a small thermally direct cell at the poles (Figure 12.28). We have shown that such a circulation is able to transfer energy and momentum toward the poles. Yet there is still one important aspect of the general circulation that we must explain. The dishpan model showed that the waves of the mid-latitudes developed as a response to faster rotation speeds. The dishpan model, however, does not explain *why* rotation leads to a wavy flow. For such an explanation, we must consider the concept of **vorticity**.

12.4 | Vorticity

Vorticity is a measure of the rate of spin in a fluid. Counterclockwise spin is defined as *positive* vorticity and clockwise spin is defined as *negative* vorticity. In the northern hemisphere, therefore, cyclones have

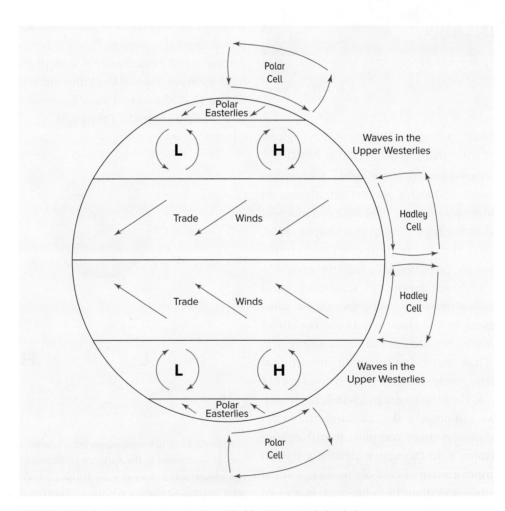

FIGURE 12.28 | A complete conceptual model of Earth's general circulation.

positive vorticity and anticyclones have negative vorticity. The reverse is true in the southern hemisphere.

There are two types of vorticity: **planetary vorticity** and **relative vorticity**. Everything on Earth has planetary vorticity. In the northern hemisphere, planetary vorticity is positive; in the southern hemisphere, it is negative. Because planetary vorticity is given by the Coriolis parameter, f_c (Equation 11.3), it will increase with latitude and is zero at the equator. Relative vorticity describes spin *relative* to Earth. Everything on Earth that spins due to its own movement has both relative and planetary vorticity. For example, air that is spinning in a cyclone has *relative vorticity* because of its own movement, and it has *planetary vorticity* because it is part of a rotating system, Earth.

Relative vorticity can arise because of curvature or shear, or both (Figure 12.29). Air parcels in curved flow will have relative vorticity because the curvature of the flow provides spin. For example, air flowing through a trough will gain cyclonic spin, thus increasing its positive relative vorticity. Conversely, air flowing through a ridge will spin anticyclonically, thus increasing its negative relative vorticity. It is a little less easy to see how shear causes vorticity. When winds blow faster on one side of an air parcel than on the other—in other words, when there is horizontal wind shear—the air parcel will spin. An analogous situation can occur in a stream. In a straight section of a stream, the flow is usually greater in the centre than it is along the banks; as a result, eddies may form in the flow. Eddies spin and, therefore, have vorticity. Both curvature and shear vorticity can be measured on upper-air charts, and these measures are used in forecasting (Section 15.5.2).

The sum of planetary vorticity, f_c, and relative vorticity, ζ_r, is called **absolute vorticity**, ζ_a.

$$f_c + \zeta_r = \zeta_a \qquad (12.2)$$

Because vorticity is expressed as angular velocity, its units are s^{-1}. Since planetary vorticity is always positive in the northern hemisphere, and since planetary vorticity is almost always greater than relative vorticity, absolute vorticity is almost always positive in the northern hemisphere. This means that even when air is spinning anticyclonically, its absolute

vorticity will likely be positive, but its absolute vorticity will be lower than the absolute vorticity of air spinning cyclonically.

Changes in the absolute vorticity of moving air occur due to convergence or divergence (see Section 9.3). This relationship is shown in Equation 12.3.

$$\left[\frac{1}{f_c + \zeta_r}\right]\left[\frac{\Delta(f_c + \zeta_r)}{\Delta t}\right] = -D \qquad (12.3)$$

The left-hand side of this equation is the normalized change in vorticity with time. D represents divergence; convergence is negative divergence. Equation 12.3 shows that convergence causes an increase in vorticity with time and divergence causes a decrease in vorticity with time.

We can use this equation to understand how air takes on cyclonic or anticyclonic spin in the atmosphere. Imagine that moving air is not noticeably spinning; it will have planetary vorticity, but no relative vorticity. It follows from Equation 12.3 that if air converges at Earth's surface, its absolute vorticity must increase. For this to happen, the air must attain positive relative vorticity; it must start to spin cyclonically, or counterclockwise, in the northern hemisphere. Because the air already had planetary vorticity, convergence did not *cause*

PLANETARY VORTICITY
Vorticity that arises due to Earth's rotation.

RELATIVE VORTICITY
Vorticity occurring due to a fluid's own motion.

ABSOLUTE VORTICITY
The sum of planetary vorticity and relative vorticity.

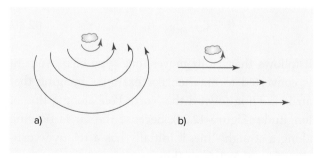

FIGURE 12.29 | In both diagrams, the cloud indicates an air parcel, viewed from above, that will develop vorticity (red arrows) as a result of the character of the flow. a) Relative vorticity develops due to curvature in the flow. b) Relative vorticity develops due to shear in the flow.

the spin; rather, it *increased* the spin. In other words, the convergence caused a *concentration* of Earth's spin (McIlveen, 2010, p. 275). If Earth did not rotate, cyclones would not form. On the other hand, when air diverges at Earth's surface, it must cause a decrease in vorticity with time. For this to happen, relative vorticity must become negative; the air must start to spin anticyclonically. In this case, the divergence has caused a *dilution* of Earth's spin (McIlveen, p. 275). Consider that as anticyclonic spin increases, absolute vorticity will approach zero, putting a limit on the spin of anticyclones. No such limit occurs with cyclones.

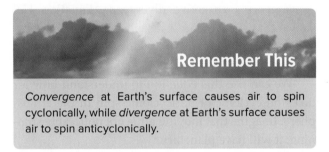

Remember This

Convergence at Earth's surface causes air to spin cyclonically, while *divergence* at Earth's surface causes air to spin anticyclonically.

When there is no convergence or divergence, D is equal to zero, as written in Equation 12.4a.

$$\frac{\Delta(f_c + \zeta_r)}{\Delta t} = 0 \tag{12.4a}$$

This equation shows that, without convergence or divergence, absolute vorticity does not change with time. In other words, absolute vorticity is *conserved*, meaning that if planetary vorticity increases over time, relative vorticity must decrease, and vice versa. This relationship is also shown by Equation 12.4b.

$$(f_c + \zeta_r)_{\text{time 1}} = (f_c + \zeta_r)_{\text{time 2}} \tag{12.4b}$$

It follows that as air moves, one type of movement is converted to another. For example, imagine that air is flowing toward the North Pole along a line of longitude (Figure 12.30). Because the air is flowing along a straight line, it initially has a relative vorticity of zero. As the air flows northward, its planetary vorticity will increase. To conserve absolute vorticity, therefore, the relative vorticity must decrease from its

initial value of zero; thus, it becomes negative. With negative vorticity, the air has anticyclonic curvature, which causes it to turn clockwise and curve back toward the equator. This example reveals that an air parcel can develop a curving flow solely because it is moving above a *spinning* planet.

Changes in vorticity are also linked to vertical motions. Because of the principle of conservation of mass, horizontal convergence and divergence in the atmosphere must be accompanied by vertical motions. When convergence occurs in the atmosphere, the column of air experiencing the convergence will vertically stretch; when divergence occurs, the column will vertically shrink. It follows that when a column of air stretches, its vorticity will increase, and when a column of air shrinks, its vorticity will decrease. These relationships are shown in Equation 12.5, where Δz is the depth of the air column.

$$\left(\frac{f_c + \zeta_r}{\Delta z}\right)_{\text{time 1}} = \left(\frac{f_c + \zeta_r}{\Delta z}\right)_{\text{time 2}} \tag{12.5}$$

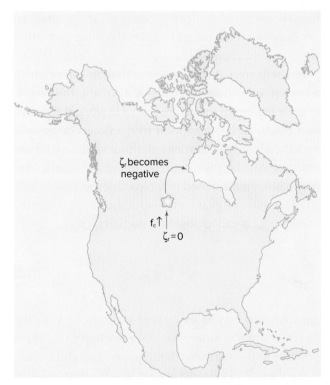

FIGURE 12.30 | As an air parcel moves toward the North Pole, it will develop negative relative vorticity, causing it to turn clockwise.

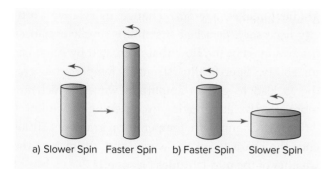

FIGURE 12.31 | a) A cyclonically spinning air column will spin faster as it stretches. b) A cyclonically spinning air column will spin slower as it shrinks.

Notice that if the depth of the air column is not changing, Δz can be eliminated from Equation 12.5, leaving Equation 12.4. When an air column shrinks with time, Δz will decrease. It follows that the absolute vorticity term must also decrease. If we assume that the latitude does not change, then the relative vorticity must decrease. Equation 12.5, therefore, shows that a cyclonically spinning air column will spin slower if it shrinks and faster if it is stretched, as shown in Figure 12.31.

12.5 | Rossby Waves

The concept of vorticity can help us understand how Rossby waves form. If we assume that there is no convergence or divergence, and no associated vertical motions, we can use Equation 12.4b. Imagine that,

initially, the northern-hemisphere jet stream is flowing directly from west to east. Under these conditions, relative vorticity is zero, so absolute vorticity is equal to planetary vorticity and, therefore, proportional to the latitude of the flow.

Next, imagine that this straight flow is deflected southward by a mountain range (Figure 12.32). As the air moves south, its planetary vorticity will decrease so, to conserve absolute vorticity, it must curve cyclonically and head north. As it moves northward, the air will gain planetary vorticity. To conserve absolute vorticity, it must curve anticyclonically and head south once again. Thus, the jet stream follows a meandering path due to the variation in the Coriolis parameter with latitude. This variation causes **barotropic instability**.

Another type of instability—**baroclinic instability**—also influences the behaviour of Rossby waves. Recall that a large decrease in temperature with *height* creates unstable conditions that lead to thermal convection in an attempt to remove the excess energy from the surface (Section 8.4). In a similar way, a large decrease in temperature with *latitude* creates baroclinic instability that influences the Rossby waves in an attempt to remove the excess energy from lower latitudes. This baroclinic

BAROTROPIC INSTABILITY
A type of instability that arises due to the rotation of Earth.

BAROCLINIC INSTABILITY
A type of instability that occurs when there are strong horizontal temperature gradients that lead to thermal advection.

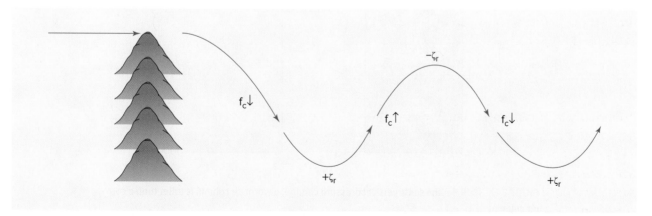

FIGURE 12.32 | The development of waves in the upper-air westerly winds of the mid-latitudes, as viewed from above. The wave motion develops because the air meanders back and forth to conserve vorticity.

instability is usually concentrated along fronts in the mid-latitudes.

The following is a simplified explanation of how temperature gradients affect the waves of the upper airflow. Recall that pressure decreases more quickly with height in cold air than it does in warm air so that, all else being equal, a warm air column will be taller than a cold air column of the same mass (Section 3.4; Figure 12.33). This means that if air is moving poleward and cooling, the column of air will shrink, whereas if air is moving equatorward and warming, the column of air will stretch. It follows that as air moves across a temperature gradient, it will be shrinking or stretching and, therefore, its vorticity will change.

Again imagine that the northern-hemisphere jet stream is moving directly from west to east and is deflected southward by a mountain range. As it moves south, not only is its planetary vorticity decreasing, it is stretching in height as it warms. For vorticity to be conserved, relative vorticity must take on a large enough positive value to compensate for the decrease in planetary vorticity *and* the stretching of the air column (Equation 12.5). This means that the air turns cyclonically and heads toward the pole.

The difference between this example and the previous one is that the addition of a temperature gradient makes a *greater* change in relative vorticity necessary. *Baroclinic* instability simply enhances the waviness of the flow that is created by *barotropic* instability. Specifically, stronger temperature gradients lead to greater-amplitude waves, which are more effective at removing excess energy from lower latitudes. Changes in temperature gradients influence the character of the waves and, ultimately, the weather of the mid-latitudes (Section 14.2).

The polar front jet streams, like the planetary circulation they are a part of, are a consequence of the equator-to-pole temperature gradient. In turn, they work to alleviate, or at least reduce, this temperature imbalance. This reduction is accomplished through the development of Rossby waves in the flow and also through the formation of cyclones and anticyclones. The Rossby waves, and the circular flow characteristic of cyclones and anticyclones, carry air north and south in the otherwise predominantly east–west flow. More intense cyclones form where there are especially strong temperature differences; by the time the cyclones die out, the temperature differences will have been reduced. Mid-latitude weather, then, is a result of the equator-to-pole temperature gradient, which manifests itself most strongly at the polar front.

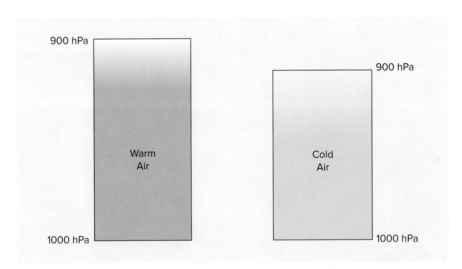

FIGURE 12.33 | For the same vertical pressure change, a warm air column is taller than a cold air column.

12.6 | Chapter Summary

1. The driving force behind the general circulation of Earth is the latitudinal imbalance in radiation. There is a surplus of radiation between roughly 40° N and 40° S, while there is deficit of radiation poleward of these latitudes. Heat is transferred poleward, preventing tropical regions from getting even warmer and polar regions from getting even colder. In addition, to transporting energy, the general circulation also transports momentum.

2. If we assume that there are no seasons and that Earth's surface is homogeneous, we can view the general circulation as a simple pressure pattern at Earth's surface. In this pattern, there is an equatorial low, and, in each hemisphere, there is a subtropical high, a subpolar low, and a polar high. These pressure patterns lead to winds. In the tropics, the winds are easterly; these are the trade winds. The zone along the equator where the trade winds from either hemisphere converge is known as the Intertropical Convergence Zone (ITCZ). Average winds in the mid-latitudes are westerly, while average winds in the polar latitudes are easterly. In the upper portion of the troposphere, there is high pressure above the equator and low pressure above the poles. This pattern results in a mostly westerly flow aloft.

3. When seasons, as well as the effects of different surface types, are considered, the simple pattern for the general circulation becomes far more complex. The pressure and wind patterns shift north and south with the seasons, and the pressure zones are interrupted due to the differential heating and cooling of land and water.

4. The tropical circulation forms a thermally direct cell known as the Hadley cell. Warm air rises in the equatorial low, flows poleward aloft, and sinks at approximately 30° N and S, where it forms the subtropical highs. To complete the cell, air flows from the subtropical highs toward the equatorial low, or ITCZ, forming the easterly trade winds. Locations influenced by the ITCZ are wet, while deserts form due to the subtropical highs.

5. Outside the tropics, the circulation is more complex and weather conditions are more variable. In the mid-latitudes, warm tropical air meets cold polar air at fronts, causing low-pressure systems to form; these systems create most of the precipitation of the mid-latitudes. These systems are carried eastward by westerly winds.

6. The equator-to-pole temperature gradient creates upper-air pressure gradients with high pressure above equatorial regions and low pressure above the poles. This pressure gradient combines with the effects of Earth's rotation to produce a westerly flow aloft over most of the planet. Two types of jet streams develop in this westerly flow: the subtropical jet streams, which occur partly to conserve angular momentum, and the polar front jet streams, which occur due to the temperature gradients associated with the polar fronts. The polar front jet streams take on a meandering pattern. The meanders are known as Rossby waves.

7. We use models to explain the general circulation. Hadley's simple model that is based on a single, thermally direct cell accounts for the equatorial low, the polar highs, and the upper-air westerlies. Ferrel's three-cell model accounts for the subpolar lows and the subtropical highs, as well as the equatorial low and the polar highs. This model also explains the surface winds over the planet: the trade winds, the mid-latitude westerlies, and the polar easterlies. A simple physical model, known as a dishpan model, shows that the Hadley cell is a good representation of the tropical circulation, where the effects of Earth's rotation are small. The model also tells us that the reason

the flow breaks down into eddies and waves in the mid-latitudes is that the effects of Earth's rotation are greater there than they are in the tropics.

8. The Hadley cells of the tropics, in combination with the waves and eddies of the mid-latitudes, are able to accomplish the necessary transfer of heat and momentum from the equator to the poles.

9. Vorticity is a measure of spin. Air spinning cyclonically has positive vorticity and air spinning anticyclonically has negative vorticity in the northern hemisphere. Planetary vorticity arises due to the rotation of Earth and increases with latitude. Relative vorticity develops when a fluid moves along a curving path or when there is shear in the flow. Absolute vorticity is the sum of planetary vorticity and relative vorticity; it increases with time in converging air and decreases with time in diverging air. It follows that convergence results in cyclonic spin, while divergence results in anticyclonic spin.

10. When there is no convergence or divergence occurring, absolute vorticity is conserved. The conservation of absolute vorticity can be applied to explain the existence and behaviour of Rossby waves.

11. Convergence and divergence are associated with vertical motions in the atmosphere; therefore, vertical motions influence vorticity. A stretching air column will spin faster, while a shrinking air column will spin more slowly. Because temperature changes can cause vertical motions, temperature gradients also influence Rossby waves. Earth's rotation alone causes waves to form, but these waves are influenced by the equator-to-pole temperature gradient.

Review Questions

1. What is the driving force behind the general circulation of Earth?

2. What four major pressure systems comprise Earth's general circulation in each hemisphere? Why do these pressure systems occur?

3. What are the three major surface winds that exist in each hemisphere? Why do these winds arise?

4. What is a Hadley cell? How is it associated with the pressure systems and winds of the tropics?

5. What is the cause of the subtropical deserts?

6. What are the two major differences between east- and west-coast climates in the mid-latitudes?

7. How does the equator-to-pole temperature gradient lead to upper-air westerly winds?

8. Why do the polar front jet streams and the subtropical jet streams exist?

9. How is the polar front jet stream of the northern hemisphere influenced by the seasons? How does this jet stream influence mid-latitude weather?

10. How can models help us explain how Earth's general circulation operates? What are the strengths and weaknesses of the models described in this chapter?

11. How does the general circulation transfer energy and momentum?

12. What is the difference between planetary vorticity and relative vorticity? What are the implications of the conservation of absolute vorticity?

13. How does the concept of vorticity explain why convergence results in cyclonic spin, while divergence results in anticyclonic spin?

14. How does the concept of conservation of vorticity explain why the Rossby waves arise?

Suggestions for Research

1.   Read some recent articles on global warming (Chapter 17) that explore why the greatest amount of warming seems to be occurring in polar regions. Find out how this uneven warming is expected to influence planetary circulation and, in particular, global patterns of precipitation.

2.   Investigate the teleconnections, caused by interactions between the atmosphere and the ocean, that can cause a change in ocean temperatures in one area to produce weather changes far away. Examples are associated with the El Niño–Southern Oscillation (ENSO) and the North Atlantic Oscillation (NAO).

References

McIlveen, R. (2010). *Fundamentals of weather and climate* (2nd ed.). New York, NY: Oxford University Press.

Strahler, Alan, and O.W. Archibold. (2011). *Physical geography: Science and systems of the human environment*, (5th Cdn. ed.). Toronto: John Wiley & Sons, Inc.

Air masses form over relatively flat, uniform surfaces—especially vast deserts, oceans, and frozen landscapes—as energy and water are transferred between Earth's surface and the air above. When air masses from warmer regions meet air masses from colder regions, they form fronts, which are the major producers of storms in the mid-latitudes.

13
Air Masses and Fronts

Learning Goals

After studying this chapter, you should be able to:

1. *describe* the characteristics of an air-mass source region;

2. *distinguish* between the five major air masses;

3. *describe* and *account* for the paths usually followed by the major air masses in North America;

4. *provide examples* of the changes in weather that can result when air masses move into a region;

5. *explain* how air masses can change as they move;

6. *distinguish* between the four types of fronts and *describe* the structure of each;

7. *list* weather changes that can occur with the passage of cold fronts, warm fronts, and occluded fronts; and

8. *explain* how our understanding of the occlusion process is changing.

In Chapter 4, you saw that heat can be transferred between the atmosphere and Earth's surface by the processes of conduction, convection, and radiation. In Chapter 7, you looked at how water vapour can be transferred from Earth's surface to the atmosphere by evaporation; then, in Chapter 10, you saw how precipitation processes operate to return water to Earth's surface. As energy and water are transferred between Earth's surface and air layers in contact with the surface, **air masses** are created. These large bodies of air can originate in the cold air of polar regions or the warm air of tropical regions. Inevitably, air masses from polar regions will meet air masses from tropical regions; this occurs in the mid-latitudes at transition zones known as *fronts* (Section 3.7). The temperature contrast between the air masses on either side of fronts provides the energy for mid-latitude cyclones, the major storms of the mid-latitudes.

Figure 13.1 depicts the relationship between air masses and fronts in a mid-latitude cyclone. Typically, such storms involve two fronts oriented in a wave-like pattern, as shown. Just as waves will form wherever fluids of different density flow past each other, waves will form along fronts where temperature contrasts create density differences in the air. In this chapter, you will learn more about the air masses and fronts associated with such storms; the storms themselves will be the focus of Chapter 14.

> **AIR MASS**
> A large body of air, thousands of square kilometres in size, throughout which the temperature and humidity are similar in the horizontal direction.

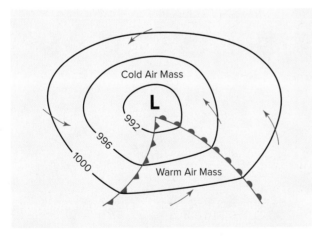

FIGURE 13.1 | A simple depiction of a mid-latitude cyclone, showing a warm front, a cold front, and a centre of low pressure. A warm, moist air mass lies in the triangular area between the warm front and the cold front, while a cold, dry air mass lies on the opposite side of these fronts. Also shown are wind arrows, and isobars labelled in hectopascals.

13.1 | Formation of Air Masses

In order for an air mass to form, the air must remain in place long enough to take on the characteristics of the surface below. As air sits stagnant, heat and water vapour are transferred between it and the surface. This process can take anywhere from several days to two weeks, with warm air masses forming more quickly than cold air masses.

Warm air masses begin to form when cool air moves over a warmer surface and, as a result, heat is transferred from the surface to the air. As air in contact with the surface is warmed by conduction, it will become less dense and rise, transferring heat upward by convection. In this unstable air mass, warm surface air is effectively carried upward. This transfer of sensible heat is driven by the temperature difference between the surface and the air (Section 6.6.2).

Alternatively, cold air masses begin to form when warm air moves over a colder surface and the surface air layer loses heat to the surface below by conduction. Cooling at the surface makes an air mass stable; the resulting lack of turbulence means that the surface cooling is not effectively carried upward. Instead, further cooling of the air mass must occur radiatively, as the air emits more radiation than it receives from the surface. The warming of an air mass by convection occurs more rapidly than the cooling of an air mass by radiation.

Likewise, water vapour can be transferred between the ground and the air above. If the surface is moist and the air is dry, the air layer closest to the ground will gain water vapour as water evaporates into it. This process of evaporation creates an upward-directed vapour pressure gradient, and water vapour is transferred into the air mass (Section 6.6.2). In this way, air that moves over oceans, large lakes, or vegetated areas will gain moisture. Recall that energy is needed for evaporation; therefore, it follows that air over *warm* tropical oceans will become moister than will air over *colder* high-latitude oceans. Air masses can lose moisture as well. For example, when a moist air mass is forced to rise over a mountain range, it may lose moisture through precipitation and descend the leeward side as a drier air mass.

13.1.1 | Air-Mass Source Regions

Air masses will form only under certain conditions; areas where these conditions are met are known as air-mass **source regions**. The best source regions are areas that have uniform surface properties stretching over thousands of square kilometres. Thus, source regions must be either land or water, not both, and they must be relatively flat. In addition, because the formation of an air mass requires that air remain stagnant for up to two weeks, winds must be light. In a cyclone, not only are winds often strong, but convergence at the surface brings together air of different properties, rather than creating the homogeneity necessary to produce an air mass. Good source regions, therefore, are the regions associated with the large, semi-permanent anticyclones of the general circulation: the polar highs and the subtropical highs (Section 12.2.1). It follows that there are four major air-mass source regions: the land and ice-covered oceans of the high latitudes, the ice-free oceans of the high latitudes, the oceans of the subtropics, and the deserts of the subtropics (Figure 13.2). The conditions necessary for an air-mass source region are not met in the mid-latitudes.

> **SOURCE REGION**
> A very large area of uniform surface type over which air can remain stagnant long enough to form an air mass.

FIGURE 13.2 | The major air-mass source regions. *Top left:* The land of the high latitudes, such as this snow-covered tundra on the Melville Peninsula, Nunavut. *Top right:* The unfrozen waters of the high latitudes, such as those off the coast of Newfoundland. *Bottom left:* The open waters of the subtropics, including the Gulf of Mexico. *Bottom right:* The deserts of the subtropics, such as those in New Mexico.

Remember This

Air masses form as

- heat is transferred between the surface and the air and
- water vapour is transferred between the surface and the air.

Air-mass source regions are

- thousands of square kilometres in area,
- of uniform surface type, and
- usually associated with semi-permanent high-pressure cells.

13.2 | Classification of Air Masses

The air-mass classification system, still in use today, was devised in 1928 by Tor Bergeron, one of the Norwegian meteorologists responsible for the polar front theory (Section 14.2). This system is purely descriptive; it provides a simple method of naming air masses based on their temperature and moisture characteristics, but it does not include any precise criteria for distinguishing one air mass from another.

The system uses two-letter symbols to name air masses. The first letter indicates the moisture content of the air. A small *c* stands for "continental" and is used to designate a dry air mass. A small *m* stands for "maritime" and is used to designate a moist air mass. The second letter indicates the temperature of the air. A capital *A* stands for "Arctic" (*AA* stands for "Antarctic"), a capital *P* stands for "Polar," and a capital *T* stands for "Tropical." (Occasionally, a capital *E* is used to designate a hot air mass.) Table 13.1 shows how these letters are used to describe five main types of air masses.

Question for Thought

Which of the five major types of air masses would be the thickest? (Consider both temperature and moisture characteristics, and recall that the concept of thickness was introduced in Section 3.5.)

There are two key purposes for classifying air masses. First, classification makes it simple to distinguish between the air masses on either side of a front. Second, classification makes it easy to keep track of an air mass's temperature and moisture characteristics. Because an air mass carries its properties with it as it moves, this sort of tracking helps weather forecasters predict how an air mass will affect other areas when it moves from its source region.

The path an air mass takes when it moves is determined by the upper airflow. Figure 13.3 shows the paths most commonly followed as the air masses

TABLE 13.1 | Air-mass classification.

Air-Mass Name	Air-Mass Symbol	Characteristics	Source Region
Continental Arctic (Continental Antarctic)	cA (cAA)	very cold and dry very stable	Arctic and Antarctic (winter only)
Continental Polar	cP	cold and dry stable in winter slightly unstable in summer	high-latitude continents and ice-covered oceans
Maritime Polar	mP	cool and moist unstable	high-latitude oceans
Maritime Tropical	mT	warm and moist unstable on west side of oceans stable on east side of oceans	subtropical oceans
Continental Tropical	cT	hot and dry very unstable	subtropical deserts (summer only in North America)

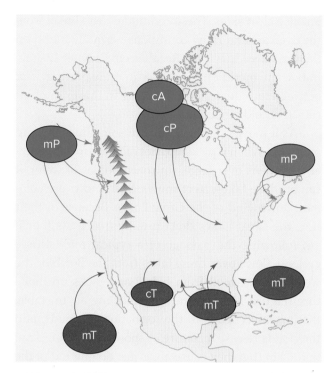

FIGURE 13.3 | The air masses that affect North America begin in polar or subtropical source regions and follow fairly regular paths (arrows) as they leave these regions. Note that the map shows annual *average* conditions; the air masses will be slightly north of these positions in summer and slightly south of these positions in winter.

affecting North America leave their source regions. As air masses move, they influence the weather in places they pass through and, in turn, they are modified by their new surroundings. For example, a cold air mass moving southeastward from northern Canada in winter will warm as it travels. As a result, it may cause a large temperature drop in southern Ontario, but it will not have such a strong effect on temperature by the time it reaches Florida.

In addition to bringing temperature changes, air masses can bring clear skies or cloudy skies. The likelihood of cloud formation depends on both the moisture content *and* the stability of the air mass (Section 8.4; Table 13.1). Stability characteristics develop as the air mass forms, but they can change as the air mass moves from its source region. An air mass that is colder than the surface over which it is travelling will be warmed from below and made unstable. On the other hand, an air mass that is warmer than the surface over which it is travelling will be cooled from below and become stable. To describe these changes in stability, a third letter is sometimes added to an air mass's symbol. A *k* means the air mass is colder than the surface below, while a *w* means the air mass is warmer than the surface below. Because *k* air masses usually become unstable, they are generally associated with the development of cumulus clouds and gusty winds. In contrast, *w* air masses become stable and are characterized by haze, fog, or the buildup of pollutants.

13.3 | Cold Air Masses

In the northern hemisphere, cold, *dry* air masses (cA and cP) form over northern Canada and Siberia, while cold, *moist* air masses (mP) develop over the northern Pacific and Atlantic oceans. In North America, southeastward moving cA and cP air masses can be experienced throughout most of the continent as far south as Florida, but mP air masses impact only a narrow zone along the west coast. In Eurasia, by comparison, cA and cP air masses are unlikely to be experienced as far south as India, but mP air masses move east over most of Europe. This comparison shows that, while the predominantly westerly flow of the midlatitudes will steer these cold air masses eastward and southward, mountains can block the movement of air masses. In North America, air masses are blocked by the north–south trending Coast Mountains and Rocky Mountains, while in Asia they are blocked by the east–west trending Himalayas.

The cold continental air masses are associated with winter: cA air masses occur *only* in winter, while cP air masses are *best developed* in winter. At this time of year, the northern latitudes experience low sun angles and short days (Section 5.8.2). Under these conditions, cold air masses will form by radiative cooling. Because air close to the surface cools faster than the air lying above, a radiation inversion will form, making the air mass very stable (Section 8.5.2). Because cold air holds little moisture, this air mass will also be very dry. For example, Table 7.1 shows that the saturation vapour pressure of air at −20°C is only about 0.1 kPa. As a result, in a cA or cP air mass, skies are usually clear, and there is little precipitation.

When cA or cP air moves from its source region, the predominantly westerly upper flow tends to move it south and east. As it moves, it is gradually warmed from below; this warming makes the air mass unstable but, as long as the air remains dry, it produces little precipitation. Under conditions of strong meridional flow, cP air, and even cA air, can cover much of Canada and the United States. In fact, fruit and vegetable crops in places as far south as Florida are occasionally threatened by such an invasion of polar, or arctic, air, an event often referred to in the media as a "Siberian express" (Figure 13.4).

LAKE-EFFECT SNOWS
Snowfall that occurs downwind of large, unfrozen lakes.

The upper airflow pattern depicted in Figure 13.4 also brought cP air toward Canada's west coast, causing a cold snap in the only region of Canada that experiences mild winters. However, because this air is warmed as it sinks westward through the mountains, it is not as cold as it is further inland to the east. As the cold air drains through the mountains and out to the west coast, it brings not only a temperature decrease, but also clear skies. These conditions are a break from the overcast and rainy conditions normally found on the west coast in winter. Recall, from Chapter 12, how important meridional flow is in transporting heat between the equator and the poles; in this example, it transfers cold air equatorward.

As it moves south, cP air can sometimes pick up moisture. When it picks up a lot of moisture, as generally happens over large, unfrozen bodies of water, it can produce **lake-effect snows**. Such snows are common downwind of the Great Lakes. Because of their size, the Great Lakes don't usually freeze. As the cP air, following the passage of a typical mid-latitude cyclone, flows over these lakes from the north or the west, it is warmed from below and picks up moisture (Figure 13.5). Warming makes the air unstable, the instability leads to the formation of clouds, and the clouds release moisture as snow. These snows are unique in that they cause areas just downwind of the Great Lakes to have the greatest snowfall accumulations of the region and in that they tend to occur *after* the passage of a storm system.

Summertime cP air masses are located farther north and are warmer and moister than their winter counterparts. With longer days and no snow, the surface warms and there is no longer an inversion at the surface, so the air is less stable. In addition, as lakes and wetlands thaw, the air picks up moisture. With more moisture and decreased stability, convective precipitation is frequent (Section 10.7).

While cP and cA air masses form over the continents and ice-covered oceans of the high latitudes, mP air masses form over the ice-free oceans of the high latitudes. These regions of the Pacific and Atlantic oceans are not ideal source regions because there are no semi-permanent high pressure cells located there, but they are extensive enough that a uniform air mass can form above them. In fact, mP air masses are sometimes referred to as *secondary air masses* because they begin as cP air masses but are modified as they move eastward from the continents over northern waters. For example, as the cP air mass from Asia moves over the Pacific Ocean, it warms and picks up moisture. Therefore, mP air is slightly warmer, and much moister, than cP air. Because the air warms

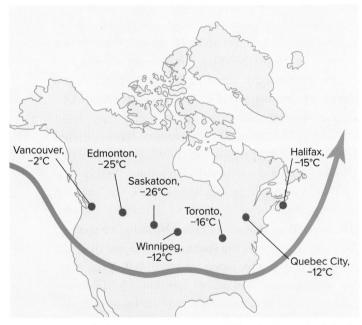

FIGURE 13.4 | The upper air flow, represented here by the path of the jet stream, that created a Siberian express on 15 January 2012. During this event, temperatures all across Canada were below freezing, as indicated by the minimum temperatures for several Canadian cities.

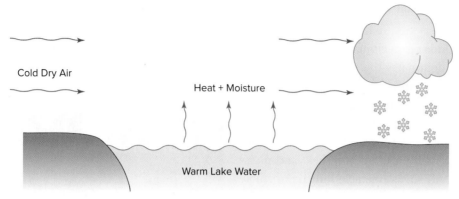

Cold Dry Air

Heat + Moisture

Warm Lake Water

FIGURE 13.5 | Lake-effect snows occur when cold, dry air warms and picks up moisture as it flows over an unfrozen lake in winter.

from below but remains cold aloft, mP air masses are unstable. This moist, unstable air can produce cumulus clouds over the ocean (Figure 13.6).

Recall from above that, in North America, mP air masses influence only the far western margin of the continent. This air brings with it the mild, moist conditions that characterize the climate of the west coast throughout the year. As an mP air mass moves east through the mountains, however, it loses its moisture as precipitation on the windward side (Section 8.3); as a result, it loses its maritime character. Because of the predominantly westerly flow of the upper air, mP air masses from the Atlantic Ocean are much less common in Canada and the United States. At times, however, such air masses can influence locations on the east coast when they are brought in from the

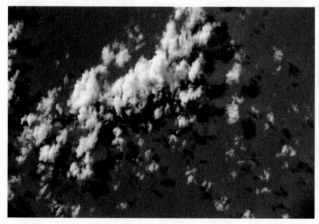

FIGURE 13.6 | A satellite image showing cumulus clouds over the Pacific Ocean. These clouds are produced by moist, unstable mP air.

northeast as part of the counterclockwise flow around mid-latitude cyclones.

Question for Thought

Why does mP air *frequently* invade the west coast of British Columbia, while cP air only *occasionally* invades Florida?

13.4 | Warm Air Masses

In both hemispheres, cT air masses form over the subtropical deserts. The North American source region, the deserts of the southwestern United States and northern Mexico, is very small compared to the source region in Africa and Asia. Over the desert surfaces of these regions, the air is strongly heated, but it picks up little moisture. As a result, it is hot, dry, and unstable.

With the exception of those that form in the North African source region, cT air masses form only in summer. Recall that thermal low-pressure systems develop over the subtropical deserts during the summer (Figure 12.3). These shallow thermal lows are capped with inversions created by the subsidence associated with the subtropical highs. This subsidence is particularly strong on the eastern sides of the subtropical highs, so the inversions are usually relatively close to the surface. Therefore, although cT air masses are very unstable, the unstable layer is shallow and the air is very dry, so these air masses do not often produce clouds or precipitation.

While cT air dominates the southwestern half of the United States in summer, mT air dominates the southeastern half. The source region for this mT air is the Gulf of Mexico and the subtropical Atlantic Ocean. Over these warm waters, mT air becomes warm and humid as it picks up both heat and moisture. Like cT air, mT air is unstable but the unstable layer at the surface is deeper than in cT air due to weaker subsidence on the western sides of the subtropical highs. After it forms, an mT air mass is steered onto the east coast of North America by the clockwise flow around the subtropical high positioned over the tropical Atlantic. This air mass brings warm and humid conditions, along with frequent thunderstorms, to the southeastern United States in the summer. As the air moves northward through the eastern portion of North America, it carries its humid conditions with it. However, as it moves, it is modified, gradually losing its humidity through precipitation. As a result, in the summer, eastern Canada is not as humid as the eastern United States. When mT air moves into the southeastern United States in winter, it is cooled from below and becomes stable. This cooling can produce advection fogs (Section 9.4.2) or, if winds are strong enough to produce mixing, stratus clouds (Section 9.3.5).

Question for Thought

In the summer, why might it *feel* much hotter in the southeastern United States (mT air) than it does in the southwestern United States (cT air)?

At least once a year, mT air will reach the west coast of North America during fall or winter. Such an event is often referred to as a "Pineapple express" because the warm, moist air originates near Hawaii (Figure 13.7). These storms bring very heavy rainfall, often resulting in over 100 mm of rain in just a few days. Extensive flooding frequently accompanies a Pineapple express, partly because of the high amounts of precipitation, and partly because the warmth causes snowmelt.

FIGURE 13.7 | A satellite image showing clouds that have formed as a result of a Pineapple express. These clouds are produced when warm, moist mT air, originating near Hawaii, reaches the west coast of North America.

During spring and summer, cT air from the southwest and mT air from the southeast frequently meet in the south-central United States. The boundary between these air masses is called the **dry line** (Figure 13.8). Because moist air is less dense than dry air (Section 3.3), the moist air will often rise over the dry air at this line, producing a line of thunderstorms (Section 14.4).

13.5 | Fronts

One air mass is separated from another by a *front*. Recall that fronts are boundaries between air of different properties. The property most commonly used to define fronts is temperature. Thus, although dry lines represent boundaries between air masses, they are not true fronts because temperatures are usually very similar on both sides of these lines.

Fronts are also associated with the uplift of air and, therefore, the formation of clouds and precipitation. This uplift occurs as warmer, less dense air rises over colder, more dense air. In addition, other changes can be expected as a front passes overhead. The uplift at fronts leads to a drop in pressure. The troughs of low pressure associated with fronts are indicated on weather maps as kinks in isobars where they cross fronts (Figure 13.1). The kink in the isobars also indicates that wind direction changes across fronts; as a front passes, winds will veer or turn in a clockwise direction. Figure 13.1 shows that winds in the warm air of the cyclone are coming from the southwest. As

the cyclone moves eastward, the cold front will pass overhead and, once it does, the winds will be coming from the northwest; thus, the wind direction will have shifted in a clockwise direction. The change in air masses associated with frontal passage means that dew-point temperature is also likely to change.

DRY LINE
A boundary separating warm, moist air from warm, dry air.

All of these changes can be used to locate fronts on weather maps. Although most of the work in compiling weather maps is now done by computers, the placement of fronts is still done by people, as it is a fairly subjective process. First and foremost,

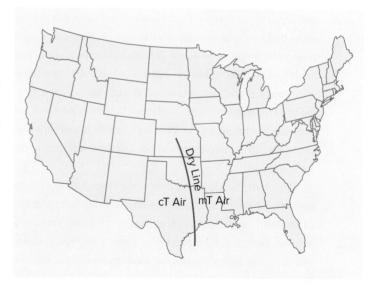

FIGURE 13.8 | The dry line in the south-central United States forms where hot, dry cT air from the southwest meets warm, moist mT air from the southeast.

In the Field

Precipitation Isotopes as Tracers for Weather Systems and Climate Variability

Shawn J. Marshall, University of Calgary

Stable water isotopes in precipitation have been used for many decades to provide insights into the hydrological cycle and climate variability. The ratio of oxygen-18 to oxygen-16 relative to "standard mean ocean water," described as $\delta^{18}O$, provides a tremendously powerful tracer of global circulation and climate conditions. Closely related proxies—such as the ratio of deuterium to hydrogen (represented as δD), and the deuterium excess, *d*, which expresses deviations in the relation between $\delta^{18}O$ and δD that arise due to kinetic (non-equilibrium) processes such as evaporation under dry, windy conditions—are available in water samples.

Fractionation processes during evaporation and condensation cause systematic, temperature-dependent variations in the isotopic character of water vapour in air masses. In general, the initial rainout from an air mass is enriched in heavy isotopes,[18]O and D, leaving residual water vapour depleted. For this reason, precipitating air masses become progressively depleted of heavy isotopes the further they travel from an evaporation source (e.g., further inland, poleward, or upslope). During cold conditions (e.g., winter months or periods of glaciation), fractionation is stronger and rainout occurs more quickly due to the reduced vapour-carrying capacity of air. Under such conditions, the rain or snow that falls inland becomes depleted in heavy isotopes, giving low, negative δ values. This distillation process is particularly well-manifest in the interior of the polar ice sheets in Greenland and Antarctica, which receive only the final dregs of moisture wrung out of a cloud. This is the basis for "isotope thermometry" in ice cores, which is perhaps the best-established application of precipitation isotopes in climate science.

The same isotopic processes that make ice cores such exceptional paleothermometers can also be exploited to understand hydrological and meteorological variability at lower latitudes. Evaporation of water from rivers, lakes, and soils and post-depositional modification of snowpacks (e.g., due to wind distribution, melting-induced homogenization, and evaporation of liquid meltwater in the pores of the snowpack) complicate the primary isotopic signal, making it complex to interpret the environmental record in such locations. However, it has recently been appreciated that isotopic records in ice cores and water bodies (e.g., lakes, groundwater) contain tremendous "secondary" information about moisture provenance and pathways, which might yield insight into past atmospheric circulation regimes and shifts in addition to local temperatures. For instance, Peruvian ice-core records provide good evidence of past El Niño cycles, due to the impact of these cycles on the tropical Pacific trade winds, convection centres, and precipitation in South America.

At mid-latitudes in the Canadian Rocky Mountains, my research group has been working to understand the isotopic signature of different air masses that transport moisture to the region. Our aim is twofold: to see whether climate insights might be gleaned from melt-affected ice cores in western Canada, and to understand the synoptic meteorological controls of the high-mountain snowpack. The latter is important for glacier mass balance and regional water resources. Because large-scale atmospheric circulation patterns are a function of the mean climate state and are reasonably well-represented in weather and climate models, this provides a potentially instructive link between projected climate change and impacts on the mountain snowpack and its glaciers.

We examine this link through characterization of $\delta^{18}O$ and δD stratigraphy in the winter snowpack prior to spring melting, and through the relation of $\delta^{18}O$ and δD values to specific synoptic snow events (i.e., individual storms that deliver moisture to the region). Several years of snowpack characterization have revealed systematic variations in $\delta^{18}O$ and δD as a function of storm trajectory and moisture source (e.g., southwesterly vs northwesterly advection from the Pacific Ocean). Low-snow years in the Canadian Rocky Mountains are associated with southwesterly flow and enriched isotopic values, which are in turn associated with northward displacement of the jet stream and persistent ridging over southwestern Canada. This pattern of circulation is common during El Niño events. Strong westerly and northwesterly storm tracks bring the opposite scenario: a heavy snowpack and depleted isotopes, due to combined source-region, trajectory, and temperature effects.

By extrapolating from these isotopic signatures of different air-mass pathways, it is possible to characterize the relative importance of different synoptic systems in contributing to the total mountain snowpack. From this, it may prove possible to forecast changes in mountain snowpack from predicted shifts in circulation under climate variability and change. This decryption key also provides a means by which ice-core isotopic values can be read to reconstruct prehistoric shifts in atmospheric circulation.

SHAWN MARSHALL is a professor and Canada research chair in climate change in the Department of Geography at the University of Calgary. He studies snow and ice in the climate system, with a focus on field-based and modelling studies of glacier response to climate change.

forecasters look for changes in air temperature, but changes in dew-point temperature and wind direction, bands of clouds and precipitation, and troughs of low pressure are also important indicators of the locations of fronts. In addition, forecasters can use these indicators, along with successive weather maps, to determine what *type* of front is occurring.

There are four main types of fronts, each of which is labelled in a specific way on weather maps (Figure 13.9). The front is a *cold front* if cold air is advancing and replacing warm air. Cold fronts are labelled on surface weather maps as lines marked with triangles that "point" in the direction in which the front is moving. The front is a *warm front* if cold air is retreating and being replaced by warm air. Warm fronts are labelled on surface weather maps as lines marked with semicircles that "point" in the direction in which the front is moving.

A **stationary front** is a front that, unlike cold and warm fronts, shows no significant movement. To

STATIONARY FRONT
A front along which there is no significant movement.

determine whether a front is stationary, a forecaster will look at a series of maps to see whether a particular front has shown movement. On weather maps, stationary fronts are labelled as lines of alternating semicircles and triangles, with triangles on the warm side of the front and semicircles on the cold side of the front.

13.6 | The Structure of Fronts

The lines drawn on weather maps imply that fronts are *sudden* temperature discontinuities but, in reality, they are large temperature gradients, or *zones*, in which temperature changes rapidly. A frontal *zone* is about 100 km across. On either side of the frontal zone are air masses in which temperature changes are small. Because Earth is unevenly heated, there will always be temperature gradients; fronts are simply zones where temperature gradients are much greater than normal. To find fronts, therefore, forecasters look

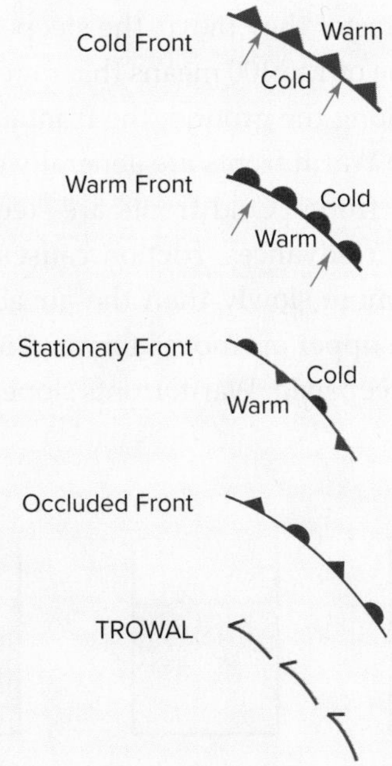

FIGURE 13.9 | The line symbols used to represent fronts on weather maps. On coloured maps, red represents warm and blue represents cold. (Occluded fronts are traditionally purple, to represent a mixture of cold and warm. These fronts, and the associated TROWALS, will be described in Section 13.8.)

for large temperature gradients. When they find significant temperature gradients, they draw the line that represents the front on the *warm* side of this temperature gradient (Figure 13.10).

Since fronts are zones of large temperature gradients, the process of formation of a front, or even the strengthening of an existing front, can be any process that increases a temperature gradient. **Frontogenesis** can occur due to convergence at the surface along a temperature gradient. This convergence might be a result, for example, of the wind slowing along the direction of flow. As convergence occurs, warm and cold air masses are pulled in closer together, thus strengthening the temperature gradient between them. In addition, any process that causes heating on the warm side of the gradient or cooling on the cold side of the gradient can strengthen the gradient and produce, or strengthen, a front. The opposite of frontogenesis is **frontolysis**. Where frontolysis is occurring, a front is either dissipating or weakening. Frontolysis can occur with surface divergence, or it can occur with cooling on the warm side of the front or warming on the cool side.

FRONTOGENESIS
A process that increases a temperature gradient and forms a front.

FRONTOLYSIS
A process that leads to the dissipation of a front.

Frontal zones usually extend upward in the atmosphere to heights of at least 5 km, but fronts are not vertical boundaries as might be expected; rather, they are *sloping surfaces*. To understand why, imagine a container in which a vertical divider separates hot water on one side from cold water on the other side (Figure 13.11a). If the divider is removed, the cold water will flow under the hot water due to the difference in their densities (Figure 13.11b); thus, the vertical boundary will become a horizontal boundary. In the atmosphere, the cold air on one side of a front will, like the water in the example, try to slide under the warmer air on the other side of the front but, as it does so, Earth's rotation comes into play. At the same time, as the cold air flows beneath the warm air, it is turned by the Coriolis force. The result is that frontal surfaces are neither horizontal nor vertical; instead, they are sloping (Figure 13.11c).

The steepest part of the slope of a frontal surface is in the air layer closest to Earth's surface, where friction has an influence on the atmosphere. In this surface layer, cold fronts slope more steeply than do warm fronts. Although the slopes of fronts are highly variable, cold fronts typically have slopes of about 1 to 100, and the faster they move, the steeper will be the slope. (A slope of 1 to 100 means that for every 100 km of distance along the ground, the frontal surface will rise by 1 km.) Warm fronts are generally about half as steep as cold fronts. Cold fronts are steeper because as the cold air advances, friction causes the surface air to move more slowly than the air above (Figure 13.12). As the upper air moves forward more rapidly, the slope is steepened. Warm fronts slope more gently

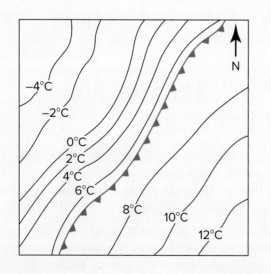

FIGURE 13.10 | The isotherms plotted on the map show a large temperature gradient. Whether it is a warm front or a cold front, the front is drawn on the warm side of the temperature gradient.

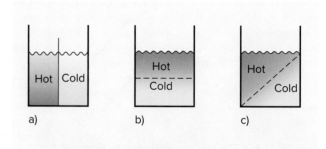

FIGURE 13.11 | a) Hot water is separated from cold water by a vertical divider. b) When the divider is removed, the cold water flows under the hot water. c) If the container were rotated, the boundary between the cold and hot water would become a sloping surface.

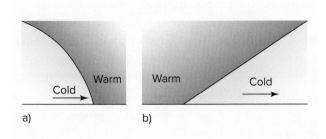

FIGURE 13.12 | a) The steep slope of a cold front. **b)** The more gentle slope of a warm front.

because as the cold air retreats, the surface air drags behind, again due to friction, while the upper air retreats more rapidly.

In both warm fronts and cold fronts, the air below the sloping frontal surface will be the colder, denser air, and the air above the sloping surface will be the warmer, less dense air. This means that frontal surfaces will be marked by inversions aloft, as the cold air below transitions to warmer air above (see Section 8.5.2). Frontal inversions can often be located on upper-air soundings and are, therefore, helpful in finding the location of a front.

13.7 | Weather Associated with Cold and Warm Fronts

So far, the changes associated with the passage of a front have been considered in a general sense. However, the *specific* changes occurring across cold fronts are usually different than those occurring across warm fronts; additionally, as you will see in the next section, changes occurring across occluded fronts differ from both those occurring across cold fronts and those occurring across warm fronts.

As a *cold* front passes overhead, both the air temperature and the dew-point temperature will fall as the colder, drier air behind the front moves into the area (Figure 13.13). In addition, the winds will veer, or rotate, in a clockwise direction with the passage of the front. With the passage of a cold front, wind direction usually shifts from south or southwest in the warm air to west or northwest in the colder air. The approach of the front will cause pressure to fall, but the pressure will rise again in the colder air once the front has passed.

At cold fronts, the advancing cold air lifts the warm air ahead of it by *convergence*. As the rising air cools to its dew-point temperature, clouds and precipitation will occur just ahead of the front (Figure 13.13b). Because the air in cold frontal zones is normally unstable, cumuliform clouds accompanied by heavy precipitation and thunderstorms are most common. Stratiform clouds can also occur at cold fronts if the lifted air is stable. Either way, the steep slope and relatively rapid movement of cold fronts means that the clouds and precipitation will be short-lived. Once the clouds have passed, the higher pressure behind the cold front can bring clear skies, but if this cold air has been made unstable by passage over a warmer surface, convective cloud is likely to occur.

When a *warm* front passes overhead, both the air temperature and the dew-point temperature will increase as warmer, more humid air moves into the

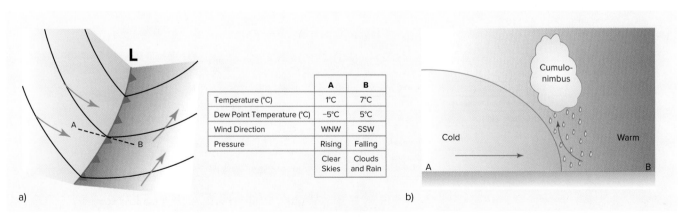

	A	B
Temperature (°C)	1°C	7°C
Dew Point Temperature (°C)	−5°C	5°C
Wind Direction	WNW	SSW
Pressure	Rising	Falling
	Clear Skies	Clouds and Rain

FIGURE 13.13 | a) An example of the difference in conditions possible on either side of a cold front. Movement of the front will cause conditions at B to become more like those at A. **b)** A cross-section through the cold front along AB.

area (Figure 13.14). Winds usually shift from east or southeast in the colder air to south or southwest in the warmer air. With the approach of the front, pressure will drop; once the front has passed, pressure will likely increase slightly.

At a warm front, warm air *overruns* the colder retreating air. This overrunning causes the air to rise slowly, cool to its dew-point temperature, and form clouds. In most cases, the air in warm frontal zones is stable, making stratiform clouds most common. It is also possible for cumuliform clouds to form at warm fronts if the air is unstable. Further, while cold fronts tend to produce narrow bands of intense precipitation, warm fronts produce wide bands of cloud accompanied by light, steady precipitation. This difference can be explained by the difference in slope of the two fronts and the slower speed of the warm front. With warm fronts, lifting occurs over a greater horizontal distance, so that the first clouds may begin to arrive as much as a day before the front passes. More often than not, the clouds can be seen to gradually lower as the front approaches. Low nimbostratus clouds and precipitation are common just before the front passes overhead.

The above is a description of the changes *most commonly* observed when fronts pass overhead. Not all fronts will bring changes just as described above; in fact, it is unlikely that any two fronts will be exactly alike. However, the changes they bring are similar enough, and occur often enough, that they are recognizable as a pattern. This pattern—in which temperature, moisture, and wind-direction changes occur in conjunction with a drop in pressure and the occurrence of clouds and precipitation—was identified after many observations had been made. Next, the *model* of fronts was *created* to provide an explanation for this pattern. Recall from Section 1.3 that scientific models are designed to simplify reality so that we can understand it; thus, they do not take into account all of the complex factors that affect processes in the real world.

Question for Thought

Why is it important to keep in mind that fronts are models?

One factor that can cause variation in the changes associated with fronts is the winds. The faster the winds are and the more they are directed toward each other, the greater will be the amount of convergence and, therefore, the strength of the uplift. The speed of movement of the front itself also influences the weather associated with the front; when fronts move slowly, the clouds and precipitation will be of longer duration than they will be when the front moves quickly. In addition, frontal weather differs by region. For example, the west coast of North America rarely experiences true cold-front weather because the Pacific Ocean causes very cold air to warm, thus reducing the contrast across the front. It is also possible for fronts to pass overhead without bringing clouds and precipitation. This could occur simply because the air being lifted at the front is dry, but the

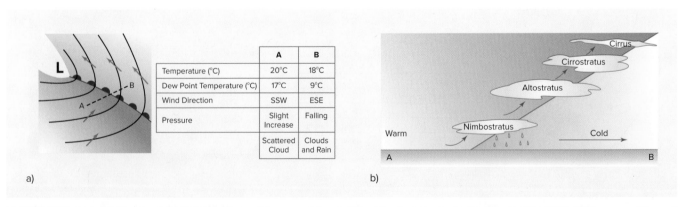

	A	B
Temperature (°C)	20°C	18°C
Dew Point Temperature (°C)	17°C	9°C
Wind Direction	SSW	ESE
Pressure	Slight Increase	Falling
	Scattered Cloud	Clouds and Rain

a)

b)

FIGURE 13.14 | a) An example of the difference in conditions possible on either side of a warm front. Movement of the front will cause conditions at B to become more like those at A. **b)** A cross-section through the warm front along AB.

shallowness of the front is a factor as well. Such is the case with the Arctic front, which separates polar air masses (usually cP) from much colder Arctic air masses (cA). Because cA air masses are very shallow, they do not usually cause enough uplift to produce clouds and precipitation.

As you can see, both warm fronts and cold fronts can display many complex variations. Yet these fronts are still far less complex than are occluded fronts.

13.8 | Occluded Fronts

Occluded fronts occur when *three* air masses come together: a cold one, a cool one, and a warm one; this combination occurs through the process of **occlusion**, which will be described below. At the surface, occluded fronts may pass by almost unnoticed, as the temperature contrast across them is small. However, because the occlusion process pushes warm air up off the surface, occluded fronts are noticeable as areas of clouds and precipitation. In general, the cloud patterns associated with occluded fronts are more complex than those associated with individual warm or cold fronts. In most cases, the approach of an occluded front will bring stratiform cloud; however, embedded within this cloud there will likely be some cumuliform cloud as well (figures 13.15c, 13.16b, and 13.16d).

Our understanding of the occlusion process is currently changing, but we will begin here with the traditional explanation, which was first proposed in 1922 as part of the Norwegian cyclone model (Section 14.2). In this model, a mid-latitude cyclone is envisioned as having a life cycle; Figure 13.1 depicts the mature stage of this cycle. After reaching its mature stage, the explanation begins, a cyclone will start to occlude as the faster-moving cold front begins to *catch up* to the warm front. It is the meeting of the air behind the cold front with the air ahead of the warm front that produces the occluded front (Figure 13.15). This front is marked on a surface weather map using the occluded front symbol depicted in Figure 13.9. At the same time, the warm air between the cold front and the warm front will be gradually pinched out and pushed up, producing a trough of warm air, or front, above the surface. As more and more of the warm air is pushed up, the occluded front will lengthen, progressively separating the low-pressure centre from the warm air. The location of the trough of warm air aloft, projected onto the surface, is marked on a weather map using the **TROWAL** (TROugh of Warm air ALoft) symbol shown in Figure 13.9. Canadian weather maps label the TROWAL rather than the occluded front.

> **OCCLUDED FRONT**
> A front that separates a cold air mass from a cool one, and the low pressure centre from the warm air in a mid-latitude cyclone.
>
> **OCCLUSION**
> A process that gradually separates the centre of low pressure from the warm air in a mid-latitude cyclone.
>
> **TROWAL**
> An acronym for "TROugh of Warm air ALoft."

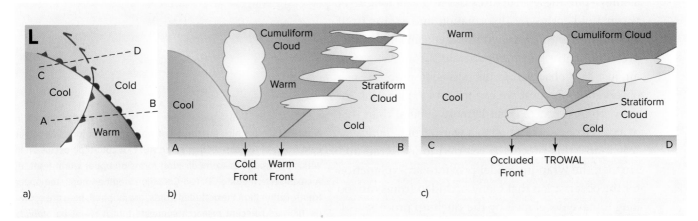

a) b) c)

FIGURE 13.15 | a) The occluded front separates the two coldest air masses at the surface, and it lies between the low-pressure centre and the warm air mass. The TROWAL represents the location, projected onto the surface, of the trough of warm air aloft. **b)** A cross-section along the line AB showing where the warm front and the cold front would be located on the surface. **c)** A cross-section along the line CD showing where the occluded front and the TROWAL would be located on the surface.

According to the traditional model, two types of occlusion are possible: **warm-type occlusions** and **cold-type occlusions**. The model explains *warm*-type occlusions as occuring when the cold air mass behind the cold front is *warmer* than the cold air mass ahead of the warm front (figures 13.15 and 13.16a). In this case, density differences cause the air behind the cold front to *ride over* the air ahead of the warm front. This structure produces an upper-level cold front in advance of the surface occluded front, as shown in Figure 13.16b. In a *cold*-type occlusion (Figure 13.16c), on the other hand, the air behind the cold front is *colder* than the air ahead of the warm front, so it *pushes under* the air ahead of the warm front, creating a structure like that shown in Figure 13.16d. In the process, an upper-level warm front is formed behind the surface occluded front.

Recent studies, however, suggest that the occlusion process may not occur as described in the Norwegian cyclone model (Schulz & Vaughan, 2011). These studies have made use of satellite imagery, detailed surface and upper-air observations, and computer models of the occlusion process. They suggest that a model based on spinning vortices might best explain how occluded fronts form. As it turns out, vortices that contain temperature gradients produce occlusion-like structures as they spin; these structures develop as a tongue of warm fluid forms by deformation and wraps itself around the centre. Through this mechanism, the low-pressure centre becomes separated from the warm air, just as is observed in the occlusion process. Such observations suggest we can give an explanation for the occlusion process that is more general than that of a cold front catching up to a warm front. Instead of "catch-up," this process is referred to as "wrap-up." Further, the wrap-up model provides an explanation for the observation that many occluded fronts are too long to have developed by the catch-up process, as it has been shown that the wrap-up process allows for such long fronts to develop.

WARM-TYPE OCCLUSION
An occlusion in which the air behind the cold front rides over the air ahead of the warm front.

COLD-TYPE OCCLUSION
An occlusion in which the air behind the cold front pushes under the air ahead of the warm front.

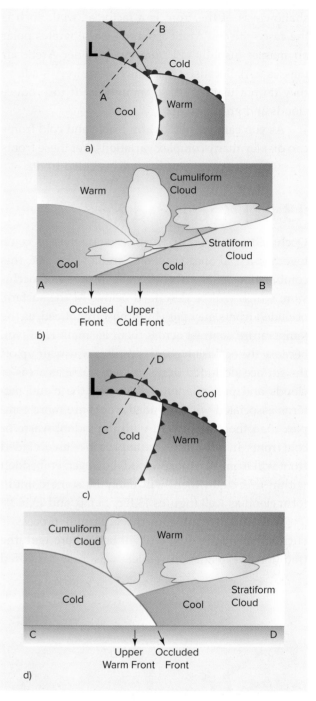

FIGURE 13.16 | a) A warm-type occlusion, as is also shown in Figure 13.15, in which the trough of warm air aloft forms an upper cold front. **b)** A cross-section along AB. **c)** A cold-type occlusion in which the trough of warm air aloft forms an upper warm front. **d)** A cross-section along CD. (On Canadian weather maps, the upper fronts, rather than the occluded fronts, are mapped; they are shown as TROWALS.) Recent research suggests that it might be *stability* differences, rather than *temperature* differences, that determine which type of occlusion will form (see Figure 13.17).

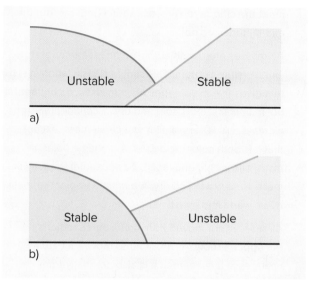

FIGURE 13.17 | a) In a warm-type occlusion, the air behind the cold front may be more unstable, rather than warmer, than that ahead of the warm front. b) In a cold-type occlusion, the air behind the cold front may be more stable, rather than colder, than that ahead of the warm front. (Compare this diagram to Figure 13.16.)

Researchers have also suggested that warm-type and cold-type occlusions may be a result of *stability* differences rather than *temperature* differences (Stoelinga, Locatelli, & Hobbs, 2002). This conclusion is based on observations showing that in warm-type occlusions, the air behind the cold front is not necessarily warmer than the air ahead of the warm front. Instead, it is more unstable (Figure 13.17). Similarly, in cold-type occlusions, the air behind the cold front is not necessarily colder than the air ahead of the warm front; it is more stable. The explanation, in both cases, is that the more unstable air will ride up over the more stable air. It follows that, because the air in cold frontal zones is usually more unstable than that in warm frontal zones, warm-type occlusions should be more common than cold-type occlusions. Indeed, recent observations of occlusions have shown that warm-types are far more common than cold-types.

13.9 | Chapter Summary

1. Air masses are very large bodies of air with similar properties throughout. Air masses form as heat and water vapour are transferred between the air and the underlying surface.

2. Air masses form in large areas known as source regions. Good source regions are large areas of uniform surface type that normally experience high pressure. The polar latitudes and the tropical latitudes

meet the criteria for air-mass source regions, but the mid-latitudes do not.

3. Air masses are classified based on their temperature and moisture content. This simple classification system gives five major types of air masses. A continental polar air mass, cP, is cold and dry. A continental Arctic air mass, cA, is very similar to a cP air mass, except that it is both colder and drier. A maritime polar air mass, mP, is cold and moist. A continental tropical air mass, cT, is warm and dry. A maritime tropical air mass, mT, is warm and moist.

4. Air masses are moved from their source regions by the upper airflow. As an air mass moves, it causes changes in the weather. It is also gradually transformed, ultimately losing its original character.

5. Cold, dry air masses (cA and cP) form over high-latitude continents and ice-covered oceans, especially in winter. These air masses are very stable because a radiation inversion usually forms at the surface where they develop. As these air masses move south, they warm from below and become less stable but, unless they also pick up water vapour, they are unlikely to produce much precipitation. In the winter, cP air can influence most of Canada and the United States. In the summer, cP air masses are less well-developed and remain farther north.

6. Cold, moist air masses (mP) form when cP air masses are modified as they move over the northern regions of the Pacific and Atlantic oceans. These air masses will be a little warmer and much moister than cP air masses. They frequently affect the west coast of North America, but they only occasionally affect the east coast of North America. The mountains of the west coast modify mP air masses, making them drier.

7. Warm, dry air masses (cT) form over the subtropical deserts in summer. This air is very unstable, but, because it is very dry and the unstable layer is shallow, precipitation is infrequent.

8. Warm, moist air masses (mT) form over the subtropical oceans. These air masses frequently invade the southeastern regions of North America. In summer, they bring heat, humidity, and frequent thunderstorms. In winter, the cooler ground makes these air masses stable, so they tend to bring fog or stratus cloud. Dry lines form where this air meets cT air in the south-central United States.

9. Fronts are narrow zones of transition between air masses. They are best represented as surfaces that slope up to at least the middle troposphere. Air temperature and dew-point temperature change rapidly across frontal zones. In addition, wind direction shifts, pressure drops, and bands of cloud and precipitation form.

10. Cold fronts occur where cold air is advancing and replacing warm air. Warm fronts occur where cold air is retreating and being replaced by warm air. Cold fronts slope more steeply than warm fronts. At cold fronts, the clouds are normally cumuliform, the precipitation is usually heavy, and thunderstorms are common. At warm fronts, the clouds are usually stratiform, and the precipitation is more moderate. These differences are thought to be due to stability differences between the cold and warm frontal zones. The clouds and precipitation at warm fronts can be expected to last longer than those at cold fronts.

11. Occluded fronts lie between the low-pressure centre and the warm air in a mid-latitude cyclone. They are noticeable at the surface as a slight temperature change and a shift in wind direction. Because the occlusion process produces a trough of warm air aloft, or an upper front, occluded fronts bring a complex variety of clouds and precipitation.

12. It is possible for two types of occluded fronts to form. If the air behind the cold front rides over the air ahead of the warm front, a warm-type occlusion forms. If the air behind the cold front pushes under the air ahead of the warm front, a cold-type occlusion forms. In the former case, an upper-level cold front is created; in the latter case, an upper-level warm front develops.

13. Our ideas about the occlusion process are changing. Instead of the catch-up process suggested by the traditional model, it is now thought that occlusions form as the result of a wrap-up process involving the deformation of a rotating fluid that contains temperature gradients. In addition, whereas it had been thought that warm-type and cold-type occlusions developed due to temperature differences, it is now thought that they result from stability differences.

Review Questions

1. How do air masses form?

2. What are the characteristics of an air-mass source region? Why is each characteristic important? What is the source region for each of the five major air masses that affect North America?

3. How do mountains influence the movement and modification of air masses?

4. Which air mass is the most stable? Why? Under what conditions does this air mass become less stable?

5. How do mP air masses form? What type of weather is associated with mP air masses? How can mP air masses be modified?

6. What factors are important determinants of whether or not an air mass is likely to produce precipitation?

7. Why is mT air likely to bring thunderstorms in summer and fog or stratus cloud in winter?

8. Why is eastern Canada more humid than western Canada in the summer?

9. How is air different on either side of a dry line? How is it the same?

10. What is a front? Why are fronts better represented as zones than as lines?

11. How are fronts identified on weather maps?

12. What are the differences between cold fronts, warm fronts, stationary fronts, and occluded fronts? What types of weather are associated with each type of front?

13. How and why do the clouds and precipitation associated with cold fronts differ from those associated with warm fronts?

14. According to the traditional model of cyclone formation, how do occluded fronts form? How and why is this thinking changing?

15. How does the *structure* of a warm-type occlusion differ from that of a cold-type occlusion? What are two possible explanations for this difference in structure?

Suggestions for Research

1. Beginning with the polar front theory, investigate how our understanding of fronts and, in particular, occluded fronts has changed over time.

2. Investigate common weather patterns in your region and explain how these patterns are associated with the movement of air masses.

References

Schultz, D.M., & G. Vaughan. (April 2011). Occluded fronts and the occlusion process: A fresh look at conventional wisdom. *Bulletin of the American Meteorological Society, 92*(4), 443–66.

Stoelinga, M.T., J.D. Locatelli, & P.V. Hobbs. (May 2002). Warm occlusions, cold occlusions, and forward-tilting cold fronts. *Bulletin of the American Meteorological Society, 83*(5), 709–21.

14

Storms

Learning Goals

After studying this chapter, you should be able to

1. *appreciate* that high- and low-pressure systems can vary in both structure and formation;

2. *outline* the stages in the life cycle of a mid-latitude cyclone, according to the polar front theory;

3. *describe* the structure of mid-latitude cyclones and anticyclones;

4. *account for* the processes at work in a mid-latitude cyclone;

5. *list* conditions favouring the formation of a mid-latitude cyclone;

6. *account* for the distribution of hurricanes;

7. *list* and *account* for each of the conditions required for hurricanes to form;

8. *account* for the weather produced by thunderstorms;

9. *distinguish* between single-cell, multicell, and supercell thunderstorms;

10. *list* the conditions required for thunderstorm formation; and

11. *compare and contrast* mid-latitude cyclones, hurricanes, and thunderstorms.

Storms can be beautiful to behold, and they can bring much-needed precipitation in times of drought, but they can also cause great damage. Thunderstorms—such as the one shown here occurring over open farmland in Saskatchewan—can be particularly damaging, as they often produce strong winds, lightning, flash floods, and even tornadoes.

In everyday usage, the word *storm* tends to conjure up images of **thunder**, **lightning**, heavy rain, and raging winds; in meteorology, however, the word *storm* is a more general term referring simply to a system that produces clouds and precipitation. Storms develop in areas of low pressure—recall from Chapter 3 that low pressure is associated with cloudy skies and high pressure is associated with clear skies. This relationship arises because, as explained in Section 11.2.3, surface *convergence* in areas of low pressure causes air to rise, while surface *divergence* in areas of high pressure causes air to sink. Also recall that rising air cools and sinking air warms (Section 8.1); and, because warmer air "holds" more moisture than does colder air, cooling can lead to saturation (Section 7.3). In short, in areas of low pressure, storms will develop as air rises, cools, and condenses to form clouds, while in areas of high pressure, air will sink and warm, resulting in clear skies (Figure 14.1).

In addition to knowing the relationship between pressure systems and weather, it is important to understand how different pressure systems form. In Section 3.6, you saw that surface heating can produce areas of low pressure. An example of such a low-pressure area is the large *thermal* low, described in Chapter 12, that rings the equator. **Tropical cyclones**, the storms that can develop into **hurricanes**, are also thermal lows (Figure 14.2). The energy for these tropical storms comes from latent heat. While *heating* is important in the formation of tropical cyclones, *upper-level divergence* is key to the formation of mid-latitude cyclones (Section 11.2; Figure 14.3). In Section 12.2.4, you also saw that mid-latitude cyclones generally form along the polar front, their energy coming from the associated temperature contrast.

Storms do not necessarily form only in areas of low pressure; sometimes, *rising* air is all that is needed. In areas of strong instability, even within a large area of high pressure, **thunderstorms** can form. As such, thunderstorms are examples of *convective* systems. They can occur at almost any latitude, although they are rare at high latitudes due to cold temperatures. In

THUNDER
The sound produced as the heat from lightning causes a sudden and rapid expansion of the surrounding air.

LIGHTNING
An electrical discharge within a cloud, between a cloud and the ground, or between a cloud and the surrounding air.

TROPICAL CYCLONE
A low-pressure area that forms in the tropics.

HURRICANE
A tropical cyclone in which sustained winds are higher than 120 km/h.

THUNDERSTORM
A convective storm that produces thunder and lightning and, usually, heavy precipitation and strong winds.

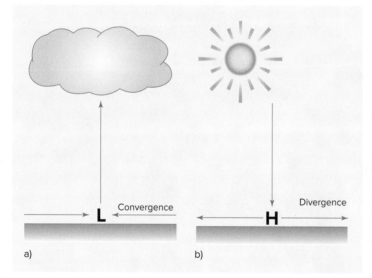

FIGURE 14.1 | a) In an area of low pressure, air converges at the surface and rises. As the rising air cools adiabatically, its relative humidity will increase until it reaches saturation at which point clouds form. b) In an area of high pressure, air diverges at the surface and sinks from above. As the sinking air warms adiabatically, the relative humidity of the air decreases; as a result, the skies remain clear.

FIGURE 14.2 | A satellite image of a tropical cyclone off the coast of Brazil.

14.1 | Characteristics of High- and Low-Pressure Systems

High- and low-pressure systems vary in both structure and mode of formation (Table 14.1). Some form due to the temperature contrasts associated with fronts; these systems are sometimes described as *baroclinic* and are characterized by temperature advection (Section 11.3). Others form in areas where there are no strong temperature contrasts; these systems are *barotropic*. Pressure systems can also be *shallow* or *deep*; shallow systems are those that weaken with height, and deep systems are those that strengthen with height. In addition, if pressure systems are colder than their surroundings, they are described as *cold-core* systems; if they are warmer than their surroundings, they are described as *warm-core* systems. Table 14.1 shows that warm-core lows and cold-core highs, which are both thermal pressure systems, are shallow. Thermal pressure systems are shallow because pressure decreases more quickly with height in cold air than it does in warm air (Section 3.6).

Another important structural distinction is that some pressure systems are *vertically stacked*, while others are *tilted*. This characteristic can be seen by identifying a system on a surface weather map and then finding it on the upper-level charts for the same time (Figure 14.4). If a system has the same position relative to Earth's surface through

SYNOPTIC SCALE
A scale of a few hundred kilometres to a few thousand kilometres.

FIGURE 14.3 | A satellite image of a mid-latitude cyclone. Notice the characteristic comma shape of this storm.

addition, thunderstorms are classed as *mesoscale* phenomenon because they are a few kilometres to a few hundred kilometres in size. Mid-latitude cyclones and tropical cyclones, on the other hand, are **synoptic-scale** systems because they range in size from several hundred to several thousand kilometres.

TABLE 14.1 | Characteristics of high- and low-pressure systems.

Characteristics					Examples
Low Pressure	barotropic	shallow	warm	vertically stacked	thermal lows (ITCZ) tropical cyclones (hurricanes)
Low Pressure	barotropic	deep	cold	vertically stacked	cutoff lows occluded lows
Low Pressure	baroclinic	deep	cold	tilted	subpolar lows mid-latitude cyclones
High Pressure	barotropic	shallow	cold	vertically stacked	polar highs
High Pressure	barotropic	deep	warm	vertically stacked (can be tilted)	subtropical highs cutoff highs
High Pressure	baroclinic	deep	warm	tilted	mid-latitude anticyclones

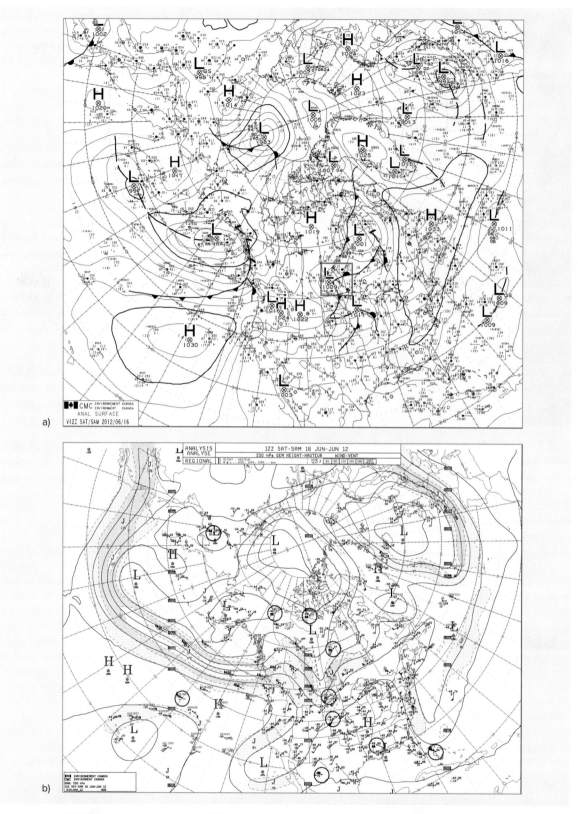

FIGURE 14.4 | Two weather maps for the same time but for different levels of the atmosphere show that mid-latitude cyclones tilt with height. a) On the surface weather map, there is a mid-latitude cyclone centred over southern Manitoba. b) This cyclone tilts upward to the northwest with height, so that it appears as a trough of low pressure over Saskatchewan on the 250 hPa chart.

Source: Adapted from Environment Canada, Weather Office, www.weatheroffice.gc.ca/analysis/index_e.html, 16 June 2012.

its entire height, it is vertically stacked. If the system gradually shifts in a certain direction, from one level to the next, it is tilted. Table 14.1 lists mid-latitude cyclones and anticyclones, both of which are deep, as tilted. If these systems did not tilt with height, they would quickly weaken because air would diverge from highs and converge into lows at all levels (Section 11.2). Note that these systems are also baroclinic, in that they form in association with fronts. On the other hand, as Table 14.1 shows, thermal systems, which are shallow, are vertically stacked. Examples of stacked systems are tropical cyclones or hurricanes; these systems are barotropic.

Question for Thought

Why do baroclinic systems develop only in the mid-latitudes?

Of the pressure systems described in Table 14.1, *mid-latitude cyclones* and *tropical cyclones* are major weather producers, or storms. As such, they will be the major focus of this chapter, starting with Section 14.2. First, however, the other systems listed in Table 14.1 deserve brief mention.

Remember This

Mid-latitude cyclones are baroclinic, deep, cold cored, and tilted. *Tropical cyclones* (hurricanes) are barotropic, shallow, warm cored, and vertically stacked.

Notice that the pressure systems of the general circulation described in Chapter 12 are listed in Table 14.1. Here, the *polar highs* are characterized as shallow, cold-cored highs, indicating that they are of thermal origin; recall that these highs form due to the cold air at the poles. The *subtropical highs* are shown to be deep, warm-cored highs, as they form where air converges and sinks at roughly 30° latitude. Thermal lows are also important in the general circulation. The *equatorial low*, or ITCZ, is a shallow, warm-cored low that forms in the warm air near the equator. Other thermal lows form over land in the summer, effectively breaking up the subtropical highs in the lower mid-latitudes (Section 12.2.2). Finally, the *mid-latitude cyclones* form the low-pressure belts of the mid-latitudes, otherwise referred to as the *subpolar lows*.

Also shown in Table 14.1 are *occluded lows*, **cutoff lows**, and **cutoff highs**. Occluded lows are low-pressure systems that remain after a mid-latitude cyclone occludes (Section 13.8). Once the storm is completely occluded, the surface fronts will no longer exist, and the air will be relatively cold throughout the system. Because the process of occlusion causes the low-pressure centre at the surface to separate from the warm air within the cyclone, the surface low shifts until it is below the upper low. As a result, the system is no longer tilted; instead, it is vertically stacked. On an upper-air chart of the northern hemisphere, such an occluded system is located to the *north* of the jet stream and surrounded by at least one closed isobar.

> **CUTOFF LOW**
> A cold upper-air low that has been cut off from the general westerly flow, so that it lies equatorward of this flow.
>
> **CUTOFF HIGH**
> A warm upper-air high that has been cut off from the general westerly flow, so that it lies poleward of this flow.

Cutoff lows and highs form when meanders in the polar front jet stream are cut off to form pools of cold or warm air, respectively. When a trough is cut off, the result is a cutoff low; in the northern hemisphere, the pool of cold air is trapped on the *south* side of the jet stream (Figure 14.5). When a ridge is cut off, the result is a cutoff high; in the northern hemisphere, the pool of warm air is trapped on the *north* side of the jet stream. Cutoff lows and highs are strongest in the upper troposphere and, although they are deep, they may or may not extend all the way to the surface. Sometimes, they will appear as weak systems on a surface chart; other times, they won't appear at all. Because cutoff highs and lows are not part of the jet stream's flow, they can remain in place for several days until they eventually die out. While it remains, a cutoff low will bring clouds and rain. A cutoff high, on the other hand, will often act as a block, forcing low-pressure systems around it; thus, a cutoff high might bring a spell of clear weather (Section 14.2.4).

FIGURE 14.5 | A 500 hPa chart showing a cutoff low over the Atlantic Ocean. The jet stream has also been highlighted on this map; notice how it flows to the north of the cutoff low.

Source: Adapted from Environment Canada, Weather Office, www.weatheroffice.gc.ca/analysis/index_e.html, 13 June 2012.

Question for Thought

Why are cutoff lows cold-cored, while cutoff highs are warm-cored?

14.2 | Mid-latitude Cyclones

14.2.1 | Polar Front Theory

The first attempt to describe the structure, formation, and development of mid-latitude cyclones was made just after World War I, by a group of Norwegian meteorologists. The model they created of the life cycle of cyclones is part of their **polar front theory**.

POLAR FRONT THEORY
A theory that states that mid-latitude cyclones form along the polar front and have a recognized life cycle.

Before this model was developed, it was known that clouds and precipitation are associated with low atmospheric pressure, and that such weather systems move. The Norwegian scientists were motivated to better understand these storms so that weather could be forecast.

With this as a goal, Jacob Bjerknes, who was only 22 when he published his first major paper on the topic, realized the need for a *dense* network of weather data. Armed with such data for Norway, Sweden, and Denmark, along with his own "simultaneous study of the sky" (Bjerknes, 1919), Bjerknes identified patterns. He recognized the existence of temperature discontinuities, which he later called "fronts" in analogy to the meeting of different armies along battle fronts during World War I. He also described the structure of cyclones by recognizing that the warm front and the cold front tend to form a characteristic wave pattern (Figure 14.6a). Further, he noted the warm air lying between the cold front and the warm front, and he identified this as the **warm-sector air** of the cyclone.

Bjerknes explained that rising air at the fronts would form clouds and precipitation. He likened the warm front to an inclined plane, but he was less certain of the vertical structure of the cold front. In his description of fronts, he noted the sequence of cloud types expected at warm fronts and indicated that different cloud types, along with more intense precipitation, form at cold fronts. He attributed the difference in cloud types and precipitation intensity to the more rapid ascent of air at the cold front. In addition, Bjerknes speculated that cyclones form as a result of the temperature contrast between polar and tropical air; he recognized this contrast as the source of potential energy for a storm. Further, he suggested that storms form with the purpose of alleviating this temperature difference.

> **WARM-SECTOR AIR**
> The warm air lying between the cold front and the warm front in a mid-latitude cyclone.

A few years after describing the structure of cyclones in his 1919 paper, Bjerknes determined that what he had described was actually the *mature* stage of a cyclone (Figure 14.6a). This realization led him to suggest that cyclones have life cycles: they are born, they mature, and they die. Subsequently, he and Halvor Solberg co-authored a paper, in 1922, in which they outlined and diagrammed the stages in a

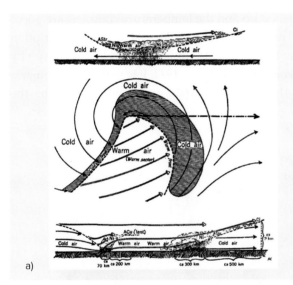

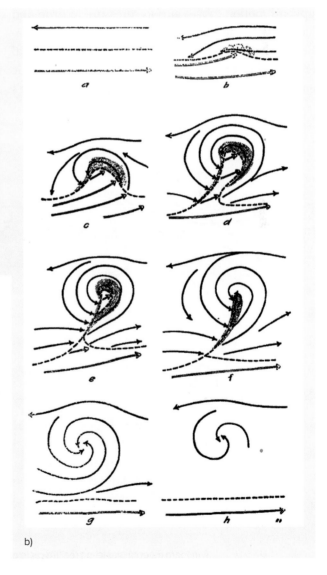

FIGURE 14.6 | a) Bjerknes and Solberg's original diagram showing the idealized structure of a mid-latitude cyclone. In the plan view of the cyclone, arrows represent airflow, and dashed lines represent fronts. (The symbols for fronts had not yet been devised.) The upper portion of the diagram is a cross section through the northern part of the cyclone. The lower portion of the diagram is a cross section through the southern part of the cyclone. **b)** Bjerknes and Solberg's original diagrams depicting the stages in the life cycle of a mid-latitude cyclone. Again, arrows represent airflow, and dashed lines represent fronts. In the first two diagrams, warmer air flows eastward while colder air flows westward along a front in which a kink begins to form. The next four diagrams depict the development of cyclonic rotation and the process of occlusion. The last two diagrams show the dying stages of the cyclone.

Source: a) A.J. Henry. (August 1922). J. Bjerknes and H. Solberg on meteorological conditions for the formation of rain. *Monthly Weather Review, 50*(8), 404. b) A.J. Henry. (September 1922). J. Bjerknes and H. Solberg on the life cycle of cyclones and the polar front theory of atmospheric circulation. *Monthly Weather Review, 50*(9), 469.

cyclone's life cycle (Figure 14.6b). (Figure 14.7 shows a more modern depiction of these stages.) The researchers noted that a cyclone can last anywhere from four to seven days and travels eastward as it evolves through its life cycle. It was also in this paper that the term *polar front* was first suggested as a name for the boundary that separates polar air from tropical air.

According to the polar front theory, a cyclone begins to form when a kink, or wave, develops in a stationary front. This kink causes warm air to begin to flow poleward and cold air to begin to flow equatorward, initiating the development of a warm front, a cold front, and cyclonic rotation (Figure 14.7b). At the same time, upward motion occurs as warm air rises over cold air at the developing fronts. This upward motion causes surface pressure to drop and leads to the development of clouds and precipitation. It also represents the conversion of the potential energy from temperature contrasts into the kinetic energy of motion. In the mature stage of the cyclone's

development, there is a clearly defined warm front and a clearly defined cold front (figures 14.7c and 14.7d). Between these fronts is the warm-sector air that Bjerknes first identified in his 1919 paper.

Toward the end of the cyclone's life, the faster-moving cold front begins to catch up to the warm front. As this happens, the warm air of the warm sector is gradually pinched out and forced aloft. In the process, an occluded front is formed (Figure 14.7e; Section 13.8). (Recall that this front is not easily noticed at the surface because the air masses it separates have similar temperatures.) Bjerknes and Solberg considered the occlusion stage to be the last stage in the cyclone's life cycle because there is no longer a source of potential energy for the storm once the warm air is pushed up and the temperature contrasts are gone.

The above is more or less the original description of the life cycle of a mid-latitude cyclone, as given by Bjerknes and Solberg in 1922. It is really quite remarkable that these scientists were able to describe the

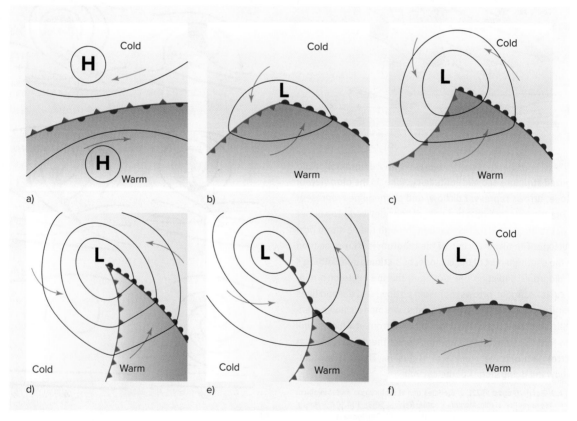

FIGURE 14.7 | A modern diagram depicting the life cycle of a mid-latitude cyclone.

structure and life cycle of cyclones so accurately—note for yourself the similarities between the original drawings (Figure 14.6) and the more modern depiction of the cyclone's stages (Figure 14.7). Their achievements appear even more outstanding when we consider how little they knew about the upper air. Aside from Bjerknes "simultaneous study of the sky," their work was based entirely on surface observations. It wasn't until the 1930s that routine upper-air observations began, and it wasn't until World War II that the polar front jet stream was discovered. Since the 1960s, satellites have provided us with a whole new view of the atmosphere.

These later observations have led to an increased understanding of the structure and development of mid-latitude cyclones. For example, whereas the original model gave no explanation for the formation of the initial kink in the polar front, we now know that this kink can be initiated by divergence in the jet stream. In addition, the newer *conveyor belt model* provides a three-dimensional view of the airflow in cyclones (Section 14.2.2.5). Recall also the relatively recent insights on the occlusion process, which were described in Section 13.8. Finally, as part of these recent studies, observations have shown that storms will often *increase* in intensity, rather than weaken or dissipate, after occlusion (Schultz & Vaughan, 2011).

Today, we still believe that mid-latitude cyclones form in association with fronts and that they have short, recognizable life cycles. On surface weather maps, these systems are still depicted just as described by the Norwegian model, and fronts often bring the expected weather. Yet we know much more about why mid-latitude cyclones form, the processes operating within them, the mechanisms through which they intensify or weaken, and the forces that move them eastward. These are, of course, the very things that can help forecasters predict when and where storms will occur.

14.2.2 | Cyclogenesis

Cyclogenesis is the name given to the processes associated with the development and strengthening of cyclones. Most cyclogenesis occurs due to temperature contrasts between air masses. Temperature contrasts are particularly marked in the air over oceans where warm currents flow past colder ones, or along coastlines where, in winter, colder continental air masses meet warmer oceanic air masses; thus, these are the kinds of regions favoured for cyclogenesis. Under these conditions, the atmosphere is baroclinically unstable because the cold air has a tendency to want to flow below the warm air, and the warm air has a tendency to want to rise above the cold air. Recall from Section 14.1 that mid-latitude cyclones and anticyclones are described as *deep, baroclinic systems that tilt with height*.

These cyclones and anticyclones are *baroclinic* in that temperature gradients, or fronts, are important in their formation. Anticyclones are baroclinic only initially because the temperature gradients are quickly obliterated through the surface divergence characteristic of high-pressure areas. The surface convergence in lows has the opposite effect, as it can bring warm air and cold air together, either producing fronts or strengthening existing fronts.

The *strength* of the temperature gradient can influence the character of the storms that develop. As the temperature difference between the air north of the jet stream and the air south of the jet stream builds up, the amplitude of the Rossby waves will increase; thus, the *meridional* component of the flow will increase (Figure 12.18; sections 12.2.5.2 and 12.5). With strong meridional flow, the temperature contrast across fronts is large, the jet stream moves very quickly, and cyclones tend to be very intense. On the other hand, under *zonal*-flow conditions, storms tend to be relatively weak. In addition, winter storms are usually more intense than summer storms because the equator-to-pole temperature gradient is greater in the winter. As recognized by Bjerknes, a storm develops with the purpose of alleviating these temperature gradients.

In addition to being baroclinic, these mid-latitude cyclones and anticyclones are *deep*, and they *tilt with height*. In cyclones, air *converges* into the region of low pressure

CYCLOGENESIS
The development or strengthening of a mid-latitude cyclone.

at the surface, *spins counterclockwise* (in the northern hemisphere), and *rises*, forming cloudy, wet weather. Cyclones appear as *closed isobars* on surface charts (Figure 14.4a) and as *troughs* of cold air on upper-air charts (Figure 14.4b). They tilt to the northwest with height, meaning that the upper-level trough is located to the northwest of the surface low. Because cyclones tilt up toward the colder air of the trough, they are cold-cored systems. (There is, of course, a small exception in the warm-sector air at the surface.) Due to this tilt, the surface low lies downstream of the upper trough, directly below an area of strong upper divergence (figures 11.13 and 14.4). In anticyclones, air *diverges* from the area of high pressure at the surface and *spins clockwise* (in the northern hemisphere); the resulting *sinking* air produces clear skies. Anticyclones appear as closed isobars on surface maps, while they appear as *ridges* of warm air on upper-air charts; thus, they are warm-cored systems. Anticyclones are *tilted* to the southwest with height, so that the surface high lies upstream of the trough, directly below an area of strong upper convergence (Figure 11.13).

Figure 14.8 shows the most fundamental processes operating within a mid-latitude cyclone. The numbering in the diagram indicates a possible sequence of events that could lead to the formation of the system. First, divergence in the mid to upper troposphere removes mass from the air column. This process lowers the surface pressure and causes air to rise. Next, the decrease in pressure causes convergence at the surface as air moves into the low-pressure area from the surroundings. Finally, as the air converges on the surface low, Earth's rotation causes the air to turn cyclonically as it rises. Remember that convergence leads to increasing vorticity (Section 12.4). Although this sequence provides us with an outline for our discussion, cyclogenesis is much more complex in reality.

To understand, and ultimately forecast, the formation and behaviour of mid-latitude cyclones, we must remember that they are *three-dimensional structures*. In addition, we must keep in mind that air is fluid, so changes at one level of the atmosphere will cause changes at other levels. Therefore, although we often consider them separately, the upper troposphere and lower troposphere are *not* separate. For example, as we have seen, convergence and divergence *aloft* lead to increases and decreases in *surface* pressure. At the same time, temperature gradients at the *surface* affect pressure patterns *aloft*. As cold air blows in behind surface cold fronts, troughs in the upper troposphere will strengthen; conversely, when warm air blows in behind surface warm fronts, ridges in the upper air will strengthen (Section 11.3). As this temperature advection causes upper troughs and ridges to strengthen, surface patterns can be further affected. Such positive feedback can strengthen cyclones, while negative feedbacks can operate to weaken them, as you will discover below.

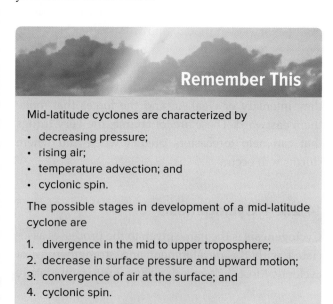

FIGURE 14.8 | This simplified diagram of a mid-latitude cyclone shows its structure and the processes operating within it. The numbers indicate a possible sequence of events in the formation of the system.

Mid-latitude cyclones are characterized by

- decreasing pressure;
- rising air;
- temperature advection; and
- cyclonic spin.

The possible stages in development of a mid-latitude cyclone are

1. divergence in the mid to upper troposphere;
2. decrease in surface pressure and upward motion;
3. convergence of air at the surface; and
4. cyclonic spin.

14.2.2.1 | Decrease in Surface Pressure

Remember that surface pressure results from the total mass of the air in a column. It follows that in order for surface pressure to decrease, mass must be removed from the air column. This could not happen if a deep, low-pressure system was vertically stacked, because convergence would occur throughout the depth of the troposphere, adding mass to the column and weakening, or **filling**, the low (Figure 14.9a). Instead, surface pressure decreases below areas of upper divergence; it is because of this that, as described above, deep low pressure systems tilt with height (Figure 14.9b). As long as the divergence aloft exceeds the convergence at the surface, there will be a net removal of mass from the air column, and the cyclone will strengthen, or **deepen**. In a similar way, if convergence aloft exceeds divergence at the surface, anticyclones will strengthen, as there is a net input of mass to the air column.

Divergence in the upper airflow is one of the most important processes in the development and maintenance of mid-latitude cyclones. In fact, it is now recognized as an important cause of the kink in the polar front first described by Bjerknes. As the divergence in the upper airflow causes surface pressure to drop, it pulls air up from below to replace the diverging air. If the troposphere is unstable, these upward motions are enhanced. In fact, it is the stability of the stratosphere that confines these vertical motions to the troposphere.

One cause of the divergence that often triggers the formation of mid-latitude cyclones is the change in wind speed associated with the gradient wind relation (Section 11.2.2). Recall that with a *constant* pressure gradient, winds blow faster around a ridge and slower around a trough than they do in straight geostrophic flow (Figure 14.10). Such changes in wind speed lead to speed divergence just downstream of an upper trough and speed convergence just upstream of an upper trough. Because it leads to divergence and convergence, the *waviness* of the upper airflow is important in the formation of cyclones and anticyclones (Figure 14.11), and this waviness is, in turn, associated with baroclinicity (Section 12.4).

Divergence and convergence in the upper air can also occur in *jet streaks*—the regions of relatively fast wind within the jet stream that are indicated on upper-air charts by the most closely spaced height contours. Because of the wind speed change, air

> **FILLING**
> An increase in the central pressure of a low-pressure system.
>
> **DEEPENING**
> A decrease in the central pressure of a low-pressure system.

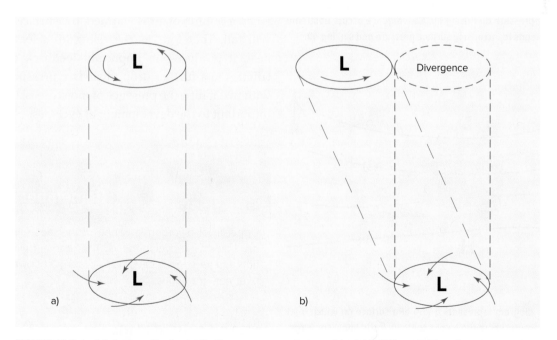

FIGURE 14.9 | a) A deep, vertically stacked low-pressure system would quickly fill because there is no process operating to remove mass from the air column. **b)** In a tilted low-pressure system, divergence occurs directly above the surface low. As long as this divergence is greater than the surface convergence, the low will deepen. (See also Figure 14.11.)

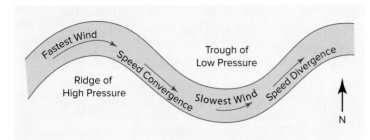

FIGURE 14.10 | Air in the mid to upper troposphere speeds up and slows down according to the gradient wind relation. This leads to convergence and divergence in the flow.

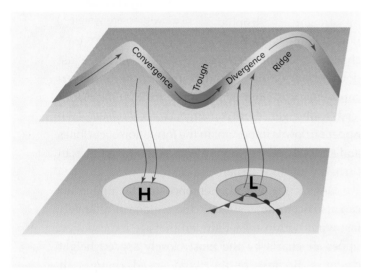

FIGURE 14.11 | Divergence occurs downstream of upper troughs and leads to decreasing surface pressure and rising air. Convergence occurs upstream of upper troughs and leads to increasing surface pressure and sinking air.

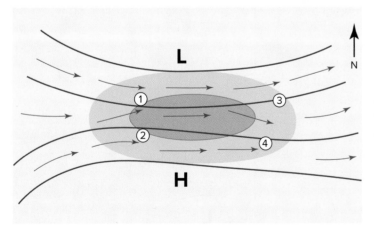

FIGURE 14.12 | This diagram represents a 250 hPa surface on which a jet streak (shaded area) occurs where the height contours (solid lines) converge. The darkest shading in the jet streak represents the fastest wind speeds. Strong directional convergence occurs at Point 1, and strong directional divergence occurs at Point 3.

moving into and out of jet streaks is temporarily put out of geostrophic balance. Air *entering* a jet streak will suddenly encounter a greater pressure gradient force. This increase in the pressure gradient force will cause it to temporarily exceed the Coriolis force and the wind will turn toward lower pressure (Figure 14.12). As a result, there will be strong directional convergence at Point 1 and weak directional divergence at Point 2. (Directional divergence and convergence were explained in Section 9.3.) The opposite happens when the pressure gradient force decreases as the air *exits* the jet streak. This decrease in the pressure gradient force will cause it to be temporarily less than the Coriolis force, and the wind will turn toward higher pressure. As a result, there will be strong divergence at Point 3 and weak convergence at Point 4.

Jet streaks are often found in troughs of the upper airflow; this makes the jet streaks curved. When jet streaks are curved, the zones of weak convergence and divergence almost disappear, so that convergence is associated with the entrance to a jet streak and divergence with the exit. Thus, the presence of jet streaks can cause divergence downstream of upper troughs and, thus, rising air.

As air rises, it cools and condenses to form the clouds characteristic of cyclones. In the process, latent heat is released; this heat warms the air. Recall that when a column of air is warmed, the surface pressure will fall. Therefore, as the rising air releases latent heat, it produces an important feedback that contributes to a further drop in surface pressure. Along with instability, the presence of moisture is, therefore, important to the development of cyclones.

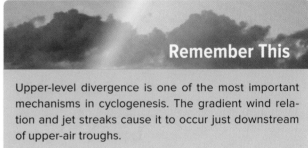

Remember This

Upper-level divergence is one of the most important mechanisms in cyclogenesis. The gradient wind relation and jet streaks cause it to occur just downstream of upper-air troughs.

14.2.2.2 | Vertical Motions

Rising air is fundamental to the development of a cyclone because it leads to the formation of clouds and precipitation. A key to forecasting the formation and strengthening of a cyclone is finding areas of divergence in the upper air, because such divergence should indicate areas of rising air. However, divergence is difficult to measure on weather maps. So, because of the relationship between vorticity and divergence (Section 12.4), meteorologists often use vorticity to find areas of divergence and, therefore, rising air.

Recall that relative vorticity is a function of curvature and/or shear in the flow. Both can be measured on upper-air charts, and vorticity can be calculated. In addition, planetary vorticity is a function of latitude, so it is easily calculated for a given location using Equation 11.3. Meteorologists can use these results to create maps of vorticity. Remember also that while convergence leads to increasing vorticity, divergence leads to decreasing vorticity (Equation 12.3). It follows that we need to find areas of decreasing vorticity in order to find areas of divergence.

As air moves through a wave at the 500 hPa level, its absolute vorticity will change according to Equation 12.3 (Figure 14.13). For the purposes of this illustration, assume that planetary vorticity remains constant. At Point 1, the air is turning anticyclonically, so it will have negative relative vorticity and, therefore, low absolute vorticity. In the straight flow at Point 2, the air will have zero relative vorticity and, therefore, a higher absolute vorticity than at Point 1. At Point 3, the absolute vorticity reaches a maximum because the air now has positive relative vorticity as a result of the cyclonic curvature in the flow. In the straight flow at Point 4, the absolute vorticity is lower than it was at Point 3. Remember that the fact that absolute vorticity is changing means that convergence and divergence must be occurring in the flow; otherwise, absolute vorticity would be conserved (Equation 12.2). Notice that Point 4, where vorticity is decreasing, is also the

area where divergence occurs. It follows that we can find areas where air is diverging and rising if we can identify areas where vorticity is decreasing, and we can find areas where air is converging and sinking if we can identify areas where vorticity is increasing.

14.2.2.3 | Temperature Advection

When an area of upper divergence is over a stationary front, the rising air and drop in pressure it creates will lead to convergence at the surface. Such convergence pulls the warm air and cold air together, forming warm and cold fronts. Sometimes there might not even be a surface front; in such cases, the surface convergence triggered by upper divergence can pull air masses together, *creating* fronts. (Recall from Section 13.6 that the term *frontogenesis* describes both the strengthening and the creation of fronts.) Figure 14.7b shows that the formation of fronts corresponds to an early stage in the life cycle of a cyclone, as outlined in the polar front theory. Behind the cold front, cold air is being advected, and behind the warm front, warm air is being advected. Cold-air advection results in sinking air because of the greater density of cold air. Warm-air advection, on the other hand, results in rising air because of the lower density of the warmer air. Thus, temperature advection enhances the vertical motions in the developing cyclone.

In addition to adding to the strength of the rising air, thermal advection can create a strong feedback that can work to intensify a storm because it will strengthen the trough-and-ridge pattern aloft.

Remember This

Where air is rising, vorticity will be decreasing in the upper half of the troposphere. Therefore, areas of rising air can be found using maps of vorticity at the 500 hPa level.

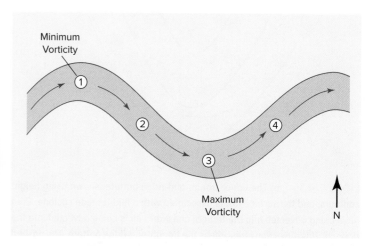

FIGURE 14.13 | The changes in absolute vorticity as air flows through a wave on the 500 hPa surface.

Figure 14.14a shows the surface fronts superimposed on the height contours of the upper air. Notice that cold air is being advected into the trough. When cold, dense air moves into this area in the lower atmosphere, its smaller thickness causes air in the upper troposphere to sink, lowering heights and deepening the upper trough. At the same time, as Figure 14.14a shows, warm air is being advected into the ridge. When warm, less dense air moves into this area in the lower atmosphere, its greater thickness pushes up on air in the upper atmosphere, increasing heights and

SHORT WAVES
Waves in the upper airflow with a wavelength of about 1000 km.

strengthening the ridge. As a result of these processes, the amplitudes of the upper waves will increase, as illustrated in Figure 14.14b. Larger amplitude waves will increase the upper divergence because it is the curvature of the waves that causes the divergence in the first place (Section 14.2.2.1). Increasing upper divergence will, in turn, intensify the surface low. Thus, a feedback loop is created, as a more intense surface low will further increase the temperature advection, and so on. Such a feedback mechanism between the upper and lower troposphere works to strengthen the cyclone. This feedback will not operate forever, of course, as mechanisms that lead to the dissipation of the storm eventually take over (Section 14.2.2.5).

Temperature advection does not occur only as a result of the surface fronts of an active cyclone; it can also develop in the upper air and *trigger* cyclone formation. A pattern in the upper airflow such as that shown in Figure 14.15 is associated with a disturbance known as a **short wave**. When a short wave moves through an upper trough, it disturbs the flow, causing thickness contours to cross height contours. Recall that such a pattern indicates baroclinic instability (Section 11.3). Short waves normally form along with, and are about the same size as, mid-latitude cyclones. However, sometimes short waves can form without cyclones. In such cases, short waves can *cause* cyclone development because they lead to temperature advection. The wind vectors in Figure 14.15 indicate that winds are backing with height and causing cold advection upstream of the trough (Section 11.3). At the same time, winds are veering with height and causing warm-air advection downstream of the trough. This temperature advection can cause fronts to develop at the surface and promote vertical motion.

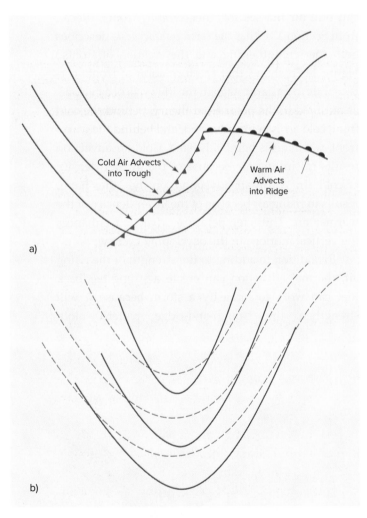

a)

b)

FIGURE 14.14 | a) The upper trough-and-ridge pattern, shown using height contours, and the surface fronts associated with a mid-latitude cyclone. Cold air is being advected into the trough, and warm air is being advected into the ridge. b) This advection strengthens the trough-and-ridge pattern. The dashed lines show the previous height contours, and the solid lines show the height contours after the change.

Cold Air Advects into Trough

Warm Air Advects into Ridge

Remember This

Temperature advection associated with *surface fronts* can act as a feedback mechanism in cyclone development by enhancing vertical motions and strengthening the upper wave pattern. Temperature advection associated with *short waves* in the upper airflow can trigger cyclone development.

14.2.2.4 | Cyclonic Rotation

In addition to forming fronts, and thereby initiating the associated temperature advection, convergence leads to cyclonic rotation. Remember that air already possesses vorticity due to Earth's rotation, but we don't notice this rotation because we are also rotating as Earth rotates. Convergence causes vorticity to increase, thus creating a *noticeable* cyclonic spin (Section 12.4). Near the equator, where planetary vorticity is zero, air will not spin even if it converges. In the mid-latitudes of the northern hemisphere, however, air converging into a low will spiral in a counter-clockwise direction. Further, the spin will increase as the rising column of air stretches (Equation 12.5).

14.2.2.5 | Dissipation of Mid-Latitude Cyclones

According to the polar front theory, mid-latitude cyclones begin to weaken and eventually die once the occlusion process begins (Figure 14.7e). Traditional reasoning held that once the warm-sector air had been pinched out and pushed above the cold air, the loss of a temperature contrast at the surface meant that there was no longer a source of potential energy for the storm. However, recent observations have shown that, in many cyclones, the pressure continues to drop *after* the storm has occluded (Schultz & Vaughan, 2011). This drop could be attributed to continued temperature advection *above* the surface, continued release of latent heat through cloud formation, or a combination of both. At some point, however, even these processes will cease to operate.

A model of the airflow in mid-latitude cyclones developed during the 1970s, known as the *conveyor belt model*, may help to explain how these storms eventually dissipate (Figure 14.16). This model provides more description of the three-dimensional structure of a mid-latitude cyclone than does the traditional cyclone model. In this model, the flow is depicted as three conveyor belts: one warm, one cold, and one dry. The warm conveyor belt represents the warm air from the warm sector rising up the slope of the warm front. This process produces the clouds along the warm front. The warm conveyor belt eventually joins the upper airflow and moves northeastward. The cold conveyor belt carries the cold air from ahead of the warm front under the warm conveyor belt and toward the northwest.

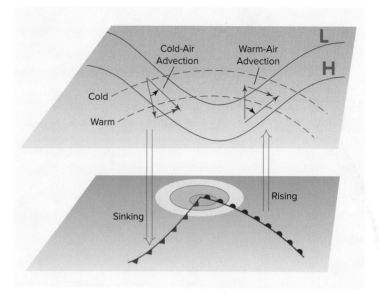

FIGURE 14.15 | A short wave has disturbed the flow at the 500 hPa level so that the thickness contours (dashed lines) cross the height contours (solid lines). As a result, cold advection is occurring upstream of the upper trough, and warm advection is occurring downstream of the upper trough. (This advection is indicated by the vector diagrams.) The short-wave disturbance can lead to the development of surface fronts and vertical motion.

Once this cold air reaches the low pressure centre, it begins to rise and rotate; like the warm conveyor belt, it ultimately joins the upper airflow, moving northeastward. Recall that mid-latitude cyclones appear on satellite images as comma-shaped clouds (Figure 14.3); the rising of the warm conveyor belt up the warm front produces the tail of the comma, while the rising and rotating of the cold conveyor belt in the low pressure centre produces the head of the comma. The dry conveyor belt carries cold, dry air from upper levels down behind the cold front, bringing the clear, dry weather that is associated with this region of the cyclone. The dry conveyor belt often appears as a strip of clear air on satellite images.

The conveyor belt model helps explain the dissipation of mid-latitude cyclones because it illustrates how the cold air gradually wraps itself around the low-pressure centre, separating it from the warm air. This separation is shown in Figure 14.7e and represents the process of occlusion; remember that there is cold air on either side of an occluded front. With the surface low wrapped entirely in cold air, it finds itself directly below the upper-level trough of cold air, so that the system is no longer tilted with height. In

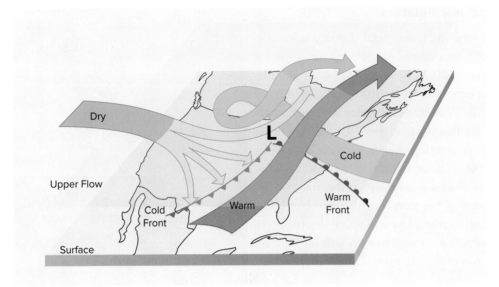

FIGURE 14.16 | The conveyor belt model of cyclone formation.

Source: Adapted from F.K. Lutgens, E.J. Tarbuck, and D. Tasa. (2009). *The atmosphere: An introduction to meteorology* (11th ed.). Upper Saddle River, NJ: Prentice Hall, p. 269.

LEE CYCLOGENESIS
The formation of a mid-latitude cyclone on the leeward side of a mountain range.

this situation, the surface low is no longer below an area of upper air divergence; therefore, as surface convergence continues, the storm will fill. Further, without temperature contrasts, temperature advection will cease.

> **Remember This**
>
> Generally, cyclogenesis begins when a large horizontal temperature gradient creates an unstable situation. The mid-latitude cyclone that results from this instability carries out the work needed to help lessen the temperature gradient by carrying cold air south and warm air north. Ultimately, the storm reduces the temperature gradient and dies out, having done its work.

14.2.3 | Lee Cyclogenesis

Sometimes, mid-latitude cyclones can form without the support of temperature gradients. Recall that Section 12.5 showed how changes in vorticity associated with airflow over a mountain range can produce waves in the upper airflow. The changes in vorticity

that result from airflow over mountains also make the lee side of mountains a favoured location for the development of cyclones, through a process known as **lee cyclogenesis**. We did not account for vertical motions in Section 12.5; here, we will account for such motions.

As an air layer is lifted over mountains, it will shrink in height as it is squeezed between the tropopause and the top of the mountains (Figure 14.17a). This shrinking results in horizontal divergence and, thus, a decrease in relative vorticity (Equation 12.5). For relative vorticity to decrease, flow that is initially straight must take on anticyclonic curvature. This is why the air turns southward (in the northern hemisphere) as it crosses the mountains (Section 12.5, Figure 14.17b). This southward turning produces a ridge in the upper flow and favours the development of anticyclones. When the air layer descends the leeward side of the mountains, it will stretch. Stretching results in horizontal convergence and, thus, an increase in relative vorticity. Recall that this increase in relative vorticity must be especially large because the air layer is both heading south and warming. To increase relative vorticity, the air must take on cyclonic rotation; thus, the lee side of mountains favours the development of cyclones. The lee side of the Canadian Rockies is a good example of such a location.

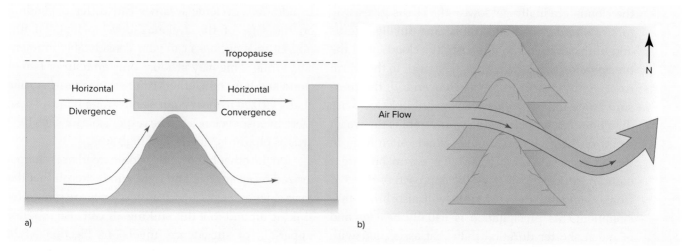

FIGURE 14.17 | a) As an air layer ascends the windward side of a mountain range, it will shrink in height, causing horizontal divergence and initiating anticyclonic spin. As the air layer descends the leeward side of the mountain range, it will stretch in height, causing horizontal convergence and initiating cyclonic spin. b) As seen from above, these processes produce an upper-air ridge over the mountain range and an upper-air trough on the lee side of the mountain range.

14.2.4 | Weather Associated with the Cyclones and Anticyclones of the Mid-Latitudes

Mid-latitude weather is characteristically variable. There is, of course, the familiar seasonal variation in temperature associated with the path of the sun in the sky (Section 5.8.2). In addition, the polar front jet stream, or *storm track*, shifts seasonally (Section 12.2.5.2, Figure 14.18) and produces more intense, and more frequent, cyclones in winter than it does in summer. As a result, most mid-latitude locations receive precipitation from mid-latitude cyclones in winter, but there are far fewer encounters with these storms in summer.

In addition to this *seasonal* variability, mid-latitude weather exhibits considerable *day-to-day* variability. This variability results because the *meandering pattern* of the jet stream is constantly changing and, as it does, it influences the formation and dissipation, as well as the positions and paths, of the travelling cyclones and anticyclones associated with it. When the jet stream is overhead, the anticyclones provide short gaps between the cloudy, wet weather associated with the cyclones. However, any mid-latitude location can receive a longer break from cloudy, wet weather when the jet stream shifts to the north or south because the warm and cold air masses where

high pressure tends to prevail lie on either side of the jet stream (Section 13.1.1).

Cyclones are associated with clouds, precipitation, and winds, and they exhibit a characteristic pattern as they move from west to east (Figure 14.19). The first sign of a cyclone approaching from the west is the cirrus clouds that appear roughly 12 to 24 hours in advance of the warm front. As the cyclone gets closer,

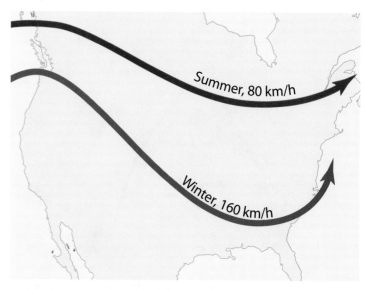

FIGURE 14.18 | The average summer and winter positions, and speeds, of the polar front jet stream.

the clouds gradually get lower; the cirrus gives way to cirrostratus, then altostratus, and finally stratus and nimbostratus. The nimbostratus clouds, and the precipitation they bring, arrive just prior to the warm front. As the warm front passes overhead, the pressure drops and the winds veer, or shift in a clockwise direction. The air behind the warm front is the warm sector of the cyclone. Cumulus clouds often form in this air because it is moist as well as warm, but larger cumuliform clouds indicate the approach of the cold front. Generally, the cold front will also bring precipitation; this precipitation is usually heavier and of much shorter duration than that associated with the warm front. The passage of the cold front brings another drop in pressure and another veer in wind direction. Once the cold front has passed, pressure rises and, particularly in winter, there is a considerable drop in temperature. The air behind the cold front is normally also dry, so, although this air may have become unstable from its travels equatorward over a warmer surface, it is unlikely to produce cloud.

Of course, few cyclones will be exactly alike. Furthermore, weather patterns can vary from one part of a cyclone to another. For example, the cloud and precipitation patterns associated with an occluded front are quite different from those associated with individual warm and cold fronts (sections 13.7 and 13.8).

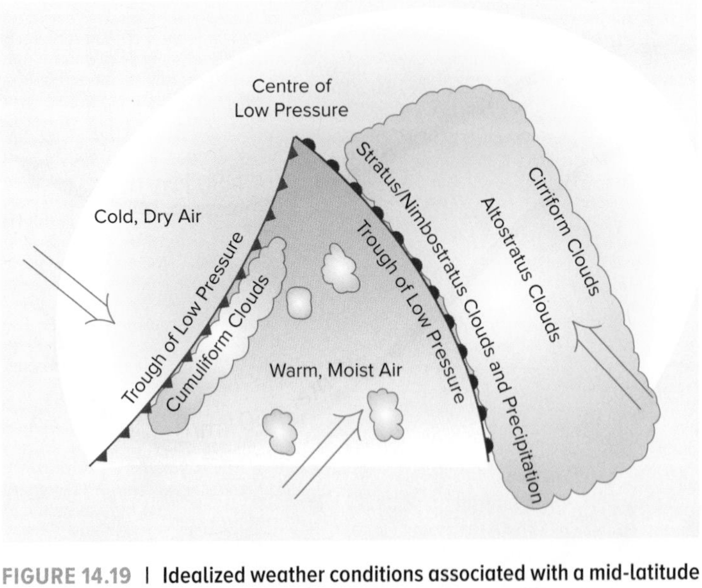

In addition, cyclonic weather can differ depending on the *stage* of the cyclone's development. Finally, mid-latitude cyclones can vary considerably in intensity. While some may produce only an hour or two of cloud and light rain, others can become raging storms characterized by heavy rains and strong winds. These more intense cyclones can be responsible for such hazards as blizzards, floods, and high seas.

On the other hand, anticyclonic weather is characterized by clear skies and light winds. Recall that the clear skies associated with anticyclones are caused by sinking air and that this sinking air can also produce upper-air, or subsidence, inversions (Section 8.5.2). Under these clear skies, anticyclones can bring very high temperatures in summer and very low temperatures in winter. In addition, recall that the reason winds are light in anticyclones is that there is a limit to the strength of the pressure gradients that can

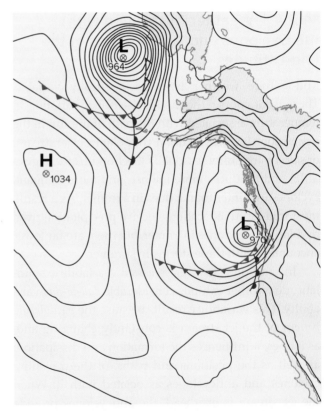

FIGURE 14.20 | This weather map for 12 March 2012 at 12 UTC shows that the pressure gradients around the two lows over the Pacific Ocean are much greater than that around the high lying just to the south of them. (The pressures labeled on the map are in hPa and the isobars are set at 4 hPa intervals.)

FIGURE 14.19 | Idealized weather conditions associated with a mid-latitude cyclone.

develop in anticyclones (Section 12.4, Figure 14.20). Light winds, combined with upper-air inversions, mean that, although the skies may be cloud-free, visibility in anticyclones can often be reduced by fog, haze, or pollution (Section 17.1.2).

As we have seen so far in this chapter, the cyclones of the mid-latitudes form as a result of divergence in the upper airflow, while the anticyclones of the mid-latitudes form below areas of upper-air convergence. These are the anticyclones that provide brief relief from the cloudy, rainy conditions normally associated with the proximity of the jet stream. However, during periods of strong meridional flow, the large anticyclones that lie on either side of the jet stream can bring longer spells of clear, dry weather. When the jet stream meanders *equatorward* of a location, the polar high brings unseasonably cold, dry weather (Section 12.2.5.2), but longer breaks can often occur when the jet stream meanders *poleward* of a location. Under such a flow regime, the subtropical high moves poleward, and unseasonably warm temperatures result. In addition, these highs are often referred to as *blocking highs* because, being *warm* highs, they extend through the troposphere (Section 14.1), blocking the paths of cyclones and, thus, forcing the cyclones to travel around them. Such a blocking pattern is known as an **omega high** (Figure 14.21). The result of such a block is that regions situated within the high, and slightly west of it, will likely experience weather that is warmer than normal, while regions to the east of the high will experience weather that is cooler than normal. In addition, because cyclones are forced to travel around the high, these areas will likely be wetter than normal, while areas within the high will be drier than normal. Because the patterns described here tend to remain in place for many days, they can cause droughts and floods.

OMEGA HIGH

A ridge of high pressure that forms in the shape of the Greek letter *omega* (Ω) in the upper airflow.

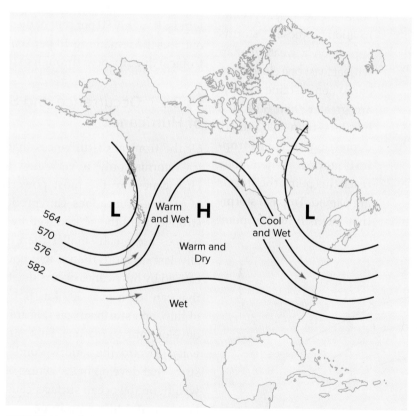

FIGURE 14.21 | **A blocking high, as it might appear on a 500 hPa chart. The height contours are in decametres, and the arrows indicate the upper airflow.**

14.3 | Tropical Cyclones: Hurricanes

Storms form differently in the tropics than they do in the mid-latitudes. To start, the tropical atmosphere is *not* baroclinic. The energy required for storms to form in the tropics comes from latent heat rather than temperature gradients. In general, these storms are known as *tropical* cyclones but, more specifically, they are named based on their intensity. **Tropical disturbances** are small areas of convection that form over tropical oceans. In these storms, sustained winds do not exceed 60 km/h. Disturbances become **tropical depressions** when they begin to develop cyclonic rotation; these storms appear on a weather map with at least one closed isobar. When sustained winds are greater than 60 km/h but less than 120 km/h, depressions become **tropical storms**. When they reach this point, the storms are named. Tropical storms that intensify to the point where sustained winds are over 120 km/h are called *hurricanes* if they form in the Atlantic or eastern Pacific, *typhoons* if they form in the western Pacific, and *cyclones* if they form in the Indian Ocean. (Here, the term *hurricane* will be used, as that is the name given to those storms that affect North America.)

Tropical disturbances often begin as troughs of low pressure in the trade wind easterlies known as **easterly waves**. Because pressure gradients are generally weak in the tropics, meteorologists plot maps of streamlines to find easterly waves. Streamlines are lines drawn parallel to the winds; easterly waves appear as ripples in the streamlines (Figure 14.22). These waves move westward over the tropical oceans. Those that travel across the Atlantic Ocean initially develop due to intense heating over the Sahara Desert. On the west side of easterly waves, the air diverges near the surface and sinks from above, and the skies are clear. East of an easterly wave's axis, however, surface convergence occurs, and the air rises to form small clusters of thunderstorms. These storms are tropical disturbances, which travel toward the west with the trade winds. Roughly 90 per cent of these storms quickly die out, but the remaining 10 per cent can intensify from tropical disturbances through to tropical storms.

14.3.1 | Occurrence and Structure of Hurricanes

Of the tropical disturbances that develop into tropical storms, only a very few become hurricanes. Hurricanes are the most powerful and destructive of storms. **Tornadoes** can produce winds that are stronger than those produced by hurricanes, but tornadoes are much smaller than hurricanes and usually last no more than a few hours. The destruction caused by hurricanes occurs due to very strong winds that can reach over 300 km/h, tremendous amounts of rain, and **storm surges** (Figure 14.23). Storm surges develop for two reasons. First, strong winds push sea water toward the land; second, the very low pressures that develop in the centre of hurricanes put less weight on the ocean surface, causing sea level to rise. For every 0.1 kPa drop in pressure, sea level can rise about 1 cm. Storm surges are particularly devastating along some of the very flat coastal regions around the Indian Ocean. Adding to the destruction, some hurricanes can also produce clusters of tornadoes.

TROPICAL DISTURBANCE
A small cluster of thunderstorms that forms in association with easterly waves over tropical oceans and in which wind speeds do not exceed 60 km/h.

TROPICAL DEPRESSION
A tropical disturbance that has begun to develop cyclonic rotation.

TROPICAL STORM
A tropical depression in which sustained winds are greater than 60 km/h but less than 120 km/h.

EASTERLY WAVE
A disturbance, in the shape of a wave, in the general easterly flow of the tropics.

TORNADO
A violently rotating column of air that can develop in association with severe thunderstorms.

STORM SURGE
A rise in water level that can cause flooding along coasts.

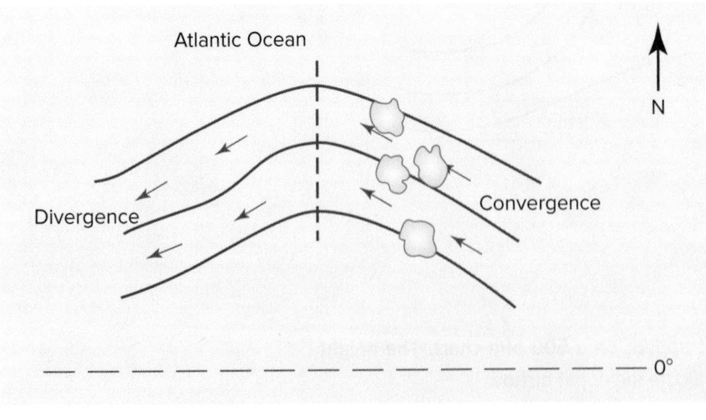

FIGURE 14.22 | Streamlines depicting an easterly wave. To the east of the wave axis, a tropical disturbance forms.

explains why we usually hear little about the hurricanes that form off the west coast of Mexico; these storms normally move west, away from land. Because hurricanes have a tendency to stay over the warmest water, some follow very erratic paths. In addition, hurricanes form only in late summer and early fall. In contrast to hurricanes, mid-latitude cyclones can form at any time of year, and their potential spatial range is far less restricted.

Mid-latitude cyclones and hurricanes are both areas of low pressure, but their *structures* are very different from one another. To start, while mid-latitude cyclones are cold-cored systems that tilt with height (Table 14.1), hurricanes are warm-cored systems that are vertically stacked. Hurricanes range in size from about 100 to about 1000 km across, which makes them considerably smaller than mid-latitude cyclones. The central pressure in hurricanes is usually about 95 kPa, but it can fall below 90 kPa. On a map, the isobars depicting a hurricane are very closely spaced, and they are circular in shape (Figure 14.25). This pattern indicates that the pressure gradients in hurricanes are much larger than those in mid-latitude cyclones and that hurricanes do not contain fronts. Recall that fronts are associated with kinks in the isobars.

Figure 14.26 shows a cross-sectional diagram through a hurricane. The cumulonimbus clouds are bands of thunderstorms that spiral counterclockwise (in the northern hemisphere) in toward a central eye. Wind speed steadily increases and pressure drops from the outer edge of the storm to the eye wall, where the storm is most intense. Cloud thickness and rainfall intensity exhibit an overall increase toward the eye wall, but both rise and fall following the banded pattern. In the circular eye of the hurricane,

FIGURE 14.23 | The devastation caused by Hurricane Katrina, in August 2005, when two levees broke and 80 per cent of New Orleans was flooded. In some areas, the water rose to 6 m above the city's surface.

The locations in which hurricanes can form are rather restricted (Figure 14.24). In general, hurricanes form over tropical waters, but they do not form within 5° of the equator, nor do they form over cold ocean currents such as those off the west coast of Africa or South America. In the northern hemisphere, most hurricanes tend to move first toward the northwest, then turn and move northeast. In general, they are directed around the subtropical highs over the oceans. This

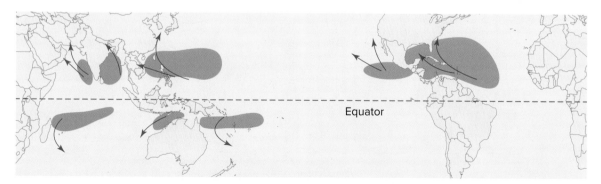

FIGURE 14.24 | Hurricanes form over tropical waters, between 5° and 20° north or south of the equator. Once they have formed, they tend to follow paths around the subtropical highs that lie over the oceans.

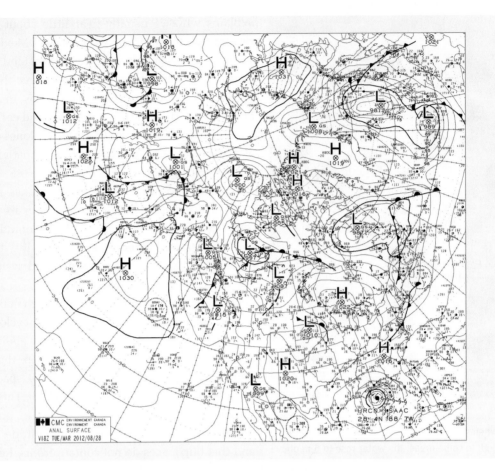

FIGURE 14.25 | A weather map for Tuesday, 28 August 2012 at 1800 UTC showing Hurricane Isaac, in the southeast corner of the map, moving toward the Gulf Coast. The latitude and longitude of the storm are labelled on the map, and the path the storm has followed is given by the small tropical storm symbols. Compare the size and shape of Hurricane Isaac to the size and shape of the mid-latitude cyclones shown to the north.

Source: Adapted from Environment Canada, Weather Office, www.weatheroffice.gc.ca/analysis/index_e.html, 28 August 2012.

which is usually about 20 to 50 km in diameter, the winds are light; the sky is fairly clear, apart from fair-weather cumulus clouds; and it is very warm due to air descending and adiabatically warming. If the widest part of the eye were to pass over an area, the calm conditions would last about an hour or two. It is not

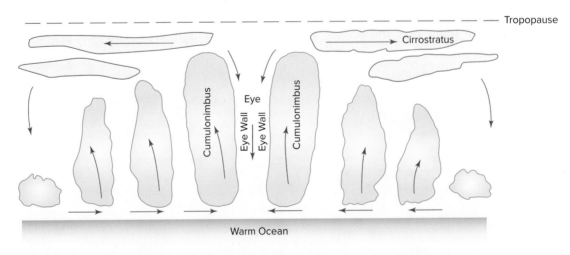

FIGURE 14.26 | A cross-sectional diagram through a hurricane.

entirely certain what causes the eye of the hurricane, but its width seems to be related to a balance between the inward-directed pressure gradient force and an outward-directed force due to the rapid rotation. In other words, it acts somewhat like a centrifuge. In more intense storms, the width of the eye is smaller, reflecting the fact that such storms are associated with a greater pressure gradient force.

Being warm-cored low-pressure systems, hurricanes have their highest temperatures in their centres. Much of this warmth comes from the release of latent heat. Remember that, because pressure decreases slowly with height in warm air, warm-cored lows are shallow. Up to a height of about 7 km, pressure in the hurricane is lower than it is in the surroundings. At about 7 km, the pressure in most hurricanes is the same as it is in the surroundings; above this height, the pressure is higher in the hurricane than it is in the surroundings. The result is that winds turn cyclonically and blow inward in the lower half of the storm, while winds turn anticyclonically and blow outward in the upper half of the storm. As the air blows outward at the top of a hurricane, cirrostratus clouds form (Figure 14.27). It is this outflow aloft that allows hurricanes to maintain their low pressures despite their stacked structure. (Recall that mid-latitude cyclones, in contrast, would fill if they weren't tilted.)

14.3.2 | Formation and Dissipation of Hurricanes

It is likely that hurricanes will grow from tropical disturbances when the release of latent heat in the thunderstorms warms the air column enough that a strong thermal low-pressure system begins to develop. The high pressure aloft, which is characteristic of such a system, leads to divergence. This divergence, coupled with the warmth, leads to further drops in the surface pressure. Eventually, the storm develops strong cyclonic spin and very fast winds. When fast winds blow over warm water, evaporation rates are very high. This evaporation supplies more water vapour and, thus, more latent heat to the storm. Although we are not certain exactly what causes a disturbance to develop into a hurricane, we have been able to identify certain requirements that are absolutely necessary.

First, hurricane formation requires a source of heat. The required heat comes from the latent heat

FIGURE 14.27 | A satellite image of a hurricane showing the eye and the cirrostratus clouds that form beyond the eye. These clouds obscure the bands of cumulonimbus clouds below. Notice that the flow around this northern-hemisphere hurricane is opposite to that for the tropical cyclone shown in Figure 14.2.

that is released when water vapour, evaporated from the ocean, condenses in the storm. Hurricanes tend to form over waters that are warmer than 26°C to a depth of about 60 m. With lower temperatures, evaporation rates would not be high enough to produce enough latent heat to sustain the hurricane; with a shallower layer of warm water, the strong winds of the hurricane would mix colder water to the surface, killing the storm. The necessity of a heat source explains why hurricanes form over tropical oceans, why they don't form where there are cold currents, and why they form in the late summer and early fall, when ocean temperatures are highest.

Second, hurricane formation requires a significant Coriolis force. Without this force, air converging into the low-pressure centre would blow *straight* into it, filling the low and killing the storm. Instead, the Coriolis force causes winds to spiral cyclonically into the low. This requirement explains why hurricanes don't form within 5° of the equator, where the Coriolis force is very weak.

Third, hurricane formation requires an unstable atmosphere. This instability is more likely to occur over the western regions of tropical oceans than it is over the eastern regions of these oceans. (Recall from Section 12.2.3 that ocean currents are warmer and the subsidence inversion associated with the subtropical highs is weaker over the western regions of tropical oceans.)

Fourth, hurricane formation requires that there be little to no vertical wind shear. Where there is wind shear, the necessary vertical circulation cannot develop. Strong vertical wind shear may explain why hurricanes rarely form off the east coast of South America, despite the ocean currents being warm (Figure 14.24). In addition, once a hurricane has started to form, the development of vertical wind shear can cause it to weaken.

Finally, hurricane formation requires high humidity throughout the troposphere. Without this humidity, the growth of clouds would be inhibited by **entrainment** of dry air.

ENTRAINMENT
A process by which air surrounding a cloud is drawn into the cloud.

Remember This

There are five requirements for hurricane formation:

1. a source of heat,
2. a significant Coriolis force,
3. instability,
4. minimal vertical wind shear, and
5. high humidity throughout the troposphere.

When all these conditions are met, a hurricane is likely to form and strengthen. Additionally, like mid-latitude cyclones, hurricanes can be self-propagating due to the development of feedbacks. As rising air cools and condenses to form clouds, latent heat is released, warming the air. Warming leads to further drops in surface pressure that strengthen convergence. With greater convergence in the lower troposphere, more air rises and cools, releasing even more latent heat.

As might be expected, these storms do not continue to intensify indefinitely. Like mid-latitude cyclones, hurricanes usually last about a week. When any of the requirements for their formation are no longer met, they will begin to dissipate.

DOWNBURST
A strong, gusty wind that forms due to downdrafts in thunderstorms.

MESOSCALE CONVECTIVE SYSTEM
A system driven by vertical circulations (i.e., convection) that is at least 100 km in one direction.

Since a continued supply of latent heat is very important, hurricanes will quickly die when they move over colder water or land. Moving over land will also weaken these storms because friction slows the winds and redirects them to blow more directly into the centre of low pressure, causing it to fill. In addition, hurricanes can weaken if they encounter vertical wind shear. This might happen, for example, as they move into the mid-latitudes, where upper winds are strong westerlies.

Question for Thought

How do tropical cyclones differ from mid-latitude cyclones?

14.4 | Thunderstorms

Thunderstorms produce thunder and lightning. In addition, they are characterized by cumulonimbus clouds, heavy rain and hail, gusty winds, and even tornadoes. Brief thunderstorms produce little, if any, damage, but longer-lasting thunderstorms can become destructive. This destruction often results from tornadoes, but thunderstorms can also produce damaging winds, **downbursts**, and flash floods. Thunderstorms have been observed extensively using radar, aircraft, and even video recordings, and they are often studied in the lab and through computer models.

Thunderstorms can occur on their own, but they are often associated with both mid-latitude cyclones and hurricanes. In mid-latitude cyclones, thunderstorms often form along the cold front. In comparison to mid-latitude cyclones, thunderstorms are much smaller and of shorter duration, they have a greater height-to-width ratio, and they involve stronger vertical motions. Hurricanes are actually very large clusters of organized thunderstorms. These clusters are known as **mesoscale convective systems**.

Thunderstorms are *convective* in nature because they form due to warm air rising in an unstable environment; as the warm air rises, cold air sinks, creating a convection cell. On average, individual thunderstorm cells might be only a few kilometres in diameter, but they can affect larger areas when they occur in a group, as clusters or lines. Because they are normally associated with strong surface heating, thunderstorms most commonly form on warm afternoons, but they can also occur at other times of the day or even during the night.

In many parts of the world, thunderstorms are the major source of rain, particularly in summer months. They are most common in the tropics, where heat and moisture are always available (Figure 14.28). In some tropical locations, they can occur almost every day. Outside of the tropics, they are most common in summer. In Canada, they occur most frequently in southwestern Ontario and least frequently north of 55° N. They are also rare on the west coast, where summers are cooler and less humid.

Thunderstorms that are able to sustain themselves are longer lasting and, therefore, more likely to become *severe*. In Canada, thunderstorms are considered to be severe when they produce winds greater than 90 km/h, hail larger than 20 mm in diameter, and/or rain with an intensity greater than 50 mm in one hour or 75 mm in two hours. Thunderstorms that are made up of a **single cell** may last for only about half an hour to an hour, as they tend to extinguish themselves relatively quickly. As a result, they are unlikely to become severe. **Multicell** and **supercell** storms, on the other hand, can sustain themselves for much longer and, thus, become severe (Figure 14.29).

In order for thunderstorms to form, three conditions are necessary. First, air near the surface needs to be both warm and moist, as indicated by a high dew-point temperature. This warm, moist air is the *fuel* for the storm. Second, the air must be conditionally unstable, meaning that rising air must be *forced* past the condensation level before it is able to rise due its own buoyancy (Section 8.4). Finally, there must be a mechanism to accomplish this forced lifting of the air to its *level of free convection* (LFC). This mechanism is sometimes referred to as a *trigger* or a *lift*, and it can be provided by further surface heating, a cold front, topography, or convergence of air at the surface. Such convergence develops, for example, in the *intertropical convergence zone* of the tropics, or in Florida where sea breezes from the Atlantic Ocean and the Gulf of Mexico converge over land. Lift can also be

SINGLE-CELL THUNDERSTORM
A thunderstorm that contains one cell.

MULTICELL THUNDERSTORM
A thunderstorm containing more than one cell, each at a different stage of development.

SUPERCELL THUNDERSTORM
A thunderstorm with one cell consisting of a strong rotating updraft.

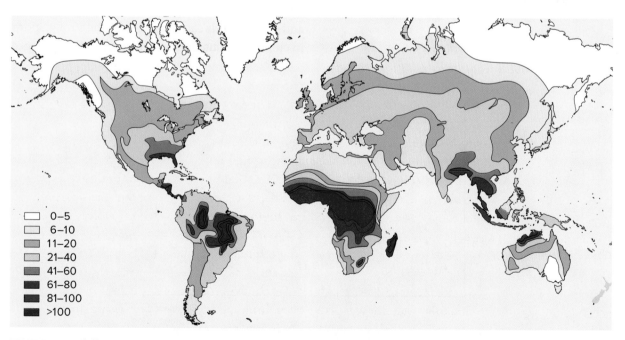

0–5
6–10
11–20
21–40
41–60
61–80
81–100
>100

FIGURE 14.28 | The average number of days per year on which thunderstorms occur.

Source: Adapted from Ahrens, C. Donald and Perry Samson (2011). *Extreme weather and climate.* California: Brooks/Cole, Cengage Learning, p. 318.

FIGURE 14.29 | A supercell storm near Grand Island, Nebraska, illuminated by lightning.

provided by divergence in the jet stream, which helps to *pull* air up.

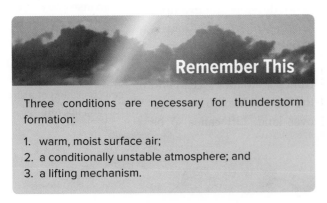

Remember This

Three conditions are necessary for thunderstorm formation:

1. warm, moist surface air;
2. a conditionally unstable atmosphere; and
3. a lifting mechanism.

14.4.1 | Life Cycle of a Thunderstorm Cell

We can think of single-cell thunderstorms as following a simple three-stage life cycle (Figure 14.30). We can also use this model to understand the processes at work in more complex multicell and supercell storms. The first stage in the life cycle of a thunderstorm cell is the *cumulus stage*; as the name implies, this stage involves the formation and growth of a cumulus cloud (Section 9.3.4). This cloud begins to develop when warm, moist air rising from the surface—the *updrafts* that fuel the storm—reaches its condensation level. Often, such a cloud will quickly evaporate as it mixes with the surrounding drier air. This process adds moisture to the air, however, so that as air continues to rise from the surface and condense, it is less likely to evaporate, and the cumulus cloud will grow in height. The release of latent heat during condensation will also add warmth to the rising air, further increasing its buoyancy and strengthening the updrafts. The updrafts involved help turn cloud droplets into raindrops by the collision–coalescence process. In addition, as the cloud rises higher, it may reach the freezing level where ice crystals can form and further aid in the formation of raindrops.

Once precipitation begins to fall, the thunderstorm cell has entered the *mature stage*. This is the stage in which very heavy rain or hail, lightning and thunder, and strong winds can be expected. The strong updrafts in cumulonimbus clouds, which typically have speeds

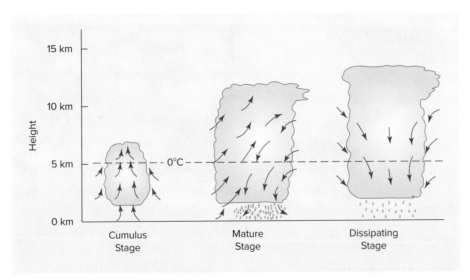

FIGURE 14.30 | The stages in the life cycle of a thunderstorm: the cumulus stage, the mature stage, and the dissipating stage.

of about 10 to 20 m/s, are responsible for the large rain-drops and the hail (sections 10.3 and 10.5). As raindrops fall, they drag air down with them, creating *downdrafts*. Downdrafts are also created by entrainment, the process by which the surrounding air is mixed into the growing cloud. As this drier air is mixed in, some of the water in the cloud evaporates, thus cooling the air. As the air cools, its density increases, and it begins to sink. The mature stage is, therefore, characterized by both updrafts and downdrafts that together create a convection cell in the cloud. It is this circulation that produces the turbulence so typical of thunderstorms.

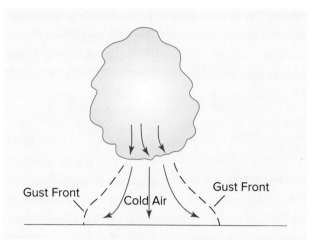

FIGURE 14.31 | **Downdrafts in a thunderstorm produce a gust front.**

Question for Thought

Why does hail form only in cumulonimbus clouds?

It is also during the mature stage that the cumulonimbus cloud develops its *anvil* top (Section 9.3.4). As the rising air reaches the tropopause, it encounters the stable air of the stratosphere; at this point, it will stop rising and spread horizontally, creating the anvil. Due to its great height, the anvil top is composed of ice crystals and, thus, has the wispy appearance of a cirrus cloud. Winds at this level also help to create the anvil. Where updrafts are particularly strong, an *overshooting top* might form as rising air penetrates the stratosphere.

The downdrafts that begin to form during the mature stage can create cool and gusty winds at the surface. If the downdrafts become concentrated, they are known as *downbursts* and can be damaging. As they flow outward below the storm, downdrafts create a **gust front** where they meet the warm, moist air at the surface (Figure 14.31). A gust front has a character that is similar to that of a cold front, but it develops on a smaller scale. Under certain conditions, the gust front can help force more warm, moist air up into the cloud and sustain the storm, but in most simple, single-cell storms, the downdrafts eventually extinguish the storm.

The third and final stage of a thunderstorm cell is the *dissipating stage*. This stage begins as the downdrafts, along with the heavy rain, fall into the updrafts, cutting off the supply of warm, moist air rising from the surface. Without its fuel source, the thunderstorm will quickly die. On the other hand, when downdrafts are prevented from falling into updrafts, dissipation does not occur, and the storm can strengthen.

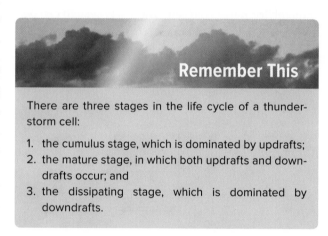

Remember This

There are three stages in the life cycle of a thunderstorm cell:

1. the cumulus stage, which is dominated by updrafts;
2. the mature stage, in which both updrafts and downdrafts occur; and
3. the dissipating stage, which is dominated by downdrafts.

14.4.2 | Types of Thunderstorms

In a *single-cell* thunderstorm, the updraft is likely to be cut off by the downdraft within half an hour to an hour. Because they form *within* an air mass and are, therefore, not associated with fronts, these storms are sometimes called *air-mass thunderstorms*. They usually form as a result of instability on a warm, humid afternoon. More importantly, they occur under conditions of little to no vertical wind shear. It is the lack of wind shear that

GUST FRONT
The boundary at which the cold air in a thunderstorm downdraft meets warm, moist air at the surface.

allows the downdrafts to fall into the updrafts and effectively extinguish the storm. Thunderstorms are likely to last longer and become severe where there is wind shear.

MESOSCALE CONVECTIVE COMPLEX
A group of thunderstorms that operate together as a system.

SQUALL LINE
A line of thunderstorms that forms at or just ahead of a cold front.

In a *multicell* storm system, many cells will be occurring together; old cells will lead to the formation of new cells, so the cells will be at various stages of development. Vertical wind shear—specifically, the shear that results from the increase of wind speed with height—causes these cells to tilt so that the downdrafts and precipitation do not fall into the updrafts (Figure 14.32). In fact, the structure of the storm allows the gust front to strengthen the updrafts by forcing warm, moist surface air to rise and form new cells. In the process, warm, moist air is drawn in from *around* the storm as well as from *below* the storm. Individual cells are still short-lived but, because new cells can form from old cells, the storm system can persist for upward of 12 hours.

When these systems of multiple cells are arranged in large *clusters*, they are known as **mesoscale convective complexes**; when they are arranged as *lines*, they are known as **squall lines**. Both arrangements are subtypes of a group of weather systems collectively known as *mesoscale convective systems*. Recall from Section 14.4 that hurricanes are also mesoscale convective systems.

Mesoscale convective complexes are circular in shape and cover areas of at least 100,000 km². These thunderstorm clusters are the major producers of summertime rain in central Canada and the United States. They are best identified using infrared satellite imagery, on which they show up as almost circular areas of very cold cloud tops (Figure 14.33). The coldness of the cloud tops is due to their great heights.

Squall lines can extend up to 500 km in length and usually form along or just ahead of cold fronts. When they form along cold fronts, it is the cold front that provides the lift. When they form ahead of cold fronts, it is possible that waves in the upper air, caused by the cold front surging forward, can help create the lift. The gust fronts associated with squall lines continually push warm, moist surface air upward as the squall line advances toward the east with the cold front. In this way, the old cells create new cells, thus sustaining the storm system. Squall lines can also form along *dry*

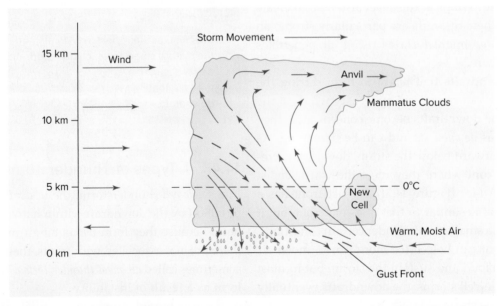

FIGURE 14.32 | The vertical wind shear in a multicell thunderstorm causes the storm to tilt so that the downdrafts do not fall into the updrafts. The mammatus clouds shown here form as air parcels sink in the very unstable air of the thunderstorm and condense, forming rounded blobs on the underside of the cloud.

FIGURE 14.33 | An infrared satellite image of a mesoscale convective complex located over a portion of the Missouri River Valley in the north central United States. The colours on this image show that the cloud tops reached temperatures of below −60°C.

lines, where less dense moist air rises over denser dry air, forming thunderstorms (Section 13.4).

Supercell storms have only one cell, which is characterized by a very strong *rotating* updraft (Figure 14.34). These cells are much larger than simple single-cell storms; they can reach heights of up to 20 km, and they can be between 20 and 50 km in width. In addition, they typically last from two to four hours. Supercells are the most severe thunderstorms and the ones that are most likely to produce tornadoes. For a supercell thunderstorm to form, there needs to be wind shear that is strong in both speed and direction. It is this strong wind shear that creates rotation; as winds at different heights move at different speeds and in different directions, the air will begin to rotate about a horizontal axis. When such horizontal spin is tilted into the updraft, it creates a *rotating* updraft known as a **mesocyclone**. This updraft is so strong that rain cannot fall through it; thus, this updraft sustains the storm. Mesocyclones can also lead to the formation of tornadoes if they lengthen and extend from the bottom of the cloud toward the ground.

> **MESOCYCLONE**
> Cyclonically spinning air in a convective system.

14.4.3 | Atmospheric Soundings and Thunderstorms

Certain conditions favour the development of thunderstorms. First, as we have seen, the air near the surface must be warm and moist to provide the

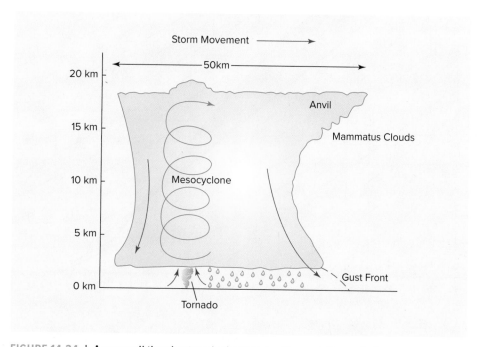

FIGURE 14.34 | A supercell thunderstorm is characterized by a rotating updraft.

fuel for the storm. Second, an inversion lying above this surface layer of air is important to help trap the warmth and moisture, allowing it to build up. In order for a thunderstorm to form, however, the air must be *forced* past this inversion until it can rise freely due to its own buoyancy. Third, the air above the inversion, and through most of the troposphere, should be cold and dry. The cold air increases the instability because it means that temperature decreases rapidly with height. The dry air is important because when it is entrained into the cloud it will cause evaporation. The cooling due to evaporation strengthens downdrafts.

The blue line on Figure 14.35 is a *hypothetical* temperature sounding used to illustrate an environment favourable for the development of thunderstorms. The

LIMIT OF CONVECTION
The height at which rising air stops rising.

red line drawn on Figure 14.35 represents an air parcel rising through this environment. Recall from Section 8.3 that a rising air parcel follows a dry adiabat until it reaches the *lifting condensation level* (LCL), which in this case occurs at 86 kPa. This is the height of the cloud base. Once in the cloud, the air parcel will continue to rise following a saturated adiabat. Notice that the air parcel needs to be *forced* to the LCL because, at this height, rising air will be colder than the surrounding air. Once the air parcel reaches 74 kPa, however, it will be warmer than its surroundings and rise on its own. Recall from Section 8.4 that this height is the *level of free convection* (LFC). From this point, the air parcel will continue to rise buoyantly until it is the same temperature as the surroundings. The height at which this occurs is the **limit of convection** (LOC); this is also the top of the cloud.

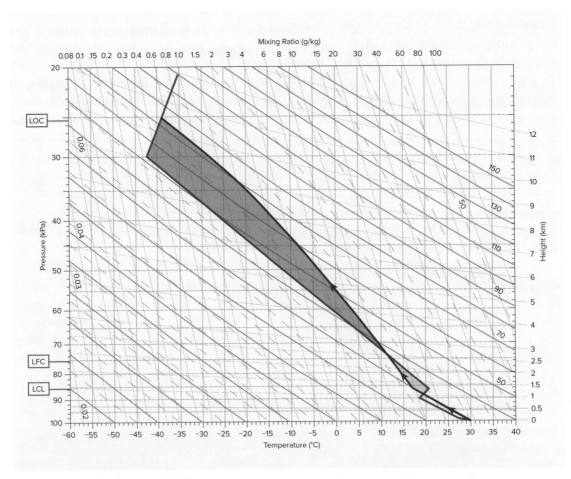

FIGURE 14.35 | A hypothetical temperature sounding, shown by the blue line, depicting a typical thunderstorm environment. The red line represents an air parcel rising through this environment. Labelled on the diagram are the heights corresponding to the LCL, LFC, and LOC. Also shown are CIN, shaded in green, and CAPE, shaded in purple.

Question for Thought

How could you use Figure 14.35 to show how the combination of significant surface moisture and cold air aloft lead to strong instability?

Figure 14.35 illustrates, through shading, two measures used to quantitatively assess thunderstorm potential. Both are calculated using atmospheric sounding data, and both are used in forecasting. The first measure is **convective inhibition** (CIN). CIN, developed in 1984, is a measure of the strength of the cap that might prevent surface air from rising (Table 14.2). CIN is expressed in joules per kilogram. It is determined by calculating the *area* below the level of free convection, between the temperature sounding and the adiabat of the rising air parcel, where the rising air parcel would be colder than the surrounding air. This value represents the size of the trigger needed to initiate thunderstorm development. The surface air can either warm enough by the afternoon to eliminate CIN, or it can be forced upward through this region by, for example, a cold front or a mountain. A larger CIN allows more warmth and moisture to accumulate near the surface, providing more fuel for a storm, but if CIN is *too* large, a storm may not be able to form at all.

The second measure is **convective available potential energy** (CAPE), also developed in 1984. CAPE is a measure of the potential intensity of a thunderstorm. Like CIN, CAPE is expressed in joules per kilogram. It is determined by calculating the *area* between the temperature sounding and the saturated adiabat of the rising air parcel, from the LFC to the LOC. The greater is CAPE, the greater will be the vertical velocity

and the intensity of the storm (Table 14.3). Figure 14.35 shows that colder air aloft can increase CAPE. In addition, CAPE will increase with a higher LOC.

14.4.4 | Lightning and Thunder

Lightning is the essential ingredient in a thunderstorm because it is lightning that creates the thunder *sound*. Lightning is a rapid and sudden flow of electricity that occurs due to a separation of electrical charges. It most commonly occurs within the cloud, but it can also occur between the cloud and the ground, between the cloud and an object on the surface (Figure 14.36), or, least commonly, between clouds. Under fair-weather conditions, the upper atmosphere is positively charged, and Earth's surface is negatively charged. During a thunderstorm, the top of the cloud typically takes on a positive charge, and the bottom of the cloud takes on a negative charge.

The reasons for such charge separation in clouds are not clearly understood, but they appear to be associated with vigorous convection. This convection carries positive charges up and negative charges down within the cloud. One of the reasons for this transfer may be that positive charges have a tendency to move from warmer surfaces to cooler surfaces. Hailstones are warmer than ice crystals because latent heat is released as supercooled water freezes onto them. When colder ice crystals collide with warmer

CONVECTIVE INHIBITION (CIN)
A measure of the strength of the cap preventing surface air from rising to produce a thunderstorm.

CONVECTIVE AVAILABLE POTENTIAL ENERGY (CAPE)
A measure of the potential intensity of thunderstorms.

TABLE 14.2 | CIN values and descriptions.

CIN (J/kg)	Description
<0	No cap
0–20	Weak capping
20–50	Moderate capping
50–100	Strong capping
>100	Intense cap; storms not likely

TABLE 14.3 | CAPE values and descriptions.

CAPE (J/kg)	Description
<300	Mostly stable, little or no convection
300–1000	Marginally unstable; weak thunderstorm activity
1000–2500	Moderately unstable; possible severe thunderstorms
2500–3500	Very unstable; severe thunderstorms; possible tornadoes
>3500	Extremely unstable; severe thunderstorms; tornadoes likely

FIGURE 14.36 | Lightning striking the CN Tower in Toronto.

hailstones, the ice crystals become positively charged, while the hailstones become negatively charged. In addition, as a supercooled water droplet freezes from the outside in, positive charges migrate to the outside. If this shell of ice is shattered, it will produce small, positively charged ice crystals. Either way, the tiny, positively charged ice crystals travel upward in the cloud, while the larger, negatively charged hailstones and water droplets fall downward (Figure 14.37).

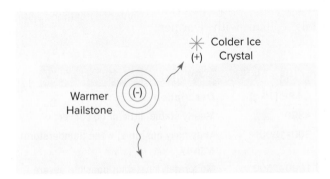

FIGURE 14.37 | When a larger, warmer hailstone collides with a smaller, colder ice crystal, the hailstone becomes negatively charged while the ice crystal becomes positively charged. As the ice crystal rises in the updrafts and the hailstone falls, they carry their charges with them.

Another reason for charge separation may be that the normal charge separation in the atmosphere induces a negative charge in the upper part of water droplets and a positive charge in the lower part. (The charge is induced because like charges repel each other.) When a large droplet is falling through the cloud, a collision with a smaller droplet will transfer negative charge to the large droplet, leaving the smaller droplet positively charged (Figure 14.38). If the droplets do not coalesce, the smaller, positively charged droplet will continue to travel up, and the larger, negatively charged droplet will continue to travel down.

No matter how it develops, the charge separation causes an electric potential, or voltage, in the cloud. In addition, because like charges repel each other, the negative charge at the base of the cloud can induce a positive charge on Earth's surface. This positive charge is particularly pronounced on objects that protrude from the surface, such as tall buildings or trees (Figure 14.39). In this way, an electric potential can also develop between the base of the cloud and Earth's surface. Because air is a poor conductor of electricity, the electric potential must be very high before lightning will occur. The objective of the lightning is

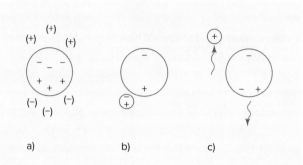

FIGURE 14.38 | a) The upper part of water droplets is negatively charged, while the lower part is positively charged. b) When water droplets collide, the larger droplet picks up the negative charge from the top of the smaller droplet. c) Small, positively charged droplets are carried up by updrafts, and larger, negatively charged droplets fall to the lower part of the cloud.

FIGURE 14.39 | The negative charge at the base of the cloud induces a positive charge on Earth's surface, especially on objects protruding from the surface. The charge separation that develops causes lightning.

to cause a current to flow and eliminate the charge separation. As this current flows, it heats the air to about 30,000°C! At this temperature, visible light is emitted; this light is what we see as a stroke of lightning. In reality, what we perceive to be a single stroke of lightning is actually several strokes, but they occur so quickly that our eyes see only one flash. If the lightning occurs between the cloud and the ground, we see a distinct stroke of lightning and usually refer to it as *fork* lightning. If, instead, the lightning occurs within the cloud, or between clouds, the light is diffused and seen as *sheet* lightning.

Not only do we *see* the lightning stroke, we also *hear* it. The very high temperature of the lightning stroke causes a rapid expansion of the surrounding air, which creates the sound we know as thunder. Since light travels almost one million times faster than sound, we see the lightning before we hear the thunder. We know that sound takes about three seconds to travel one kilometre, so we can determine the distance to the storm by counting the seconds between the lightning and the thunder, and dividing this number by three.

In the Field

Understanding the Connection between Mid-latitude Cyclogenesis and Weather
Kaz Higuchi, York University

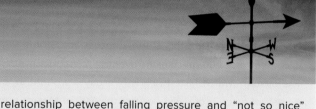

As we watch and listen to weather reports and forecasts, we become aware of the fact that inclement weather is usually associated with falling atmospheric pressure. In fact, people have recognized the relationship between falling atmospheric pressure and an approaching storm since the invention of the barometer, over 300 years ago. The observed relationship between falling pressure and "not so nice" weather is a physical property characteristic of a cyclone (sometimes referred to by weather forecasters simply as a *synoptic system*).

As can be seen on any weather map, a mid-latitude cyclone is a low-pressure structure composed of a centre of

(continued)

low pressure (quite often less than 1000 hPa) and a wave-like frontal system. The frontal system separates the cold air from the warm air, with warm air sandwiched in the upside-down V–shaped region between the warm and cold fronts. Cyclones form as the atmosphere transports energy (mainly in the form of heat) from the low-latitude subtropical region to the high-latitude polar regions, so that warm air pushes poleward while cold air pushes equatorward.

Associated with the formation of a cyclone is an upward transport of warm air over cold air along the frontal system. The upward movement of warm air along the warm front is relatively gentle, and it often results in the formation of various stratus-type clouds and associated precipitation. In contrast, the upward movement of warm air along the cold front is much more rapid and forceful, and it generally results in the formation of convective-type clouds, such as cumulonimbus, which can produce a large amount of rain—as well as strong wind—in a very short period of time. Severe weather events, such as strong wind gusts, tornadoes, excessive rainfall (or snowfall), and extreme freezing rain are usually associated with cyclonic storms. Here are a couple of examples.

In the winter season, with a high-pressure system sitting over northern Quebec, a synoptic cyclonic system approaching southern Ontario and Quebec can bring warm air and moisture from the southern United States and the Gulf of Mexico. With the intrusion of warm air overriding the surface cold air advected from the Quebec high in front of the warm front, the atmosphere can produce a meteorological condition ahead of the warm front that can cause freezing rain to occur. The severity of a freezing-rain event depends on the amount of moisture transported in by the cyclonic system and how quickly the system moves.

A particularly severe example of such an event occurred in early January 1998 over southern Ontario, Quebec, and the Maritimes; in fact, this storm, dubbed "Ice Storm '98," was the worst natural-disaster event in Canadian history. The system, which lasted nearly five days, brought in an excessive amount of moisture from the Gulf of Mexico and the western tropical Atlantic Ocean. During our research to understand the storm, we found that a variation in the regional atmospheric circulation pattern (called the North Atlantic Oscillation, or NAO) over the North Atlantic Ocean blocked the usual west-to-east flow of air across Canada and the northern United States, allowing the northward transport of excessive moisture along eastern North America, from the Gulf of Mexico. The North Atlantic Oscillation is a see-saw oscillation of atmospheric mass between the subtropical and subarctic North Atlantic Ocean. Meteorologists look for changes in these kinds of hemispheric circulation patterns on their weather maps to predict the direction and speed of synoptic systems.

In the summer season, lines of severe thunderstorms, called squall lines, can form along or in front of an advancing cold front. Tornado occurrences associated with squall lines have been observed in Canada, particularly in the prairies and in southern Ontario and Quebec. Meteorologists can detect occurrences of tornadoes by looking for hook-like echoes on radar displays. The echoes are a reflection of rain and hail circling around the mesocyclonic flow inside a supercell thunderstorm. One of the more memorable extreme squall-line events, which involved several F2 tornadoes, occurred in August 2009, just north of Toronto, Ontario.

In our research, we have used a mathematical model based on the physics of meteorological dynamics to understand and predict the processes by which extreme winds and precipitation are produced. Although the model, which has a spatial resolution of 2.5 km, is not able to resolve the occurrences of individual tornadoes, it is able to achieve realistic simulations of events such as the August 2009 squall line. Numerical models constitute an essential tool in predicting which mid-latitude cyclones will result in the production of extreme weather.

KAZ HIGUCHI is an adjunct professor in the Faculty of Environmental Studies and the Graduate Program in Geography at York University and an executive member of IRIS (York's Institute for Research and Innovation in Sustainability). His current work involves the development of socio-economic adaptation strategies for extreme weather and climate events, and complex systems research to investigate human interactions with natural systems within the context of climate change. He has also been a weather forecaster at the Arctic Weather Centre, Environment Canada, in Edmonton.

14.5 | Chapter Summary

1. Mid-latitude cyclones are deep, cold-cored, baroclinic systems that tilt with height. Tropical cyclones are shallow, warm-cored, barotropic systems that are vertically stacked. Both produce clouds and precipitation due to rising air.

2. The polar front theory describes the stages in the life cycle of a mid-latitude cyclone. The storm is born when a disturbance forms along a stationary front. This disturbance leads to the development of cyclonic circulation, and separate warm and cold fronts. In the mature stage, the fronts form an open wave pattern, and there is a sector of warm air between the fronts. In the final stages, the warm-sector air is separated from the low-pressure centre by an occluded front. The storm dies once temperature contrasts are eliminated.

3. Mid-latitude cyclones form below the downstream portion of an upper-air trough because this tends to be an area of divergence. As long as this upper-air divergence is greater than surface convergence, mass is removed from the air column and the cyclone will intensify. Anticyclones form below the upstream portion of an upper-air trough because this is a region of convergence.

4. Cyclogenesis usually occurs due to large temperature gradients, or baroclinic instability; such conditions are concentrated along fronts in the mid-latitudes. Mid-latitude cyclones are characterized by decreasing pressure, rising air, temperature advection, and cyclonic spin. Pressure decreases, and air rises, as a result of divergence in the jet stream. We can identify areas of divergence and rising air on upper-air charts by finding areas where vorticity is decreasing. Temperature advection occurs at fronts and enhances vertical motions. Temperature advection can also strengthen the upper-air wave pattern. This strengthening creates a feedback because the greater waviness of the upper airflow leads to an increase in the upper-air divergence, and the increase in upper-air divergence, in turn, strengthens the cyclone. Cyclonic spin is caused by increasing vorticity as air converges into the low-pressure centre at the surface. Mid-latitude cyclones dissipate when they become separated from the region of upper-level divergence and the temperature contrasts are wiped out. Cyclogenesis can also occur on the leeward side of a mountain range.

5. Hurricanes are very severe storms that form between 5° and 20° latitude over oceans warmer than 26°C. They tend to form in the late summer and early fall. Because they produce very strong winds, torrential rain, and storm surges, hurricanes are the most devastating of storms. A distinctive feature of hurricanes is their central eye, where winds are calm, clouds are fair-weather cumulus, and temperatures are high. These storms are warm-cored low-pressure systems characterized by surface convergence and upper-air divergence. Hurricanes can form from tropical disturbances when all of the following conditions are met: there is a source of latent heat, the Coriolis force is large enough that convergence produces rotation, the atmosphere is unstable, there is minimal vertical wind shear, and humidity is high throughout the troposphere.

6. Thunderstorms are convective storms that produce thunder and lightning; heavy rain and/or hail; strong, gusty winds; and, occasionally, tornadoes. They form when air near the surface is warm and moist, the atmosphere is conditionally unstable, and some mechanism works to lift the air. The life cycle of single-cell thunderstorms begins with the cumulus stage. In this stage, updrafts are dominant. The mature stage begins when downdrafts develop. The combination of updrafts and downdrafts makes the air very turbulent,

so the most intense weather occurs during this stage. In the dissipating stage, the storm is dominated by downdrafts that weaken the storm. Multicell and supercell thunderstorms can become severe because they are able to sustain themselves. Multicell storms sustain themselves as old cells create new ones, while very strong rotating updrafts sustain supercell storms.

7. Atmospheric soundings can be used to assess thunderstorm potential. Warm, moist air at the surface provides the fuel for the storm. An inversion creates a cap, allowing the warmth and moisture to accumulate. Warm, moist surface air combined with cold air aloft produces unstable conditions. If a trigger allows the warm, moist air to move past the inversion, a thunderstorm will form. Convective inhibition (CIN) is a measure of the strength of the cap, while convective available potential energy (CAPE) is a measure of thunderstorm potential.

Review Questions

1. How is a mid-latitude cyclone different from a tropical cyclone?

2. What are the stages in the life cycle of a mid-latitude cyclone?

3. How would you characterize the tilt with height of mid-latitude cyclones and anticyclones? What would happen if these systems did not tilt?

4. Consider a mid-latitude cyclone. What causes the pressure to drop? What causes the air to rise? Why does temperature advection occur? What causes the air to spin?

5. What causes convergence and divergence in the jet stream? Under what conditions can this convergence and divergence strengthen?

6. What is the relationship between vertical motions and vorticity? How do meteorologists make use of this relationship to find areas of vertical motion?

7. What role does thermal advection play in the development and strengthening of mid-latitude cyclones?

8. Why can cyclogenesis occur on the lee side of mountain ranges?

9. What causes the eventual dissipation of mid-latitude cyclones?

10. Considering that they are vertically stacked, how are hurricanes sustained?

11. Consider the five requirements necessary for a hurricane to form. Why is each one important?

12. How do hurricanes eventually weaken and die?

13. How and why are the paths generally followed by mid-latitude cyclones different from the paths followed by hurricanes?

14. What are the three stages of a single-cell thunderstorm?

15. What are the differences between single-cell thunderstorms, multicell thunderstorms, and supercell thunderstorms? Why are multicell and supercell thunderstorms able to become severe?

16. What are CIN and CAPE? What do they indicate about the conditions that favour the formation of thunderstorms?

17. In general, how do storms sustain themselves?

Suggestions for Research

1. Find out, in detail, how tornadoes form, and describe the various methods that have been used to detect them. List factors that make accurate prediction of the paths, intensity, and life span of tornadoes difficult.

2. Investigate methods for suppressing hurricanes, and explore the degree to which these methods are successful. Comment on how climate change might influence the frequency and intensity of hurricanes.

References

Bjerknes, J. (February 1919). On the structure of moving cyclones. *Monthly Weather Review, 47*(2), 95–99.

Schultz, D.M., & G. Vaughan. (April 2011). Occluded fronts and the occlusion process: A fresh look at conventional wisdom. *Bulletin of the American Meteorological Society, 92*(4), 443–66.

We rely on a wide array of technologies and tools to help us forecast the weather. The Doppler on Wheels—shown here scanning a supercell thunderstorm—is one such tool that researchers can use to track, analyze, and predict the behaviour of severe weather systems.

15

Weather Forecasting

Learning Goals

After studying this chapter, you should be able to

1. *distinguish* between various methods of making a weather forecast;

2. *appreciate* how advances in both science and technology have been important in weather forecasting;

3. *explain* why weather observations are the first step in making a weather forecast;

4. *describe* how satellites and radar are used in weather forecasting;

5. *list* ways in which weather maps are used in the analysis and diagnosis steps of weather forecasting;

6. *use* weather maps to make simple forecasts;

7. *describe* the process of numerical weather prediction;

8. *appreciate* why it is likely impossible to produce perfectly accurate weather forecasts; and

9. *explain* how long-range forecasts are made.

Weather forecasting is the art and science of predicting the weather. It is an *art* because it is a partially subjective process in which weather forecasters make judgements, based on their knowledge and experience, about how the atmosphere will evolve from its present state. It is a *science* because predicting how the atmosphere will change also requires an understanding of the physics that govern the behaviour of the atmosphere, and the mathematical equations that describe these processes.

CHANCE OF PRECIPITATION
The chance that measurable precipitation will fall on any random point of the forecast region during the forecast period. *Measurable precipitation is defined as 0.2 mm of rain or 0.2 cm of snow.*

Canadian weather forecasts are issued by the Meteorological Service of Canada (MSC). These forecasts can be found on Environment Canada's Weather Office website (www.weatheroffice.gc.ca). The forecast period is divided into two parts: regular and extended. The *regular* forecast provides detailed coverage of the first 48 hours, while the *extended* forecast covers days three to seven in brief. Forecasts are presented in an icon format and a more detailed text format (Figure 15.1). The *icon* forecast normally indicates the expected sky conditions, the maximum and minimum temperatures, and the type of precipitation. If the **chance of precipitation** (COP) is between 30 and 70 per cent, it is also included in the icon forecast. For

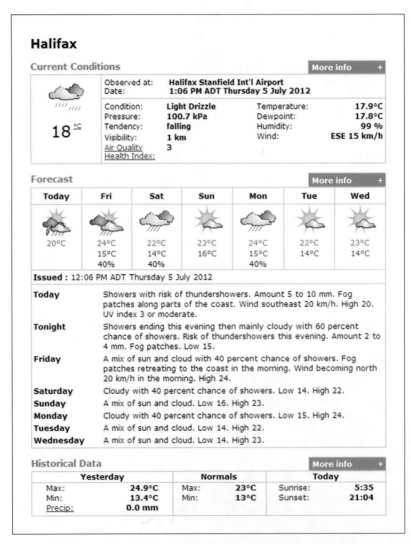

FIGURE 15.1 | A weather forecast for Halifax, produced by the Meteorological Service of Canada.

Source: Environment Canada, Weather Office, www.weatheroffice.gc.ca, 5 July 2012.

the regular forecast period, the *text* forecast provides a predicted precipitation amount, rough information about the time that the predicted events might occur, and the humidex or wind chill, if applicable. This level of detail is not included in the extended part of the text forecast. If strong winds are expected during the forecast period, they will be noted in both the regular and the extended text forecasts.

As the name Meteorological *Service* of Canada implies, weather forecasting provides a service. In most countries of the world, this service has come to be not only expected, but essential. Consider that weather influences nearly every aspect of our lives, from our safety and the availability of our food and water, to our recreational activities. Marine forecasts are issued regularly to provide boaters with information about the location and severity of storms. Before a flight, pilots are briefed on the weather so that they can avoid hazardous conditions such as those associated with thunderstorms. For farmers, accurate weather forecasting might help prevent crop damage due to frost, very heavy rain, or hail. Weather forecasts help the operators of ski resorts plan the best times for snow-making (Section 10.6). Accurate forecasts allow utility companies to plan for unusual power demands that might result from severe cold in the winter or excessive heat in the summer. The MSC also issues *weather warnings* for any weather events that could result in dangerous conditions that might pose a risk to the public. These events include high heat coupled with humidity, heavy rain, freezing rain, blizzards, strong winds, thunderstorms, and hurricanes.

Although forecasters do their best to provide consistently accurate forecasts, forecasts do go wrong. In fact, we all love to joke about how unreliable forecasts can be. Yet, instead of criticizing forecasters for being "wrong" *some* of the time, we should appreciate that they are "right" as often as they are, especially when we consider the complexity of the processes that create and influence the weather (Figure 15.2).

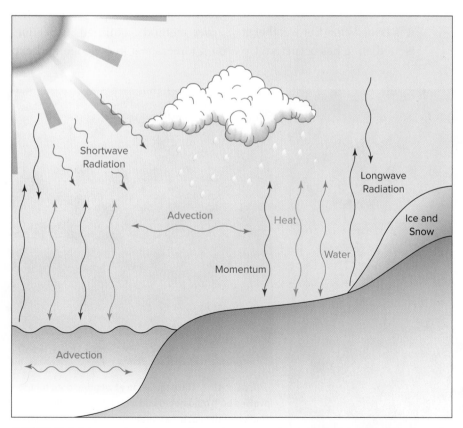

FIGURE 15.2 | Interactions between flows of energy, moisture, and momentum make the processes influencing weather complex and the task of weather forecasting difficult.

15.1 | The Forecasting Process

Although the practice of forecasting is complex, the idea behind it is simple: forecasting is the process of projecting current weather conditions into the future. There are four steps in this process, the first three of which are mostly based on acquiring information about, and understanding, present conditions. All forecasts, therefore, begin with *observation* (Figure 15.3). This step provides a picture of the current state of the atmosphere and, thus, the initial conditions for the forecast. Weather observations include surface observations over both land and water, upper-air observations, satellite imagery, and **radar** data (Section 3.7).

During the second step of the forecasting process, forecasters *analyze* the observations. This step involves compiling and processing huge amounts of data, much of which is used to produce weather maps. This enormous task must be done very quickly—after all, a forecast isn't much good if it is issued *after* the weather it is predicting has occurred! Up until the 1950s, researchers drew weather maps by hand (Figure 15.4); since then, computers have greatly speeded up the process (Figure 15.5). Today, a variety of weather maps are produced. Some portray variations in the *observed* values of pressure, temperature, moisture, and winds, at the surface and at various levels of the atmosphere (Section 3.7), while others show *computer-derived* values such as thickness, vorticity, and vertical motion.

As forecasters produce and study weather maps, they develop a good understanding of the processes operating in the atmosphere and how these processes are changing. Through analysis, therefore, a forecaster is piecing together a *diagnosis*. Diagnosis, although not entirely separable from analysis, is considered to be the third step in a forecast. While preparing the diagnosis, forecasters also likely begin the fourth stage of the process, which is *prognosis*.

Prognosis involves determining how the present state of the atmosphere, determined through the analysis and diagnosis steps, will change with time. Traditionally, forecasters have relied on a few *subjective* methods, outlined in Section 15.2, that draw on a forecaster's knowledge and experience. Today's

RADAR
A system that transmits radio waves and receives their reflection.

PROGNOSIS
A weather forecast.

FIGURE 15.3 | The acquisition of weather data for use in forecasting often involves both automated and manual observations. *Left:* An automated weather station at a meteorological facility in Britain; researchers can collect data from this station remotely. *Right:* One of several Environment Canada meteorological experts on site at the 2010 Vancouver Olympic Games collecting data at a weather station.

15.2 | Forecasting Methods

The simplest traditional method for forecasting the weather is the *persistence* method. The idea behind forecasting by persistence is simply that what is happening now will continue into the future. We all make persistence forecasts. If a steady rain is falling in the morning, we might decide that it will rain all day and cancel outdoor activities. On the other hand, on sunny mornings, we usually assume that the skies will remain clear, and we leave our umbrellas behind. Persistence forecasts, by their very nature, are good only until something changes, and a forecaster must attempt to anticipate when that change is likely to occur. In the mid-latitudes, for example, we know that a steady winter rain will usually continue for at least a few *hours*, while a clear, dry spell in the summer will likely persist for several *days*.

A second method is called *trend* forecasting. This method is based on the assumption that changes that are occurring will likely continue in the same way. Home barometers provide simple trend forecasts; when the pressure is steadily increasing, the needle moves toward "Fair." The trend method is based on the observation that weather systems often move in the same direction and at the same speed as they have been moving. This trend normally holds as long as the upper-air long-wave pattern doesn't change. Using the trend method, forecasters can make simple

FIGURE 15.4 | A hand-drawn weather map from 1899.

forecasts are more *objective* in nature, as they are largely based on the output of computer models that *calculate* the future state of the atmosphere. However, forecasters continue to use the traditional, subjective methods to make short-term predictions or evaluate computer-generated forecasts.

Remember This

There are four steps in a weather forecast:

- observation, which involves acquiring vast quantities of data that describe the current state of the atmosphere;
- analysis, which involves compiling and mapping the data that describe the current state of the atmosphere;
- diagnosis, which involves developing an understanding of the processes at work in the atmosphere and how they are changing; and
- prognosis, which is the generation of a forecast.

FIGURE 15.5 | Computers feature prominently in modern weather-forecasting offices, as they can help forecasters collect and analyze a vast amount of data in a relatively short period of time.

forecasts using a series of weather maps. The maps make it possible to determine the speed and direction of a moving weather system and, thus, to predict when it will arrive at a given location. In the same way, the trend method can also be useful for forecasting the severe weather associated with thunderstorms. However, as in persistence forecasting, forecasters must anticipate *changes*.

A third forecasting method is known as the *analog*, or pattern-recognition, method. (The word *analog* is used because this method involves *comparing* a current weather pattern to weather patterns of the past.) Prior to the introduction of computer forecasts, this was the most commonly used forecasting method. It is based on the assumption that certain weather patterns will repeat themselves. For example, snow is not nearly as common on the west coast of Canada as it is in most of the rest of the country. However, when an upper-air ridge lies over the eastern Pacific and an upper-air trough lies over British Columbia, snow usually arrives on the west coast. This upper-air pattern puts cold Arctic air over the province, while the jet stream travels southward along the coast (Figure 15.6). If a low-pressure system develops in association with the jet stream, precipitation will fall

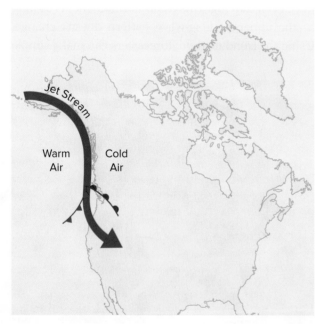

FIGURE 15.6 | **The conditions that commonly bring snow to Canada's west coast.**

through cold air and reach the ground as snow. When forecasters observe this pattern on a weather map, they will usually issue a forecast of snow.

A fourth method involves using *"rules of thumb."* This method is also called *empirical* forecasting because the "rules" it relies on have been developed based on observations. For example, forecasters have observed that winter precipitation will usually be snow north of the 5400 m contour line on the 1000 to 500 hPa thickness chart, while precipitation south of this line will almost always be rain. Observations have also shown that this rule works only for locations at or near sea level. Other rules of thumb are described in Section 15.5.

A fifth forecasting method makes use of *climatology*. This is the only method that does not require knowledge of current conditions, as it is based instead on weather statistics compiled over many years. These forecasts are based on probabilities. For example, Winnipeg is one of Canada's sunniest cities, particularly in summer. It follows that a forecast of sunshine on a July day in Winnipeg has a high probability of being accurate.

Using only the methods described above, our ability to make accurate, detailed forecasts that extend beyond 24 hours or so is limited. To improve on accuracy, provide more detail, and make longer-term forecasts, we now rely heavily, but not entirely, on computers. This forecasting method, known as *numerical weather prediction* (NWP), involves using numerical models programmed into computers. In comparison to the forecasting methods described above, this method is entirely objective, in that the forecast is *calculated*. The sheer volume of calculations necessary to produce a forecast makes it impossible to apply this method *without* computers. Today we use *supercomputers*, which can perform billions of calculations per second, to make forecasts by the numerical weather prediction method. Even on such high-speed computers, a single run of a forecast model can take upward of an hour to complete.

Since the 1950s, forecasting skill has improved considerably. Consider that today's 72-hour forecasts are as accurate as 36-hour forecasts were in the 1970s (NOAA, 2007). This improvement is largely due to the use of computers and the increase in computing power over the years, but it is also due to our steadily improving knowledge of atmospheric processes. In fact, the history of forecasting provides an interesting story of how advances in science have increased our understanding of the atmosphere, while advances in technology have made it possible to apply this understanding.

Remember This

Forecasts can be made using the following methods:

- persistence,
- trend,
- analog,
- rule of thumb,
- climatology, and
- numerical weather prediction.

15.3 | The History of Weather Forecasting

Several of the early scientific advances that eventually made forecasting possible have already been discussed in this book. Chapter 4 introduced the concept of *latent heat*, which was first discovered in the eighteenth century by Joseph Black, the same scientist who also did some important initial work on *specific heat*. In 1802, John Dalton described the atmosphere as a mixture of gases, including water vapour, that each contribute a partial pressure to the total atmospheric pressure (Section 7.3). The work of Robert Boyle and Jacques Charles that eventually led to the formulation of the *ideal gas law* in 1834 was described in Section 3.2. In addition, the *laws of thermodynamics*, also developed during the nineteenth century, have allowed us to understand the relationship between heat, work, and temperature changes in the atmosphere (Section 4.3). Even the development of *calculus* in the seventeenth century is significant because it is this branch of math that provides the differential equations that are the basis of computer-based forecasting models.

By the early part of the nineteenth century, it had been observed that storms are associated with low atmospheric pressure and that they move. Toward the middle of the nineteenth century, scientists were beginning to recognize the *structure* of these storms. They noted that wind blows around them in a counterclockwise direction, in the northern hemisphere, and that the rotation at the surface is directed slightly inward toward the lowest pressure. It was recognized that this inward motion results in convergence of air and, thus, the uplift that produces clouds and precipitation.

It was also in the early part of the nineteenth century that weather observation networks were being established in some countries. The idea was that if weather systems could be mapped, they could be better understood and possibly even forecast. However, in order to produce weather maps of large areas—and, ultimately, to produce weather forecasts—the rapid communication of weather observations is needed. Such communication was made possible by the telegraph, beginning in the mid-1800s.

The Meteorological Service of Canada was established in 1871. At the time, this organization started a weather office in Toronto and set up a network of observing stations in eastern Canada. Initially, this weather office issued only weather warnings, but it began producing weather forecasts in 1876. During the 1880s, the Canadian weather observation network expanded westward along with the construction of the Canadian Pacific Railway and its telegraph lines. By the end of the nineteenth century, weather forecasts were being provided for most of Canada.

At the global scale, it was recognized that weather forecasting would require international co-operation; for this reason, the International Meteorological Organization (IMO) was founded in 1873. This organization became the United Nation's World Meteorological Organization (WMO) in 1951; today, the WMO is made up of 189 countries.

Early forecasts were usually based on a forecaster's experience and the application of the *trend* or *analog* methods. As you might expect, these forecasts were not usually accurate. Recall that it was the desire to be able to accurately forecast the weather that led to

Bjerknes's detailed study of mid-latitude cyclones and the development of the Norwegian cyclone model in 1919 (Section 14.2.1). Shortly after this time, however, most meteorologists were beginning to realize that they needed to know what was happening in the upper atmosphere in order to produce more reliable forecasts. To this end, they began using unmanned balloons and kites to measure temperature, pressure, and water vapour above the surface. By the 1930s, instrument packages containing radiosondes (Section 3.7) were being launched on a regular basis; the data they provided made it possible for meteorologists to produce upper-air maps. During World War II, pilots first encountered the jet stream, and Carl-Gustaf Rossby discovered the waves of the upper troposphere (Section 12.2.5.2). Since then, we have learned more and more about this upper airflow and how it influences the development and movement of surface weather systems.

By expanding on the work of Bjerknes, mathematician Lewis Fry Richardson attempted to develop a method for *calculating* weather forecasts. His initial idea was simple: he planned to use a set of equations, which would describe the physics of the atmosphere, to project the current state of the atmosphere forward in time, step by step. The *implementation* of his idea, however, was not simple. In fact, at the time, it was impossible. In 1922, Richardson attempted to calculate a forecast using a slide rule. Because of the number of calculations required, it took him six weeks to calculate a forecast for six hours into the future. Even at that, he calculated only how pressure would change and, unfortunately, his result was far from correct. Despite his lack of success, Richardson envisioned that a large number of people might be able to work together to produce timely forecasts. Although his idea wasn't implemented at the time, it led to the development of the method we know as *numerical weather prediction* (sections 15.2 and 15.6).

In the latter half of the 1940s, the first electronic computers were produced; this technological advancement made the practice of numerical weather prediction a real possibility. The first computer-based weather forecast was made in 1950. By the late 1950s, technology had advanced to such a degree that computers were being used to draw weather maps. At the same time, the first reasonable numerical forecasts

were being produced, but these forecasts were for the 500 hPa surface only. In 1962, Canada acquired its first computer for weather forecasting; it was located in the Canadian Meteorological Centre (CMC) in Dorval, just outside Montreal (Figure 15.7). By the mid-1960s, computers had begun to produce fairly reliable *surface* forecasts. Routine use of computers for forecasting followed, in most industrialized countries, in the late 1960s or early 1970s. In 1983, Canada acquired its first supercomputer; the increased capacity of this computer allowed for the implementation of many modelling improvements. For the most part, as our understanding of atmospheric processes improved and our models became more complex, computer power kept pace and we made full use of it. In general, computer models now produce forecasts that are much more accurate and longer-term than those produced by earlier models.

Along with computers, radar and satellites have proved to be invaluable tools of weather forecasting. During World War II, it was discovered that certain wavelengths of radiation transmitted by radar systems are strongly reflected by raindrops. Since then, radar has provided vital information about everything from light rain showers to severe storms (Figure 15.8). In the 1960s, the first satellites were launched. Satellites provide a complete global view of atmospheric conditions such as cloud cover, temperature, and atmospheric moisture. Satellite imagery helps

FIGURE 15.7 | The Canadian Meteorological Centre in Dorval, Quebec, is a hub for Environment Canada's network of weather and climate forecasting offices and home to Environment Canada's supercomputer.

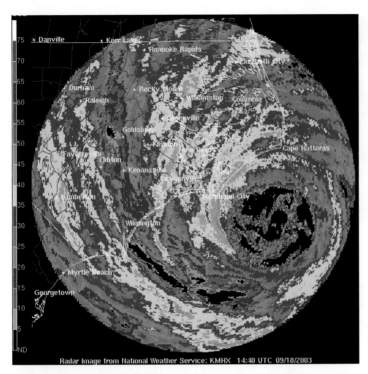

Radar Image from National Weather Service: KMHX 14:40 UTC 09/18/2003

FIGURE 15.8 | A Doppler radar image of hurricane Isabel on the east coast of the United States in September 2003. Green, yellow, and orange correspond with areas of heavy rainfall; areas with the lightest rainfall are indicated by blue and black. This image clearly shows the banded structure of a hurricane, as described in Section 14.3.1.

fill in the gaps in surface observations, and it is particularly valuable for viewing weather systems in areas where observations are sparse, such as over the oceans and in polar regions.

15.4 | Observations

As noted above, all weather forecasts begin with observation; this step provides the initial conditions for the forecast. To get a complete picture of atmospheric conditions, observations are needed from all over the world. To this end, weather observations are coordinated by the WMO. All observations are made at the same time, based on Coordinated Universal Time (UTC, Section 3.7). In addition, the instruments used for the observations are standardized, and the conditions under which the observations are taken must be the same everywhere. The WMO also ensures that both this observational data and the output of forecasting models are shared among countries. Without international co-operation, reliable weather forecasting would not be possible.

In Section 3.7, we saw that surface weather observations are made over land and at sea. In addition, radiosondes provide information about the upper air. These upper-air observations are supplemented by observations made by equipment on commercial aircraft. Because aircraft tend to favour the tailwinds often provided by the jet stream, these important upper-air winds are well sampled. Observational data are also provided by satellites and radar.

15.4.1 | Satellites

There are two types of orbits used for weather satellites: *geostationary* and *polar orbiting* (Figure 15.9). Geostationary satellites are set in orbit at a height of about 36,000 km above the equator. These satellites orbit with Earth and are thus always above the same spot. Polar-orbiting satellites, on the other hand, orbit at altitudes of only about 800 km. In addition, they observe different locations because they orbit Earth along longitudinal lines, travelling over the poles, and scanning areas farther west with each successive orbit. As a result, polar-orbiting satellites pass over any location twice a day. Both types of orbit are needed because geostationary satellites do not provide a good view of polar regions.

Most satellite imagery for use in weather forecasting is of three types: visible, infrared, and water vapour (Figure 15.10). Visible imagery depicts visible light from the sun that has been reflected by Earth's surface (Section 5.1). Thus, visible imagery is most like a photograph. Surfaces with high albedos—like clouds, snow, and ice—will appear bright on visible imagery. Further, thick clouds appear brighter than thin clouds because they are more reflective. Cumulonimbus clouds generally have the highest reflectances because they tend to be the thickest clouds, while cirrus clouds generally appear less bright than other clouds because they tend to be thin. The darkest areas on visible imagery are water and forests. An advantage of visible imagery is that it is easy to interpret because it is similar to what we see, but a disadvantage is, of course, that it is available only during the day.

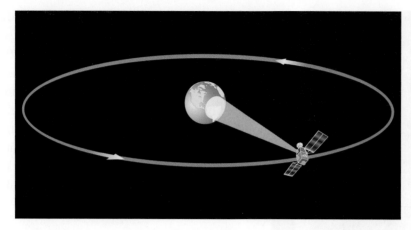
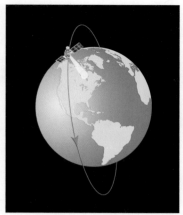

FIGURE 15.9 | *Left:* A satellite in geostationary orbit. *Right:* A polar-orbiting satellite.

Infrared imagery usually depicts radiation at wavelengths between 10 and 11 μm. Radiation in these wavelengths falls in the range of those emitted by Earth and by clouds (Section 5.5). The specific wavelengths between 10 and 11 μm are used because they fall in the atmospheric window and can thus travel through the atmosphere to the satellite. Recall that the amount of radiation emitted depends on temperature, with warm objects emitting more radiation than cold objects emit (Section 5.2.1). On infrared

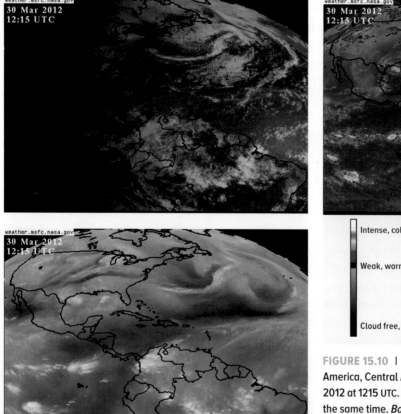

Intense, cold cloud tops, high altitude

Weak, warm cloud tops, low altitude

Cloud free, very warm surface temperature

FIGURE 15.10 | *Top Left:* Visible satellite imagery for eastern North America, Central America, and northern South America on 30 March 2012 at 1215 UTC. *Above:* Infrared satellite imagery produced for the same time. *Bottom Left:* Water vapour imagery produced for the same time. Notice that the western portion of the visible image is dark, while this same portion of the other two images is not.

imagery, the brightness scale is reversed, so that cold objects appear bright, and warm objects appear dark. Because the tops of low clouds are much warmer than the tops of high clouds, this imagery allows us to distinguish between high and low clouds. Yet it is sometimes difficult to see thin, high clouds such as cirrus clouds because they tend to disappear in contrast to warmer clouds below.

Together, visible and infrared imagery allow forecasters to determine the location, type, shape, and height of clouds. Based on these characteristics, forecasters can identify certain weather processes. For example, they can identify a mid-latitude cyclone by the characteristic comma-shaped cloud that it creates. In addition, they can locate the jet stream by finding the line of cirrus clouds that is characteristic of its northern edge (Figure 15.11, Section 9.3). Forecasters can also use satellite imagery to locate fronts, which often appear as the distinct edge of a mass of cloud, or even as a distinct change in cloud type. Forecasters can even use successive satellite images to monitor the *movement* of storms and, thus, determine their speed and direction.

FIGURE 15.11 | Jet stream cirrus over the Atlantic, photographed from the unmanned Apollo 6 spacecraft in 1968.

Question for Thought

Why would cumulonimbus clouds appear bright on both visible and infrared imagery?

To produce water vapour imagery, satellites sense radiation with a wavelength of 6.7 μm, which is strongly absorbed by water vapour, particularly in the middle to upper troposphere. Where the upper half of the troposphere is moist, the water vapour absorbs this radiation; where it is dry, the radiation is not absorbed, so it continues through the troposphere and can be measured by the satellite. On the resulting image, moist areas appear bright because, as on infrared imagery, the brightness scale is reversed. Radiosonde data are also useful for finding areas of the atmosphere where there is water vapour, but water vapour imagery helps to fill in the gaps and provide a more complete picture of the distribution of water vapour.

15.4.2 | Radar

Radar systems transmit and receive pulses of radio waves. If the pulses of transmitted energy are reflected by objects in the surroundings, they will be received back at the radar unit. At its simplest, radar can determine the distance to objects by measuring the speed at which the reflected pulse is returned. Remember that one reason radar is useful in forecasting is that radio waves are strongly reflected by precipitation. The time taken for the reflected pulse to return indicates the distance to the storm, and the character of the reflected pulse indicates the type and intensity of precipitation.

Doppler radar has been widely used in meteorology since the 1990s. Because Doppler radar can detect motion, it is particularly useful in forecasting thunderstorms (Figure 15.12). Doppler radar can not only find areas of heavy rain and hail in the storm, it can also detect the rotation patterns that lead to the formation of tornadoes. Thus, the use of Doppler radar has increased the lead time for tornado warnings.

FIGURE 15.12 | Doppler radar—such as this one in New Underwood, South Dakota—can determine how distant the supercell is as well as the type and intensity of precipitation being produced—in this case, hail. It may also be able to determine whether or not a tornado is likely to form.

In the Field

Probing Storms to Reveal Their Inner Secrets

John Hanesiak, University of Manitoba

Atmospheric scientists, such as I, who study storms can do so using a variety of technologies or "tools of the trade," including computer models, satellite products, ground-based and airborne remote sensing, and direct in-situ measurements. The real atmosphere is our lab, and Mother Nature tells us whether we got it right or not! There are many aspects of storms and severe weather that continue to elude our complete understanding, partially due to our inability to observe the smaller-scale controlling processes and phenomena. Technologies that map out temperature,

humidity, and wind variations over short time scales (minutes) are needed to help answer questions such as "Why do thunderstorms first form where and when they do?" or "Why do some storms produce tornadoes while others do not?"

A large network of scientists, each with his or her own expertise, have to come together to try and answer these kinds of complex questions. A common approach is for the scientists to conduct large-scale field experiments that utilize the "tools of the trade." These kinds of research projects are very exciting, not only for the scientists, but also for the students involved in them. Students get the chance to experience, first hand, how technology can be used to study the atmosphere. Some typical tools include weather balloons and ground meteorological instrumentation mounted on tripods. However, there are always newer technologies available to us that can enhance or even completely change the way we study and monitor the atmosphere. That is the focus here—highlighting some of the newer technologies and tools that are used to better understand storms and severe weather.

At the ground, we can measure temperature, humidity, wind speed and direction, pressure, and location (via GPS) using an automated mobile meteorological observation system (AMMOS), which is mounted to the roof of a vehicle. The concept of the AMMOS dates back to the mid-1990s, but the newer AMMOS has the advantage of being able to provide accurate locations of real-time meteorological measurements every one to two seconds while the vehicle is moving. This data can be used to study, for example, the spatial characteristics of fronts and dry lines, where thunderstorms first initiate, as well as thunderstorm downdraft characteristics that are believed to be critical for tornadogenesis in many cases.

Although not all severe storms owe their initiation and perpetuation to planetary boundary layer (PBL) processes, many of them do. This makes meteorological PBL observations of temperature, humidity, and wind critical. The trouble is that no single instrument provides all three quantities. Thus, we must use multiple instruments to make observations in order to gain a full picture of what is happening in the atmosphere. Some examples of newer measurement technologies include atmospheric emitted radiance interferometers (AERIs), profiling passive microwave radiometers (PMRs), and lidars. The primary advantage of these systems is that they can take samples every one to ten *minutes*; by contrast, traditional weather balloons take samples every couple of *hours*. PMRs and lidars observe the atmosphere in such high temporal resolution that they allow us to see small-scale phenomena that have never before been directly observed at these scales. Such observations can provide many exciting new insights into severe weather processes. The main drawback to all of these systems is their inability to probe into storms due to cloud and/or heavy precipitation; hence, they are generally used to assess the pre-storm environment or conditions ahead of or around the main storm system.

To observe conditions *within* storms, we use radar systems. Ground-based weather radar systems have been in existence for many years, providing the location, intensity, and local wind characteristics of storms (via Doppler radar). These radars have been instrumental in helping us to better understand internal storm structures, dynamics, and other processes. All of the US and Canadian operational radars will soon have polarization capabilities, which will help us discern precipitation types and even identify when powerful tornadoes touch down, via detection of debris. Polarization data may also provide new insights into microphysical processes within storms and how these processes relate to updraft and downdraft intensity. Radars have also been deployed on mobile trucks, called Doppler on Wheels (DOWs); the information gathered from these devices has transformed tornadogenesis theories. Newer rapid-scan DOWs are capable of scanning storms much more quickly in order to help us better understand tornadic storm physics. New radar technology such as phased-array systems (fixed and mobile) are now providing better spatial resolution and quicker scanning capabilities for thunderstorm initiation prediction and severe weather applications. Some new and existing radars are also capable of taking advantage of refractivity measurements in order to spatially "map out" water vapour variations in the PBL that can be critical for convection initiation research and prediction. New radar systems onboard aircraft and satellites are also very valuable for studying storms over broader domains; however, they offer only "snap shots" of the atmosphere at any particular time. Examples of newer satellite radars include CloudSat (www.nasa.gov/mission_pages/cloudsat/) and GPM (Global Precipitation Measurement mission; http://pmm.nasa.gov/GPM).

As mentioned, a team of scientists is needed to address cross-cutting scientific atmospheric problems. The same goes for instrumentation. That is, we typically require a suite of different technologies and tools to tackle big issues. Large field projects use a combination of instruments and tools to better understand the inner workings of severe storms and the severe weather they produce. The need for such elaborate studies will continue into the future.

JOHN HANESIAK is a professor in the Department of Environment and Geography and the Centre for Earth Observation Science at the University of Manitoba. He has been a weather forecaster with Environment Canada. His research focuses on storms and severe weather, convection initiation processes, and surface–atmosphere interactions.

15.5 | Analysis and Diagnosis

Once the observations have been made, the next step in a forecast is the analysis and diagnosis of the data that have been collected. A major part of this step is the production and study of weather maps. Up until the 1950s, maps were plotted by hand; since then, maps have been plotted using computers. These maps are known as *synoptic weather maps* because they give us a visual synopsis of the weather conditions that are occurring at a moment in time (Section 3.7).

We can gain the most from weather maps when we look at surface maps in combination with upper-air maps. Recall, after all, that weather systems are three dimensional. For example, divergence in the upper troposphere leads to dropping surface pressure, while thermal advection at the surface can influence the waves of the upper atmosphere (Section 14.2.2). Thus, it is by looking at maps of processes occurring at different levels that we can best anticipate how systems are changing—what direction they are travelling in, how fast they are going, and, most importantly, whether they are gaining or losing strength.

15.5.1 | Surface Maps

Surface maps are normally drawn for 0000, 0600, 1200, and 1800 UTC each day. The computer plots the surface observations as *weather station symbols* (Figure 15.13, Appendix), draws isobars to show pressure patterns, and labels areas of high and low pressure according to the isobar pattern. Forecasters then draw fronts, based again on the isobar pattern, but also using information provided by the weather station symbols (Section 13.5).

The placement of fronts is more subjective than the drawing of isobars. By convention, forecasters usually place fronts on the warm side of large temperature gradients. Because air rises at fronts, they are locations of low pressure; thus, kinks in the isobars often indicate the presence of a front. Forecasters often draw fronts onto maps according to the Norwegian cyclone model (Section 14.2.1). A trough of low pressure extending to the south and west of a low usually indicates a cold front, while a trough of low pressure extending east from the low indicates a warm front. Other indicators of fronts are clouds, wind shifts, and changes in

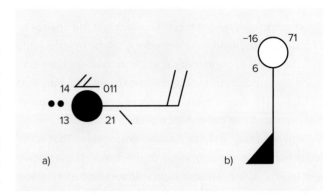

FIGURE 15.13 | a) A surface weather station symbol indicating that the skies are overcast with nimbostratus clouds and light rain. The temperature is 14°C, and the dew-point temperature is 13°C. The winds are from the east at 37 to 45 km/h. The pressure is 1001.1 hPa and has dropped steadily by 2.1 hPa since the last observation. **b)** An upper-air weather station symbol, for a 500 hPa surface, indicating that the temperature is -16°C, and the dew-point depression is 6°C. The winds are from the south at 92 to 100 km/h. The height of the pressure surface is 571 dam. (See the Appendix for a key to these symbols.)

dew-point temperature. As mentioned above, satellite imagery is also useful for finding fronts. In addition, forecasters can use thickness charts to find areas of thermal advection, which often indicate the presence of a front (Section 15.5.2).

On its own, a surface map is most useful in helping us to determine where high-and low-pressure systems are currently located. In other words, it can tell us where the skies are clear and where they are cloudy. In addition, a *series* of surface maps can help us determine the direction and speed at which pressure systems are moving; we can do so by using either the persistence method or the trend method of forecasting. To determine whether such systems are intensifying or weakening, however, we must consult upper-air charts.

Remember This

Surface weather maps show the following features:

- weather station symbols,
- isobars,
- centres of high and low pressure, and
- fronts.

15.5.2 | Upper-Air Maps

Upper-air maps are also produced based on observational data. Unlike surface maps, these charts are produced only twice daily, for 0000 and 1200 UTC, and they are maps of *pressure* surfaces. In Canada, maps are produced for the 250, 500, 700, and 850 hPa pressure surfaces. Unlike the drawing of surface maps, the drawing of upper-air maps is done completely by computer. The drawing of these maps begins with the plotting of weather station symbols. Based on this information, height contours, and certain secondary fields, are drawn (Table 3.2). In addition to these observed values, several computer-derived values—such as vorticity, thickness, and vertical motion—are also usually mapped. As the following discussion illustrates, each type of upper-air map has its own unique value in analysis and diagnosis.

15.5.2.1 | The 250 hPa Chart

The 250 hPa pressure surface is located near the tropopause, at an average height of 10,000 m. Maps of this surface, such as the one shown in Figure 15.14, show height contours as well as **isotachs**. At this height, there is no strong radiative heating, little evaporation, and little friction. As a result, 250 hPa charts tend to show a wave pattern that is much simpler than the more complex flow patterns that we see on maps representing the surface or lower troposphere. The

waves are, of course, the Rossby waves (Section 12.2.5). These waves are important in forecasting because they trace out the flow of the *polar front jet stream*, and because the wave troughs are large regions of cold air, while the wave ridges are large regions of warm air.

In the last few chapters, you have seen how important the upper-air wave pattern is to weather. Not only do the upper-air winds determine the paths of surface storms, the general *pattern* of these waves determines the intensity and duration of these storms. With *zonal* flow (Figure 12.18), storms are weaker and move more quickly; as a result, weather changes more frequently. With large-amplitude waves, or *meridional* flow, storms are more intense. There are times when upper-air patterns are constantly changing, but there are also times when they are very stable and a particular pattern can remain in place for several days. Since the upper-air wave pattern determines weather, it is important to be able to forecast changes in this pattern.

One way to diagnose change on the 250 hPa chart is to look for *jet streaks* (Section 14.2.2). Jet streaks are very fast upper-air winds created by large temperature contrasts in the lower atmosphere. They appear on the 250 hPa chart as elongated concentric isotachs that lie in the axis of the jet stream (Figure 15.14). Recall that jet streaks are regions where areas of convergence and divergence will develop because the flow is temporarily out of geostrophic balance. Forecasters often use a conceptual model to show where convergence and the associated sinking air, and divergence and the associated rising air, are likely to occur in relation to a jet streak (Figure 15.15). When a jet streak is located behind, or westward, of a trough, it will cause the trough to strengthen because air will rise and cool where divergence is occurring. The cold air will further lower the height of the trough. When a jet streak is located ahead, or eastward, of a trough, the trough will weaken because air will sink and warm where convergence is occurring. The warm air will raise the height of the trough. The same rule holds for ridges.

ISOTACHS
Lines that connect points of equal wind speed.

15.5.2.2 | The 500 hPa Chart

The 500 hPa surface is located roughly midway through the troposphere, at an average height of about

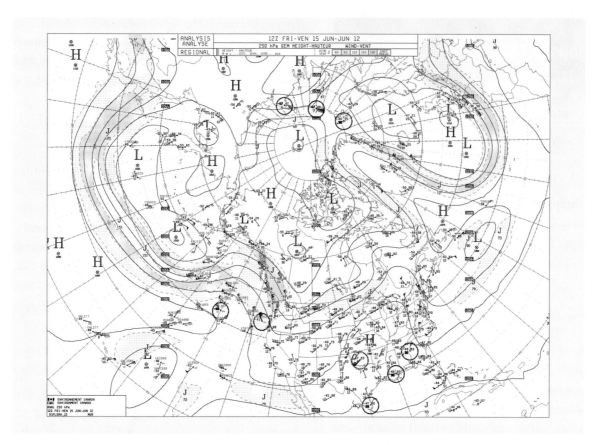

FIGURE 15.14 | A 250 hPa chart. Shading is used to highlight areas where wind speeds are above 60 knots, or about 100 km/h. These areas of very fast winds are jet streaks.

Source: Environment Canada, Weather Office, www.weatheroffice.gc.ca, 15 June 2012.

5500 m. As Figure 15.16 shows, the waves of the upper troposphere are still apparent at this height, but the baroclinic zones, or large temperature contrasts, more characteristic of the lower atmosphere also begin to

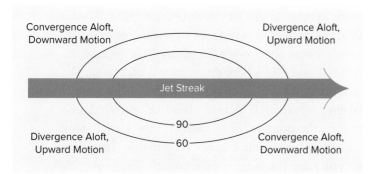

FIGURE 15.15 | A conceptual model for determining convergence and divergence around a jet streak. The curved lines are isotachs labelled in knots. Travelling in the direction of the jet streak, the convergence and divergence are strongest on the left. (See also Figure 14.12.)

Source: Adapted from Vasquez, T. (2011). *Weather analysis and forecasting handbook.* Garland, TX: Weather Graphics Technologies, p. 67.

appear. It is for this reason that, in addition to height contours, this chart includes thickness contours for the 1000 hPa to 500 hPa layer. Remember that thickness values are good indicators of the *average* temperature of a layer, with larger thicknesses indicating warmer temperatures and smaller thicknesses indicating cooler temperatures (Section 3.5).

There are a few forecasting *rules of thumb* that apply to the 500 hPa chart. The first is that surface low-pressure systems tend to follow the direction of winds at this level and move at about half their speed. The second is that for the west coast of North America, precipitation is more likely north of the 5640 m contour line than it is south of this line. The third rule of thumb, mentioned as an example above, is that precipitation will normally be *snow* north of the 5400 m thickness line and *rain* south of this line.

The 500 hPa chart, like the 250 hPa chart, is commonly used to look for changing atmospheric conditions. Because this map includes thickness contours,

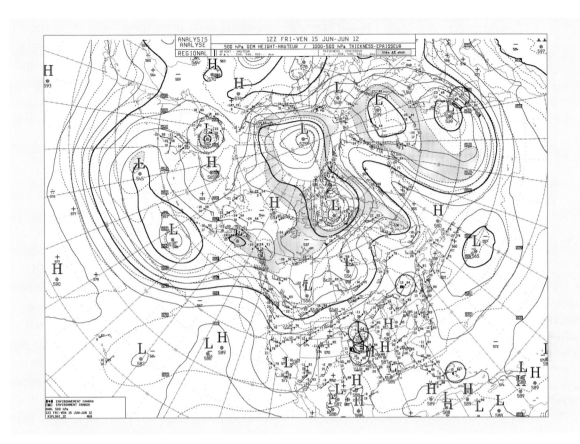

FIGURE 15.16 | A 500 hPa chart. Notice that the area between the 5400 m thickness line and the 5340 m thickness line is shaded. Locations north of this area will likely experience any precipitation as snow, while locations south of it will likely experience any precipitation as rain.

Source: Environment Canada, Weather Office, www.weatheroffice.gc.ca, 15 June 2012.

it can be used to find areas of *thermal advection*, which are associated with *short waves*. Recall that these are areas where thickness contours cross height contours (Section 14.2.2.3). Forecasters often use 500 hPa charts to assess thermal advection using the *box method*, in which they draw boxes where thickness contours cross height contours (Figure 15.17). Since winds parallel height contours, these boxes indicate that winds are crossing the thickness contours and, thus, advecting warm or cold air. Where boxes are small, thermal advection is strong; where boxes are large, thermal advection is weak; where there are no boxes, thermal advection is not occurring. Remember that warm advection leads to rising air and cloudy, wet weather, while cold advection leads to sinking air and clear skies. In addition, thermal advection influences the long-wave pattern. Upper-air troughs will be strengthened by cold-air advection, because the cold air will further lower their heights. On the other

hand, upper-air ridges will be strengthened by warm-air advection, because the warm air will further raise their heights. Thermal advection, therefore, is an indicator of changing conditions.

Another indicator of change on 500 hPa charts is *vorticity advection*. Forecasters have been using vorticity advection to identify areas of vertical motion since the 1960s. For this purpose, the 500 hPa chart must include lines of constant absolute vorticity. Recall that absolute vorticity is a function of the Coriolis parameter, shear, and curvature (Equation 12.2). Vorticity values are calculated, and plotted on charts, by computers. As with thermal advection, vorticity advection can be assessed using the box method. Where height contours cross vorticity contours, vorticity is being advected. Positive vorticity advection is occurring where winds blow from high vorticity to low vorticity. Remember from Section 14.2.2.2 that divergence and, thus, rising air are occurring in areas where

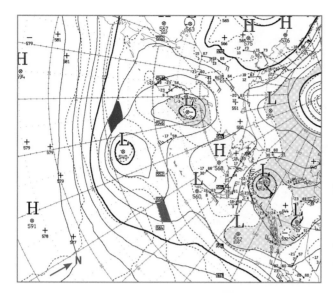

FIGURE 15.17 | Thermal advection is illustrated on this section of a 500 hPa chart by the box method. The blue box indicates that cold advection is occurring, while the red box indicates that warm advection is occurring. (Refer back to Figure 11.23b to remind yourself how this was determined using the thermal wind concept.)

Source: Adapted from Environment Canada, Weather Office, www.weatheroffice .gc.ca/analysis/index_e.html, 13 June 2012.

vorticity decreases. Therefore, forecasters have associated areas of positive vorticity advection with areas of clouds and precipitation.

However, vorticity advection is not always a reliable way to assess vertical motion because other processes also cause vertical motion. For example, if in an area of the atmosphere there is positive vorticity advection in the upper troposphere and cold advection in the lower troposphere, these processes are likely to cancel each other out. This is because positive vorticity advection is associated with rising air, while cold advection is associated with sinking air. Of course, it follows that vorticity and thermal advection could also enhance each other where positive vorticity advection occurs together with warm advection. A way to take *both* vorticity advection and thermal advection into account is to use a rather complicated equation known as the *omega equation*, which was developed in 1972. Using the omega equation, vertical velocities can be calculated by computer and plotted on a chart for analysis.

The 500 hPa chart is the best one for recognizing omega highs (Figure 14.21). (Note that there is no connection between omega highs and the omega

equation, aside from the similarity in their names.) It is important to be able to recognize such a pattern because these highs can lead to droughts and floods, but it is also important to be able to determine when such a pattern might change. This can be done using the thickness contours to find areas of thermal advection, as described above.

15.5.2.3 | The 700 hPa Chart

The 700 hPa surface lies on average at an altitude of about 3000 m above sea level (Figure 15.18). This chart shows isotherms as well as height contours. In addition, shading is used on this chart to provide information about dew-point depression (Section 7.5.2). Lighter shading indicates areas with dew-point depressions of less than or equal to 5°C, and darker shading indicates areas of dew-point depressions of less than or equal to 2°C. Dew-point depression is plotted on this chart because 700 hPa represents the height of the middle of most clouds that are associated with large areas of low pressure. A forecasting rule of thumb is that clouds are likely in the areas where the dew-point depression is 5°C or less, and precipitation is likely in areas where the dew-point depression is 2°C or less.

This chart is also useful for forecasting summer thunderstorms because winds at this height tend to steer air-mass thunderstorms (Section 14.4). In addition, temperatures at this height help define the *cap* that can either cause thunderstorms to form later in the day, after heat and humidity have had a chance build up, or prevent them from forming altogether. In fact, another forecasting rule of thumb is that if temperatures on this chart are above 14°C, there will be no thunderstorms because air will not be able to rise past such a strong cap. Like the 500 hPa chart, the 700 hPa chart is also used to find temperature advection and vorticity patterns.

15.5.2.4 | The 850 hPa Chart

The 850 hPa chart shows conditions at about 1500 m above sea level (Figure 15.19). Except in high-altitude areas, the 850 hPa surface lies just above the planetary boundary layer (Section 11.1.4). Thus, for most areas of Earth, this surface is above the effects of surface friction, and heating and cooling. In addition to height contours, 850 hPa charts show isotherms. Where the

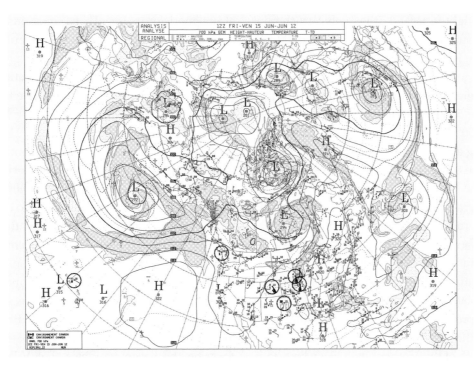

FIGURE 15.18 | A 700 hPa chart. The shaded areas indicate regions where dew-point depression is low and, therefore, clouds and precipitation are likely.

Source: Environment Canada, Weather Office, www.weatheroffice.gc.ca, 15 June 2012.

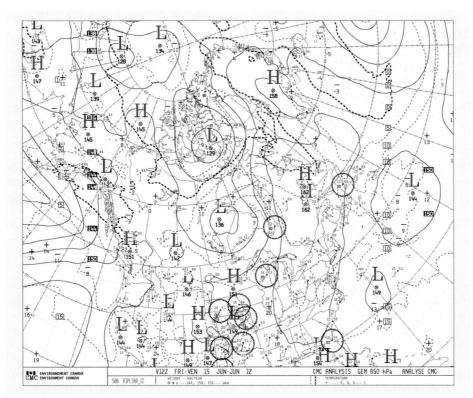

FIGURE 15.19 | An 850 hPa chart.

Source: Environment Canada, Weather Office, www.weatheroffice.gc.ca, 15 June 2012.

isotherms are tightly packed, it is likely that there is a front at the surface. As with the 500 and 700 hPa charts, this one is also used for finding temperature advection. In general, these three charts are used because temperature differences become less pronounced above 500 hPa, while temperature patterns below about 850 hPa are complicated by diurnal patterns of heating and cooling.

Because the 850 hPa surface lies above the effects of daily surface heating and cooling, this chart can be used to forecast maximum temperatures. For this, the forecasting rule of thumb is that in the winter, the maximum surface temperature will be 9°C above the 850 hPa temperature, while in summer, the maximum surface temperature will be 15°C above the 850 hPa temperature. The 850 hPa chart can also be used to determine whether precipitation will be rain or snow. Areas to the north of the –5°C isotherm will likely have snow, while areas south of this isotherm will likely have rain.

15.6 | Numerical Weather Prediction

After carrying out a thorough analysis and diagnosis, a forecaster should have a good understanding of the processes operating in the atmosphere. In most cases, she or he will also have a general idea about what the forecast might be in the short term. It is at this point that a forecaster might seek guidance from computer forecasts generated using the *numerical weather prediction* method (Figure 15.20).

The calculations necessary in numerical weather prediction are done by supercomputers that are programmed with models that use equations to mimic atmospheric processes. These equations are known as the *primitive equations* because they represent the *fundamental* laws governing atmospheric physics. Simpler versions of the first five of these equations have been introduced throughout this book. The first two equations are *equations of motion* (Chapter 11) that represent the horizontal movement of air. These equations are based on Newton's second law, and they account for the pressure gradient force, the Coriolis force, and friction. The third equation is the *hydrostatic equation* (Section 3.4); it is used in the models to determine vertical motions. The fourth equation is the *ideal gas law* (Section 3.3), which links pressure, temperature, and density. The fifth equation is based on the *first law of thermodynamics*, and it is used in the models to determine temperatures (Section 4.3). The sixth equation is for the *conservation of mass*, and the final equation is for the *conservation of moisture*. These seven equations describe the seven variables that Vilhelm Bjerknes recognized as those that control the weather: winds in the north–south direction, winds in the east–west direction, winds in the vertical direction, pressure, temperature, air density, and moisture.

The model equations are in the form of nonlinear partial differential equations. These equations cannot be solved exactly, so computer models use numerical methods to find approximate solutions. Because of their nonlinearity, these equations are very sensitive to initial conditions. This means that slight differences in initial conditions can produce quite different forecasts. It also means that any errors in initial conditions will be amplified over time, which is why forecasts become increasingly inaccurate with time.

Behaviour, like that of the weather these equations describe, that is extremely sensitive to initial

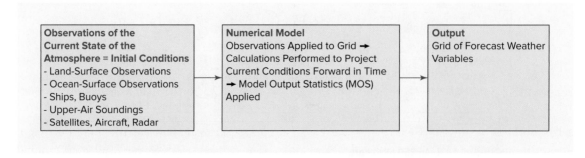

FIGURE 15.20 | The steps involved in numerical weather prediction.

conditions is known as **chaos**. Chaos arises because of the complexity that characterizes the behaviour of fluids, such as the atmosphere. Phenomena that are chaotic are difficult to predict. Imagine trying to predict the future location of a leaf that is floating down a river. It is highly probable that the leaf will continue to move downstream, but there is also a small probability that it will get caught on a rock and stop altogether. As the leaf gets farther from its starting point, its position becomes increasingly harder to predict. It is partly because of chaos that there is a limit to how far into the future weather forecasts can be made. Forecasts out to 24 hours are usually accurate, and forecasts from two to five days are usually fairly accurate, but the accuracy of forecasts beyond seven days rapidly decreases.

We have not always recognized chaos. In fact, Bjerknes and Solberg's successful development of their cyclone model made them quite confident that we would soon be producing reliable weather forecasts (Section 14.2.1). This confidence continued into the 1950s. During those early days of the development of numerical weather prediction models, it was believed that all that stood in the way of reliable forecasts was the need for more powerful computers. In the early 1960s, however, Edward Lorenz discovered chaos in weather. Once his ideas caught on, we came to the realization that the existence of chaos places a limit on our ability to make reliable weather forecasts.

The computer models used in numerical weather prediction represent the atmosphere as a three-dimensional grid, based on latitude, longitude, and altitude. The size of the grid boxes varies considerably from one model to another; it ranges from tens to hundreds of *kilometres* in the horizontal direction, and from tens to hundreds of *metres* in the vertical direction. In general, models with smaller grid boxes have a higher resolution and should produce more accurate results, but the tradeoff is that they will take longer to run. Increasing computer power over the years has made it possible to run increasingly high-resolution models.

Because the observational data that are input to the models will not correspond exactly to the grid points, and because there are vast areas of Earth, such as the oceans and the polar regions, where observations are sparse, the computer must begin a model run by assigning values to all the grid points. It does this by using what might best be described as rather sophisticated interpolation techniques. The "interpolated" values represent the initial conditions, or the starting point, for the model run; they will be projected forward in time to produce the forecast. This process of assigning values to grid points is one of the ways that computer models are forced to *approximate* the real world.

Models are either *global*, covering the entire Earth, or *regional*, applying only to a region of interest—for example, North America. Regional models can have higher resolutions than global models can have because their smaller area of coverage means they require less time to run on a computer. This higher resolution should make regional models more accurate than global models, but regional models quickly decrease in accuracy after about 48 hours. This is because these models cannot consider weather systems outside their domains, and such weather systems will be moving into the domain after about 48 hours. Regional models are, therefore, used for shorter-term forecasts, and global models are used for longer-term forecasts.

Once it has assigned values to the grid points, the computer begins to perform the calculations that will move these values ahead in time by discrete steps, known as *time steps*. The length of each time step varies from one model to another, but it is usually on the order of tens of minutes for global models, and minutes for regional models. As with smaller grid-box sizes, shorter time steps usually produce more accurate results, but again the compromise is that the computer will need more time to process the model. This need is compounded by the fact that smaller grid squares *require* shorter time steps. When the model starts up, the first set of calculations advances the initial values forward in time by one time step. These new values are then used to calculate the values for the next time step, and so on. As the model progresses, it will show weather systems that already exist beginning to dissipate, and it will depict new weather systems beginning to form. It will produce

CHAOS
Behaviour that is extremely sensitive to initial conditions and very difficult to predict.

forecasts at regular intervals, usually every six or twelve hours. This process will continue, time step by time step, until the desired forecasting time is reached. At present, 16 days is about the longest forecasting period of any model.

In Canada, we use a model known as the Global Environmental Multiscale (GEM) model, which was developed by the Meteorological Service of Canada (MSC). This model is run on the supercomputer located in the Canadian Meteorological Centre (CMC) in Dorval (Figure 15.7). The GEM model is run as both a global model and a regional model. At the present time, the resolution of the regional model is 15 km, and that of the global model is 0.3° latitude by 0.45° longitude. The regional models are run four times a day, while the global models are run twice a day. Both versions of the GEM model produce forecasts at 12-hour intervals. For example, a model run that is initialized using data observed at 0000 UTC will produce its first forecast for 1200 UTC. The regional model is run for 48 hours into the future, and the global model is run out to, at most, 16 days.

As described so far, computer models are best at forecasting large-scale weather features, particularly those of the upper air. In addition, the results for a given grid box are *averages* for a relatively large area. In other words, these models do not produce *detailed* forecasts of surface weather conditions that take into account local variations such as topographical features and water bodies. Because forecasts would be of little value without such information, forecasters use a statistical method known as **model output statistics** (MOS) to refine the large-scale model output for local conditions. MOS are created by comparing years of numerical weather prediction results to the actual weather in an area. From this, regression equations are developed to adjust the results of the numerical weather prediction forecast to local conditions. These regression equations, or statistical models, are added

MODEL OUTPUT STATISTICS (MOS) A statistical method that adjusts the output of a weather forecasting model for local conditions.

ENSEMBLE FORECASTING A forecasting technique that involves running forecasting models several times, with slight variations in initial conditions and/or the way the model operates.

to the numerical weather prediction models to refine the large-scale results for given locations. MOS provide forecasts of such things as the maximum and minimum temperature, the probability of precipitation, the amount and type of precipitation, the extent of cloud cover, and winds.

Question for Thought

How is numerical weather prediction fundamentally different from model output statistics?

In an attempt to deal with the chaotic nature of weather, and thus improve forecasts, meteorologists have developed a technique known as **ensemble forecasting**. This technique involves running a model several times for the same forecast period. In each run, there will be slight changes in either the initial conditions, the workings of the model itself, or both. Because it requires large amounts of computer power to run a model once, let alone several times, this method has been in use only since the 1990s. Instead of producing *one* possible outcome, an ensemble forecast produces a *range* of possible outcomes. The *average* of these outcomes might be used as the forecast, but the information can also be used to define the *uncertainty* associated with a forecast. The longer the models are run, the more different the results should get. It follows that we can have the greatest confidence in a forecast when the results of the runs are most similar. The variability of results can be seen on maps that are referred to as *spaghetti plots* (Figure 15.21). On these maps, only one or two different height contours are shown, but they are shown several times, once for each run of the model. Ensemble forecasting has allowed us to extend the period over which we can make reliable forecasts.

The output of a model run is a set of weather variables for each grid point. These variables are normally presented in map form (Figure 15.22). Such forecasting maps are called *prognostic charts*, or *progs* for short. Because there are different models, developed at forecasting centres around the world, several different progs will be available for each forecast period. Models differ

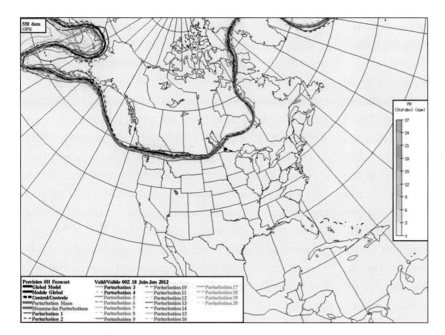

FIGURE 15.21 | An ensemble forecast, or spaghetti plot, for the 5580 m height contour. This forecast was produced by running the GEM model 20 times; each run represents a different perturbation. The heavy black line is the output of an unperturbed model run, while the heavy red line is the average of the ensemble runs.

Source: Environment Canada, Weather Office, www.weatheroffice.gc.ca/ensemble/index_e.html, 15 June 2012.

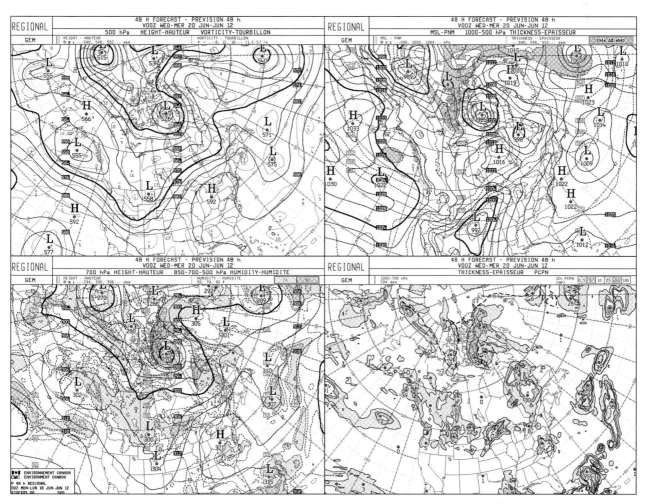

FIGURE 15.22 | An example of a four-panel set of forecast maps produced by Canada's regional GEM model.

Source: Environment Canada, Weather Office, http://www.weatheroffice.gc.ca/model_forecast/index_e.html, 20 June 2012.

in resolution, length of time step, the way in which the primitive equations are solved, and the **parameterizations** that are used.

As part of the final step in issuing a forecast, a forecaster will interpret and compare the outputs from different models in order to decide on the final forecast that should be issued. To carry out this step, forecasters generally need a good knowledge of the various models. They should know which models tend to perform best in certain situations, and they should know the general strengths and weaknesses, or *biases*, of each model. At times, forecasters might choose the model output they think is best based on their previous *diagnosis* of the current conditions and how these conditions seem to be changing. Sometimes, they use the analog, or pattern-recognition method to choose. At other times, they may decide to issue a forecast that is a *consensus* of the model outputs.

It is relatively easy to issue a forecast during a period when weather conditions are steady, such as during a sunny spell in the summer, because the progs will likely all be quite similar. On the other hand, forecasting decisions are much more difficult to make when weather conditions are changing, as they are when a storm is approaching, because model outputs may differ greatly. For example, as a storm approaches the Pacific Northwest, one output may show that the storm will bring heavy rain to Vancouver, while another output shows that Vancouver will remain dry but Seattle will receive heavy rains. Thus, when conditions are highly changeable, weather prediction becomes a much more challenging task.

Numerical weather prediction is often described as one of the greatest achievements in the science of meteorology. However, as was noted above, it is unlikely that we will ever be able to make perfectly accurate forecasts. There are three reasons for this. First, the weather observations on which the models rely are far from perfect, or even complete. Despite our best efforts, observations inevitably contain errors, and there are large gaps in observational networks. Second, the models cannot represent reality perfectly.

> **PARAMETERIZATION**
>
> An approximation used in a model either when we don't understand a process well enough to represent it with equations or when the process is so complex that it would require too much computer time to adequately represent it.

The grid system used in these models is only a crude approximation of the real world; for example, it cannot account for features such as thunderstorms that are smaller than the resolution of the model. Further, our understanding of atmospheric processes is far from complete, as is clear from our need to use parameterizations to represent certain complex processes such as convection or the formation of clouds. In addition, it is very likely that there is a limit to computer power, so we may never be able to *fully* represent natural processes through models, even if we do come to understand these processes completely. Third, because weather is chaotic by nature, we will likely never be able to predict the exact sequence of events that will cause weather to change.

Remember This

It is likely impossible that we will ever be able to make weather forecasts that are 100 per cent accurate because

- there will always be gaps and errors in weather observations,
- it is impossible to make a model that represents reality perfectly, and
- weather is chaotic.

Despite these limitations, the numerical method described in this section can often provide fairly accurate *short-* and *medium-range forecasts*. Short-range forecasts usually extend to about 72 hours, while medium-range forecasts usually extend to about seven days. Although these seven-day forecasts likely represent close to the limit of what we can forecast, people are often interested in what conditions might be like over an even longer range.

15.7 | Long-Range Forecasting

Most long-range forecasting is based on climatology and statistical analysis rather than the modelling of physical processes. As a result, instead of predicting *specific* temperatures or weather patterns, long-range forecasting offers *probabilities* that temperatures or precipitation will be above or below normal. In Canada, both monthly and seasonal long-range forecasts are issued (Figure 15.23).

Today, these long-range forecasts are based mostly on the connections that researchers have discovered between atmosphere–ocean interactions and worldwide weather patterns. The best known of these atmosphere–ocean interactions is the El

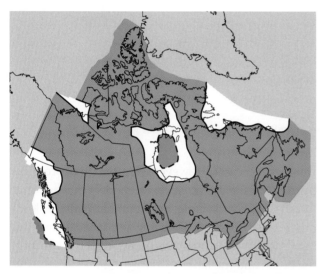

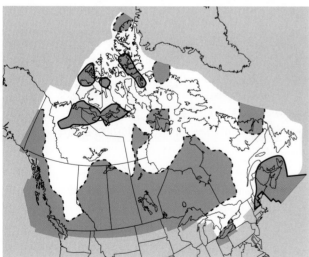

FIGURE 15.23 | *Top:* This map shows where *temperatures* are expected to be above normal (shaded in red) or below normal (shaded in blue) over the three-month period covering June, July, and August of 2012. *Bottom:* This map shows where *precipitation* is expected to be above normal (shaded in red) or below normal (shaded in blue) over the same time period.

Source: Adapted from Environment Canada, Weather Office, www.weatheroffice.gc .ca/model_forecast/index_e.html, 31 May 2012.

Niño–Southern Oscillation (ENSO), which refers to atmospheric pressure reversals and the associated changes in ocean temperatures in the equatorial Pacific Ocean. Normally, there is a cold current, and relatively high atmospheric pressure, in the eastern Pacific off the coast of Peru, while there is lower atmospheric pressure in the western Pacific. This pressure pattern helps to drive the trade winds westward. An El Niño refers to a warming of the Peru Current and a weakening of the high pressure. Under these conditions, the trade winds weaken, and a large area of the eastern Pacific warms. During the opposite conditions, known as a La Niña, the Peru Current is colder than normal, the high pressure in the area strengthens, and the trade winds strengthen. Under these conditions, large areas of the Pacific become colder.

Because of interactions between the ocean and the atmosphere, the warming and cooling of the Pacific influences global wind patterns—and, thus, weather around the world—in what are known as **teleconnections**. During El Niño or La Niña events, the paths of jet streams often change, so that places experience conditions that are warmer or colder than normal. For example, El Niño events can make much of western Canada *warmer* than normal in winter, while La Niña events can make much of western Canada *colder* than normal in winter (Figure 15.24). Shifts in the upper airflow can also cause places to be wetter or drier than normal; it is common to hear of floods or droughts during strong El Niño episodes. Not only that, but the strength of hurricanes can be affected, and monsoon patterns can change. In addition to ENSO, researchers have been discovering other such atmosphere–ocean interactions. Two important ones appear to be the North Atlantic Oscillation (NAO), which can influence the weather of Europe, and the Pacific Decadal Oscillation (PDO), which appears to influence the strength of El Niño and La Niña events. The exact causes and effects of these atmosphere–ocean interactions are not well understood, but the weather changes associated with these interactions have been carefully tracked and documented over the years.

TELECONNECTIONS
Linkages that result because atmosphere–ocean interactions in one part of the world affect the large-scale atmospheric circulation in another part of the world.

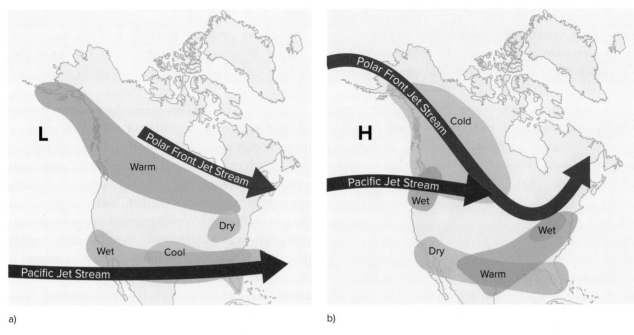

a)

b)

FIGURE 15.24 | a) El Niño and **b)** La Niña events can influence weather in North America as the paths of the jet streams shift.

This documentation has provided useful statistics on the relationship between atmosphere–ocean interactions and weather; in turn, these statistics have been used to create *statistical* models. These models are *analog* in that they make use of past patterns to predict future patterns and, thus, produce long-range weather forecasts. However, scientists have recently begun to make use of *numerical* models in long-range forecasting. For example, Environment Canada's long-range forecasts are now being produced using two numerical models that incorporate both the effects of the ocean and the land.

15.8 | Chapter Summary

1. Weather forecasts are made by first observing current weather conditions, then projecting these conditions forward in time using our understanding of atmospheric processes. The steps involved in making a forecast are observation, analysis, diagnosis, and prognosis.

2. Weather forecasters use a variety of methods to make forecasts. The persistence method is based on the assumption that present conditions will not change. The trend method involves identifying how the atmosphere is changing and assuming that this change will continue into the future. The analog method involves recognizing patterns from the past and assuming that they will produce the same kind of weather as they have before. The "rule of thumb" method relies on rules that have been formulated based on past observations. The climatology method involves using climate statistics to make forecasts based on averages. Numerical weather prediction is a forecasting method that relies on computers to do billions of calculations to determine future conditions based on initial conditions.

3. The invention of the telegraph, which made instant communication possible, is often regarded as the point at which weather forecasting became possible. Forecasts remained short term, and fairly unreliable, until the computer made numerical weather prediction

possible. Subsequent increases in computer power and our expanding knowledge of atmospheric processes have allowed us to make increasingly sophisticated models for forecasting the weather.

4. Weather observations are required to establish the initial conditions for a forecast. The acquisition of weather data from around the world is coordinated by the World Meteorological Organization. The goal is to provide as complete coverage as possible. Satellite imagery helps us fill in the gaps left by other observational data. Visible and infrared satellite imagery provides information about the location and nature of cloud coverage. Water vapour satellite imagery provides information about the distribution of atmospheric water vapour. Radar data provide information about local storms.

5. In the analysis and diagnosis steps of weather forecasting, the forecaster makes use of weather maps for the surface and four layers through the troposphere. These maps display weather observations as well as such computer-derived values as vorticity, thickness, and vertical motion. Together, these maps allow a forecaster to develop an understanding of the weather and how it is changing. Forecasters can use maps for the upper half of the troposphere to determine if, and when, a storm system might reach a location; they can also use these maps to determine whether temperatures over a large area might be above or below normal. They can also assess jet streaks on the 250 hPa chart, and thermal advection and vorticity advection on maps of the lower half of the troposphere, to determine whether weather systems are strengthening or weakening. The 700 hPa chart is useful for finding areas of moisture in the atmosphere. A variety of forecasting rules of thumb can also be applied on these charts.

6. Numerical weather prediction is a forecasting method in which computers project current conditions into the future by performing calculations. These calculations make use of seven equations that mimic the behaviour of the atmosphere. Weather observations are used to assign values to a three-dimensional grid representing the atmosphere. Calculations are done to advance these initial values forward in time, time step by time step. Statistics are used to refine the model output for local conditions. Ensemble forecasting is an attempt to deal with chaos and extend the forecasting period. In issuing a forecast, forecasters normally consult the output of several different models for guidance. It will likely never be possible to make forecasts that are completely reliable or extend much further into the future than they do now.

7. Long-range forecasting relies on statistical and numerical models that are based on relationships between atmosphere–ocean interactions and weather. Long-range forecasts provide probabilities that temperatures or precipitation will be above or below normal, over periods of months or seasons.

Review Questions

1. What are the four steps in the weather forecasting process? Why is each step important?

2. What methods do researchers use to make forecasts? How is climatological forecasting different from the other methods? How is numerical weather prediction different from the other methods?

3. For each statement below, which forecasting method is most likely being used?
   a) Thickness contours on the 500 hPa chart suggest that the precipitation will be snow.
   b) We should get about 20 mm of rain before noon tomorrow.
   c) The clouds will continue to thicken as a low-pressure system moves in to the area.
   d) This steady rain should continue for a few more hours.
   e) Based on past experience, we can expect it to rain almost every day this month.
   f) When the jet stream comes from the southwest, our city usually gets a lot of rain.

4. What was the significance of each of the following technological advances to the science of weather forecasting?
   a) the telegraph
   b) weather instruments
   c) radio transmission
   d) computers
   e) supercomputers
   f) radar
   g) satellites

5. What was the significance of each of the following scientific advances to the science of weather forecasting?
   a) the formulation of the Norwegian cyclone model
   b) the development of the primitive equations
   c) the discovery of the Rossby waves

6. What do bright areas indicate on
   a) visible satellite imagery?
   b) infrared satellite imagery?
   c) water vapour satellite imagery?

7. How is radar useful in forecasting?

8. How can each of the four upper-level charts be used in forecasting?

9. What are the steps involved in numerical weather prediction? What are the possible sources of error in this process?

10. How have we been able to extend the period over which weather forecasts can be reliably made? Why is there likely a limit to the period over which a reliable weather forecast can be made?

11. What are teleconnections? How is an understanding of teleconnections useful in long-range forecasting?

Problem

Imagine that you are a weather forecaster producing a weather map from the set of observations plotted as weather station symbols on the map below. Begin by drawing isobars on the map. Use an interval of 4 hPa, and begin with the 996 hPa isobar. Next, draw fronts on the map, using the wind direction information on the weather station symbols to determine which front is the cold front and which front is the warm front.

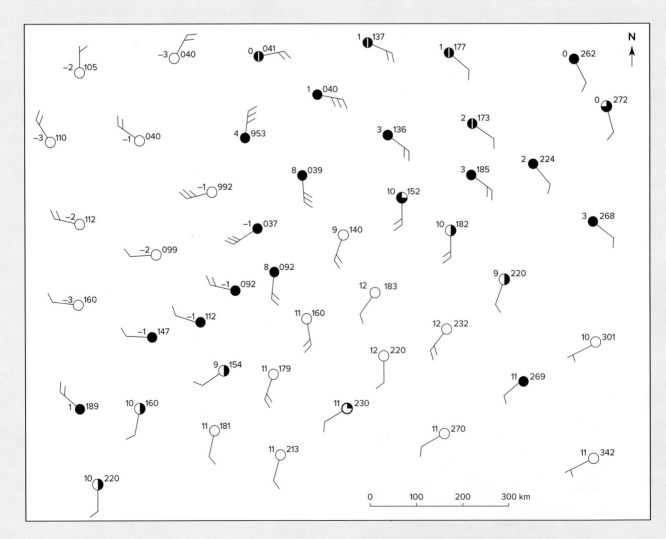

Suggestions for Research

1. Explore the history of weather forecasting, and describe the major technological advancements that have allowed meteorologists to improve the accuracy of their forecasts. Comment on the value of specific technologies that are used today.

2. Find the seven-day forecast for your city on the website for Environment Canada's weather office, and record the details of the prediction. Revisit the website each day for the following week, and note how the forecast changes. List factors that could account for the changes you observe.

References

NOAA. (June 2007). The history of numerical weather prediction. Retrieved from http://celebrating200years.noaa.gov/foundations/numerical_wx_pred

16

Global Climate

Learning Goals

After studying this chapter, you should be able to

1. *appreciate* the role of each of the six climate controls in producing the diversity of climates around the world;

2. *relate* the world's five major vegetation biomes to climate;

3. *distinguish* between the empiric and genetic methods of classifying climate;

4. *describe* Köppen's method of classifying climate;

5. *associate* the climate types of the world with their locations; and

6. *distinguish* between the various climate types in terms of both their characteristics and their causes.

One of the most visible indicators of an area's climate is natural vegetation. The grasses and widely spaced trees that are typical of the savanna biome, for example, have evolved to survive the long periods of drought characteristic of wet–dry tropical climates.

In Chapter 1, climate was defined as "average weather." As this definition suggests, the climate of a place can be described quantitatively using statistics. Since such statistics also include extremes, the variability of weather at a place can be included as part of the description of its climate. Today, most countries have reliable climate data going back at least 100 years. In Canada, climate averages are based on 30-year periods and can be found on Environment Canada's Weather Office website (www.weatheroffice.gc.ca).

Throughout this book, we have explored the processes responsible for our day-to-day *weather*. Because the same weather patterns repeat themselves year after year, certain types of weather will characterize a place, creating the *climate* of that place. Climates are least variable near the equator, where most places are hot and wet all year. In these regions, sun angles are always high, and the rising air of the Intertropical Convergence Zone (ITCZ) is always close by (Section 5.8.2 and 12.2.3). On the other hand, high-latitude places located in the interiors of large continents experience large seasonal variations, with warm summers and bitterly cold winters. Coastal locations are characterized by less seasonal variation, as they are influenced by the slow thermal response of water (Section 6.6.3). The climates of coastal locations are also strongly affected by ocean currents. Windward mountainous coasts are some of the wettest places on Earth; however, under certain conditions, it is possible for deserts to form along coasts (Section 10.7).

Latitude, distribution of land and water, ocean currents, high- and low-pressure systems, prevailing winds, and mountains are all **climate controls**. Because certain combinations of these controls repeat themselves, we see patterns repeating themselves in the global distribution of climate. For example, climates with mild, wet winters and hot, dry summers tend to occur on west coasts in the lower mid-latitudes; California has such a climate, as do the countries surrounding the Mediterranean Sea. Systems of climate classification have been created based on identifying such patterns. The most well-known of these systems was created by Wladimir Köppen and will be described in this chapter.

> **CLIMATE CONTROLS**
> Characteristics of a location that combine to create the climate of that location.

Climate controls create climate types that, in turn, influence other characteristics of a location. Agriculture, forestry, recreation, and tourism are some activities that are strongly influenced by climate. In most parts of the world, homes and buildings are constructed with climate in mind. In the natural world, the weathering of rock, the formation of soils, the development of landforms, and the native vegetation are all, to varying degrees, responses to climate. In fact, the strong connection between climate and vegetation is the basis of Köppen's climate classification system (Figure 16.1).

16.1 | Climate Controls

16.1.1 | Latitude

The most important climate control is *latitude*. Through its control on sun angle and day length, latitude is a major determinant of temperature. This

FIGURE 16.1 | What climate types are suggested by the vegetation shown in the photos above?

was recognized by the ancient Greeks when they called the tropics the *torrid zone*; the mid-latitudes, the *temperate zone*; and the polar regions, the *frigid zone*. Latitude also controls seasonality. While temperatures are consistently high in the tropics (Figures 16.2 and 16.3), mid-latitude regions are characterized by large annual temperature ranges and, thus, warm and cold seasons. In the polar regions, the large annual variation in the length of daylight leads to large temperature ranges, but it is always cold because sun angles are always low. We can see the effect of latitude on temperature by comparing temperature characteristics in Yellowknife, which is located at about 62° N, to those in Calgary, which is located at about 51° N. In Yellowknife, the *mean annual temperature* is −5.6°C, and the *annual temperature range* is close to 45°C; by contrast, Calgary's *mean annual temperature* is 3.4°C, and its *annual temperature range* is about 27°C.

16.1.2 | Distribution of Land and Water

The *distribution of land and water* accounts for differences in climate between marine and continental locations. Marine climates are moderated by water bodies and, thus, exhibit smaller daily and annual variations in temperature than do continental climates. For example, Calgary and Victoria are located at about the same latitude but, due to the latter's coastal location, the annual temperature range of Victoria—about 13°C—is half that of Calgary. Proximity to the water can also produce wetter climates, as illustrated by the fact that Victoria receives considerably more precipitation throughout the year than does Calgary. This is not always the case, however, as coastal deserts are not uncommon. The Atacama Desert, along the coasts of Peru and Chile, is an example of such a desert.

16.1.3 | Ocean Currents

Warm and cold *ocean currents* can affect both temperature and precipitation. For example, the North Atlantic Drift, which is a warm current that crosses the north Atlantic and continues northward up the northwest coast of Europe, makes parts of Scandinavia warmer than they would be without it (Figure 16.2). Bergen, Norway, located at about 60° N, has a mean annual temperature of about 8°C and an average January temperature of about 1.5°C. In comparison, Anchorage, Alaska, located at approximately the same latitude, has an average annual temperature of about 2°C and an average January temperature of only −9°C. Both towns are located on a coast, but the warm North Atlantic Drift significantly impacts temperatures in Bergen. In addition, because the warmer waters make the air above moist and unstable, Bergen receives over five times more precipitation than does Anchorage. Cold currents have the opposite effect. For example, the Atacama Desert, mentioned above, is strongly influenced by the cold Peru Current, which travels northward along South America's west coast. Although this desert is not *caused* by the cold current, it is *strengthened* by it because as air moves over the cold water, it is cooled from below and becomes more stable. These conditions produce a coastal desert where it almost never rains but is often foggy.

16.1.4 | Pressure Systems

Average *pressure patterns* control precipitation. Recall from Chapter 12 that the general circulation is made up of major pressure systems: the equatorial low (ITCZ), the subpolar lows (mid-latitude cyclones), the subtropical highs, and the polar highs. Thus, on average, pressure is low near the equator and in the mid-latitudes, and it is high in the subtropics and at the poles. As a result, equatorial and mid-latitude regions are wet, while the subtropical and polar regions are dry. In fact, if Earth were of uniform surface type, either all water or all land, there would be alternating belts of wet and dry regions circling the planet. Because so many other factors influence precipitation, the real pattern is much more complex (Figure 10.17). You also learned, in Chapter 12, that these pressure patterns shift with the seasons, meaning that some regions of Earth experience high pressure in one season and low pressure in another. For example, some tropical locations are wet in the summer, when they are affected by the ITCZ, and dry in the winter, when they are affected by a subtropical high.

16.1.5 | Prevailing Winds

Prevailing winds can affect both temperature and precipitation. The prevailing winds of the tropics are the easterly trade winds, while the prevailing winds of the mid-latitudes are westerlies (Chapter 12). As a result of prevailing westerly winds, locations on Canada's west coast are more strongly influenced by the effects of the ocean than are places on Canada's east coast. Consider, for example, that Sydney, Nova Scotia, has an annual temperature range much closer to that of Calgary than that of Victoria. Prevailing winds also determine the locations of rain-shadow deserts. Recall, for example, that in the tropics of South America, where prevailing winds are easterly, the Atacama Desert lies on the *western* side of the Andes, but in the mid-latitudes of South America, where prevailing winds are westerly, the Patagonian Desert lies on the *eastern* side of the Andes (Figure 12.8).

16.1.6 | Mountains

One way that *mountains* control climate is by acting as barriers that force air to rise and release moisture as precipitation on their windward sides; having lost its moisture, the air that sinks on the leeward sides of mountains produces deserts. The impact of mountains is illustrated by the large amount of precipitation that Canada's west coast receives, compared to the smaller amount of precipitation that Europe's west coast receives (Figure 10.17). Although the two locations are at similar latitudes, and they both experience prevailing westerly winds off the oceans, Canada's west coast is wetter because of the presence of mountains. The Coast Range of British Columbia is also largely responsible for preventing the invasion of cold Arctic air into the coastal regions of the province during the winter. Mountains also control climate simply because temperature decreases with altitude.

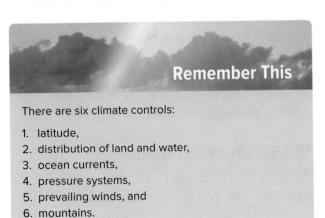

Remember This

There are six climate controls:

1. latitude,
2. distribution of land and water,
3. ocean currents,
4. pressure systems,
5. prevailing winds, and
6. mountains.

16.2 | Vegetation and Climate

The natural vegetation that has evolved in an area is largely a result of temperature, precipitation, and the seasonal distribution of each. Vegetation needs warmth and moisture; where either or both of these requirements are limited, plants have developed a variety of adaptations. Apart from being shaped by climate, vegetation is also affected by such things as soil type, topography, conditions of extreme exposure to sun or wind, and even the history of fire in a location. There are five major vegetation groups, or *biomes*, each of which can be linked to climate.

Trees are the main plant form of the *forest* biome. This biome requires the most moisture of any of the major vegetation groups. Although most forests can tolerate a dry season, *rain*forests thrive only where there is no dry season, or where the dry season is very short. The tropical rainforests enjoy abundant precipitation year round. For the temperate rainforests of Canada's west coast, there is a significant drop in summer precipitation, but the climate is never really dry. In addition, forests require a warm season, and they will not develop at all where the average monthly temperature is always below 10°C. The boreal forests of northern Canada are needleleaf forests that require a warm season of only about three or four months. The temperate deciduous forests, common in southern regions of eastern Canada, are broadleaf forests that adapt to the cold winter season by shedding their leaves.

As the name suggests, the *grassland* biome is dominated by grasses and other non-woody plants, or herbs. Although grasses can tolerate drier conditions than can forests, a moderate amount of precipitation is required. Trees and shrubs are absent from grasslands except in the slightly wetter areas that might occur along the banks of rivers. The amount of precipitation determines the height of the grasses; in wetter areas, the grasses are taller. Grasslands tend to occur in regions with warm summers and large annual temperature ranges. The climate of the Canadian prairies is suited to grassland vegetation.

The *savanna* biome is sometimes considered to be transitional between forests and grasslands because it consists of grasslands with isolated trees. Savanna vegetation is limited to areas of the tropics that have seasonal precipitation. In the wet season, the grasses and leaves thrive; in the dry season, the grasses wither and the trees shed their leaves. The total annual precipitation associated with savanna is similar to that for temperate grasslands; the major difference is that there must be a drought period lasting several months in order for savanna to develop. It is believed that this drought is important in maintaining savannas because fires that occur during the dry season help limit the number of trees.

The *desert* biome is made up of plants that are adapted to dry conditions. Not only are precipitation amounts very low in desert regions, they are also extremely variable. In addition, temperatures in tropical deserts can be higher than anywhere else on Earth; these high temperatures increase evaporation rates, reducing moisture availability even further. Desert plants exhibit a variety of adaptations that allow them to survive under conditions where moisture is extremely limited. They tend to grow far apart, so that they do not have to compete with other plants for water; they often have widespread, shallow root systems, which allow them to take in water from a large area; and they usually have small or thick-skinned leaves that open their stomata only at night, to minimize the loss of water through transpiration. In hot deserts, plants often have shiny leaves to help reflect away harsh solar radiation. In addition, the leaves of some plants, such as the sagebrush that is common in the semiarid deserts of the southern interior of British Columbia, are covered in fine hairs that help to reduce water loss and keep the leaves cool.

Plants of the *tundra* biome are adapted to cold conditions. The long, severely cold winters of tundra regions limit the growing season to, at most, a few months. In addition to being able to resist the cold, plants of the tundra must be adapted to strong winds and poorly drained swampy soils, which develop when the upper layer of **permafrost** melts in the summer. Tundra plants are small and compact in order to conserve warmth. They consist mostly of lichens, mosses, grasses, and small woody shrubs. Canada's entire northern coast supports tundra vegetation.

PERMAFROST
A subsurface layer of soil and rock material that remains frozen throughout the year.

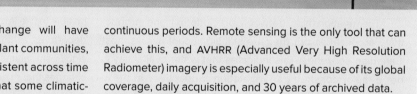

In the Field

Monitoring Vegetation Responses to Climatic Variation
Yuhong He, University of Toronto Mississauga

Scientists know that global climate change will have impacts on the distribution and health of plant communities, but changes to vegetation will not be consistent across time and space. Research strongly suggests that some climatically induced changes in vegetation are already occurring in Canada. For example, northerners and scientists in the tundra across the Canadian Arctic have noted that shrubs are getting taller with increasing temperatures.

My research seeks quantitative answers to important questions such as *How will plants respond to climate change? Which climate variables will drive the geographic shifting of vegetation?* and *Will the timing of events in seasonal cycles, such as leaf emergence, change with increased temperature?*

To address these questions, we need accurate and consistent vegetation information for a large geographical area. This information must be collected at specific times and over continuous periods. Remote sensing is the only tool that can achieve this, and AVHRR (Advanced Very High Resolution Radiometer) imagery is especially useful because of its global coverage, daily acquisition, and 30 years of archived data.

We can use AVHRR data to calculate the Normalized Difference Vegetation Index (NDVI), which is an indicator of vegetation productivity. Several studies have taken this approach to investigate the responses of vegetation in Canadian ecosystems to climate conditions. For example, Li and Guo (2012) found that the grassland growing season exhibits earlier green-up and later *senescence* (that is, biological aging) as temperature increases.

In our study on the response of ecosystems to climate in Canada, I and my co-authors analyzed AVHRR NDVI/productivity data and climate data over a 23-year period (Figure 1).

(continued)

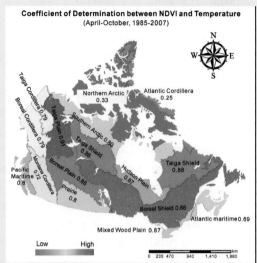
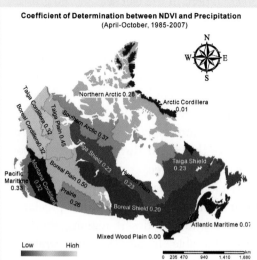

FIGURE 1 | On the *temperature* diagram (left) the warmer colours indicate stronger relationships between temperature and vegetation productivity. On the *precipitation* diagram (right) the lighter colours indicate stronger relationships between precipitation and vegetation productivity.

Source: Y. He, Z. Guo, P. Dixon, & J. Wilmshurst. (2012). NDVI variation and its relation to climate in Canadian ecozones. *The Canadian Geographer*, Vol. 56, 4, 492-507.

We found very strong relationships between temperature and productivity across Canadian ecozones; in other words, we found that temperature variability during the growing season is one of the most important climatic parameters influencing vegetation growth. This is expected because warmer temperatures during the summer growing season typically enhance photosynthesis and, therefore, productivity. Spatially, we found lower correlations between temperature and productivity in the Arctic Cordillera and Northern Arctic ecozones, likely because of the shorter growing season.

The relationships between precipitation and productivity are weaker, but they are still significant across most Canadian ecozones. This indicates that water availability is not a major constraint to vegetation productivity in these ecozones. Adding more precipitation does not supplement a limiting element; rather, it reduces a limiting element (temperature) for those plants that are temperature-limited.

Agreeing with other studies, we found that productivity has increased over time in Canadian ecozones; this increase corresponds to increasing temperatures across the country. However, annual precipitation changes were not uniform across Canada; rather, annual precipitation has increased in some ecozones but decreased in others. One of the greatest concerns related to the future of vegetation in Canada is that the anticipated minor increases in precipitation will be insufficient to offset the much higher evapotranspiration

rates to be expected in a warmer climate (Zoltai, Singh, & Apps, 1991). This could lead to significantly drier soils and more severe droughts.

The conclusions from our study, derived from remote sensing data, provide an overview of how climate affects Canadian ecosystems. What is the next step? Using available ground information and high spatial resolution remote sensing data, our future efforts will focus on establishing a remote sensing baseline to define vegetation conditions in each terrestrial ecosystem in Canada. In practice, land managers will be able to use this baseline as a standard to determine whether departures exhibited by modern ecosystems are unusual or significant; ultimately, this information will help them identify changes that may require their specific attention.

REFERENCES

Li, Z., & Guo, X. (2012). Detecting climate effects on vegetation in Northern mixed prairie using NOAA AVHRR 1-km time-series NDVI data. *Remote Sensing, 4*, 120–134.

Zoltai, S.C., Singh T., & Apps, M.J. (1991). Aspen in a changing climate. In S. Navratil and P.B. Chapman (Eds), *Aspen management for the 21st century* (pp. 143–152). Edmonton, AB: Forestry Canada, Northwest Region.

YUHONG HE is an assistant professor at University of Toronto Mississauga. Her research focuses on the use of remote sensing techniques, spatial analysis, climate data, and ecosystem modelling in studies of natural or managed systems (grassland, forest, wetland, and agriculture), and on the links between observed changes and environmental and anthropogenic driving factors at multiple spatial and temporal scales.

16.3 | Climate Classification

The variety of possible climates on Earth is almost endless, and the distribution of climates over Earth's surface is complex. It is partly for these reasons that we attempt to classify climate. In general, classification reduces the complexity of reality by putting items into categories that make it easier for us to describe, compare, and understand them. In particular, climate classification allows us to prepare *maps* of climate, a task that would likely be impossible without some form of classification system.

Individual aspects of climate, such as precipitation or temperature, can be more easily mapped.

For example, as you saw in Figure 10.17, we can map the global distribution of total annual precipitation. Temperature can be a bit less straightforward to represent. We often map temperature *distributions*—generally on two maps, one for January temperatures, and one for July temperatures (Figure 16.2)—and annual temperature *ranges* (Figure 16.3). In addition, we can map patterns related to temperature to provide information, such as the length of the growing season (Figure 16.4). Other phenomena such as number of days with thunderstorms (Figure 14.28), frequency of fog, annual snowfall, and number of sunshine hours can also be mapped to provide information about various aspects of climate.

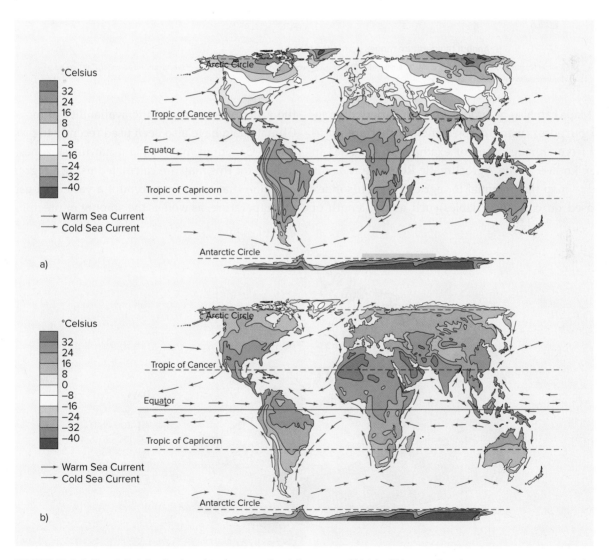

FIGURE 16.2 | The global distribution of temperature for a) January and b) July. This map also shows warm ocean currents in red, and cold ocean currents in blue.

Source: Adapted from Q. Stanford. (2008). *Canadian Oxford world atlas* (6th ed.). Don Mills, ON: Oxford University Press Canada, p. 125.

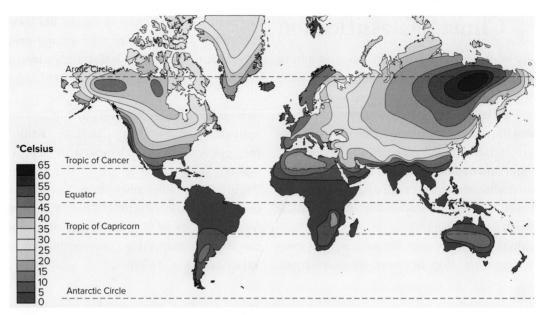

FIGURE 16.3 | Annual temperature ranges across the globe.

Source: Adapted from Strahler, Alan, and O.W. Archibold. (2011). *Physical geography: Science and systems of the human environment*, (5th Cdn. ed.). Toronto: John Wiley & Sons, Inc, p. 86.

Maps can also be produced, for a variety of practical purposes, using derived indices that are *calculated* from climate data. For example, in Canada, an index known as the Agroclimatic Resource Index is used to map the potential for agriculture. This index is based on climate variables such as the length of the growing season and the availability of moisture. Climate data have also been used recently to produce a new Canadian map of plant hardiness zones. The indices for this map are calculated using a regression equation that takes into account a variety of aspects of temperature, as well as the amount of precipitation

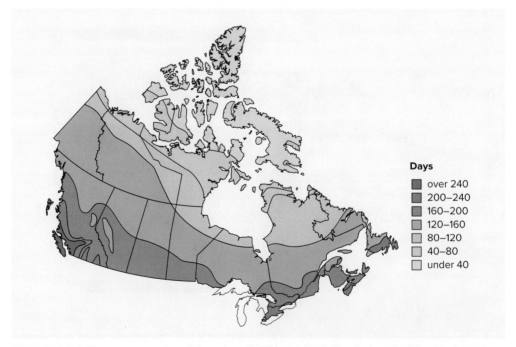

FIGURE 16.4 | The average number of days above 5.6°C is used to define the length of the growing season in this map of Canada.

Source: Adapted from Q. Stanford. (2008). *Canadian Oxford world atlas* (6th ed.). Don Mills, ON: Oxford University Press Canada, p. 18.

during the growing season, snow depths, and even windiness. Another example is the Canadian climate zones map created for use with the Energy Star energy conservation program (Figure 16.5). This map is based on *heating degree days* (HDD). Heating degree days are normally calculated using a base temperature of 18°C because buildings don't normally need to be heated when temperatures are above this value. For each day of the year that average temperature is *below* 18°C, the temperature is subtracted from 18°C. When all such values are obtained, they are added together to give the HDD for a particular location. The larger is this number, the greater are the heating requirements.

Because they are created for a specific use, the above are examples of *applied* climate classifications. Two other types of climate classification systems are *empiric* and *genetic*. In empiric systems, climates are grouped based on observed data, usually temperature and precipitation. The classification system devised by Wladimir Köppen, which will be discussed in Section 16.3.1, is an empiric system. In genetic systems, climates are grouped based on their causes. For example, the constant hot, wet conditions near

the equator are *caused* by the consistently high sun angles and the close proximity of the ITCZ. In genetic systems, this climate is normally called the *wet equatorial* climate. Notice that two of the climate controls listed in Section 16.1, latitude and pressure systems, are used as the basis for explaining the characteristics of this climate type. In a genetic system, climate types can be defined by different combinations of the climate controls.

A genetic classification scheme is shown in Table 16.1. The three major groups—tropical, mid-latitude, and polar—are based on latitude, because latitude is the *cause* of the three main temperature regimes. Low-latitude climates are always hot, mid-latitude climates have warm and cold seasons, and high-latitude climates are always cold. Climates are further subdivided based on whether they are always wet, always dry, or have a wet season and a dry season. The precipitation regime is usually determined by the fourth climate control, the dominant pressure system in the region.

Each of the 12 climate types that result from this genetic climate classification system is given a

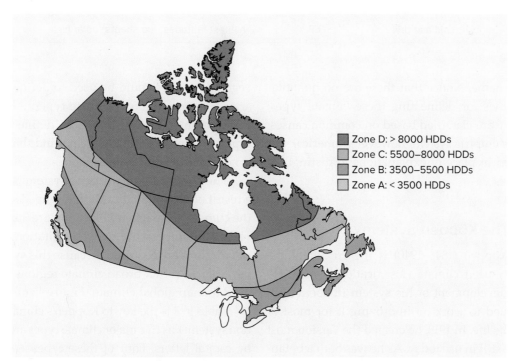

FIGURE 16.5 | Canada has four Energy Star zones, which are identified based on the number of heating degree days (HDDs) in each area. These zones are used to categorize products such as windows and doors as "Energy Star qualified." Products are labelled to indicate the zone(s) for which they are Energy Star qualified.

Zone D: > 8000 HDDs
Zone C: 5500–8000 HDDs
Zone B: 3500–5500 HDDs
Zone A: < 3500 HDDs

Source: Adapted from Natural Resources Canada, "Climate zones—Energy Star," http://oee.nrcan.gc.ca/equipment/windows-doors/1371, 30 June 2011.

TABLE 16.1 | A genetic climate classification system. The symbols for the Köppen climate system have been included (see Table 16.2).

	Descriptive Characteristics	Köppen Symbol	Causes
Tropical			
Wet Equatorial	hot and wet	Af	low latitudes; ITCZ always near
Wet/Dry Tropical	hot and wet/dry (dry in low-sun season)	Aw	low latitudes; ITCZ in high-sun season, subtropical high in low-sun season
Tropical Monsoon	hot and wet/dry (dry in low-sun season)	Am	low latitudes; seasonal reversal of winds
Subtropical Desert/ Steppe	hot and dry	BWh, BSh	low latitudes; subtropical high all year
Mid-latitude			
Moist Subtropical	warm/cold and wet	Cfa	lower mid-latitudes, east coast; moist, unstable side of subtropical high in summer, mid-latitude cyclones in winter
Marine West Coast	warm/cold and wet	Cfb, Cfc	mid-latitudes, west coast; mid-latitude cyclones all year
Moist Continental	warm/cold and wet	Dfa, Dfb, Dwa, Dwb	mid-latitudes, continental, east coast; mid-latitude cyclones all year
Subarctic	warm/cold and wet	Dfc, Dfd, Dwc, Dwd	higher mid-latitudes, continental, east coast; mid-latitude cyclones all year
Mediterranean	warm/cold and wet/dry (dry in summer)	Csa	lower mid-latitudes, west coast; dry, stable side of subtropical high in summer, mid-latitude cyclones in winter
Mid-latitude Desert/ Steppe	warm/cold and dry	BWk, BSk	mid-latitudes; rain shadow or far inland
Polar			
Polar Tundra	cold and dry	ET	high latitudes; coastal regions; polar high
Polar Icecap	cold and dry	EF	highest latitudes; ice sheets; polar high

descriptive name. Notice that there are no quantitative guidelines for delineating these climate types; instead, they are classified based on common causes. In Köppen's empiric system, however, numerical values are used in place of explanations to distinguish climate types.

16.3.1 | The Köppen System

Wladimir Köppen (1846–1940) is the creator of the most widely used climate classification system. He began the development of his system in about 1900 and continued to work on improving it for most of the rest of his life. In 1918, he created the version most like the one still in use today. As he was both a botanist and a climatologist, Köppen based his system on the assumption that the type of vegetation in a region is closely linked to the climate of that region; thus, he believed that the boundaries between different

vegetation types could be used to define the boundaries between different climate types. Although this assumption allowed Köppen to define his climate types, it is important to keep in mind that vegetation is not influenced by climate alone.

The endurance of Köppen's system is most likely a result of its simplicity. All that is needed to classify the climate of a particular location are mean monthly values of temperature and precipitation; little to no calculations are required. Because this system allows us to define rather broad climate regions, it is widely used to map global climate (Figure 16.6).

Table 16.2 is the key to Köppen's climate classification system. His five major climate types are designated by capital letters. Four of these types—A, C, D, and E—are defined by temperature; of these four, three—A, C, and D—are considered warm and moist enough to support the growth of forests. Group A climates are *tropical rainy climates*, in which the average temperature

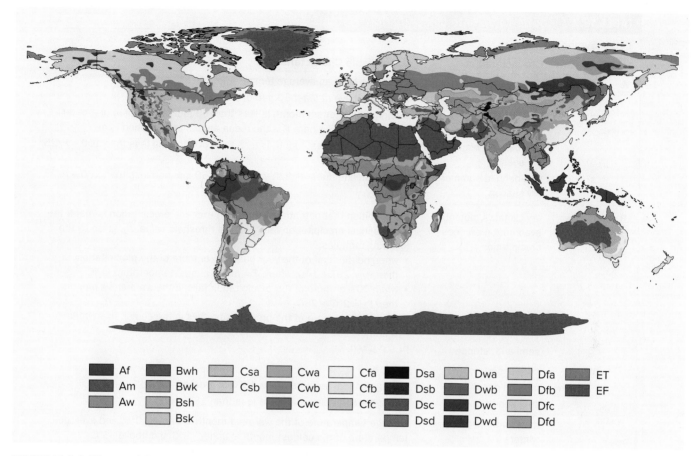

FIGURE 16.6 | Climates of the world mapped according to Köppen's climate classification system.

Source: Adapted from M.C. Peel, B.L. Finlayson, & T.A. McMahon. (2007). "Updated world map of the Köppen–Geiger climate classification." *Hydrology and Earth Systems Science 11*, 1633–44.

of every month is 18°C or above. Köppen based this category on the distribution of tropical rainforests; for the most part, these forests occur only where this temperature criterion is met. Groups C and D are *mid-latitude rainy climates*. The distinction between these two groups is that group C climates are characterized by *mild* winters, while group D climates are characterized by *severe* winters. Group E climates are neither warm enough nor wet enough to support forests. These are the *polar climates*. In order for a climate region to fall in the E group, the average temperature of the warmest month must be below 10°C. This temperature boundary was chosen because it is generally associated with the northern limit of tree growth. Where no months have temperatures above 10°C, forests give way to tundra. This boundary is visible as the tree line in northern Canada. Polar climates are divided between *polar tundra climates*, designated ET, and *polar ice cap climates*, designated EF. While polar tundra climates are warm

enough to support tundra vegetation, polar ice cap climates are permanently frozen, so they cannot support vegetation at all.

The fifth major climate type, B, covers the *dry* climates—those in which there is a moisture deficit for most of the year. This group is further subdivided, based on the *degree* of dryness, or aridity. The most arid climates are designated BW and are named *desert* climates. The semiarid climates are designated BS and are named *steppe* climates. Köppen's dry climates can be warm or cold, but they are too dry to support the growth of forests.

Because dryness is not a function of precipitation alone, B climates are more difficult to define than are the other four. For moisture deficits to occur, **potential evaporation** must exceed precipitation. Potential evaporation depends

POTENTIAL EVAPORATION
The amount of evaporation that would occur if moisture were unlimited.

TABLE 16.2 | Köppen's climate classification system.

Letter Symbol 1st	2nd	3rd	Descriptive Characteristics	Quantifiable Characteristics
A			hot and moist	all months have an average temperature of 18°C or higher
	f		wet all year	all months have at least 60 mm of precipitation
	m		short dry season	precipitation in driest month is less than 60 mm but equal to or greater than $100 - P/25$, where P is the mean annual precipitation in mm
	w		well-defined winter dry season	precipitation in driest month is less than 60 mm and less than $100 - P/25$
	s		well-defined summer dry season	precipitation in driest month is less than 60 mm and less than $100 - P/25$
B			dry climates, potential evaporation exceeds precipitation	the condition that potential evaporation exceeds precipitation is met if the mean annual precipitation falls below a threshold value, P_T, given by the following formulas: • when neither half of the year has 70% or more of the precipitation, then $P_T = 20T + 140$, where T = mean annual temperature in °C • when 70% or more of the precipitation falls in the warmer six months, then $P_T = 20T + 280$ • when 70% or more of the precipitation falls in the cooler six months, then $P_T = 20T$
	S		semi-arid; steppe	$P_T > P > \frac{1}{2}P_T$
	W		arid; desert	$P < \frac{1}{2}P_T$
		h	hot and dry	average annual temperature is 18°C or greater
		k	cool and dry	average annual temperature is less than 18°C
C			moist with mild winters	average temperature of the warmest month is above 10°C, and average temperature of the coldest month is under 18°C and above −3°C
	w		dry winters	average precipitation of the wettest summer month is at least ten times as much as in the driest winter month
	s		dry summers	average precipitation of the driest summer month is less than 40 mm, and average precipitation of wettest winter month is at least three times as much as in the driest summer month
	f		wet all year	criteria for w or s cannot be met
		a	hot summers	average temperature of warmest month is over 22°C, and average temperature of at least 4 months is over 10°C
		b	warm summers	average temperatures of all months are below 22°C, and average temperature of at least 4 months is over 10°C
		c	cool summers	average temperatures of all months are below 22°C, and average temperature of one to three months is above 10°C
D			moist with severe winters	average temperature of warmest month is above 10°C, and average temperature of coldest month is −3°C or below
	w		dry winters	same as under C
	s		dry summers	same as under C
	f		wet all year	same as under C
		a	hot summers	same as under C
		b	warm summers	same as under C
		c	cool summers	same as under C
		d	very cold winters	average temperature of the coldest month is −38°C or below
E			polar	average temperature of the warmest month is below 10°C
	T		tundra	average temperature of the warmest month is greater than 0°C and less than 10°C
	F		ice cap	average temperature of the warmest month is 0°C or below

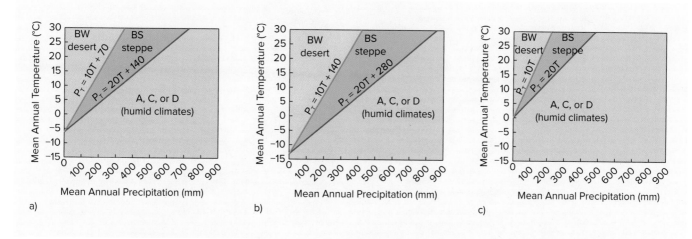

FIGURE 16.7 | Graphical depiction of Köppen's formulas for defining dry climates when a) precipitation is distributed evenly throughout the year, b) precipitation is concentrated in the warmest six months (wet summer), and c) precipitation is concentrated in the coolest six months (wet winter). Notice that, for a given amount of precipitation, climates become increasingly arid as the temperature increases.

largely on energy availability, or temperature; as a result, potential evaporation is much higher in hot climates than it is in cold climates. Therefore, to determine whether a climate is dry or not, both temperature *and* precipitation must be taken into account. With this requirement in mind, Köppen derived simple formulas that make it possible to determine whether or not a climate is dry and, if it is dry, whether it is a desert or a steppe. These formulas are given as equations in Table 16.2 and depicted graphically in Figure 16.7.

For the three moist climate types, a second letter is used to indicate the seasonal distribution of precipitation. These letters are assigned based on whether the region has a summer dry season, a winter dry season, or an even distribution of precipitation throughout the year. The timing of precipitation is an important determinant of the amount of moisture available for vegetation. If precipitation falls mostly in the warm season, more of it will evaporate than if it falls mostly in the cold season. Thus, in mid-latitude areas, where the warm season is the major growing season, a lack of rain at this time is particularly stressful for plants.

For the B, C, and D climate groups, there is also a third letter. In the case of the B climate group, the third letter is used to distinguish between hot and cold deserts. In general, most deserts in tropical regions have high mean annual temperatures and are designated as hot deserts, while most deserts in mid-latitude

regions have lower mean annual temperatures, due to cold winters, and are designated as cool deserts. For both the C and D groups, the third letter is used to provide more detail about seasonal temperatures.

Remember This

- Genetic climate classification systems group climates based on their common causes. The resulting climate types are given descriptive names.
- Köppen's climate classification system groups climates using temperature and precipitation data. The resulting climate types are given letter symbols.

Section 16.3 has described a genetic climate classification system, in which climates are grouped based on common causes, as well as Köppen's climate classification system, in which climate types are defined using temperature and precipitation values. As you can see from tables 16.1 and 16.2, there is, in most cases, a close fit between the two systems. Building on what has been covered here, the next three sections will take a more detailed look at each of the world's 12 main climate types. For your reference, Figure 16.8 shows the locations of places that will be used as examples of the various climate types.

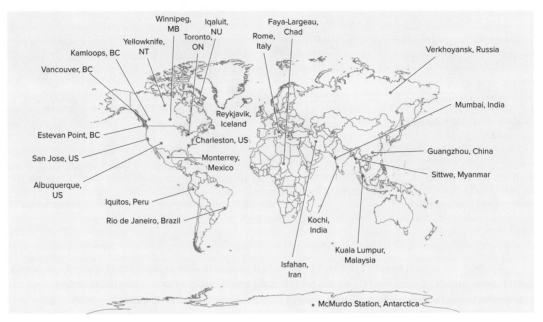

FIGURE 16.8 | This world map shows the locations of places used as examples of the various climate types discussed in sections 16.4 through 16.6.

16.4 | Tropical Climates

Most areas with tropical climates lie between the latitudes of 30° N and 30° S. At these low latitudes, sun angle is always high and day lengths fluctuate by only, at most, about four hours over the course of a year. The high sun angles, together with the lack of variability of both sun angles and day lengths, means that tropical climates are always hot. In fact, in most tropical climates, temperature will change more over the course of a day than it will over the course of a year. The annual temperature range generally increases with distance from the equator.

The pressure systems associated with the tropics are, of course, the equatorial low, or ITCZ, and the subtropical highs. Recall that the former can cause regions to be very wet, while the latter can cause regions to be very dry; thus, because of the influence of these systems, tropical regions are characterized by some of the wettest and driest places on Earth. In addition, due to the prevailing easterly trade winds, tropical east coasts are likely to be wetter than tropical west coasts.

The four main climate groups of the tropics are all hot. Of these, one is always wet, two have wet and dry seasons, and one is always dry.

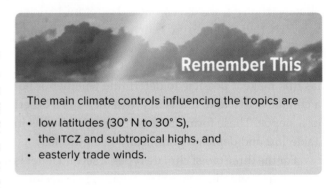

Remember This

The main climate controls influencing the tropics are

- low latitudes (30° N to 30° S),
- the ITCZ and subtropical highs, and
- easterly trade winds.

16.4.1 | Wet Equatorial Climates (Af)

Characteristically hot and wet all year round, *wet equatorial climates* are the least variable climates on Earth. They are normally located close to the equator, between about 10° N and 10° S. The largest areas of wet equatorial climates occur in the Amazon Basin of South America, the Congo Basin of Africa, and the East Indies. Both Iquitos, Peru, and Kuala Lumpur, Malaysia, are examples of locations with wet equatorial climates that also fit the criteria of Af climates in Köppen's system (Figure 16.9). In both cases, the annual temperature range is less than 2°C, and mean monthly temperatures fluctuate around 26 or 27°C. Total annual precipitation is over 2000 mm and, although monthly precipitation totals do vary, they are always high. In most wet equatorial climates, the rain falls mostly as afternoon thundershowers.

The consistently high temperatures in these regions are associated with being close to the equator, where the sun angle is high all year and the day lengths are always close to 12 hours. However, these climates are not the hottest on Earth because their skies are frequently cloudy, and the clouds reduce the amount of sunlight that reaches the surface. In addition, the high amounts of moisture mean that much energy is used for evaporation, leaving less for heating. The abundant precipitation associated with wet equatorial climates is due to the close proximity of the ITCZ for all, or most of, the year. The variation in monthly rainfall totals is associated with the movement of the ITCZ (Figure 16.10). When the ITCZ is closest, precipitation is highest.

The native vegetation of the wet equatorial climate is the *tropical rainforest* (Figure 16.11). In these hot, wet conditions, there are virtually no limitations on the growth of vegetation; thus, tropical rainforests are composed of mostly broadleaf evergreen plants. Broadleaf plants have leaves with a large surface area; these leaves allow the plants to cool easily through transpiration. In addition, the plants are evergreen, as there is neither a cold season nor a dry season. Under these warm, wet conditions, plants tend to grow very close together, making tropical rainforests the densest forests on Earth. These forests also support a greater diversity of life than does any other ecosystem.

Under certain conditions, climates very similar to the wet equatorial type can develop in places that are farther away from the equator. These climates are sometimes called *trade wind coastal climates* because they develop where the trade winds blow toward mountainous coasts. The resulting uplift of moist air produces large amounts of rain during the high-sun season when the ITCZ is nearby. When the ITCZ moves away, precipitation amounts drop more than they do in the wet equatorial climates. Thus, trade wind coastal climates display slightly more seasonality of precipitation than is the norm for wet equatorial climates. Examples of trade wind coastal climates occur along the eastern, or windward, sides of both Central America and Madagascar. Such climates normally classify as Af in Köppen's system. In a genetic system, however, they would be a separate group, as mountains

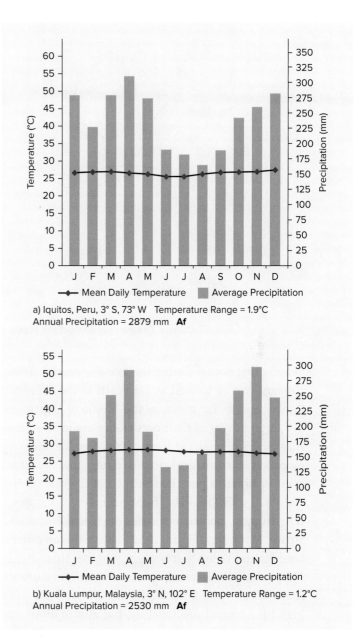

a) Iquitos, Peru, 3° S, 73° W   Temperature Range = 1.9°C
Annual Precipitation = 2879 mm   **Af**

b) Kuala Lumpur, Malaysia, 3° N, 102° E   Temperature Range = 1.2°C
Annual Precipitation = 2530 mm   **Af**

FIGURE 16.9 | a) Iquitos, Peru, and b) Kuala Lumpur, Malaysia, have wet equatorial climates.

and prevailing winds, in addition to low latitude and pressure systems, play a role in producing them.

16.4.2 | Tropical Monsoon Climates (Am)

Tropical monsoon climates are also always hot, but they are distinguished from wet equatorial climates by their seasonal precipitation regime. Locations with monsoon climates have a dry season of three to six

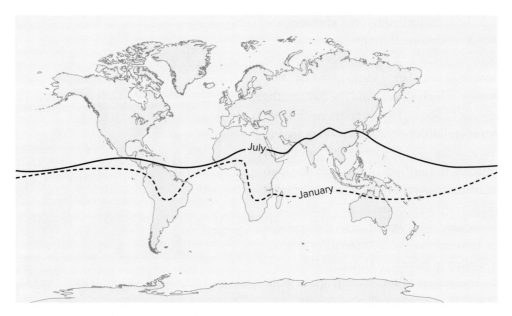

FIGURE 16.10 | The average limits of the annual migration of the ITCZ.

months that occurs during the low-sun season. These climates can be located as far north as the Tropic of Cancer and as far south as the Tropic of Capricorn. The largest areas of their occurrence are in Southeast Asia and India, along west coasts. Smaller areas of monsoon climate are located along the northeast coast of Brazil, along the south coast of West Africa, and in the Philippines. Two examples of locations with tropical monsoon climates are Kochi, India, and Sittwe, Myanmar (Figure 16.12). Sittwe has an annual temperature range that is slightly higher than that of wet equatorial climates because of its greater distance from the equator. Both Kochi and Sittwe have very high total annual precipitation, but almost all of this precipitation falls in just over half the year. For the rest of the year, precipitation is very low.

The seasonal pattern of precipitation characteristic of monsoon climates is associated with coastal locations in the tropics where there is a seasonal reversal of winds (Section 12.2.2). For example, the wet season in both India and Southeast Asia occurs when the ITCZ moves north of these locations (Figure 16.10). This shift produces a *moist* onshore flow from the southwest that brings rain during the high-sun season. During the low-sun season, the ITCZ lies over the Indian Ocean and produces a *dry* offshore flow from the northeast and very little rain. The length of the wet season varies from place to place, but it is usually

FIGURE 16.11 | A tropical rainforest in Tortuguero National Park, Costa Rica, with characteristic broadleaf plants.

longer than the dry season. Although monsoon climates are hot all year, they are usually a little warmer in the dry season than in the wet season. Once the rains arrive, the clouds reduce the amount of solar radiation that reaches the surface and, as a result, temperatures are slightly cooler.

Monsoon climates produce the highest amounts of precipitation on Earth. In some locations, total annual precipitation can be over 11,000 mm. Considering that this rainfall is concentrated in seven or eight months, that is a lot of rain! It is the abundant precipitation characteristic of monsoon climates that allows them to support rainforests, despite having dry seasons. Because these climates receive less than 60 mm of rain in the driest month but still receive enough

precipitation annually to support rainforests, they are designated Am in Köppen's system.

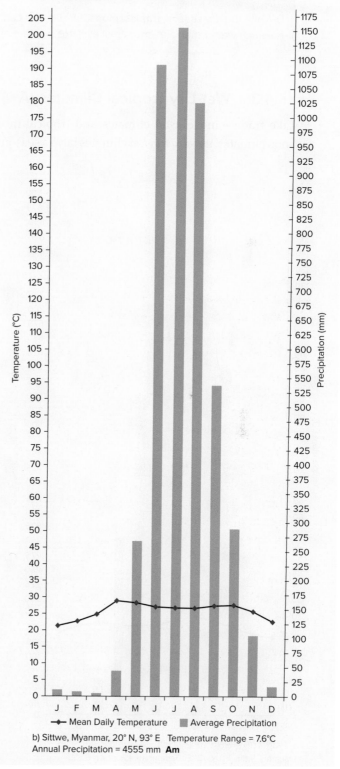

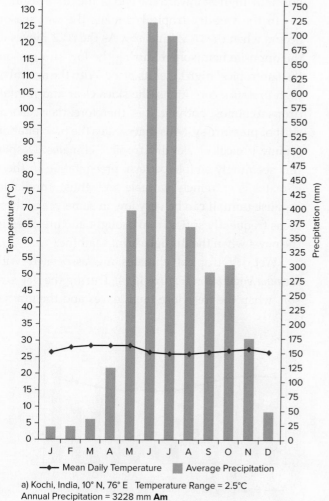

a) Kochi, India, 10° N, 76° E Temperature Range = 2.5°C
Annual Precipitation = 3228 mm **Am**

b) Sittwe, Myanmar, 20° N, 93° E Temperature Range = 7.6°C
Annual Precipitation = 4555 mm **Am**

FIGURE 16.12 | a) Kochi, India, and b) Sittwe, Myanmar, have tropical monsoon climates.

Remember This

Despite the dry season, tropical monsoon climates can be much wetter than wet equatorial climates.

16.4.3 | Wet-Dry Tropical Climates (Aw)

Like trade wind coastal climates and tropical monsoon climates, *wet-dry tropical* climates have a seasonal precipitation regime in which the wet season is the high-sun season, but the distinguishing feature of wet-dry tropical climates is that the wet season is usually short and brings much less precipitation. Regions of wet-dry tropical climates are generally found between the latitudes of 5° to 20° N and S. Figure 16.6 shows that the largest areas with wet-dry tropical climates lie poleward of the wet equatorial climates in both Africa and South America. There are also small regions of wet-dry tropical climates in northern Australia, Southeast Asia, and southern Mexico. The climates of Mumbai, India, and Rio de Janeiro, Brazil, are representative of wet-dry tropical climates (Figure 16.13). In comparison to both the wet equatorial and the tropical monsoon climates, these climates have lower annual precipitation totals. Although the temperature changes very little in these climates, it is normally highest toward the end of the dry season.

In the wet-dry tropical climate, the wet season occurs when the ITCZ is nearby. As the ITCZ moves to the opposite hemisphere during the low-sun season, the subtropical high takes its place. With the arrival of high-pressure conditions, the skies clear and rainfall drops to almost nothing. It is, therefore, the seasonal shift of pressure systems that creates the precipitation regime typical of wet-dry tropical climates. Despite this seemingly simple pattern, precipitation in these climates is extremely variable and, thus, unreliable. Because rainfall can be very low in some years, these areas frequently suffer from drought and are often in the news when the drought brings famine.

Wet-dry tropical climates are associated with *savanna* vegetation (Figure 16.14). During the dry season, when the trees lose their leaves and the grasses

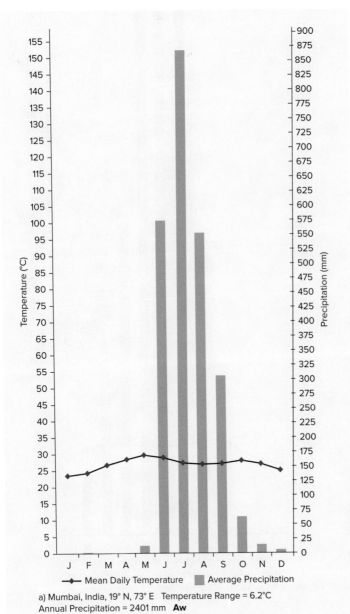

a) Mumbai, India, 19° N, 73° E   Temperature Range = 6.2°C
Annual Precipitation = 2401 mm   **Aw**

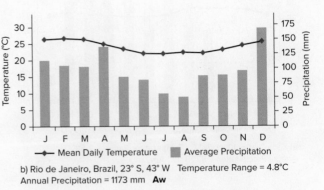

b) Rio de Janeiro, Brazil, 23° S, 43° W   Temperature Range = 4.8°C
Annual Precipitation = 1173 mm   **Aw**

FIGURE 16.13 | a) Mumbai, India, and b) Rio de Janeiro, Brazil, have wet-dry tropical climates.

FIGURE 16.14 | Savanna vegetation on the Serengeti plains, Tanzania, with widely spaced acacia trees.

wither, the grazing animals that are dependent on the savanna migrate in search of food. Although it is not certain, it seems that fires occurring during the drought season are what maintain this type of vegetation. In Köppen's system, wet-dry tropical climates are classed as Aw because the precipitation of the driest month is less than 60 mm, but the total annual precipitation is not enough to support forests. The Aw climate type is the second most common climate type, covering 11.5 per cent of Earth's land surface (Peel, Finlayson, & McMahon, 2007).

There are at least two climate types that are similar to wet-dry tropical climates but that don't fit the Aw designation. The first type occurs in places where the dry season occurs during the time of *high* sun rather than *low* sun; this type is designated As in Köppen's system. These climate types are rare; an example is Honolulu, Hawaii. The second type occurs in tropical locations that are at high altitudes. Because these climates have dry seasons at the time of low sun but are too cold to be classed as tropical climates, they are designated Cwa or Cwb. An example of a place with a Cwa climate is Guadalajara, Mexico, which is located at about 21° N, at an altitude of just over 1500 m. An example of a place with a Cwb climate is Cuzco, Peru, which is located at 13½° S, at an altitude of 3400 m.

16.4.4 | Subtropical Desert and Steppe Climates (BWh, BSh)

The *subtropical desert and steppe climates* are hot and dry year round. Larger annual temperature ranges make these climates a bit more variable than the wet

Question for Thought

How do you think the Cwa and Cwb climate types associated with high-altitude tropical locations could be included in the genetic climate classification system outlined in Table 16.1?

equatorial climates. These dry climates lie between 10° and 30° N and S, with the steppes often forming transitional zones surrounding the drier deserts. Together, the dry desert and steppe climates of the subtropics and the mid-latitudes cover about one-third of Earth's land area (Figure 16.6). In fact, BWh is the most common climate type, as it covers 14.2 per cent of Earth's land surface (Peel, Finlayson, & McMahon, 2007). The world's largest area of desert climate extends from the Sahara in northern Africa all the way through the Middle East and includes the mid-latitude deserts of central Asia. Subtropical deserts also occur in Mexico and the southwestern United States, southern Africa, and most of Australia. Finally, the Atacama Desert, described above, is situated in the tropics along South America's west coast.

Faya-Largeau, Chad, located in the middle of the Sahara Desert, is an example of a subtropical *desert*, while Monterrey, Mexico, is an example of a subtropical *steppe* (Figure 16.15). Both climates have annual temperature ranges that are larger than those we have seen for any other tropical climate, due mostly to their higher latitudes. Faya-Largeau receives, on average, only 18 mm of rain per year. Monterrey receives considerably more—approximately 383 mm per year—but it remains a dry climate because of its high temperatures and the fact that most of its precipitation falls in summer. Using Köppen's formulas, Faya-Largeau is classified as a BWh climate, while Monterrey, with lower temperatures and more precipitation, is classified as a BSh climate. Both Faya-Largeau and Monterrey are designated with the third letter *h* because their mean annual temperatures are above 18°C.

Notice that a pattern is beginning to emerge. Regions of wet equatorial climates are wet all year due to their proximity to the ITCZ. Regions of wet-dry tropical climates have a wet season when the ITCZ is near and a dry season when they are influenced by

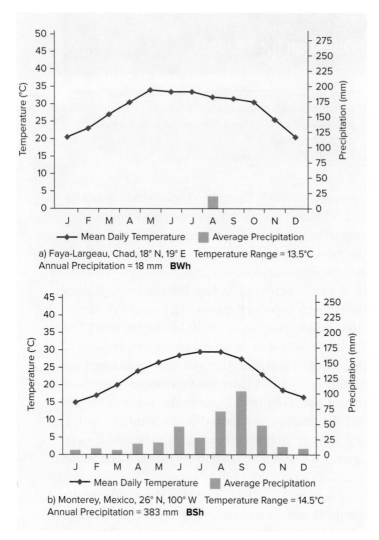

a) Faya-Largeau, Chad, 18° N, 19° E   Temperature Range = 13.5°C
Annual Precipitation = 18 mm   **BWh**

b) Monterey, Mexico, 26° N, 100° W   Temperature Range = 14.5°C
Annual Precipitation = 383 mm   **BSh**

FIGURE 16.15 | a) Faya-Largeau, Chad, has a subtropical desert climate, and b) Monterrey, Mexico, has a subtropical steppe climate.

the subtropical high. Regions of subtropical dry climates are influenced by the subtropical high all year. The small amount of rain that does fall in these dry climates is usually convectional in origin and, as a result, sporadic and unreliable. Steppe climates tend to receive their precipitation either from the ITCZ or from mid-latitude cyclones. The summertime peak in precipitation in Monterrey is due to the close proximity of the ITCZ. In subtropical steppe climates that are located farther poleward, the rain falls mostly in winter because it is more likely to be associated with mid-latitude cyclones.

Summer afternoon temperatures in subtropical deserts are the highest on Earth. In fact, daytime maximum temperatures in these regions can reach 45°C. With high sun angles and clear skies, high amounts of solar radiation reach the surface; because the surface is dry, little energy is used for evaporation. In addition, the clear skies and low humidity mean that absorption of longwave radiation leaving the surface is minimal, so temperatures drop rapidly overnight. As a result, not only do deserts record the highest temperatures, they also exhibit the greatest *daily range* in temperature of any climate type. As described in Section 16.2, desert vegetation must adapt to these extreme conditions of heat and aridity (Figure 16.16).

All deserts are dry, but not all deserts have clear skies and high temperatures, even in the subtropics. Where cold currents flow along west coasts, subtropical deserts of a different character will form. One

FIGURE 16.16 | Subtropical desert vegetation in the Atacama Desert, Chile.

such desert is the Atacama, along the west coast of South America; another is the Namib Desert, along the west coast of southern Africa. In these locations, air is cooled from below as it moves over the cold ocean currents; this cooling makes the air very stable and, thus, unlikely to rise and form clouds and rain. However, at the same time, the air can be cooled to its dew-point temperature and form advection fog. These processes produce foggy deserts with temperatures that are considerably lower than temperatures in other deserts at comparable latitudes.

One of these coastal deserts, the Atacama, holds the distinction of being the world's driest desert. There are places in the Atacama Desert where rainfall has never been recorded. This extreme aridity is due not only to the influence of the subtropical high and air that is further stabilized by the cold current, but also to the desert's position on the leeward side of the Andes Mountains. In recent years, people from a village in the Atacama Desert have devised an ingenious method of collecting water from the fog. This method, which they call *camanchaca*, involves hanging fine-meshed nets over troughs. As the fog rolls through the nets, the moisture is captured and drips into the troughs below. In this way, the village is supplied with several thousand litres of water every day.

16.5 | Mid-latitude Climates

Areas with mid-latitude climates typically fall between 30° and 60° N and S. Unlike the tropics, these regions experience both warm *and* cold temperatures. The mid-latitudes experience seasonal temperature variations because both sun angle and day length are highly variable in these locations (Section 5.8.2). Throughout the mid-latitudes, the noon sun angle fluctuates 47° over the course of a year and, in the higher mid-latitudes, day length can vary by as much as 12 hours over the same period. Mid-latitude locations can experience summer temperatures that are as warm as those in the tropics, while their winter temperatures are considerably colder than those of the tropics (Figure 16.2). This is because the poleward *decrease* in sun angle is offset by the poleward *increase* in day length in summer, while both sun angle *and* day length decrease with increasing latitude in winter.

It follows that annual temperature ranges are considerably greater in the mid-latitudes than they are in the tropics (Figure 16.3). *Within* the mid-latitudes, however, there is considerable variation in annual temperature range due to the combined effects of continentality and prevailing winds.

Day-to-day weather is also far more variable in the mid-latitudes than it is in the tropics. This variation results from the influence of mid-latitude cyclones, which form where warm air from the tropics meets cold air from the poles (Section 14.2). As these storms move eastward in the prevailing westerly flow, they cause the weather to fluctuate between stormy and calm, cloudy and clear. Weather in the mid-latitudes is also influenced by the subtropical highs and the polar highs. In fact, the only pressure system unlikely to influence mid-latitude weather is the ITCZ.

There are six mid-latitude climate types. The first three are moist climates with *mild* winters; they are designated C in Köppen's system. The fourth and fifth are moist climates with *severe* winters, designated D in Köppen's system. The sixth type comprises the dry climates of the mid-latitudes. As you will see, vegetation types don't correlate as well with climate types in the mid-latitudes as they do in the tropics.

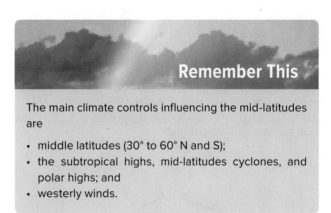

Remember This

The main climate controls influencing the mid-latitudes are

- middle latitudes (30° to 60° N and S);
- the subtropical highs, mid-latitudes cyclones, and polar highs; and
- westerly winds.

16.5.1 | Mediterranean Climates (Csa, Csb)

Mediterranean climates have moderate annual temperature ranges with hot summers and mild winters, but the distinguishing feature of these climates is their dry summers. Climates with dry summers are less common than climates with wet summers or climates with even annual distributions of precipitation. Areas of Mediterranean climate are found between about 30° to 45° N and S, on *west* coasts. As might be expected, the largest area with Mediterranean climates occurs around the Mediterranean Sea. These climates are also found along the coast of California, in small regions on the west coast of South America, and along the southwest coasts of Africa and Australia.

The climate graphs for Rome, Italy, and San Jose, United States, show the characteristic features of a Mediterranean climate (Figure 16.17). Lower mid-latitude west coast locations give these climates their hot summers, mild winters, and relatively low annual temperature ranges. Precipitation drops significantly in the summer, when these areas are influenced by the dry, stable side of the subtropical highs (Section 12.2.3). In the winter, the subtropical highs weaken and shift equatorward, leaving the mid-latitude cyclones to bring precipitation to Mediterranean climates. With their mild, wet

SCLEROPHYLL
Woody plants with small, waxy leaves.

winters and hot, dry summers, both Rome and San Jose are classified as Csa in Köppen's system. However, because San Jose is considerably drier, it comes very close to being classified as a steppe climate, or BSk. Notice that it is only because most of San Jose's precipitation falls in the cool season that it is *not* a BSk climate.

The hot, dry summers put severe limitations on the native vegetation in regions with Mediterranean climate types. Because these areas are hot and dry at the same time, they develop severe moisture deficits. The vegetation found in these regions is known as **sclerophyll**, meaning "hard leaves" (Figure 16.18). This vegetation type consists of woody trees and shrubs with small, waxy leaves that reduce the loss of moisture through transpiration. Dominant tree species in these regions include olive, oak, and eucalyptus.

16.5.2 | Moist Subtropical Climates (Cfa)

The main difference between Mediterranean climates and *moist subtropical climates* is that the latter do not have a dry season; thus, while Mediterranean

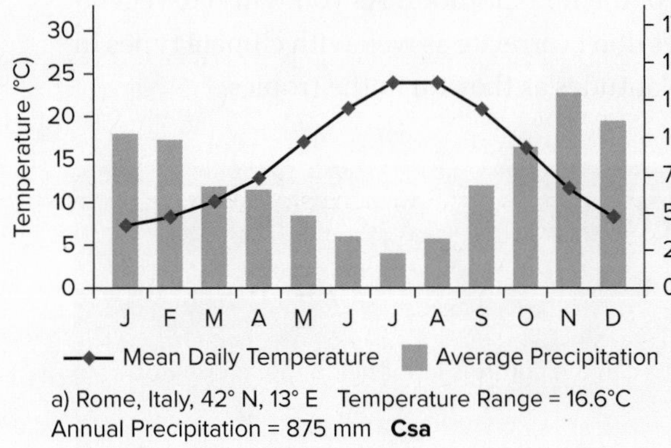

a) Rome, Italy, 42° N, 13° E   Temperature Range = 16.6°C
Annual Precipitation = 875 mm **Csa**

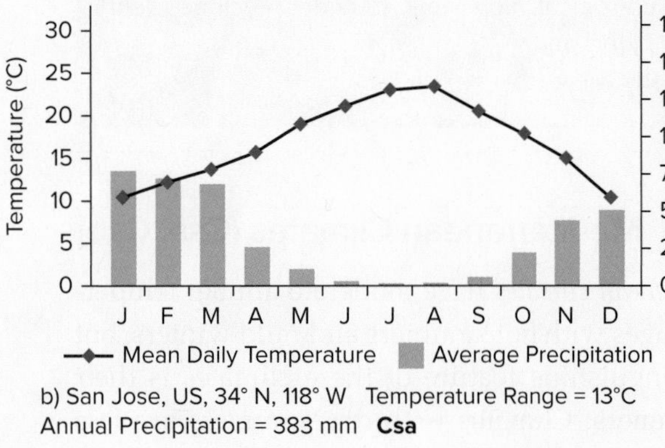

b) San Jose, US, 34° N, 118° W   Temperature Range = 13°C
Annual Precipitation = 383 mm **Csa**

FIGURE 16.17 | a) Rome, Italy, and b) San Jose, United States, have Mediterranean climates.

FIGURE 16.18 | A sclerophyll forest of eucalyptus near Pemberton, Australia.

climates are classified as Csa in Köppen's system, moist subtropical climates are classified as Cfa. Areas with moist subtropical climates are located in roughly the same latitude range as those with Mediterranean climates, although moist subtropical climates extend a little farther equatorward in some places. In addition, they occur on the eastern, rather than the western, sides of continents. The largest areas of moist subtropical climates are found in the southeastern United States, southeastern China, southern Japan, and southeastern South America. There are also small areas of these climates on the east coast of southern Africa and the northeast coast of Australia.

Charleston, United States, and Guangzhou, China, are both examples of locations with moist subtropical climates (Figure 16.19). The annual temperature range in these regions tends to be slightly higher than those in Mediterranean climates as these regions tend to have summers that are slightly warmer and winters that are slightly cooler than those of Mediterranean climates. This greater variation in temperature occurs because these locations are situated in the eastern portions of continents, where prevailing winds have travelled over land rather than water. Total precipitation amounts are relatively high in these climates, and precipitation falls mostly in summer. Despite being located at roughly the same latitudes as areas with Mediterranean climate, areas with moist subtropical climates do not experience the dry summers associated with the Mediterranean climate type because they are located on the moist, unstable side of the subtropical highs. As a result, summers are wet and often very humid. The summer rain is usually in the form of thundershowers, but these locations can receive extreme amounts of rain from hurricanes toward the end of the summer. As in Mediterranean climates, moist subtropical climates receive winter precipitation from mid-latitude cyclones.

Although considered to be a moist subtropical climate and classified as Cfa in Köppen's system, Guangzhou is less typical of this climate type than is Charleston. Guangzhou has both a lower annual temperature range and a less even distribution of precipitation. The former is due to its lower latitude, while the latter is due to its location in *southeast* Asia, where it is influenced by the monsoon circulation.

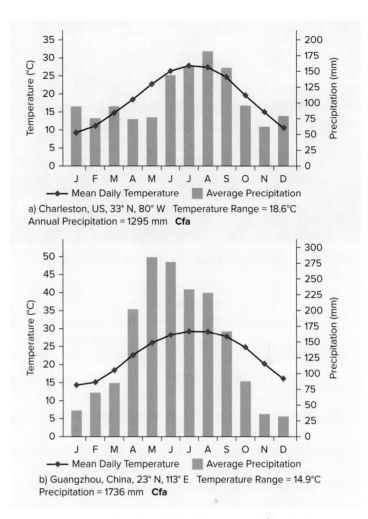

a) Charleston, US, 33° N, 80° W Temperature Range = 18.6°C
Annual Precipitation = 1295 mm **Cfa**

b) Guangzhou, China, 23° N, 113° E Temperature Range = 14.9°C
Precipitation = 1736 mm **Cfa**

FIGURE 16.19 | a) Charleston, United States, and b) Guangzhou, China, have moist subtropical climates.

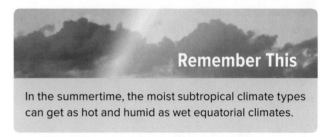

Remember This

In the summertime, the moist subtropical climate types can get as hot and humid as wet equatorial climates.

With relatively abundant precipitation throughout the year, particularly in the warm season, moist subtropical climates can support *broadleaf evergreen forests* similar to, but less dense than, those of the tropical rainforests (Figure 16.20). However, native vegetation in this climate type is quite varied. For example, *needleleaf evergreen forests* can be found in the southeastern United States, and *mid-latitude deciduous forests* gradually become dominant in the northern and western areas of this climate region.

FIGURE 16.20 | A subtropical broadleaf evergreen forest in Danxiashan, China.

There are also regions in southeast Asia in which climates are designated Cwa. These regions are similar to those designated Cfa, but they have wet summers and dry winters that are even more strongly influenced by the monsoon circulation in the area than is Guangzhou. As a result, these regions are grouped with the high-altitude regions of the tropics, which were discussed in Section 16.4.3. In Köppen's system, therefore, the Cwa and Cwb of the tropics and the Cwa of the lower mid-latitudes have the same designation, yet their origins are quite different.

Question for Thought

How do you think the Cwa climate type associated with the monsoon circulation in southeast Asia could be included in the genetic climate classification system outlined in Table 16.1?

16.5.3 | Marine West Coast Climates (Cfb, Cfc)

Marine west coast climates are similar to moist subtropical climates in that they have no dry season, but they differ from these climates as well as Mediterranean climates in that they are typically colder. Most often, they are located on west coasts between about 40° and 60° N and S. In North America, the marine west coast climate occurs in a narrow strip that runs along the west coast, from northern California to Alaska. This climate type also occurs in northwestern Europe, New

Zealand, and southern Chile. In addition, marine west coast climates occur, somewhat anomalously, on the *east* coasts of southern Africa and Australia.

Estevan Point, Canada, and Reykjavik, Iceland, are two examples of locations with marine west coast climates (Figure 16.21). Estevan Point is classified as a Cfb climate, while Reykjavik is a Cfc climate because of its colder summers. Due to their locations on windward coasts, these two places have very mild winters compared to other places at similar latitudes. This is particularly true of Reykjavik, which is located at 64° N! In most cases, precipitation in marine west coast climates is quite high, and more of it falls in

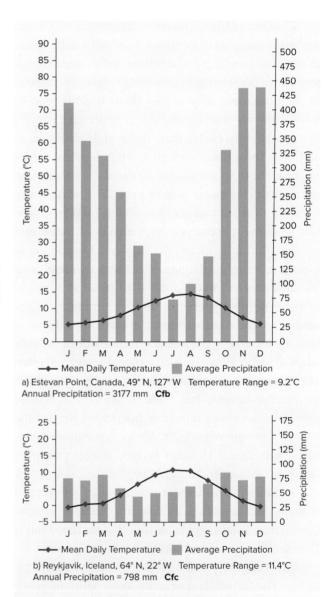

a) Estevan Point, Canada, 49° N, 127° W   Temperature Range = 9.2°C
Annual Precipitation = 3177 mm   **Cfb**

b) Reykjavik, Iceland, 64° N, 22° W   Temperature Range = 11.4°C
Annual Precipitation = 798 mm   **Cfc**

FIGURE 16.21 | a) Estevan Point, Canada, and b) Reykjavik, Iceland, have marine west coast climates.

winter than in summer. This year-round precipitation comes from being in the path of eastward-moving mid-latitude cyclones, while the drop in precipitation during the summer results from the poleward shift of the subtropical highs at this time. Being on west coasts, these locations are influenced by the dry stable sides of these highs, in the same way that Mediterranean climates are influenced by the stable side of these highs. In fact, some marine west coast locations remain dry enough in the summer that they are classified as Csb in Köppen's system. An example is Vancouver, which has a precipitation regime similar to that of a Mediterranean climate, although its more northerly location makes it both cooler and wetter than typical Mediterranean climates (Figure 16.22).

Not only is the Pacific Northwest region of North America a windward coast, it is also a *mountainous* coast. As a result, marine west coast climates of North America are generally wetter than those of Europe, where there are no comparable mountain ranges. In fact, the Pacific Northwest is one of the wettest places on Earth. To see this, compare the total annual precipitation at Estevan Point to that of the wet equatorial and tropical monsoon climates discussed earlier.

It is largely this difference in precipitation amounts that accounts for the differences in vegetation between the marine west coast climates of western North America and those of Europe. In most of the marine west coast climate regions of Europe, the natural vegetation is *temperate deciduous forest* (Figure 16.23). These

Remember This

Some marine west coast climates can be as wet as wet equatorial and tropical monsoon climates.

Question for Thought

How do you think the climates of Canada's west coast could be included in the genetic climate classification system outlined in Table 16.1?

forests are made up of trees, such as oaks, birches, elms, maples, and beeches, that adapt to winter conditions by shedding their leaves. In contrast, the natural vegetation of North America's marine west coast climate regions is *temperate rainforest* (Figure 16.24). These needleleaf evergreen forests contain trees such as Douglas fir, western red cedar, Sitka spruce, and western hemlock, which are among the largest trees on Earth. The *needleleaf* evergreen trees are better adapted to colder temperatures than are the *broadleaf* evergreen trees of the tropical rainforests. However, some broadleaf trees, such as bigleaf maples, thrive in the temperate rainforests by shedding their leaves in the winter. The wet conditions of these forests also allow for a lush growth of such plants as mosses, lichens, and ferns.

16.5.4 | Moist Continental Climates (Dfa, Dfb, Dwa, Dwb)

Moist continental climates are distinguished from the three mid-latitude climates discussed so far by their severe winters and, thus, their larger annual temperature ranges. This climate type is found between 40° and 60° N. In North America, a large region of moist continental climate extends from the Rockies, through southern Canada and the northern United States, to the east coast. Thus, most Canadians live in regions associated with this climate type. In addition, moist continental climates occur in eastern Europe, parts of southwestern Russia, and northeastern China. These climates are not found in the southern hemisphere because there are no large continents at these latitudes.

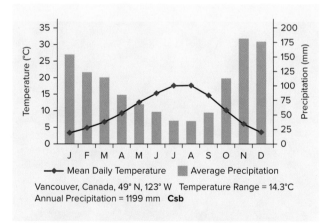

FIGURE 16.22 | According to Köppen's system of classification, Vancouver is classified as Csb, indicating that, similar to Mediterranean climates, its summers are much drier than its winters. But, overall, Vancouver's climate is both cooler and wetter than a Mediterranean climate.

Examples of moist continental climates include the climates of Toronto and Winnipeg (Figure 16.25). Using Köppen's system, Toronto is classified as having a Dfa climate, while Winnipeg is classified as having a Dfb climate because it experiences slightly

Remember This

There are no "D" climates in the southern hemisphere because there are no large continents between the latitudes of 40° and 60° S.

cooler summers. The typical annual temperature range for moist continental climates is higher than that for any other climate we have considered so far. This wide range in temperatures arises because moist continental climate regions, being mid-latitude continental or east coast locations, are not affected by the moderating influence of the oceans. (Remember that, because of the prevailing westerly winds, east coasts in the mid-latitudes tend to be influenced more by continental air than by maritime air.) In addition, the annual temperature range within moist continental climates increases with distance inland and toward the north.

Places with moist continental climates receive moderate amounts of precipitation, with precipitation coming mostly from mid-latitude cyclones. They have no dry season, but summer precipitation is sometimes higher than winter precipitation due to convectional showers. Moving westward toward the interior of the North American continent, and also northward, there is less precipitation overall, and more of the winter precipitation is snow. This is because these places are

FIGURE 16.23 | Temperate deciduous forests in summer (*top*) and winter (*bottom*).

FIGURE 16.24 | Temperate rainforest in Carmanah Valley, on Vancouver Island, with characteristic big trees, mosses, and ferns.

farther from the moist air that originates in the south-eastern portions of the continents, and closer to the cold, dry air that invades from the north in winter. For example, Toronto, located at 44° N, has a total annual precipitation of almost 830 mm, and this precipitation is distributed fairly evenly throughout the year; at the same time, Winnipeg, located at 50° N and far-ther west, receives just over 500 mm of precipitation, most of which falls in summer. In addition, Toronto receives snow as well as a moderate amount of rain in the winter, while Winnipeg receives almost all of its winter precipitation in the form of snow. In far eastern Asia, there are regions of moist continental climate where winters are so dry that these regions are classi-fied as Dwa and Dwb in Köppen's system. These very dry winters are associated with the cold, dry air of the Siberian high, while the much wetter summers result from the monsoon circulation in this area.

In the southern regions of the moist continental climate type, the vegetation is predominantly *tem-perate deciduous forest*. In the drier western regions of this climate type, the forests give way to grasslands. As conditions get drier and colder toward the north, *needleleaf* forests become dominant. In Canada, this type of forest is known as the *boreal forest*; in Eurasia, it is called the *taiga* (Figure 16.26). Like the temperate rainforests, the boreal forests are mostly needleleaf evergreen trees, but that is where the similarity ends. The trees of the boreal forests are much smaller and more widely spaced than those of the temperate rain-forests. A few varieties of spruce, pine, and fir grow in these forests, as do larches, which are *deciduous* needle-leaf trees. Although boreal forests do occur toward the northern limits of moist continental climates, they are better associated with subarctic climates.

16.5.5 | Subarctic Climates (Dfc, Dfd, Dwc, Dwd)

There is no distinct difference between the *subarctic climate* type and the moist continental climate type, but areas of subarctic climate have shorter, cooler summers and much colder winters than do moist con-tinental climates. As such, the third letter for this cli-mate type is either *c* or *d* in Köppen's system. Subarctic climates are generally found in regions lying between 50° and 70° N. The southern limit of this climate region

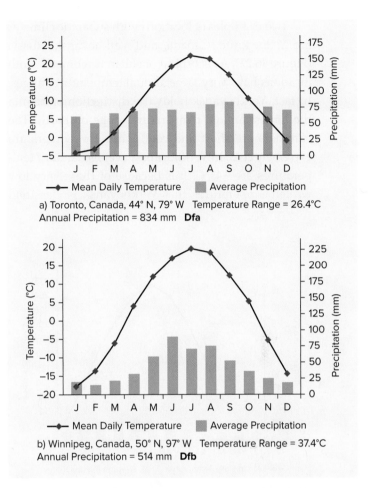

a) Toronto, Canada, 44° N, 79° W   Temperature Range = 26.4°C
Annual Precipitation = 834 mm   **Dfa**

b) Winnipeg, Canada, 50° N, 97° W   Temperature Range = 37.4°C
Annual Precipitation = 514 mm   **Dfb**

FIGURE 16.25 | a) Toronto, Canada, and b) Winnipeg, Canada, have moist continental climates.

is much farther north in the west than it is in the east, in both North America and Eurasia. This difference can be attributed to the prevailing westerly winds bringing milder winter conditions to the west coast.

FIGURE 16.26 | A spruce tree boreal forest in the Yukon. Note how the evergreens are widely spaced in this subarctic climate.

Two examples of locations with subarctic climates are Yellowknife, Canada, and Verkhoyansk, Russia (Figure 16.27). As is typical of subarctic climates, both locations have very large annual temperature ranges. In fact, Verkhoyansk holds the distinction of having the largest annual temperature range on Earth. The summers are short and cool, while the winters are extremely cold. In Verkhoyansk, average January temperatures are −45°C! It is because of these very low winter temperatures that Verkhoyansk is classified as Dfd in Köppen's system, while Yellowknife is Dfc. Precipitation in subarctic climates is low due to both the cold and the dominance of high-pressure conditions. Precipitation falls mostly in the summer and comes from mid-latitude cyclones. It is in these cold, dry conditions that the boreal forests of North America and the taiga of Eurasia are found.

16.5.6 | Mid-latitude Desert and Steppe Climates (BWk, BSk)

Like the desert and steppe climates of the subtropics, the *mid-latitude desert* and *steppe* climates are dry all year; unlike their subtropical counterparts, however, they have cold seasons. In Köppen's system, the mid-latitude deserts and steppes are designated BWk and BSk, respectively. Most of these climates are located between about 35° and 50° N, and they cover extensive regions of the northern hemisphere, particularly in Asia. In the southern hemisphere, they are limited to areas of Argentina and Chile.

Isfahan, Iran, and Albuquerque, United States, both have mid-latitude *desert* climates, while Kamloops, Canada, has a mid-latitude *steppe* climate (Figure 16.28). Because of their higher latitudes, mid-latitude dry climates have cooler temperatures, and larger annual temperature ranges, than do the dry climates of the subtropics. The low precipitation in areas of mid-latitude dry climates is due to their position on either the leeward side of mountain ranges or far from major moisture sources. Kamloops, for example, lies in the rain shadow of the Coast Mountains of BC, while Isfahan lies in the rain shadow of the Himalayas. In the southern hemisphere, the Patagonian Desert and Steppe of Argentina lie in the rain shadow of the Andes. Precipitation regimes in mid-latitude dry climates show considerable variation from one place to another. In Isfahan, precipitation is very low, and it falls mostly in the winter due to the passage of mid-latitude cyclones at this time. On the other hand, precipitation is somewhat higher in Albuquerque, and it is mostly convectional summer precipitation. In fact, it is *because* most of Albuquerque's precipitation falls in summer that this location has a desert climate rather than a steppe climate. In Kamloops, the small summer peak is associated with convectional

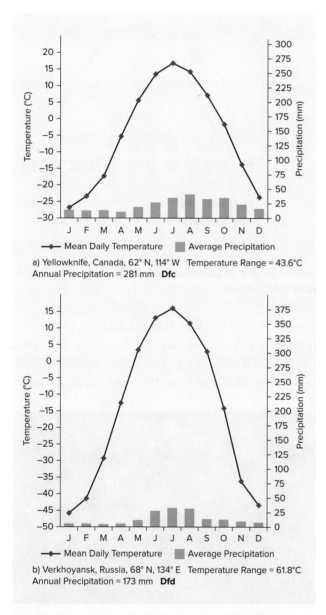

a) Yellowknife, Canada, 62° N, 114° W Temperature Range = 43.6°C
Annual Precipitation = 281 mm **Dfc**

b) Verkhoyansk, Russia, 68° N, 134° E Temperature Range = 61.8°C
Annual Precipitation = 173 mm **Dfd**

FIGURE 16.27 | a) Yellowknife, Canada, and b) Verkhoyansk, Russia, have subarctic climates.

precipitation, while the small winter peak is associated with cyclonic precipitation.

Because they are colder, the mid-latitude dry climates are not usually associated with the extreme aridity that is characteristic of some of the subtropical deserts; thus, conditions for vegetation are somewhat

less limiting. Under the most arid conditions, vegetation in dry mid-latitude climates consists of small woody shrubs, such as sagebrush (Figure 16.29, top). In regions of semi-arid climate, the vegetation is usually grassland (Figure 16.29, bottom). The natural vegetation of much of the south-central Canadian Plains is mixed prairie, a grassland type that is made up of

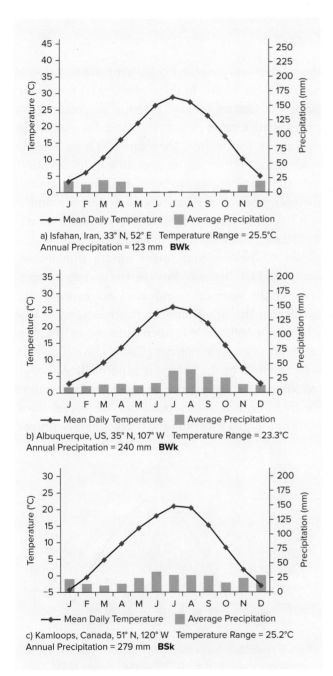

a) Isfahan, Iran, 33° N, 52° E   Temperature Range = 25.5°C
Annual Precipitation = 123 mm   **BWk**

b) Albuquerque, US, 35° N, 107° W   Temperature Range = 23.3°C
Annual Precipitation = 240 mm   **BWk**

c) Kamloops, Canada, 51° N, 120° W   Temperature Range = 25.2°C
Annual Precipitation = 279 mm   **BSk**

FIGURE 16.28 | a) Isfahan, Iran, and b) Albuquerque, United States, have mid-latitude desert climates. c) Kamloops, Canada, has a mid-latitude steppe climate.

FIGURE 16.29 | *Top:* Sagebrush is found in the Great Sandhills near Sceptre, Saskatchewan. *Bottom:* Big bluestem is among the tall native wild grasses found in the Tall Grass Prairie Preserve in Tolstoi, Manitoba.

short- and medium-height grasses. Short-grass prairie grows toward the west in the rain shadow of the Rockies, while tall-grass prairie thrives toward the east in the wetter regions of southern Manitoba.

16.6 | Polar Climates

Polar climates generally occur from about 60° N and S to the poles, but they can be found as far south as almost 50° N in eastern Canada. Polar climates are always cold; in Köppen's system, they are defined as those climates in which the average temperature of the warmest month is below 10°C. Despite the long summer days, temperatures remain low because sun angles are always low, and because much of the sun's radiation is reflected by ice and snow. Further, the large variation in day lengths causes large annual temperature ranges.

Because this region of the world is dominated by the polar high, the skies are usually clear and precipitation is low. The dryness is also due to the cold: cold air holds very little moisture and provides little energy to fuel storms. In fact, the poles are the least stormy areas on Earth. Although polar climates have low precipitation, the cold ensures that evaporation rates will be even lower, so there are no moisture deficits. Polar climates are divided into two types—polar tundra climates, in which temperatures are above freezing for a few months, and polar ice cap climates, in which the average monthly temperature is never above freezing.

16.6.1 | Polar Tundra Climates (ET)

In *polar tundra* climates, there is a short, cool summer and a long, cold winter. This climate type occurs along Canada's northern coast from about 65° N in the west to just north of 50° N in the east, and it extends from this coast through the Arctic islands to the North Pole. It is also found on the coasts of Greenland, the north coast of Iceland, and the north coast of Eurasia. In the southern hemisphere, the southern tip of South America and the Antarctic Peninsula are the only regions with polar tundra climates.

Iqaluit, Canada, is an example of a location with a polar tundra climate (Figure 16.30). Although

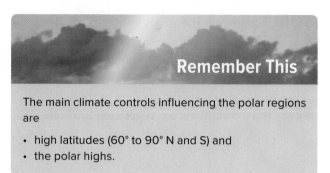

Remember This

The main climate controls influencing the polar regions are

- high latitudes (60° to 90° N and S) and
- the polar highs.

the winters are long and cold, temperatures do not normally fall as low in these climates as they do in subarctic climates because places with polar tundra climates are located close to the coast. Even when the oceans are ice-covered, they can still have a slight moderating effect on temperatures. In addition, because temperatures are below freezing for most of the year, regions of polar tundra climate are underlain by permafrost. Precipitation tends to fall mostly in summer, when the most moisture is available.

In Köppen's system, polar tundra climates are designated ET. In these climates, the average temperature of the warmest month does not exceed 10°C. Recall that the significance of this boundary is that where no months have temperatures above 10°C, forests will not grow. In northern Canada, this transition is represented by the *tree line* (Figure 16.31). Although trees may continue to grow in small patches north of the tree line, it is at this point that the boreal forest more or less gives way to tundra vegetation. As noted

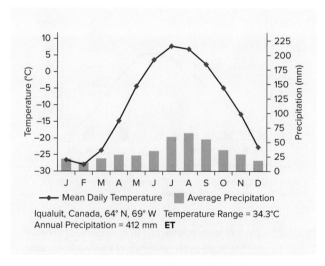

Iqaluit, Canada, 64° N, 69° W   Temperature Range = 34.3°C
Annual Precipitation = 412 mm   **ET**

FIGURE 16.30 | Iqaluit, Canada, has a polar tundra climate.

FIGURE 16.31 | The tree line in Tombstone Territorial Park, Yukon. To the south are boreal forests, and to the north is tundra.

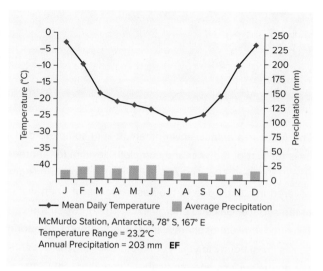

McMurdo Station, Antarctica, 78° S, 167° E
Temperature Range = 23.2°C
Annual Precipitation = 203 mm **EF**

FIGURE 16.33 | McMurdo Station, Antarctica, has a polar ice cap climate.

in Section 16.2, above, tundra plants are low-growing plants that are adapted to a very short growing season, extreme cold, saturated soil conditions, and harsh winds (Figure 16.32).

16.6.2 | Polar Ice Cap Climates (EF)

Areas of *polar ice cap* climate are found in the interior of Greenland and in Antarctica. McMurdo Station, Antarctica, is an example of a place with a polar ice cap climate (Figure 16.33). In this climate type, the average temperature of all months is below freezing. Such climates are designated EF in Köppen's system. The total annual precipitation in polar ice cap climates is usually very low, and precipitation is normally snow. Because temperatures remain below freezing, the precipitation simply accumulates on the glaciers, and these climates cannot support vegetation (Figure 16.34).

Question for Thought

What are the strengths and weaknesses of Köppen's climate classification system?

FIGURE 16.32 | Tundra vegetation on southern Victoria Island, Nunavut, with the low-growing plants typical of this climate.

FIGURE 16.34 | The Greenland ice sheet.

In the Field

Understanding and Monitoring Arctic Environments

John Iacozza, University of Manitoba

The Arctic is a unique environment. It is in complete darkness for part of the year and complete daylight for another part of the year. During the daylight period, large areas are covered in clouds due to the vast expanse of open water at the surface. Gas and energy exchange between the surface and the atmosphere varies over different temporal scales (from hourly to yearly), and this exchange is dependent on the surface and atmospheric conditions. In Canada, this area is also quite remote and lacks a significant human presence over large areas. These factors make studying the Arctic using *in situ* measurements (such as a meteorological tower) difficult; thus, remote sensing plays a key role in understanding and monitoring the Arctic system.

Remote sensing tools for studying Arctic meteorology can be classified into two groups: ground-based tools and satellite-based tools. Ground-based radars are currently used to examine parameters such as precipitation events and amounts. These systems are typically located at weather stations found close to Arctic communities. The radar system sends out a signal composed of radio waves or microwaves and records how much of the signal is scattered and how much is sent back to the receiver by the liquid droplets in the air. It operates using the Doppler effect, which uses the change in frequency or phase of the return signal to estimate the amount of precipitation in a general area. The primary advantage of this technique is that it provides information on short temporal scales (for instance, it can obtain precipitation estimates every 15 minutes or less) as well as short spatial scales (for instance, it can show variability in precipitation at the metre scale). However, the logistics of this system limit its usefulness for Arctic meteorology. Because this system is ground-based, it must be located near a weather station where it can be maintained. As well, this system can provide information for only a limited area (typically less than 500 km). Therefore, only a small number of ground-based radars are operated in the Arctic.

Satellite-based remote sensing tools are used more often than ground-based systems because they offer a more practical method for studying the Arctic. Researchers have been studying and observing meteorology from space since the early 1960s. For the Arctic region, polar-orbiting remote sensing systems orbit around the globe, essentially from pole to pole, collecting a swath of information. Unlike ground-based systems, the sensors onboard the satellites provide information for large areas and can cover the entire Arctic region on a weekly or even daily basis. In addition, since the sensors orbit Earth, the remoteness of the Arctic does not pose a hindrance on the acquisition of the information. However, these systems cannot provide the high spatial or temporal resolution that is characteristic of ground-based radar systems.

Over the past decade or so, a number of polar-orbiting satellites have been placed in orbit to examine various aspects of the planet's atmosphere, including the atmosphere of the Arctic region. For instance, NASA's CloudSat satellite was launched in 2006 to examine clouds, precipitation, and cloud–climate feedbacks. This radar sensor is 100 times more sensitive than existing ground-based radars and is able to detect liquid water particles in the atmosphere that are smaller than rain drops. Satellite-based remote sensing also provides information on climatological and meteorological parameters other than clouds. For instance, CERES (Clouds and Earth's Radiant Energy System) measures both solar-reflected and Earth-emitted radiation from the top of Earth's atmosphere to the surface, providing energy budget variables for the Arctic region as well as cloud fraction and optical depth. MOPITT (Measurements of Pollution in the Troposphere) is an instrument designed to measure tropospheric carbon monoxide and other pollutants. Such measurements allow researchers to investigate factors related to changes in the Arctic climate, such as transportation of carbon monoxide from industrialized areas and land use changes. MODIS (Moderate Resolution Imaging Spectroradiometer) uses visible and infrared radiation to provide daily information on various meteorological parameters, such as surface temperature, water vapor, and aerosols. Unlike the previous sensors that use the visible and infrared portions of the electromagnetic spectrum to investigate Arctic meteorology, QuikSCAT (Quick Scatterometer) uses microwaves to provide information on near-surface winds over much of the Arctic region. These various sensors and instruments illustrate the functionality of using satellite-based remote sensing to examine Arctic, and global, meteorology and climatology.

JOHN IACOZZA is a senior instructor in the Clayton H. Riddell Faculty of Environment, Earth, and Resources at the University of Manitoba. His research focuses on the physical controls of habitat selection of Arctic marine mammals and involves using remote sensing technologies and geographic information systems to study the Arctic environment, in order to understand climate change impacts on the Arctic marine system.

16.7 | Chapter Summary

1. Six factors control the distribution of climate types around the world. Latitude influences temperatures and seasonality through its control over sun angle and day length. Effects of the distribution of land and water account for the fact that coastal locations have lower daily and annual temperature ranges than do inland locations; these differences arise because water has a moderating effect on temperature. Ocean currents control temperatures and precipitation because warm ocean currents warm, moisten, and destabilize the air above, while cold ocean currents have the opposite effect. The major pressure systems of the Earth's general circulation control precipitation. The equatorial low (ITCZ) and the subpolar lows (mid-latitude cyclones) bring precipitation; the subtropical highs and the polar highs bring clear skies. Prevailing winds also control climate; as a result, windward coasts are wetter, and have lower annual and daily temperature ranges, than leeward coasts. Finally, mountains have a strong control on the distribution of precipitation, as they force air to release its moisture on their windward sides, leaving their leeward sides dry. Mountains also control temperature because temperature decreases with altitude.

2. Forests require warmth and abundant moisture. Grasslands grow where moisture is more limited. Savanna grows in the tropics in places with a seasonal drought that lasts several months. Desert vegetation is adapted to conserve moisture. Tundra vegetation occurs at high latitudes where temperatures are too low to support forests.

3. There are three types of climate classification systems. *Applied* systems group climates for specific purposes. *Genetic* systems group climates based on their common causes. *Empiric* systems group climates based on climatic data, usually temperature and precipitation. Köppen's system is an example of an empiric climate classification system that sets boundaries for climate regions based on mean monthly temperature and precipitation values.

4. Tropical climates are hot due to their low-latitude locations. The precipitation regimes of these climates are influenced by the movement of the ITCZ and the subtropical highs. Places that are always affected by the ITCZ are wet year round, places that are always affected by the subtropical highs are dry year round, and places that are affected by both have wet and dry seasons. Tropical monsoon climates have a very wet season and a short dry season due to seasonal shifts in the winds. In the tropics, prevailing easterly winds make east coasts wetter than west coasts. The hottest, the wettest, and the driest places on Earth are located in the tropics.

5. Mid-latitude climates have warm and cold seasons. Continental locations experience greater seasonal differences in temperature than do marine locations. Most precipitation comes from mid-latitude cyclones. Some locations experience dry summers due to the poleward shift of these storms at that time. Prevailing westerly winds make west coasts wetter and milder than east coasts. Most dry mid-latitude climates form on the lee sides of mountains. The largest annual temperature ranges on Earth occur in the mid-latitudes.

6. Polar climates are always cold and have large annual temperature ranges. Precipitation is low because the air is cold and the pressure is normally high.

Review Questions

1. What are some examples of how each climate control influences climate?

2. What are the differences between the five vegetation biomes? What type of climate is associated with each biome?

3. How do genetic climate classification systems differ from empiric climate classification systems?

4. How are the differences between wet equatorial climates, wet/dry tropical climates, and dry subtropical climates associated with the movement of the two pressure systems that influence tropical climates?

5. How are wet-dry tropical climates different from tropical monsoon climates?

6. How do the climate controls associated with the trade wind coastal climates differ from those associated with the wet equatorial climates?

7. What causes the variation of annual temperature ranges in tropical climates?

8. How are Mediterranean climates, moist subtropical climates, and marine west coast climates similar to and different from one another?

9. What are the two sets of conditions that can produce Cwa and/or Cwb climates?

10. What climate controls produce "D" climates?

11. How are the origins and characteristics of dry subtropical climates similar to those of dry mid-latitude climates? How do these dry climates differ from one another?

12. In what climate types are prevailing winds important?

13. Which climate type has the distinction of having
    a) the highest temperatures?
    b) the lowest temperatures?
    c) the highest precipitation?
    d) the lowest precipitation?
    e) the smallest annual temperature ranges?
    f) the largest annual temperature ranges?
    g) a summer dry season?
    h) a winter dry season?

14. What climate types occur in Canada?

Suggestions for Research

1. Investigate climate classification systems other than Köppen's, and describe the benefits and drawbacks of each one.

2. Choose a climate type described in this chapter, and investigate how it might be affected by anthropogenic climate change. Find historical climate data for a region in which the climate type occurs, and note evidence that the climate of the region is changing.

References

Peel, M.C., B.L. Finlayson, & T.A. McMahon. (2007). "Updated world map of the Köppen–Geiger climate classification." *Hydrology and Earth Systems Science 11*, 1633–44.

17
The Changing Atmosphere

Learning Goals

After studying this chapter, you should be able to

1. *list* the major air pollutants and *describe* the main sources of each one;

2. *describe* conditions that influence air quality;

3. *explain* how the thinning of the ozone layer is linked to human activities;

4. *appreciate* how conditions over Antarctica have led to the formation of the ozone hole;

5. *identify* causes of both natural and anthropogenic climate change;

6. *list* the greenhouse gases that are increasing in the atmosphere and *explain* why each is increasing;

7. *describe* the various feedbacks associated with anthropogenic climate change;

8. *explain* how radiative forcing and climate sensitivity are used to make predictions about climate change;

9. *describe* sources of uncertainty associated with anthropogenic climate change; and

10. *evaluate* action taken to curb anthropogenic climate change.

Human-generated pollutants are drastically changing the composition of our atmosphere. Indeed, research suggests that anthropogenic climate change—driven in large part by the burning of fossil fuels—may be the greatest challenge we, as a species, have ever faced.

Since the beginning of the Industrial Revolution, humans have been inadvertently changing the composition of the atmosphere through their actions. At first, we noticed problems at the local scale, in the polluted air of our cities. *Air pollution* became a major concern in the 1960s, and since then the development of cleaner technologies has helped reduce this problem to some degree. In addition, because most of the pollutants found in the air of cities have relatively short residence times, air pollution tends to remain a *local* problem.

More recently, we have become increasingly aware that some of our emissions *do not* have such short residence times. In fact, these long-lived gases can be mixed so thoroughly throughout the atmosphere that their influence can have global consequences. During the 1980s, we made the terrifying discovery that certain long-lived synthetic gases—the chlorofluorocarbons (CFCs)—were *destroying the ozone layer*. Facing this threat to life on Earth, the nations of the world successfully banned the production and use of these gases. As a result, the ozone layer should eventually recover.

Like CFCs, most greenhouse gases also have relatively long residence times in the atmosphere. Since the early 1990s, there has been mounting evidence that our increasing emissions of these gases are beginning to impact global climate and that these climate changes will likely accelerate. Unlike CFCs, however, greenhouse gases are linked to almost every aspect of our lives, from our industries and transportation to our agricultural practices. Further, the consequences of *climate change* are far less clear than those of *ozone depletion*. For these reasons, we have had little success in dealing with this problem. Thus, *anthropogenic climate change* is possibly the greatest challenge we have ever faced.

AIR POLLUTANT
Any gaseous, liquid, or solid substance present in the atmosphere in sufficient quantity to negatively impact the health of people and animals, or harm plants or materials.

PRIMARY POLLUTANT
A pollutant emitted directly into the atmosphere.

SECONDARY POLLUTANT
A pollutant that forms in the atmosphere.

PARTICULATE MATTER
Atmospheric pollutants in solid or liquid form.

ACID RAIN
Rain with a pH less than about five.

17.1 | Air Pollution

17.1.1 | Types of Air Pollutants

Some **air pollutants** are emitted directly into the atmosphere; these are known as **primary pollutants**. Others form in the atmosphere as a result of chemical transformations; these are known as **secondary pollutants**. Pollutants of concern in most large urban and industrial areas are carbon monoxide, sulphur dioxide, nitrogen oxides, volatile organic compounds (VOCs), ozone, and **particulate matter**, or aerosols.

Question for Thought

Do the gases listed as air pollutants also occur naturally?

Carbon monoxide (CO) is produced by the incomplete combustion of carbon-based fuels; thus, motor vehicles are a major source of carbon monoxide. When it is cold, or oxygen levels are low, combustion is less efficient and more carbon monoxide is produced. Carbon monoxide is a toxic gas that, in high enough concentrations, can cause asphyxiation and death. Through the widespread use of catalytic converters in motor vehicles, carbon monoxide emissions have been reduced in recent years (Figure 17.1).

Sulphur dioxide (SO_2) comes from burning fossil fuels that are high in sulphur. Coal-fired power generators that burn high-sulphur coal are significant sources of sulphur dioxide. Other sources include smelters and pulp mills. This gas is also produced naturally by volcanic eruptions. Sulphur dioxide is directly harmful to both plants and animals. Further, in moist air, sulphur dioxide can form droplets of sulphuric acid (Section 2.9.1), which either settle out of the atmosphere as acid deposition or rain out of the atmosphere as **acid rain**. Because of concerns about acid rain, sulphur dioxide emissions were capped, in both Canada and the United States, as of 1994. The resulting reductions, achieved mostly through pollution-control technologies, are shown in Figure 17.1.

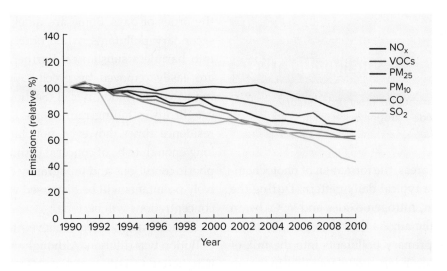

FIGURE 17.1 | The declining trend in the emissions of some major air pollutants in Canada, as a percentage of 1990 emissions.

Data source: Environment Canada, "Air pollution emissions data," www.ec.gc.ca/indicateurs-indicators, June 2012.

The two most common nitrogen oxide (NO$_x$) gases are nitric oxide (NO) and nitrogen dioxide (NO$_2$), both of which form as the nitrogen and oxygen in the atmosphere react during high-temperature combustion, such as in gasoline engines. Above certain concentrations, nitrogen oxides can lead to heart and lung problems. In addition to being directly harmful, nitrogen oxides react in the atmosphere to form droplets of nitric acid (Section 2.9.1), which, like sulphuric acid, can contribute to acid deposition and acid rain. In Canada, emissions of nitrogen oxides have recently begun to decrease, mostly as a result of emission controls on motor vehicles (Figure 17.1). Nitrogen oxides, along with volatile organic compounds (VOCs), are also precursors to **photochemical smog** (Figure 17.2).

VOCs, which were first described in Section 2.9.1, enter the atmosphere through incomplete fuel combustion in motor vehicles, the evaporation of fuels and solvents, and the emissions of vegetation. Figure 17.1 shows that VOC emissions have been decreasing; again, this decrease is a result of emission controls. Many VOCs are believed to have negative health effects. In addition, they are a concern because of their role in the formation of photochemical smog.

> **PHOTOCHEMICAL SMOG**
>
> A mixture of toxic gases that forms when gases emitted by motor vehicles react in sunlight.

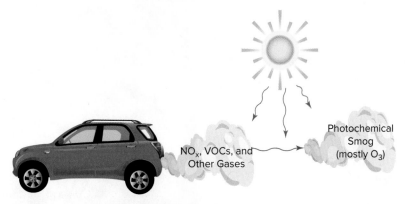

FIGURE 17.2 | Some of the pollutants emitted from the tailpipes of cars react in the presence of sunlight to form photochemical smog.

Remember This

The three major gaseous pollutants from motor vehicles are carbon monoxide, nitrogen oxides, and VOCs.

In large urban areas, the formation of photochemical smog follows a typical daily pattern. During the morning rush hour, nitrogen oxides and VOCs begin to accumulate in the air. As the hours pass, sunlight transforms these primary pollutants into the mix of ozone and other secondary pollutants that make up photochemical smog. Ozone is a toxic gas that irritates the eyes and lungs, and it is damaging to vegetation. By the afternoon, levels of these secondary pollutants will be high. Due to the important role of sunlight in its formation, photochemical smog is most common on clear days when the sun is high in the sky.

Particulate matter associated with anthropogenic activities can be either emitted directly or formed in the atmosphere (Section 2.9.1). Major *primary* sources of particulates include dust from roads and agriculture, smoke and soot from burning biomass or fossil fuels, and some industrial processes. *Secondary* sources are mostly the gas-to-particle conversions of sulphur and nitrogen gases, and VOCs. Particulate matter is classified as PM_{10}, that is, aerosols smaller than 10 µm in diameter; or $PM_{2.5}$, that is, aerosols smaller than 2.5 µm in diameter. Particulate matter can cause adverse health effects because it can be inhaled deep into the lungs and cause respiratory problems. The very small size of $PM_{2.5}$ makes it a greater health risk than PM_{10}. In addition, particulate matter can scatter sunlight and reduce visibility. This is why skies are often very pale blue, or even white, over polluted cities. Figure 17.1 shows that, like most of the other pollutants, emissions of both $PM_{2.5}$ and PM_{10} are decreasing in Canada.

17.1.2 | Conditions Affecting the Dispersal of Air Pollutants

First and foremost, pollutant concentrations will depend on the amounts, and residence times (Section 2.2), of the various substances emitted. Most air pollutants have relatively short residence times, on the order of days. Some are quickly transformed to secondary pollutants, while others are transformed into harmless substances. Further, many pollutants are easily removed by precipitation, either by acting as condensation nuclei (Section 9.1), or by being caught up in falling rain drops. Despite their short residence times, however, pollutants usually linger long enough to be of concern. Ultimately, both atmospheric conditions and topography control how effectively pollutants will be dispersed and how high their concentrations will become.

Traditionally, it was believed that the *solution* to pollution was dilution. Although we now realize that it is better to reduce pollutants at their source, we continue, to a large extent, to rely on the atmosphere to mix pollutants into larger volumes of air, thereby diluting them. Winds tend to control mixing in the *horizontal* direction, while atmospheric stability tends to control mixing in the *vertical* direction (Figure 17.3).

FIGURE 17.3 | Emissions from an oil refinery smokestack in Saskatchewan. Judging by the direction of the billowing smoke, is mixing being controlled by winds or by atmospheric stability?

In general, faster wind speeds provide for greater dispersal of pollutants for two reasons. First, faster winds transport pollutants downwind more quickly than do slower winds. Figure 17.4 illustrates this concept. Imagine that pollutants are emitted from a smokestack in "puffs." With faster winds, the distance between puffs will quickly increase and, therefore, pollutant concentrations will quickly decrease. Second, faster winds create more turbulence than do slower winds. The vertical motions associated with turbulence allow for more mixing.

The nature of the winds is also important. In circulating wind systems, such as *land–sea breeze circulation systems* (Section 11.4), pollutants are generally poorly dispersed. This is because, instead of being removed from a region, the pollutants are simply recirculated as they blow one way during the day and the opposite way at night. For example, in Vancouver, daytime sea breezes carry air pollutants eastward, through the Fraser Valley, and nighttime land breezes blow these pollutants back westward (Figure 17.5).

Also important for the dispersal of pollutants are the vertical motions influenced by atmospheric stability. Recall from Chapter 8 that in an unstable atmosphere, vertical motions are favoured, while in a stable atmosphere, vertical motions are suppressed.

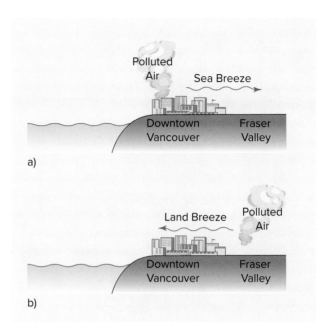

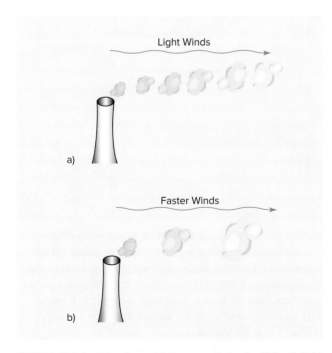

FIGURE 17.4 | a) The distance between pollution "puffs" with light winds. b) The distance between pollution "puffs" with faster winds.

FIGURE 17.5 | a) During the day, sea breezes carry pollutants eastward through the Fraser Valley. b) At night, land breezes carry pollutants back toward the city.

Thus, when conditions are unstable, pollutants emitted at the surface will rise and, therefore, be effectively dispersed in the vertical direction (Figure 17.6a). Conversely, when conditions are stable, pollutants emitted at the surface will tend to stay near the surface, where they can accumulate to potentially dangerous levels (Figure 17.6b).

The influence of atmospheric stability on air quality is far more complex than is implied by Figure 17.6. After all, not all pollutants will be emitted at the ground, and the environmental lapse rate is very unlikely to be constant with height. In addition, the temperature and exit velocity of the pollutants themselves can vary. A variety of possible outcomes can be illustrated by imagining what would happen to pollution plumes from smokestacks under different stability conditions (Figure 17.7).

Recall that neutral stability results when the ELR is equal to the DALR (Figure 17.7a). Under these conditions, a rising or sinking air parcel will be at the same temperature as the surrounding air, so that vertical and horizontal motions are equally likely. As a plume of pollutants from a smokestack is blown downwind, therefore, it will spread equally in the vertical and horizontal directions, producing a pattern described as *coning*.

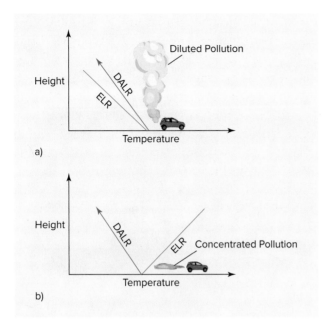

FIGURE 17.6 | a) When the ELR is greater than the DALR, the atmosphere will be unstable, and polluted air will rise from the surface with ease because it will remain warmer than the surrounding air. **b)** When there is a surface inversion, the atmosphere will be extremely stable, and polluted air will not be able to rise because it will be cooler than the surrounding air. (In these diagrams it is assumed that there is no wind.)

Figure 17.7b illustrates unstable conditions, in which vertical motions are favoured over horizontal motions; such conditions were also shown in Figure 17.6a. In the case of the smokestack, however, the pollutants are released at a distance from the surface. As a result, the plume will exhibit a *looping* pattern as it is blown downwind in the unstable atmosphere. Under these conditions the plume can be carried to the surface before it is substantially diluted, bringing dangerously high levels of pollutants to ground level. However, the strong mixing associated with an unstable atmosphere is generally good for dispersing pollutants as they are carried farther downwind.

Figure 17.7c shows a surface inversion like the one illustrated by Figure 17.6b. Again the difference is that here the pollutants are released farther above the ground. With the stable conditions of an inversion, the pollutants will spread horizontally but not vertically as they are blown downwind, producing a fan shape if viewed from above. With a *fanning* plume, air pollution will be kept away from the surface. A comparison of figures 17.7c and 17.6b indicates that the

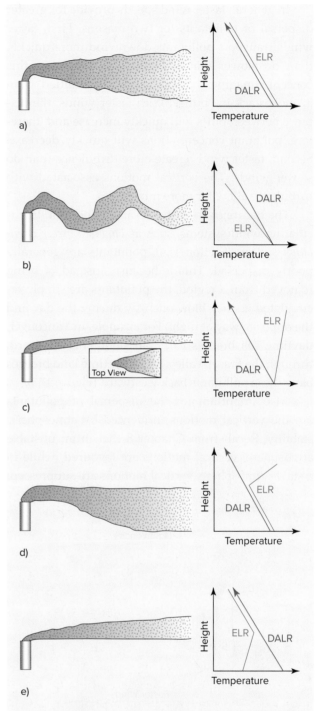

FIGURE 17.7 | Patterns of smokestack plumes under different stability conditions. **a)** With neutral stability, the plume will form a *cone* shape. **b)** With unstable conditions, the plume will form a *looping* pattern. **c)** With stable conditions, the plume will look like a *fan* from above. **d)** When a mixed layer is capped by a subsidence inversion, the effect on the plume is described as *fumigation*. **e)** When a neutral layer lies above an inversion, the resulting plume is described as *lofting*. (In these diagrams, it is assumed that the wind is blowing from left to right.)

height at which pollutants are emitted can influence pollutant concentrations at the surface. This explains why smokestacks are now often built to extend quite high into the atmosphere.

Recall that inversions can also form *above* the surface as *subsidence inversions*; these inversions develop as air sinks and warms under anticyclonic conditions (Section 8.5). Figure 17.7d shows a subsidence inversion above a *mixed layer*. The effect on a pollution plume in this situation is aptly named *fumigation*, as it can lead to high pollution concentrations near the surface. As the pollutants are blown downwind, they will be thoroughly mixed throughout the mixed layer, but the inversion will create a *cap*, preventing pollutants from escaping through the top of the mixed layer. The shallower the mixed layer, the greater will be the pollutant concentration. If anticyclonic conditions persist for an extended period, air quality can gradually deteriorate as more and more pollutants are added each day. The polluted air throughout the mixed layer will be seen from a distance as a reddish-brown layer of air with a clearly defined top (Figure 17.8).

Very different conditions lead to the *lofting* of pollutants (Figure 17.7e). This condition, produced when there is a surface inversion with a neutral layer aloft, is common in clear weather as the surface begins to cool in the evening. If pollutants are emitted above the height of the top of the inversion, they will be mixed upward as they are blown downwind with little risk that they will reach ground level because they will be prevented from doing so by the stable layer below.

Anticyclonic conditions can create the poorest air quality, especially if they occur in summer. With clear skies and warm temperatures, photochemical smog is likely to form. These same clear skies mean that there will be no precipitation to remove pollutants from the atmosphere. At the same time, the subsiding air will produce an inversion that traps pollutants beneath it. Finally, anticyclones are associated with light winds and even circulation systems such as land–sea breeze systems, so pollutants will not be blown far from their source.

The effects of topography on air pollution are rather straightforward. Cities or industrial regions located in valleys or basins are most likely to experience episodes of severe pollution. Under these conditions,

cold air drainage down the sides of the mountains can lead to the formation of inversions that trap pollutants in the valley. In addition, mountains can act as barriers to wind movement. For example, in Vancouver, the effect of land–sea breeze systems on air pollutants is exacerbated by the surrounding mountains. These mountains form a gradually narrowing basin (Figure 17.9a). Sea and land breezes move in an east–west direction, back and forth through this basin, while significant north–south movement of air is prevented. In addition, the subsidence inversions that form in Vancouver in the summer are generally lower in altitude than the mountaintops. The result is that pollutants are trapped beneath the inversion and between the mountains (Figure 17.9b).

FIGURE 17.8 | **A polluted mixed layer lying over the Gardiner Expressway in Toronto. Nitrogen dioxide is responsible for the reddish-brown colour of the smog.**

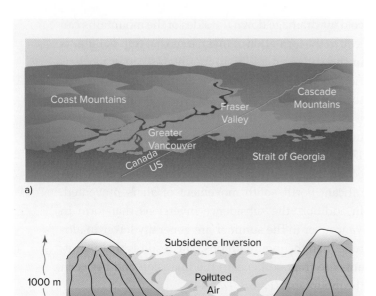

a)

b)

FIGURE 17.9 | a) The Lower Fraser Valley airshed. **b)** Pollutants become trapped between the mountains and below the subsidence inversion.

To report to the Canadian public about the *health risk* associated with certain levels of air pollutants, the Canadian government uses its **Air Quality Health Index (AQHI)**, which is calculated based on the amounts of pollutants such as ozone, PM_{10}, $PM_{2.5}$, and nitrogen dioxide present in the atmosphere. Of these pollutants, ozone and $PM_{2.5}$ are the ones associated with the greatest health risks. The AQHI is determined by measuring the concentrations of these pollutants, assigning a numerical value to the health risk associated with each one, and calculating a total value for the combined risk associated with the level of pollutants in the atmosphere at a given time. The higher the number of the index, the poorer is the air quality and the greater are the health risks (Table 17.1).

The Canadian government has also developed *a set of environmental indicators* as part of the Canadian Environmental Sustainability Indicators (CESI) program. The purpose of this program, which was launched in 2004, is to use these indicators to monitor the state of the Canadian environment and assess the impact of human activities on this environment. The environmental indicators for *air quality* report the concentrations of ground-level ozone, $PM_{2.5}$, sulphur dioxide, nitrogen dioxide, and VOCs. These indicators are determined for the national, regional, and local levels and are invaluable to Canadian decision-makers developing programs and policies to improve air quality.

AIR QUALITY HEALTH INDEX (AQHI)
An index used to report the health risk associated with certain levels of air pollutants.

An area in which the air is confined by topography and atmospheric conditions is sometimes referred to as an *airshed* in the context of air pollution management. The airshed described here is referred to as the *Lower Fraser Valley airshed*. In the past, when high-pressure conditions have persisted for several weeks, as they often do in this area in summer, air pollution advisories have been issued for the Lower Fraser Valley airshed.

17.1.3 | Air Quality Assessment

In order to alert the public to poor air quality conditions, we need to monitor pollutant levels and be aware of the levels at which various pollutants will begin to produce ill effects. To this end, there are roughly 300 air pollution monitoring stations across Canada; most are located in urban areas (Figure 17.10). The airborne pollutants measured at these stations include carbon monoxide, sulphur dioxide, nitrogen oxides, VOCs, ozone, particulate matter, and heavy metals.

FIGURE 17.10 | Air quality monitoring stations use air intakes, such as the ones shown here, to monitor the levels and types of pollutants present in the air.

TABLE 17.1 | Air Quality Health Index categories and messages.

Health Risk	Air Quality Health Index	Health Messages	
		At Risk Population[a]	**General Population**
Low	1–3	**Enjoy** your usual outdoor activities.	**Ideal** air quality for outdoor activities.
Moderate	4–6	**Consider reducing** or rescheduling strenuous activities outdoors if you are experiencing symptoms.	**No need to modify** your usual outdoor activities unless you experience symptoms such as coughing and throat irritation.
High	7–10	**Reduce** or reschedule strenuous activities outdoors. Children and the elderly should also take it easy.	**Consider reducing** or rescheduling strenuous activities outdoors if you experience symptoms such as coughing and throat irritation.
Very High	Above 10	**Avoid** strenuous activities outdoors. Children and the elderly should also avoid outdoor physical exertion.	**Reduce** or reschedule strenuous activities outdoors, especially if you experience symptoms such as coughing and throat irritation.

[a] People with heart or breathing problems are at greater risk. Follow your doctor's usual advice about exercising and managing your condition.
Source: Environment Canada, "Air Quality Health Index categories and messages," www.ec.gc.ca/cas-aqhi, 13 December 2011.

The national ground-level ozone indicator shows that concentrations of this gas increased about 10 per cent from 1990 to 2010 (Figure 17.11a). The regional indicators show that this increasing trend is due mostly to increases in Ontario and the Prairies. These regions also tend to be those with the highest ozone concentrations in Canada (Figure 17.12a). The $PM_{2.5}$ indicator (Figure 17.11b) shows no trend in $PM_{2.5}$ concentrations over the 10 years for which this indicator has been determined. However, there was a significant increase from 2009 to 2010, which was most likely due to the major forest fires that occurred during the warm, dry summer of 2010. As with concentrations of ground-level ozone, concentrations of $PM_{2.5}$ tend to be highest in areas that are heavily populated and/or heavily industrialized (Figure 17.12b). Unlike ozone and $PM_{2.5}$, the other three air quality indicators have shown significant decreases in concentration since 1996 (figures 17.11c, d, and e).

Remember This

The monitoring of air pollutants is important for

- issuing health warnings to the public,
- identifying locations and/or times with poor air quality,
- recognizing trends in air quality, and
- assessing the effectiveness of mitigation measures.

In addition to showing trends and regional patterns within Canada, these indicators can be used to compare air quality in Canada to that in cities around the world. Table 17.2 shows that, in general, levels of both $PM_{2.5}$ and ozone are relatively low in Canada compared to levels in other cities around the world.

17.2 | Depletion of Stratospheric Ozone

As motor vehicle exhaust *creates* ozone at ground level, other anthropogenic emissions are *destroying* ozone in the stratosphere. The irony is that ground-level ozone is *detrimental* to life because ozone is toxic, while stratospheric ozone is *essential* to life because ozone absorbs harmful ultraviolet radiation from the sun. If the amount of ozone in the lower troposphere were to reach the levels that exist in the stratosphere, the AQHI would be far above 10. Stratospheric ozone can be destroyed by chlorine (Cl), bromine (Br), nitrogen oxides (NO_x), and hydroxyl (OH). All of these occur naturally in the atmosphere but, due to human activities, amounts of atmospheric chlorine, bromine, and nitrogen oxides have increased.

For many years, the largest source of chlorine in the stratosphere has been the synthetically produced gases collectively known as chlorofluorocarbons (CFCs). As their name suggests, CFCs contain chlorine, as well as fluorine and carbon. For example, two common CFCs are CFC-11 ($CFCl_3$) and CFC-12 (CF_2Cl_2);

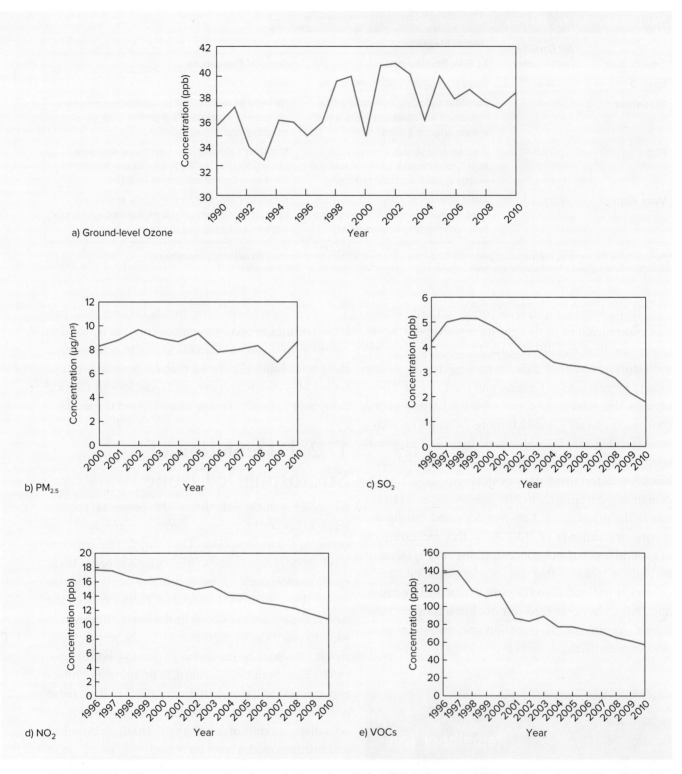

FIGURE 17.11 | a) The national ground-level ozone indicator from 1990 to 2010. **b)** The national PM$_{2.5}$ indicator from 2000 to 2010. **c)** The national sulphur dioxide indicator from 1996 to 2010. **d)** The national nitrogen dioxide indicator from 1996 to 2010. **e)** The national volatile organic compounds indicator from 1996 to 2010.

Data source: Environment Canada, "Ambient levels of air pollutants," www.ec.gc.ca/indicateurs-indicators, 18 June 2012.

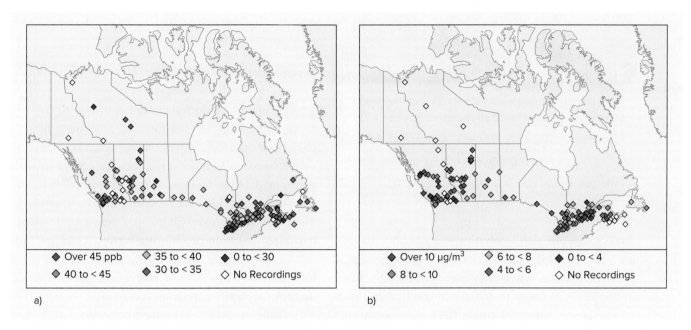

FIGURE 17.12 | a) Warm-season average ground-level ozone concentrations in ppb. **b)** Warm-season average PM$_{2.5}$ concentrations in μg/m³. (The values shown on these maps have not been population-weighted.)

Source: Adapted from Environment Canada, "Ambient levels of ground-level ozone at monitoring stations, Canada, 2010" and "Ambient levels of fine particulate matter at monitoring stations, Canada, 2010," www.ec.gc.ca/indicateurs-indicators, 23 April 2012.

TABLE 17.2 | PM$_{2.5}$ and ground-level ozone averages for selected cities around the world.

City	Average Concentration of PM$_{2.5}$ (μg/m³)	Average Concentration of Ground-Level Ozone (ppb)	City	Average Concentration of PM$_{2.5}$ (μg/m³)	Average Concentration of Ground-Level Ozone (ppb)
Amsterdam	17.63	24.73	MONTREAL	9.80	32.54
Barcelona	17.16	34.48	OTTAWA/ GATINEAU	6.61	34.00
Berlin	21.09	35.01	Paris	18.49	31.13
Boston	10.01	30.95	Phoenix	8.03	47.93
Brussels	19.96	26.84	Pittsburgh	12.58	44.18
Bucharest	23.81	23.37	Prague	18.47	32.40
CALGARY	9.82	30.61	Rome	18.75	35.85
Chicago	12.59	38.04	Seattle	5.68	25.79
Denver	7.78	45.84	Sofia	20.06	39.40
EDMONTON	13.62	28.35	Stockholm	7.88	32.03
Hamburg	15.92	33.96	TORONTO	6.45	33.60
Houston	12.33	38.07	VANCOUVER	4.01	26.52
London	14.00	25.41	Warsaw	30.86	32.99
Lyon	22.24	34.44	Washington	11.00	43.11
Madrid	12.39	37.56			

Source: Adapted from Environment Canada, "Annual average concentrations of fine particulate matter for selected Canadian and international cities, 2010" and "Annual average concentrations of ground-level ozone for selected Canadian and international cities, 2010," www.ec.gc.ca/indicateurs-indicators, 18 June 2012.

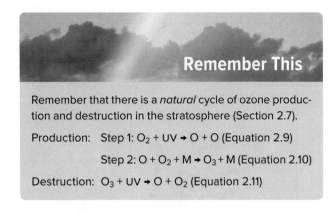

Remember This

Remember that there is a *natural* cycle of ozone production and destruction in the stratosphere (Section 2.7).

Production: Step 1: O_2 + UV → O + O (Equation 2.9)

Step 2: $O + O_2 + M → O_3 + M$ (Equation 2.10)

Destruction: $O_3 + UV → O + O_2$ (Equation 2.11)

these gases are also known as *Freon*. Although they are now banned (see Section 1.3), CFCs generated great excitement when they were first synthesized, in the 1930s, because they are non-reactive and non-toxic. These properties made CFCs safe substitutes for ammonia and sulphur dioxide, which were being used as refrigerants. CFCs also came to be used extensively as propellants in aerosol sprays, as agents for producing Styrofoam, and as solvents.

Being non-reactive, the only *sink* for CFCs is breakdown by exposure to intense radiation. As a result, CFCs have residence times in the atmosphere of, on average, about 100 years (Table 2.2). This is long enough for them to become thoroughly mixed through the troposphere, where they now occur in measurable, though very small, quantities (Table 2.1). From the troposphere, CFCs can enter the stratosphere through breaks in the tropopause, or by way of the strong convection associated with thunderstorms. Once in the stratosphere, CFCs encounter ultraviolet radiation; this radiation has the intensity to break them down. When CFCs are broken down, chlorine is released. Equation 17.1 shows this reaction for CFC-11.

$$CFCl_3 + UV → CFCl_2 + Cl \qquad (17.1)$$

Once released, chlorine can react with ozone, producing chlorine monoxide (ClO) and molecular oxygen (O_2); in the process, the ozone is destroyed (Equation 17.2).

$$Cl + O_3 → ClO + O_2 \qquad (17.2)$$

Because chlorine can be regenerated when chlorine monoxide reacts with an oxygen atom (Equation 17.3),

a single chlorine atom can destroy hundreds of thousands of ozone molecules.

$$ClO + O → Cl + O_2 \qquad (17.3)$$

Fortunately, chlorine cannot continue to destroy ozone indefinitely. In the form of chlorine atoms or chlorine monoxide, chlorine is *reactive* and can destroy ozone. Eventually, however, chlorine atoms react with methane (CH_4) to form hydrogen chloride (HCl), and chlorine monoxide reacts with nitrogen dioxide to form chlorine nitrate ($ClONO_2$). Locked up in either hydrogen chloride or chlorine nitrate, chlorine cannot destroy ozone. These compounds are *reservoirs* for chlorine; without these reservoirs, chlorine could destroy far more ozone than it does.

Nitrogen oxides also destroy ozone, and they do so in a similar way to chlorine. First, nitric oxide reacts with ozone (Equation 17.4), then nitric oxide is regenerated to destroy more ozone (Equation 17.5).

$$NO + O_3 → NO_2 + O_2 \qquad (17.4)$$

$$NO_2 + O → NO + O_2 \qquad (17.5)$$

Like chlorine, however, nitric oxide eventually winds up in a reservoir molecule. For this to happen, the nitrogen dioxide reacts with hydroxyl, forming nitric acid (HNO_3).

Human activities have increased the amount of nitric oxide in the atmosphere in two ways. First, nitrogen fertilizers add nitrogen to the soil; this nitrogen is transformed into nitrous oxide (N_2O) by soil bacteria (Section 2.3), and it eventually travels to the stratosphere, where it forms nitric oxide. Second, supersonic jets flying in the stratosphere emit nitric oxide in the same way as motor vehicles do (Section 17.1.1). Fortunately, such flights do not occur on a regular basis.

In addition to chlorine and nitric oxide, bromine can also destroy ozone. The amount of bromine in the stratosphere is increasing due mostly to the use of **halons** in fire extinguishers. Like CFCs, halons are artificially produced, and they are stable enough to reach the stratosphere, where they are broken apart to release bromine. Eventually, bromine ends up in the reservoir compounds hydrogen bromide (HBr)

> ## Remember This
>
> The following reactions form chlorine reservoirs:
>
> $$Cl + CH_4 \rightarrow HCl + CH_3$$
>
> $$ClO + NO_2 + M \rightarrow ClONO_2 + M$$
>
> The following reaction forms a nitric oxide reservoir:
>
> $$NO_2 + OH + M \rightarrow HNO_3 + M$$
>
> (Recall that M represents a molecule that is needed to carry away excess energy; this can be any other gas molecule in the atmosphere.)

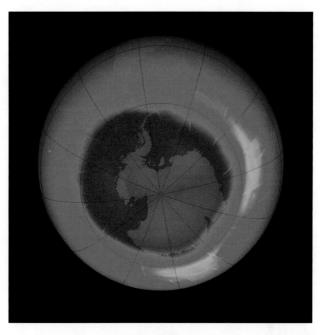

FIGURE 17.13 | A satellite image of the Antarctic ozone hole in 2009. What is the cause of the ozone hole?

and bromine nitrate ($BrONO_2$). Bromine is potentially more destructive to ozone than is chlorine because a smaller proportion of it forms reservoir compounds, but chlorine is much more abundant in the stratosphere than is bromine.

> ## Remember This
>
> Chlorine, nitric oxides, and bromine all destroy ozone, and their levels have been increasing in the stratosphere due to human activities. The source of chlorine is the breakdown of CFCs. The source of nitric oxide is nitrous oxide from fertilizer use and high flying jets. The source of bromine is the breakdown of halons.

17.2.1 | The Ozone Hole

After Canada, the US, and several other countries banned the use of CFCs in aerosol sprays in 1978, concern about the ozone layer began to fade because there was no proof that it was being destroyed. In 1985, however, the discovery of the now infamous **ozone hole** over the South Pole was announced (Figure 17.13). The ozone hole is a roughly circular area centred over, and somewhat bigger than, the continent of Antarctica.

The ozone hole is best developed in October, during the southern-hemisphere spring. Figure 17.14 shows that the hole began to form in about 1970. The loss of ozone continued to increase each year until about 1990, when it began to level off. The discovery of this region of drastically reduced ozone came as a great surprise because it had been thought that if indeed the ozone layer were thinning, the thinning would be gradual. Clearly, the ozone hole showed otherwise.

This startling, and frightening, discovery led to extensive research to uncover the *cause* of the ozone hole. In September 1987, evidence to connect chlorine with ozone depletion was finally found. Measurements made during a flight through the ozone hole revealed that in the hole, ozone levels were very low, while chlorine monoxide levels were very high (Figure 17.15). Outside the hole, ozone levels were more than double the values in the hole, while chlorine monoxide levels were about ten times less than the values in the

HALONS

A type of halocarbon that contains carbon, bromine, and other halogen atoms.

OZONE HOLE

A roughly circular area in the ozone layer where the amount of stratospheric ozone is about 50 per cent less than it is in the surroundings.

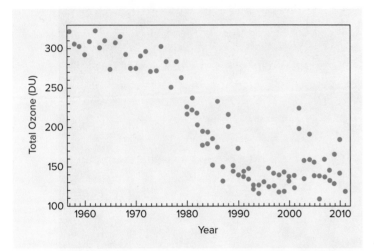

FIGURE 17.14 | The decrease in ozone over Antarctica. The units used to measure the depth of ozone in an atmospheric column are Dobson units. (One Dobson unit is equal to a layer of ozone 0.001 cm thick at normal sea-level pressure.) Normal amounts of ozone in the mid-latitudes are about 300 Dobson units.

Source: Adapted from NASA, Ozone hole watch, "Ozone facts: History of the ozone hole," http://ozonewatch.gsfc.nasa.gov/facts/history.html, 18 November 2009.

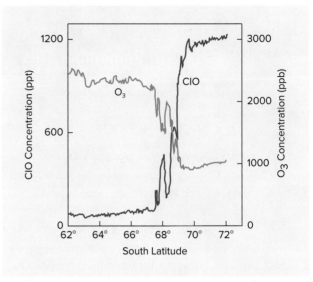

FIGURE 17.15 | Measurements of ozone and chlorine monoxide concentrations made as a research aircraft was flown through the ozone hole in September 1987. You can see where the ozone hole begins, around latitude 69° S, by the abrupt decrease in ozone concentrations and the abrupt increase in chlorine monoxide concentrations.

Source: J.G. Anderson, W.H. Brune, and M.H. Proffitt. (1989). Ozone destruction by chlorine radicals within the Antarctic vortex: The spatial and temporal evolution of ClO-O3 anticorrelation based on *in situ* ER-2 data. *Journal of Geophysical Research, 94*(D9), pp. 11, 465–79.

hole. The conclusion was made that the chlorine was responsible for destroying the ozone and producing the hole.

Since that time, we have come to understand the complex chemistry responsible for the creation of the ozone hole. The process begins during the Antarctic winter. At this time of little to no sunlight, upper-air winds form a strong vortex of spinning air called the *polar vortex*. An important effect of the polar vortex is that it separates the air over Antarctica from the rest of the atmosphere. As a result, the air in the vortex becomes very cold, and there is no exchange of gases between it and the surroundings. When temperatures in the stratosphere drop below about −80°C, certain atmospheric gases begin to condense and form clouds known as *polar stratospheric clouds* (PSCs). These clouds are crucial to the formation of the ozone hole because they provide *surfaces* on which chemical reactions can occur (Figure 17.16). In fact, it is due to the presence of PSCs that ozone depletion is considerably higher over Antarctica than it is anywhere else. (PSCs are much less likely to form over the Arctic, where the vortex is weaker and the air does not get quite as cold. As a result, ozone depletion over the Arctic is less than it is over the southern pole, but greater than in the mid-latitudes.)

The net result of the reactions within the PSCs is a *decrease* in the amount of nonreactive chlorine, HCl and ClONO₂, and an *increase* in the amount of reactive chlorine, Cl and ClO. Equation 17.6 illustrates the

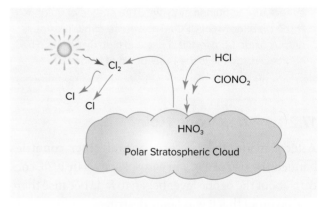

FIGURE 17.16 | The process by which the chlorine reservoir compounds are converted to nitric acid, HNO₃, and molecular chlorine, Cl₂, on PSCs. The nitric acid becomes part of the cloud and, once the sun reappears in the spring, the molecular chlorine eventually forms reactive chlorine.

reaction in which chlorine is released from its reservoir compounds as molecular chlorine (Cl_2).

$$ClONO_2 + HCl \rightarrow Cl_2 + HNO_3 \qquad (17.6)$$

When the sun returns to Antarctica toward the end of September, the molecular chlorine is photodissociated to form *reactive* atomic chlorine. The condensed nitric acid (HNO_3) becomes part of the cloud preventing the chlorine from reforming chlorine nitrate. Thus, the reactive chlorine is left available to react with the ozone. The resulting reactions occur very rapidly, producing the ozone hole by October. As the sun rises higher in the sky and the days get longer, the polar vortex begins to break down. Ozone-rich air from the mid-latitudes mixes with the ozone-poor air over Antarctica, and the ozone hole disappears for another year.

The ozone hole showed such a dramatic, and terrifying, thinning of the ozone layer that action was taken very quickly to prevent further damage. In September 1987, an international agreement on the issue, called the Montreal Protocol on Substances that Deplete the Ozone Layer, was ratified. Initially, this agreement called for a 50 per cent reduction in CFC production, based on 1986 levels, by 2000. At about the same time as this agreement was put in place, however, the evidence that chlorine was almost certainly the cause of the ozone hole was announced. This "proof" led to several revisions to the treaty that ultimately saw CFCs, halons, and a few other chlorine-containing substances banned completely by 1996. To date, the Montreal Protocol has been signed by almost 200 countries.

Since the mid-1990s, atmospheric concentrations of CFCs have, for the most part, either decreased or levelled off (Figure 17.17), and it is predicted that the ozone layer should fully recover by 2050 if this trend continues. With so much at stake, governments and industries around the world have been eager to come up with less harmful substitutes for CFCs. In many cases, CFCs have been replaced with hydrochlorofluorocarbons (HCFCs) and hydrofluorocarbons (HFCs). Yet these substitutes are not without their problems. Like CFCs, HCFCs and HFCs are both powerful greenhouse gases, although their residence times in the atmosphere are much shorter than the residence time of CFCs. In addition, HCFCs do contain chlorine, but they are far more likely than CFCs to break down in the troposphere, far from the ozone layer. Under the terms of the Montreal Protocol, HCFCs must be completely phased out by 2030.

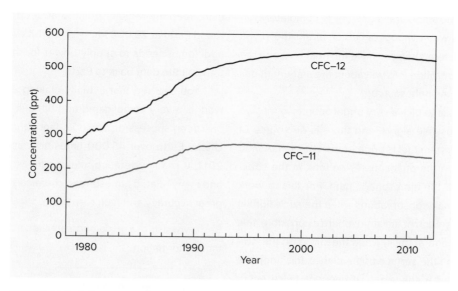

FIGURE 17.17 | Concentrations of the two major CFCs are beginning to show a slight decline. Considering that these gases were banned in 1996, why have they not shown a greater decrease?

Data source: NOAA, "Chlorofluorocarbon-11 (CCl_3F)—Combined Data Set" and "Chlorofluorocarbon-12 (CCl_2F_2)—Combined Data Set," www.esrl.noaa.gov, 1 June 2012.

In the Field

Measuring Atmospheric Ozone from Space

James Drummond, Dalhousie University

How do you measure a few molecules in a billion, 1000 km away from a spacecraft moving at about 25,000 km/h? This is the very great challenge that I found myself facing as part of the very large team that designed and built the Canadian SciSat satellite, which launched on 12 August 2003.

The aim of the project was to measure ozone and related chemicals in the atmosphere from space. Changes in these chemicals—particularly the ozone of the ozone layer—could have a serious effect on life on the surface. Although we have regulations in place to control the chemicals that affect ozone, we are still figuring out whether these regulations are having the right effect. Such research involves a thorough knowledge of science mixed with some state-of-the-art engineering.

One of the ways that we can find these atmospheric constituents is to shine a light through the atmosphere and then use a spectrometer to break the light energy up into its separate wavelengths and look at them individually. Nearly all atmospheric gases absorb certain wavelengths of light, particularly in the visible and infrared regions, and these absorption patterns are as unique as a person's signature. For example, ozone's spectral signature is completely different from that of water vapour or carbon monoxide. If you find the right signature, then you have the right gas, and the strength of the absorption tells you about the amount of gas between you and the light source.

The next step is to pick a very bright source so that we get the largest possible signal, and there is no source of light in the solar system that is brighter than the sun. As the sun shines through the atmosphere, we look at the spectrum at the end of the atmospheric path. For this to work well, we need to make observations when the sun's light is shining through as much of the atmosphere as possible; this happens at dawn and at dusk, when the sun is on the horizon. For people on the surface of the planet, this happens twice a day, but for a spacecraft, it happens twice every orbit. In addition, the position of the spacecraft is such that its instruments can measure sunlight passing through different parts of the atmosphere. If measurements are made

as quickly as one measurement every two seconds, we can get measurements for different altitudes from a single sunrise or sunset. And, finally, as the satellite comes round into the full sunlight, it can measure the sunlight without any atmosphere between it and the sun. This gives us an important check on our measurements, since we know that there is no atmosphere in the path and that our results must reflect this.

Measuring all the individual wavelengths of light quickly enough is an engineering challenge, as we need a spectrometer that is sensitive enough, fast enough, light enough, and strong enough to survive on a spacecraft. This is possible using a combination of Fourier transform and grating spectrometers, but only if we can make some of the detectors very cold—about 80 K (–190°C). Space gives us a hand here, as deep space is even colder—about 2.7 K (–270°C). With a very efficient energy radiator, we can radiate heat away from the detectors to space and cool them to the operational temperature.

Finally, we need a spacecraft that can keep the spectrometer accurately pointed at the sun and, at the same time, keep the radiator pointed at deep space while going round Earth at 25,000 km/h (one orbit every 90 minutes). In addition, it needs to supply power for the instruments and transmit the data back to Earth.

You can see that actually making a space instrument work is a very complicated and difficult task. Yet our work has been successful. Since its launch in 2003, SciSat has circled Earth over 45,000 times and, as of the summer of 2012, it is still operating, making measurements of ozone and many related atmospheric chemicals twice an orbit with great accuracy and high resolution.

See www.asc-csa.gc.ca/eng/satellites/scisat/default.asp for more information.

JAMES DRUMMOND is Canada Research Chair in Remote Sounding of Atmospheres and a professor in the Department of Physics and Atmospheric Science at Dalhousie University. His current research is in the remote sounding of atmospheres on Earth.

17.3 | Climate Change

As mentioned in the introduction to this chapter, the relatively short residence time of most air pollutants means that air pollution tends to remain a *local* problem. On the other hand, ozone depletion is a *global* problem because of the much longer residence times of the gases involved. Also global in scale is the problem of anthropogenic climate change. This problem is an example of how human activities have contributed to atmospheric changes that interfere with the planetary energy balance (Section 6.4) and, thus, affect climate.

Most people are at least familiar with the link between rising levels of greenhouse gases and global warming—the increase in Earth's temperature caused by increasing concentrations of greenhouse gases. Yet the connections between other human-induced changes and climate are generally less well known and not as well understood. Some of these changes amplify global warming while others lessen it, but the *net* effect remains one of warming. For example, increasing levels of aerosols are offsetting the warming caused by greenhouse gases. The problem becomes even more complex when it is taken into account that warming will lead to further consequences, both direct and indirect. For example, a direct and easily anticipated consequence of rising temperatures is the melting of the polar ice caps. Other consequences, such as possible changes in ocean currents, are less direct and, thus, much more difficult to anticipate. The wide-reaching and complex nature of this problem is the reason that it is often referred to generally as *climate change*, but it is important to remember that *global warming* best describes the direction, and main driver, of the changes. In many ways, anthropogenic climate change is a far more difficult problem to both understand and deal with than is ozone depletion.

As part of our effort to understand and deal with climate change, the Intergovernmental Panel on Climate Change (IPCC) was formed in 1988. To maintain an interdisciplinary and international approach, this panel is composed of physical, natural, and social scientists from around the world. Their purpose is to compile and evaluate climate change research. In particular, they assess the following aspects of climate change: the science of the processes involved,

observed and predicted impacts, possible adaptations to these impacts, and actions for reducing emissions. As part of this process, the IPCC has produced four reports, the latest of which was released in 2007.

17.3.1 | Natural Climate Change

Climate change is *natural* for planet Earth, and one of the best ways we have to understand present climate change is to study climate changes of the past. Throughout Earth's long history, there have been periods that were colder than today, and there have been periods that were warmer today; these alternating periods of warmth and cold are sometimes referred to as *icehouses* and *greenhouses*, respectively (Figure 17.18). We currently live in a relatively warm period of an icehouse era. Compared to today's temperatures, past temperatures have been warmer more often than they have been colder. In addition, there have not always been ice sheets at the poles, as there are now.

Much of the fluctuation in Earth's temperature throughout the planet's history can be linked to changing concentrations of carbon dioxide. On time scales of roughly 100 million years, the amount of carbon dioxide in the atmosphere seems to have been largely controlled by plate tectonic activity. Recall that, in relation to plate tectonics, an important *source* for carbon dioxide is volcanic eruptions, while an important *sink* is chemical weathering (Section 2.5). During times of rapid seafloor spreading, higher amounts of volcanic activity gradually add carbon dioxide to the atmosphere, increasing the atmospheric concentration of this gas. In turn, when seafloor spreading rates are slow, carbon dioxide concentrations decrease. Plate tectonic activity can also provide fresh rock for chemical weathering, increasing the sink for carbon dioxide, and lowering its concentration.

As outlined in Section 2.5, the last cold spell, before the one we are experiencing today, occurred about 300 million years ago, during the geologic period known as the Carboniferous Period. Recall that this was a time of high rates of organic carbon burial and, thus, low levels of atmospheric carbon dioxide. About 250 million years ago, this relatively cold period ended, and the age of the dinosaurs began. From then until about 65 million years ago, the dinosaurs

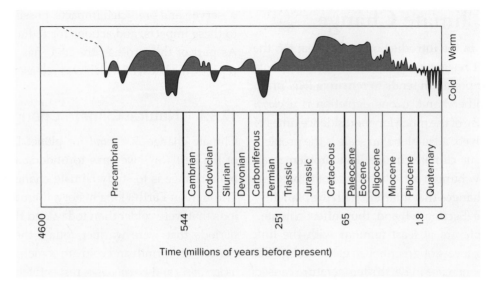

FIGURE 17.18 | Throughout its history, Earth has experienced both warm periods and cold periods as a result of natural processes.

enjoyed one of the warmest times our planet has experienced (Figure 17.18). It has been estimated that during this time, temperatures were up to 10°C higher than they are today, and carbon dioxide concentrations were about four times higher than they are today. The higher carbon dioxide levels were due to faster seafloor spreading rates.

When seafloor spreading rates slowed about 100 million years ago, carbon dioxide levels began to drop, and the planet began to cool (Figure 17.19). Then, about 30 million years ago, the Indian subcontinent collided with Asia, beginning the uplift of the Tibetan Plateau. This event provided abundant fresh rock for weathering and, thus, further accelerated the rate of decrease in atmospheric carbon dioxide and the cooling of the planet. As a result of this cooling, an ice

sheet began to form at the South Pole about 40 million years ago, and another began to form at the North Pole about 10 million years ago. For the last 2 million years, Earth has been in an ice age during which the climate has cycled between glacials and interglacials. During the coldest part of this period, temperatures were likely 5 to 6°C colder than they are today.

Glacial cycles are affected by both variations in Earth's orbit and changes in the atmospheric concentration of carbon dioxide. Due to the gravitational pull of the moon and the other planets, the shape of Earth's orbit, the tilt of its axis, and the direction in which the

Remember This

- 360 to 250 million years ago: cold period with CO_2 levels about the same as today
- 250 to 65 million years ago: warm period with CO_2 levels up to four times higher than today
- 65 to 2 million years ago: period of dropping temperatures and dropping CO_2 levels
- 2 million years ago to present: cold period

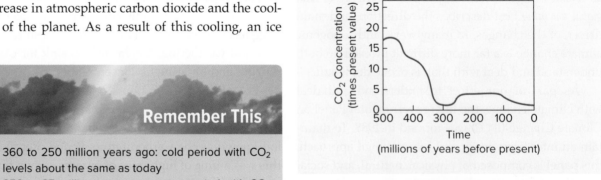

FIGURE 17.19 | Changes in carbon dioxide concentrations, relative to today's value, over the last 500 million years. (These values were estimated using models.)

Data source: Estimated based on R.A. Berner and Z. Kothavala. (2001). "GEOCARB III: A revised model of atmospheric CO_2 over Phanerozoic time," *American Journal of Science, 301*, pp. 182–201.

north end of its axis points all change. These cycles occur over time periods of 100,000 years, 41,000 years, and 26,000 years, respectively. Although the 100,000 year cycle does cause a very small change in the *total* amount of solar radiation received by Earth, these orbital changes mostly just affect the *seasonal distribution* of solar radiation. When summers get cooler and winters get warmer, glaciers can grow because less snow melts in the summer and more snow falls in the winter. Orbital changes, however, are not enough to account for the *thickness* of the ice sheets that develop. Another mechanism is involved: the fluctuation in carbon dioxide levels.

The changes associated with the growth of ice sheets seem to influence the carbon dioxide content of the atmosphere, producing a positive correlation between carbon dioxide concentrations and temperature (Figure 17.20). Several possible positive feedback mechanisms are believed to link changes in ice sheets and changes in the concentration of carbon dioxide (Kump, Kasting, & Crane, 281–7). In one such feedback, the drop in sea level associated with the growth of ice sheets exposes the nutrient-rich continental shelves. As rivers wash these nutrients into the oceans, the oceans' biological productivity increases and, as a result, so does photosynthesis. Photosynthesis removes carbon dioxide from the atmosphere; when the photosynthesizing organisms die, the carbon is carried into the deep water, where it is stored. This is a positive feedback mechanism because it is the growth of ice sheets that leads to the accelerated rates of carbon dioxide removal that, in turn, lead to further growth of ice sheets. Eventually, glaciers recede due to orbital changes. As they do, the feedbacks should operate in the reverse, causing carbon dioxide levels to rise and help warm the planet once again. In glacial cycles, therefore, orbital variations seem to *drive* the climate changes, while carbon dioxide levels seem to respond to these changes and *amplify* them.

Certain processes also cause natural climate change over the very short term. One such process is the variation in the sun's output due to sunspot activity. The more sunspots there are, the more energy the sun emits, and vice versa. For example, over the 70-year period from 1645 to 1715, sunspots were very rare and, according to historical records, European winters were unusually long and harsh. The resulting

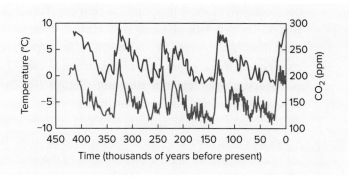

FIGURE 17.20 | Carbon dioxide concentrations (blue line) and temperature changes (red line)—relative to today's temperature—are closely correlated over the last 400,000 years. The amount of carbon dioxide in the atmosphere is determined using air bubbles in ice cores; temperatures are determined using oxygen isotopes.

shorter growing seasons caused widespread famine. This unusual cold period has been named the Maunder Minimum.

Another process that affects short-term natural climate change is the eruption of volcanoes. In addition to adding carbon dioxide to the atmosphere, volcanic eruptions also add sulphur dioxide gas to the atmosphere. Recall that this gas reacts with water to form droplets of sulphuric acid that are very effective at scattering solar radiation. Also recall that if sulphur dioxide gas reaches the stratosphere, the resulting haze of acid droplets can remain there for a few years, scattering back the sun's radiation and, thus, causing cooling (Section 2.9).

Question for Thought

Considering that both carbon dioxide and sulphur dioxide are emitted by volcanic eruptions, why do you think the former is associated with long-term climate change while the latter causes much shorter-term climate change?

The above is a very brief history of *natural* climate change. It shows that climate change can occur on a variety of time scales, from hundreds of millions of years for changes associated with plate tectonics, to a few years for changes associated with sunspots and volcanic eruptions. It also shows that variations in atmospheric carbon dioxide concentrations are closely correlated with most temperature changes. Based on this history, it would seem that climate change is most *directly* connected to changes in solar output and changes in the composition of the atmosphere. The planetary energy balance diagram (Figure 6.6) can be used to see how each of these factors could influence climate; changes in either would clearly alter flows of energy in the Earth-atmosphere system. In addition, Figure 6.6 shows that changes in the surface albedo could influence climate. In general, therefore, we can list three causes of climate change: changes in solar output, changes in atmospheric composition, and changes in the albedo of Earth's surface. Of these, both atmospheric composition and surface albedo are being influenced by human activities.

17.3.2 | Anthropogenic Climate Change

As far back as 1896, Svante Arrhenius, a Swedish scientist, speculated that the burning of coal associated with the Industrial Revolution *might* be increasing the amount of carbon dioxide in the atmosphere, and that this increase *might* lead to a warming of Earth. Prior to this, work had been done to establish that carbon dioxide, like water vapour, can absorb Earth's longwave radiation, creating a *greenhouse effect*. Arrhenius later won the Nobel Prize for his work. During his time, however, there were no measurements of atmospheric carbon dioxide concentrations, so it was impossible to tell whether amounts of this gas were, in fact, increasing.

In 1957, American scientist David Keeling began a program to monitor the amount of carbon dioxide in the atmosphere by establishing an observatory on Mauna Loa in Hawaii. This location, high on a mountaintop in the middle of the Pacific Ocean, is far from any sources of contamination. Observations made from this observatory show that atmospheric carbon dioxide levels have increased from just under 320 ppm in 1958 to close to 390 ppm in 2010 (Figure 2.1). There is now mounting evidence that temperatures are also increasing, just as Arrhenius predicted.

Increasing the amount of carbon dioxide in the atmosphere is not the only way we are influencing the atmosphere on a global scale. We are also adding other greenhouse gases, as well as aerosols, and we have introduced a new cloud type, jet contrails. In addition, through large-scale changes in land use,

which often involve deforestation, we are altering the albedo of Earth's surface. These changes impact climate, as they alter the flows of energy in the Earth-atmosphere system.

17.3.2.1 | Greenhouse Gases

Figure 17.21 shows the increase in the greenhouse gases carbon dioxide, methane, and nitrous oxide over the last two thousand years. This figure indicates that carbon dioxide levels were fairly consistent, at about 280 ppm, before they began to rise in the late nineteenth century. For this reason, 280 ppm is used to represent the pre-industrial level of carbon dioxide, or the level prior to 1750, the year generally associated with the start of the Industrial Revolution (Table 17.3). Given that carbon dioxide concentrations have now reached 389 ppm, we can calculate that the level of this gas in Earth's atmosphere has increased by about 39 per cent. The bulk of this increase has occurred since the 1970s. Recent research indicates that this increase is linked to human activities rather than natural processes.

Carbon dioxide levels are increasing due to fossil-fuel burning, deforestation, and, to a much lesser degree, cement production; fossil-fuel burning has

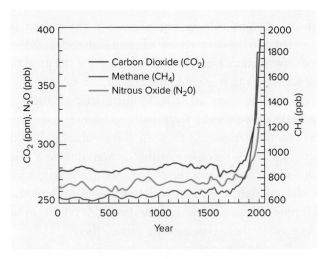

FIGURE 17.21 | The increasing concentrations of three greenhouse gases over the last 2000 years. The last 50 years' or so worth of data are based on direct measurements, such as those recorded at Mauna Loa, while earlier data are estimated from ice cores.

Source: Adapted from P. Forster, et. al. (2007.) Changes in atmospheric constituents and in radiative forcing. In S. Solomon et al. (Eds.) *Climate change 2007: The physical science basis. Contribution of working group I to the fourth assessment report of the intergovernmental panel on climate change* (pp. 129–234). Cambridge, UK: Cambridge University Press, p. 135.

TABLE 17.3 | Concentrations of carbon dioxide, methane, and nitrous oxide in 2010 and before 1750 (estimated).

	2010 Concentration	Pre-1750 Concentration	Change
Carbon Dioxide (ppm)	389	280	39%
Methane (ppb)	1808	700	158%
Nitrous Oxide (ppb)	323	270	20%

contributed about twice as much carbon dioxide as has deforestation. Remember that fossil fuels represent a *long-term* storage of carbon. Under natural conditions, such carbon eventually returns to the atmosphere when the sedimentary rocks containing the fossil fuels are exposed at the surface and weathered. Burning fossil fuels speeds up the process of returning carbon dioxide to the atmosphere (Section 2.5). Equation 17.7 illustrates the chemical reaction that occurs in the combustion of fossil fuels; it is written here for methane (CH_4), a very common natural gas.

$$CH_4 + 2O_2 \rightarrow CO_2 + 2H_2O \qquad (17.7)$$

Deforestation increases atmospheric carbon dioxide because trees store carbon over the *short* term. The cut trees can be a *source* for carbon dioxide if they are burned, or as they decompose. However, deforestation is most significant because it represents the loss of an important *sink* for carbon dioxide. Most deforestation is for agricultural purposes, and the crops that replace the trees do not remove nearly as much carbon dioxide from the atmosphere as do trees.

Not all of the carbon dioxide produced by the combustion of fossil fuels and deforestation has ended up in the atmosphere; almost half of it has either been dissolved in the oceans or taken up by organic matter on land. (An unfortunate side effect of increased ocean absorption of carbon dioxide is the increasing acidity of the oceans.) It seems there is no simple relationship between the amount of carbon dioxide we emit and the amount that ends up in the atmosphere. In addition, it is very likely that as we continue to produce carbon dioxide, the sinks for carbon will weaken or even reverse and, thus, become sources.

The amount of methane in the atmosphere has more than doubled since pre-industrial times (Table 17.3). Compared to carbon dioxide, methane is a much more powerful greenhouse gas, but it has a much shorter residence time (Table 2.2). Methane is produced naturally, primarily by anaerobic decomposition in wetlands (Equation 2.4) and, to a slightly lesser extent, through the emissions of termites. Today, however, anthropogenic sources of methane—such as rice paddies, grazing animals, landfills, and the extraction of fossil fuels—produce more methane than do natural sources.

Atmospheric concentrations of nitrous oxide have increased about 20 per cent since pre-industrial times (Table 17.3); this increase is due mostly to the use of nitrogen fertilizers. Nitrous oxide is of particular concern because it has a long residence time (Table 2.2), and because it is partly to blame for the depletion of stratospheric ozone, as described earlier.

Remember This

Atmospheric concentrations of carbon dioxide are increasing due to fossil fuel burning and deforestation. Atmospheric concentrations of methane are increasing due to the large numbers of grazing animals, the proliferation of garbage dumps, the extraction of fossil fuels, and the large areas of land used to grow rice. Atmospheric concentrations of nitrous oxide are increasing due to the use of nitrogen fertilizers.

In addition to the above, several of the **halocarbons**, including synthetically produced CFCs, HCFCs, and HFCs, are also greenhouse gases. In fact, some of these gases are

HALOCARBONS
Substances containing carbon and one or more halogen atoms (i.e., chlorine, bromine, fluorine, iodine).

particularly powerful greenhouse gases. Concentrations of CFCs were increasing steadily in the atmosphere until these substances were banned in 1996 (Figure 17.17), but CFCs are still present in the atmosphere in significant concentrations due to their long residence times (tables 2.1 and 2.2). On the other hand, the concentrations of HCFCs and HFCs are rapidly increasing, as these substances are being used as substitutes for CFCs. However, these two gases are of slightly less concern due to their shorter residence times.

The gases described above—carbon dioxide, methane, nitrous oxide, and the halocarbons—are known as the *long-lived greenhouse gases*. Because these gases differ in their capacity to cause warming, the IPCC developed an index known as the **Global Warming Potential (GWP)** that allows us to compare the relative strengths of the various gases. The GWP of a gas is calculated as a ratio of the amount of warming that would be caused by the emission of a kilogram of that gas compared to the amount of warming that would be caused by the emission of a kilogram of carbon dioxide (Table 17.4). This index is a complicated function of the effectiveness of the greenhouse gas *and* its atmospheric residence time. In addition, because the greenhouse gases have different residence times, the GWPs have to be expressed based on different time horizons. For example, methane is a much more effective greenhouse gas than carbon dioxide is, but methane has a much shorter residence time. Because of this, methane is 72 times more powerful than carbon dioxide over 20 years, but only 25 times more powerful over 100 years.

As you saw in sections 17.1 and 17.2, amounts of another greenhouse gas, ozone, are also being influenced by human activities. While ozone levels are *increasing* in the troposphere, they are *decreasing* in the stratosphere. Therefore, the changes in ozone

GLOBAL WARMING POTENTIAL (GWP)
A measure of the capacity of a greenhouse gas to impact global temperatures, relative to the warming potential of carbon dioxide.

concentrations could lead to either warming or cooling. With less stratospheric ozone, less ultraviolet radiation is being absorbed in the stratosphere, and more is reaching the surface. Based on this information, it might be predicted that the stratosphere should cool, while the lower atmosphere should warm. It turns out that because the stratosphere is, in fact, cooling, it is emitting less radiation to the troposphere, so that the net effect of decreasing stratospheric ozone is cooling. On the other hand, the increase in tropospheric ozone is causing warming. Overall, the warming due to increased tropospheric ozone is greater than the cooling due to decreased stratospheric ozone, so that the *net* effect of changes in ozone concentration is warming. Yet it is important to remember that ozone has a relatively short residence time (Table 2.2), so its influence on global temperatures is generally of less concern than is that of the long-lived greenhouse gases described above.

17.3.2.2 | Other Changes: Aerosols, Jet Contrails, and Surface Albedos

In addition to adding greenhouse gases to the atmosphere, we are making other changes that have been linked to climate change. First, as discussed in Section 2.9, we are increasing the amount of aerosols in the atmosphere. Most aerosols scatter the sun's radiation, producing a cooling effect, but black carbon (or soot) absorbs radiation, causing warming. Although there is still a lot we don't understand about the radiative properties of aerosols, we currently think that, overall, aerosols have a *net* cooling effect.

A particularly large amount of uncertainty is associated with the role aerosols play in influencing the radiative properties of clouds. Remember from Section 9.1.2 that aerosols provide surfaces upon which water vapour condenses. When more aerosols are available, the number of droplets in clouds *increases*, while the size of the droplets *decreases*. A greater amount of smaller droplets provides more surface area, which seems to increase the reflectivity of clouds. Recent observations support this relationship between aerosols and the reflectivity of clouds. For example, a satellite study of clouds in the Amazon indicated that the clouds' albedos increased during the burning season (Kaufman & Fraser, 1997). It has also been shown,

TABLE 17.4 | The Global Warming Potential (GWP) of long-lived greenhouse gases for a variety of time horizons.

Gas	Residence Time (Years)	Global Warming Potential		
		20 years	100 years	500 years
CO_2	100	1	1	1
CH_4	12	72	25	7.6
N_2O	114	289	298	153
CFC-11	45	6730	4750	1620
CFC-12	100	11,000	10,900	5200

Data source: IPCC. (2007). *Fourth assessment report*. Geneva, Switzerland: IPCC.

again through satellite observation, that emissions from ships have been increasing the reflectivity of clouds over the oceans (Wallace & Hobbs, 2006, p. 218). Changes in the radiative properties of the atmosphere due to anthropogenic aerosols clearly have important, and as yet uncertain, implications for climate change; thus, this is currently an area of active research.

Jet contrails (Section 9.3.6) are also linked to climate change. Because they form high in the troposphere, jet contrails are most similar to cirrus clouds. Recall that clouds both warm *and* cool Earth, and that high, thin clouds tend to have a net warming effect because of their relatively low albedos (Section 6.4). It is likely, therefore, that jet contrails also have a warming effect, although the effect of jet contrails on climate is not well understood.

Finally, large-scale land use changes likely influence climate because such changes alter Earth's albedo. Of these changes, deforestation probably has the greatest impact. Because forests have low albedos (Table 5.2), deforestation over large areas can significantly *increase* the planetary albedo. With a higher albedo, more sunlight is reflected, and the result is cooling.

Remember This

Anthropogenic changes that impact climate include

- increasing levels of the long-lived greenhouse gases: carbon dioxide, methane, nitrous oxide, and halocarbons;
- increasing levels of tropospheric ozone;
- decreasing levels of stratospheric ozone;
- increasing amounts of aerosols;
- the creation of jet contrails; and
- changes in surface albedo.

17.3.2.3 | Radiative Forcings

In order to quantify the climatic impacts of the changes described above, we use the concept of *radiative forcing* (Chapter 6). Recall that a radiative forcing is a change in the flows of radiation in the Earth-atmosphere system that will upset the energy balance and, thus, *force*

climate to change. Because they are energy-balance terms, forcings are expressed in W/m^2, with a positive forcing representing an energy surplus, and a negative forcing representing an energy deficit. Figure 17.22 provides estimates of the radiative forcings, from 1750 to 2005, resulting from the anthropogenic changes described above. This table also includes the estimated radiative forcing of the sun over this same time period. Another natural factor that has influenced climate over this period is volcanic eruptions, but the effects of these eruptions are not considered here because they are very short term. The effects of some of the listed changes are much better understood than are others; notice that the only terms for which we have a high level of scientific understanding are the long-lived greenhouse gases. Radiative forcing values are calculated for each gas based on its effectiveness as a greenhouse gas and its increase since pre-industrial times. (See Example 17.1.)

Radiative forcings provide us with a method for comparing the effects of the various changes that impact climate. Figure 17.22 shows that the total radiative forcing of all the long-lived greenhouse gases is 2.64 W/m^2, and that more than half of this forcing is due to carbon dioxide. Carbon dioxide is associated with the largest forcing because, despite being the weakest greenhouse gas, it has shown the greatest increase. The forcing of tropospheric ozone is comparable to that of the other greenhouse gases, but more uncertainty is attached to this value than to the others. Human activities are linked to three other small positive forcings: a decrease in the albedo of snow due to black carbon, an increase in stratospheric water vapour due to the breakdown of methane, and an increase in jet contrails. Notice that the latter two effects are both very small and poorly understood. Finally, since 1750, there has been a small increase in radiation coming from the sun; thus, the sun contributes a small positive forcing. Notice, however, that this *natural* forcing is considerably smaller than are the combined *anthropogenic* forcings.

Offsetting the positive forcings are several negative forcings. The largest negative forcing comes from aerosols, which account for about 80 per cent of the total negative forcing. Their forcing includes both their *direct* effect on the scattering of solar radiation and

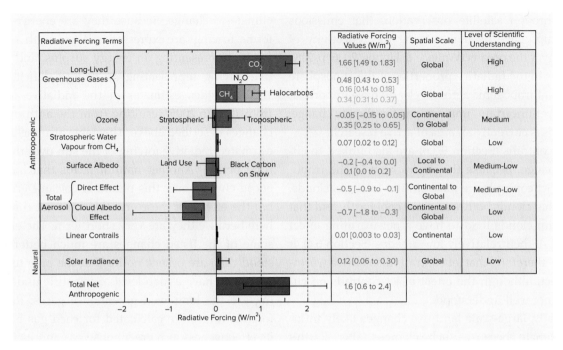

Radiative Forcing Terms		Radiative Forcing Values (W/m²)	Spatial Scale	Level of Scientific Understanding
Long-Lived Greenhouse Gases	CO₂	1.66 [1.49 to 1.83]	Global	High
	N₂O CH₄ Halocarbons	0.48 [0.43 to 0.53] 0.16 [0.14 to 0.18] 0.34 [0.31 to 0.37]	Global	High
Ozone	Stratospheric Tropospheric	−0.05 [−0.15 to 0.05] 0.35 [0.25 to 0.65]	Continental to Global	Medium
Stratospheric Water Vapour from CH₄		0.07 [0.02 to 0.12]	Global	Low
Surface Albedo	Land Use Black Carbon on Snow	−0.2 [−0.4 to 0.0] 0.1 [0.0 to 0.2]	Local to Continental	Medium-Low
Total Aerosol — Direct Effect		−0.5 [−0.9 to −0.1]	Continental to Global	Medium-Low
Total Aerosol — Cloud Albedo Effect		−0.7 [−1.8 to −0.3]	Continental to Global	Low
Linear Contrails		0.01 [0.003 to 0.03]	Continental	Low
Solar Irradiance		0.12 [0.06 to 0.30]	Global	Low
Total Net Anthropogenic		1.6 [0.6 to 2.4]		

FIGURE 17.22 | Radiative forcings associated with anthropogenic change (and the sun's output) from 1750 to 2005.

Source: Adapted from P. Forster, et al. (2007). Changes in atmospheric constituents and in radiative forcing. In S. Solomon et al. (Eds.). *Climate change 2007: The physical science basis. Contribution of working group I to the fourth assessment report of the intergovernmental panel on climate change* (pp. 129–234). Cambridge, UK: Cambridge University Press, p. 203.

their *indirect* effect on the albedo of clouds. Although the latter is the largest contributor to the negative forcing, it is associated with the greatest amount of uncertainty. A smaller negative forcing comes from the increase in albedo due to land use changes, and an even smaller negative forcing is associated with the decrease in stratospheric ozone.

To determine the total radiative forcing, we cannot simply add up the individual forcing values because the uncertainty associated with some of them is not evenly distributed. However, taking the skewed uncertainty into account provides a total anthropogenic forcing from 1750 to 2005 of 1.6 W/m². Notice that without the negative forcings, the total forcing would be almost double what it is. This has an important consequence. If all emissions were to stop suddenly, the total radiative forcing would actually *increase* before it started to decrease. The following explains why. Since ozone has a very short residence time, tropospheric ozone would be removed from the atmosphere almost immediately, reducing the positive forcing by a small amount. Aerosols have longer residence times, but they would settle out of

the atmosphere much more quickly than would the long-lived greenhouse gases, so they would be gone within a few years. The *net* effect of these two changes would be to decrease the negative forcing, leaving the positive forcing associated with the long-lived greenhouse gases.

Remember This

Certain changes linked with human activity have led to a *total* radiative forcing of 1.6 W/m² since 1750. The largest *positive* forcing comes from long-lived greenhouse gases, while the largest *negative* forcing comes from aerosols.

The positive radiative forcing of 1.6 W/m² is very likely associated with the *observed* warming of close to 0.8°C over the last 100 years or so. It would be nice if there were a simple relationship between radiative forcing and temperature change that could be used for making predictions. Unfortunately, there is no

Example 17.1

The radiative forcing attributable to a gas can be approximated as the product of the *radiative efficiency* of the gas in $W \cdot m^{-2} \cdot ppm^{-1}$ and the increase in atmospheric concentration of the gas in ppm.

From 1750 until 2005, carbon dioxide increased from 280 ppm to 379 ppm, methane increased from 700 ppb to 1774 ppb, and nitrous oxide increased from 270 ppb to 319 ppb. (These data for 2005 are taken from the IPCC's 2007 report.) Using these values and the radiative forcings given in Figure 17.22, estimate the radiative efficiency of each of these three greenhouse gases.

Carbon Dioxide:

$$\frac{1.66 \ W/m^2}{99 \ ppm} = 0.017 \ W \cdot m^{-2} \cdot ppm^{-1}$$

Methane:

$$\frac{0.48 \ W/m^2}{1.07 \ ppm} = 0.45 \ W \cdot m^{-2} \cdot ppm^{-1}$$

Nitrous Oxide:

$$\frac{0.16 \ W/m^2}{0.049 \ ppm} = 3.3 \ W \cdot m^{-2} \cdot ppm^{-1}$$

These calculations show that carbon dioxide is the least effective greenhouse gas. However, it has the greatest radiative forcing because its concentration in the atmosphere has increased much more than the concentrations of the other two gases over the period from 1750 to 2005.

Question for Thought

In what ways are anthropogenic climate change and ozone depletion linked?

such relationship; each $1 \ W/m^2$ increase in forcing will not necessarily produce the same increase in temperature. This nonlinear relationship is due to the complexity of the climate system and, in particular, to the operation of feedback effects (Section 1.2).

17.3.2.4 | Feedback Effects

In the context of global warming, a positive feedback effect will *increase* the warming, and a negative feedback effect will *decrease* the warming. Although many of the possible feedbacks associated with climate change have been identified, and are quite straightforward to explain and understand, many uncertainties remain. First, it is very difficult to estimate the *strengths* of these feedbacks; as a result, it is also very difficult to attach radiative forcings to these feedbacks. Second, it is hard to predict when any particular feedback might begin to operate. It is likely that there are *thresholds* associated with feedbacks, meaning that a certain amount of warming must occur before the feedback will begin to operate. Examples of some of the feedbacks that have been identified so far are described below.

Because they increase the rate of warming, positive feedback effects have a *destabilizing* effect. A strong, and fairly well understood, positive feedback is the *water vapour feedback*. As the temperature increases, the evaporation of water vapour should also increase. In fact, since the 1980s, observations have shown that the amount of water vapour in the atmosphere is increasing. Because water vapour is also a greenhouse gas, its increase will lead to further warming. The radiative forcing of this positive feedback is estimated to be about $1.8 \ W/m^2$. The fact that this forcing is greater than the forcing due to carbon dioxide shows that the consequences of feedback effects are not trivial. Another important positive feedback effect that is believed to be operating is the *ice feedback*. As the temperature increases, the rate of ice melt will increase; with less ice, the planetary albedo will decrease. With a lower albedo, Earth will absorb more of the sun's radiation, and more warming will occur. This will lead to further melting of ice, and so on. The radiative forcing of this feedback effect is estimated to be about $0.3 \ W/m^2$.

Some feedback effects don't influence warming rates *directly*; instead, they influence greenhouse gas concentrations. For example, because temperatures have risen, permafrost has begun to melt and, as it does so, stored carbon dioxide and methane are released. While this positive feedback increases the *sources* for greenhouse gases, other feedbacks may cause decreases in the *sinks* for these gases, and they may even eventually turn sinks into *sources*. Such a feedback effect seems to be operating in relation to the oceans and the land, both of which act as sinks for carbon dioxide. These sinks are very powerful; in fact, it has been estimated that they have absorbed almost 50 per cent of the carbon dioxide we have already emitted. As the oceans warm, however, their ability to absorb carbon dioxide will decrease because warm water holds less carbon dioxide than does cold water. It is also thought that climate change might drive changes in ocean circulation that are likely to reduce carbon dioxide uptake. As the land warms, it is probable that biological activity will increase, leading to an increase in carbon dioxide emissions from soils. If both of these powerful sinks—the oceans and the land—were to be lost, atmospheric concentrations of carbon dioxide could increase dramatically.

In contrast to positive feedbacks, negative feedbacks are *stabilizing*. A potentially important negative feedback is the *cloud feedback*. In Section 5.5, we saw that the radiative properties of clouds allow them to both warm and cool Earth, while in Section 6.4, we saw that clouds are believed to have, at present, a net *cooling* effect on Earth. As the planet warms, it is likely that there will be more clouds because there will be more water vapour in the air. What remains very uncertain, however, is what these clouds will be like and, therefore, just what their effect on temperatures might be. Current studies show that the increased cloudiness will likely have a cooling effect. This makes the cloud feedback a negative one, and the size of the forcing associated with it is estimated to be anywhere between −0.3 and −1.8 W/m².

Other negative feedbacks may act to decrease the amount of greenhouse gases in the atmosphere. It is thought that both increased temperatures and increased amounts of atmospheric carbon dioxide might increase the rates of plant growth. The resulting higher rates of photosynthesis would, in turn, increase the sink for carbon dioxide. This feedback remains an area of active research. Another possibility is that the residence time of methane may

CLIMATE SENSITIVITY
A term that provides the relationship between a radiative forcing and the resulting temperature change.

decrease as a result of warming. A warmer atmosphere may lead to greater storminess and, thus, more lightning. The increase in lightning means there will be more hydroxyl (OH), which reacts with methane (Equation 2.1), effectively removing it from the atmosphere (Dawson & Spannagle, 2009, p. 95).

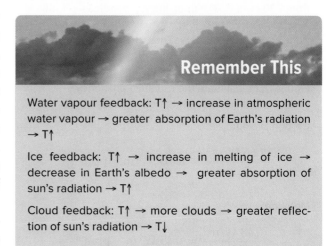

Remember This

Water vapour feedback: T↑ → increase in atmospheric water vapour → greater absorption of Earth's radiation → T↑

Ice feedback: T↑ → increase in melting of ice → decrease in Earth's albedo → greater absorption of sun's radiation → T↑

Cloud feedback: T↑ → more clouds → greater reflection of sun's radiation → T↓

Based on our present understanding of feedback effects, it appears most likely that the positive ones will outweigh the negative ones. This is bad news because it means that feedbacks will likely accelerate global warming. An alarming aspect of positive feedbacks is that they can potentially lead to rapid, escalating climate change. Further, the uncertainty associated with feedbacks makes it particularly difficult to make predictions about the magnitude and timing of climate changes.

17.3.2.5 | Consequences: Predictions and Evidence

As noted above, although radiative forcings allow us to assess the *relative* impact of various changes to the Earth-atmosphere system, they do not easily translate to temperature changes. Therefore, we rely on another term, **climate sensitivity**, to predict how much the temperature should change for a given radiative forcing. The idea is simple: a certain amount of energy should produce a given temperature change, as shown by Equation 17.8.

$$\text{Temperature Change} = \text{Climate Sensitivity} \times \text{Radiative Forcing} \quad (17.8)$$

The units of climate sensitivity are °C/W·m⁻². A high climate sensitivity means that a small forcing can cause a large temperature response, whereas a low climate sensitivity means that a large forcing is required to produce even a small temperature change.

Unfortunately, this simple idea is not easily applied to something as complex as an entire planet. Despite the challenges involved, attempts have been made to determine Earth's climate sensitivity, based on natural climate changes of the past. For example, the radiative forcing that brought about the roughly 5°C temperature decrease associated with the last ice age has been estimated to be about 7.1 W/m². Based on this radiative forcing and using the equation given above, we can estimate that Earth's climate sensitivity is about 0.7°C/W·m⁻² (Wallace & Hobbs, 2006, pp. 445–6).

In general, Earth's climate sensitivity is estimated to be between 0.4°C/W·m⁻² and 1.0°C/W·m⁻². With the inclusion of positive feedback effects, climate sensitivity is closer to the high end of this range. On the other hand, the storage of heat in the oceans *reduces* climate sensitivity in the *short* term because the large specific heat of the oceans produces a time lag between the forcing and the final temperature change, or response (Section 4.1.1). As Earth warms, the oceans will warm more slowly than the air. This slower rate of warming in the oceans will, in turn, slow the warming of the air. (Remember that when two substances are in contact heat will be transferred between them. If both are warming but one is warming more slowly than the other, then heat will be transferred from the substance that is warming more quickly to the substance that is warming more slowly.)

If climate sensitivity is taken to be 0.7°C/W·m⁻², then the radiative forcing of 1.6 W/m² that we have experienced since 1750 should produce a temperature change greater than the 0.8°C increase we have observed so far. It hasn't, because some of the energy associated with the forcing is being absorbed by the oceans, and some is being used to evaporate water and melt ice. As noted above, the heat stored in the oceans is a temperature change waiting to happen; the implication here is that even if all greenhouse gas emissions were to stop immediately, the temperature would continue to rise.

Question for Thought

Although the warming we have experienced so far seems quite small, why is it urgent that we act now to reduce emissions?

Despite the complications of feedbacks and the lag time associated with the oceans, climate sensitivity can be used in a simple model to illustrate how predictions about future temperatures are made. In climate forecasting, we usually make predictions based on a doubling of **carbon dioxide equivalent** over pre-industrial levels, as shown in Example 17.2. As of 2007, all the long-lived greenhouse gases had reached a carbon dioxide equivalent of 455 ppm, which is approximately 155 ppm over pre-industrial levels. Based on current projections of emission rates, it is estimated that a doubling of carbon dioxide equivalent will have occurred before 2050.

The simple model used in Example 17.2 allows us to *predict* that a doubling of carbon dioxide equivalent

CARBON DIOXIDE EQUIVALENT
The amount of carbon dioxide that would have the same global warming potential as a given amount of another greenhouse gas. For example, the carbon dioxide equivalent of 1 tonne of methane is 25 tonnes.

Example 17.2

Pre-industrial values of carbon dioxide equivalent are estimated to be about 300 ppm. Using a mid-range climate sensitivity value of 0.7°C/W·m⁻², how much should the temperature increase if carbon dioxide equivalent doubles?

Use the radiative efficiency for carbon dioxide of 0.017 W·m⁻²·ppm⁻¹ that we calculated in Example 17.1 in Equation 17.8.

$$0.7°C/W·m^{-2} \times 0.017\ W·m^{-2}·ppm^{-1} \times 300\ ppm = 3.6°C$$

This simple model shows that a doubling of carbon dioxide equivalent should result in a temperature increase of 3.6°C.

Question for Thought

Example 17.2 showed that a doubling of carbon dioxide equivalent could cause a warming of 3.6°C. What could cause the actual warming to be less than this estimate? What could cause the actual warming to be more than this estimate?

could produce a warming of 3.6°C. Such a calculation provides a very basic idea of the way in which climate modelling works. Climate scientists normally use very sophisticated models known as *general circulation models* (GCMs). In the context of anthropogenic climate change, these models can be used to predict distributions of temperature, and other climatic variables, from a set of initial conditions that are based on possible future concentrations of greenhouse gases.

These models are very similar to those used to forecast the weather; they are based on equations that mimic the physics of the atmosphere, and they divide the atmosphere—and, in this case, the oceans—into three-dimensional grid boxes (Section 15.6; Figure 17.23). While the inputs to weather forecasting models are current conditions, the inputs to GCMs are future scenarios of, for example, the concentrations of greenhouse gases. As you have seen throughout this book, the atmosphere does not act independently from the oceans, the land, ice, or life. Thus, today's models attempt to include as many interactions between the atmosphere and the other spheres as possible. Even the detailed outlines of continents and their topography are built into the most sophisticated models. Although all of this information is complex and difficult to incorporate into computer models, one of the biggest uncertainties in all GCMs is their handling of clouds. Clouds remain far too complex to be adequately represented

Question for Thought

In what ways is the atmosphere linked to the other three spheres of the Earth-atmosphere system?

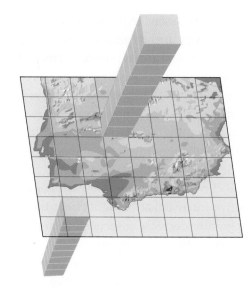

FIGURE 17.23 | The three-dimensional structure of a climate model. The model divides the atmosphere and the oceans into 3-D boxes, and it accounts for the influence of such factors as vegetation, ice, and terrain. It also accounts for the fact that the fluid within each box interacts both horizontally and vertically with the fluids in adjoining boxes.

by a set of equations. However, climatologists generally have confidence in GCMs because these models are good at replicating climate changes of the past.

According to the 2007 IPCC report, it is predicted that a doubling of carbon dioxide over pre-industrial levels, to a level of about 560 ppm, will *likely* produce a temperature increase in the range of 2 to 4.5°C, with a best estimate of 3°C. The report states that it is *very unlikely* that this value will be less than 1.5°C. A temperature increase of 2°C would be associated with a sea level rise of about half a metre; a temperature increase of more than 3°C would make Earth warmer than it has been for over three million years. These projected temperature increases are defined by the IPCC as the *sensitivity of the climate to a doubling of carbon dioxide*. As was noted above, we had reached a carbon dioxide *equivalent* of 455 ppm as of 2007, and it is expected that a doubling of carbon dioxide *equivalent*, to a level of approximately 600 ppm, will occur before the middle of this century. Because the climate system is slow to respond, the full amount of predicted warming may not be realized for several decades after the doubling is reached.

There is a general consensus that once Earth's temperature has increased by more than 2° or 3°C, dangerous impacts are likely (Figure 17.24). In Canada, it is expected that glaciers in the mountains of BC will lose half their volume, summer Arctic sea ice will be reduced to half its former extent, the risk of desertification in the prairies will significantly increase, water resources will be impacted due to both flooding and drought, many ecosystems will be severely disrupted, and much more. In addition, it is believed that if temperature increases are higher than 3°C, a threshold could be reached at which runaway feedback effects will begin to operate. These feedback effects might continue to drive climate change even without further increases in greenhouse gases.

To limit global warming to 2°C, it is thought that carbon dioxide levels must not exceed 450 to 500 ppm. Recently, some scientists have even suggested that levels must be *reduced* to 350 ppm—roughly the level of atmospheric carbon dioxide in the mid-1980s—in order to prevent dangerous consequences. The reasoning here is that even if the amount of carbon dioxide were stabilized at today's level of close to 390 ppm, temperatures would continue to rise. Meeting any of these targets would require drastic, and almost immediate, reductions in our emissions. Whether or not this can be achieved depends on society's response, which is equally as difficult to predict as the amount of warming that might result from increasing greenhouse gas concentrations.

As mentioned above, there is evidence that Earth's average temperature has already begun to increase. Careful analysis of temperature records shows an increase of about 0.8°C over the last

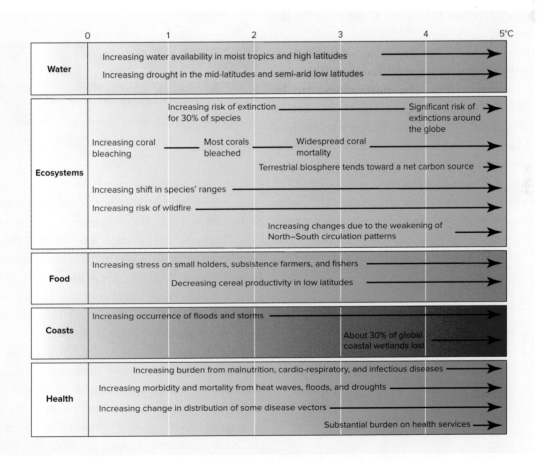

FIGURE 17.24 | Some actual and predicted impacts associated with increases in global temperatures, relative to 1980–99 temperatures. Note that the impact is predicted to begin where the description begins and increase as temperatures increase.

Source: Adapted from IPCC. (2007). *Climate change 2007: Synthesis report. Contribution of working groups I, II, and III to the fourth assessment report of the Intergovernmental Panel on Climate Change.* Geneva, Switzerland: IPCC, p. 51.

century (Figure 17.25). Most of this increase has been in the last 50 years. So far, 1998, 2005, and 2010 are tied for the warmest year on record. In the Arctic, temperature increases have been almost double the global average.

Although Figure 17.25 shows considerable fluctuation in the temperature record, the overall upward trend is clearly evident. The fluctuations can be accounted for by volcanic eruptions, El Niño–La Niña cycles, sunspot activity, and the inherent variability of climate. Much work has been done to remove from the temperature record any inconsistencies due to instrument errors and measurement methods. In addition, because cities are generally warmer than their surroundings, any warming trend that might have been

caused by the growth of cities around measurement stations has been eliminated.

We have already observed a variety of changes that can be linked to warming. For example, sea level is currently rising at a rate of about 3 mm per year due to both the melting of ice and the thermal expansion of the warming oceans. Indeed, glaciers have lost 25 m of water equivalent since 1945, and areas covered by Arctic sea ice and underlain by permafrost have been decreasing in size. There has also been an increase in weather-related natural disasters, including floods, droughts, hurricanes, heat waves, and wildfires. In addition, precipitation patterns have shifted, so that some parts of the world are receiving more precipitation than they did in the past, while others are receiving much less. Finally, evaporation rates have increased, leading to an increase in the salinity of the oceans.

Question for Thought

Why might global warming increase the frequency and intensity of hurricanes?

As you have seen, we have evidence that greenhouse gas concentrations are rising due to human activities and that global temperatures are increasing. We also know that there is a connection between the level of greenhouse gases in the atmosphere and global temperatures. Based on this evidence and understanding, is it possible for us to conclude that *our actions are causing Earth's climate to change?* There is, of course, always the possibility that the observed warming has *natural* causes. Over the years, however, the IPCC has collected more and more evidence—and has, thus, become increasingly confident—that anthropogenic greenhouse gases are indeed causing the observed warming. Such growing certainty has been reflected in each successive report. In 1995, the IPCC report stated that "the balance of evidence suggests a discernible human influence on global climate" (p. 22). In 2001, the IPCC report stated that "most of the observed warming over the last 50 years is *likely* to have been

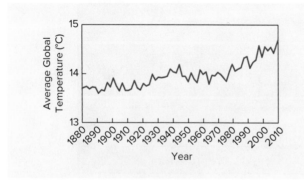

FIGURE 17.25 | Average global temperature increase from 1880 to 2011. What is the rate of increase in temperature shown in this graph? How might this *rate* change?

Data source: Adapted from NASA, "GISS surface temperature analysis." http://data.giss.nasa.gov/gistemp/graphs_v3, 12 July 2012.

due to the increase in greenhouse gas concentrations" (quoted in IPCC, 2007a, p. 39). ("Likely" indicates a 66 per cent probability.) Most recently, in 2007, the IPCC strengthened its statement to say that it was "very likely" that humans are influencing the climate. ("Very likely" indicates a 90 per cent probability.) In a sense, this increasing certainty that we are indeed causing the observed temperature increases is good news. The 2007 IPCC report has not only generated a much greater awareness of the problem, but also created a greater feeling of urgency that action is needed.

Question for Thought

When it comes to climate change, what are we certain about, and what are we not certain about?

17.3.2.6 | Action on Climate Change

The first formal international action on climate change was taken in June 1992, with the establishment of the United Nations Framework Convention on Climate Change (UNFCCC). This convention was signed at the United Nations Conference on Environment and Development in Rio de Janeiro. Its stated goal is "to achieve . . . stabilization of greenhouse gas concentrations in the atmosphere at a level that would prevent dangerous anthropogenic interference with the climate system" (UN, 1992, p. 4). Because we don't know exactly what this level is, developed nations made somewhat arbitrary commitments to stabilize their emissions at 1990 levels by 2000. These commitments were voluntary. It was agreed that there would be annual conferences, beginning in 1995, to continue the work of carrying out the goals of the convention.

The most significant of these annual meetings took place in 1997 in Kyoto, Japan, and led to the establishment of an international agreement known as the Kyoto Protocol. As part of this protocol, it was agreed that developed nations would collectively reduce their greenhouse gas emissions by 5 per cent below 1990 levels by 2012. To meet this goal, some countries—including Canada, the United States, and most western

European nations—agreed to reduce their emissions by slightly more than 5 per cent, while others agreed to a reduction of less than 5 per cent. In order to meet their targets, developed countries could reduce emissions, trade emission rights, or collect emission rights through the development of clean technology. No targets were set for the developing countries.

The Kyoto Protocol went into force in February 2005, when the ratification requirements were finally met. For those countries that ratified the protocol, the commitments are legally binding. Unfortunately, the United States—the country with the highest emissions—refused to sign on, arguing that there was little point in doing so if the developing countries were not making cuts.

Although the developed countries, as a group, managed to meet their targets by 2012, this success was not a result of great efforts to reduce emissions. Rather, emissions were reduced due to the decline in economic activity in the countries of Eastern Europe and the former Soviet Union. In fact, emissions continued to increase in most other developed nations. Canada, which withdrew from the protocol late in 2011, is one of the worst offenders; its present emissions are more than 50 per cent *above* 1990 levels. At the same time, emissions from the less developed countries are rapidly increasing. As a result, *global* emissions have increased significantly since 1990.

Nonetheless, we should not lose hope. Kyoto was intended to be only the first step; further cuts in greenhouse gas emissions have always been part of the plan, and we now know that the cuts need to be much greater. Moving forward, all developed countries—including Canada and the United States—must actively work together to reduce emissions, and the developing countries—particularly those that are developing most rapidly—must also join in and do what they can.

Question for Thought

Why do you think we have had much more success dealing with ozone depletion than we are having dealing with climate change?

In the Field

Using Global Climate Model Output for Future Scenarios of Climate Change

Adam Fenech and Neil Comer, University of Prince Edward Island, Climate Collaborating Unit

Our present understanding of the climate system and how it is likely to respond to increasing concentrations of greenhouse gases in the atmosphere would be impossible without the use of global climate models (GCMs). GCMs are powerful computer programs that incorporate physical processes to simulate, as accurately as possible, the functioning of the global climate system in three spatial dimensions and in time.

The results from 24 GCMs were used in the deliberations of the Intergovernmental Panel on Climate Change's (IPCC) Fourth Assessment Report (4AR), which was released in 2007. Each of the modelling centres provided future projections for at least two, sometimes three, emission scenarios (scenarios that describe how greenhouse gas emissions could evolve over the next 100 years). Thus, the report was able to take into account about 72 possible future outcomes for a location's climate (figures 1 and 2).

While the models are all in agreement on the direction of temperature change, results between models can vary widely, and models each contain their own inherent biases. The differences in results exist because of the differences between each GCM's model resolution, model formulations, and model parameterization. Differences also arise depending on which emission scenario of future greenhouse gases is selected.

There are several caveats to using GCMs for future projections. The resolution of the models varies and is

completely determined by the modelling centre; that is, there is no "standard" grid size or projection method. The output from the model represents an average of the entire grid cell area—a grid cell with a large resolution of about 250 by 250 km. This approximation means that even the distribution of land/water grid cells differs between models. For many locations, this can have important implications because many climates are influenced significantly by the moderating effect of proximity to large water bodies. Regardless of these caveats, models remain the best option available for creating future climate projections—versus, for example, a simple extrapolation of historical trends.

Three approaches have been developed to provide some direction for determining which of the many future projections of climate should be used in planning: the extremes (max./min.) approach; the ensemble approach; and the validation approach. The *extremes (max./min.) approach* suggests that it is best to plan within the full range of possibilities that the GCMs present. It takes the projection for the maximum change as well as the projection for the minimum change and uses both as the range of considerations when planning. The *ensemble approach* suggests that it is best to plan for the average change of all the models. It uses a mean or median of all the models (or many models) to reduce the uncertainty associated with any individual model. The *validation approach* suggests that only those

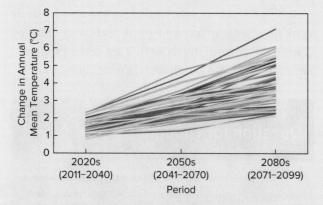

FIGURE 1 | Future projections of annual mean temperature changes at the City of Toronto grid cell in the 2020s, 2050s, and 2080s using all 24 models contributing to the IPCC 4AR (change in degrees Celsius from 1961–90 baseline period).

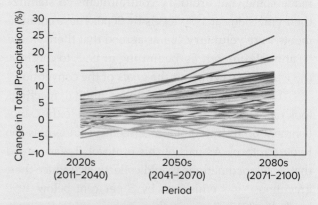

FIGURE 2 | Future projections of annual total precipitation changes at the City of Toronto grid cell in the 2020s, 2050s, and 2080s using all 24 models contributing to the IPCC 4AR (change in percentage from 1961–90 baseline period).

models that compare well to historical climate observations should be used for planning. It takes the historical climate observations over a 30-year period from a global gridded dataset (for example, that of the National Centers for Environmental Prediction, NCEP) and compares this against all models to see which ones reproduce the values best. Subsequently, only the four or five best-agreement models are used to produce the validated projections for planning (usually within one standard deviation).

As an example, all three approaches are presented for the future projections of climate for the grid cell that contains the City of Toronto, Ontario (figures 3 and 4). Impact or adaptation studies may require the application of additional downscaling methods to provide more precise results, but what is presented here is how the GCMs project the future magnitude of annual mean climate change. Notice that the validated and ensemble methods show very similar average results.

ADAM FENECH is the director of Climate Collaborating Initiatives at the University of Prince Edward Island and was previously a senior climatologist with Environment Canada. He was awarded the 2007 Nobel Peace Prize as a member of the Intergovernmental Panel on Climate Change.

NEIL COMER is a research climatologist working in Climate Collaborating Initiatives at the University of Prince Edward Island and was previously manager of the Canadian Climate Change Scenarios Network for Environment Canada.

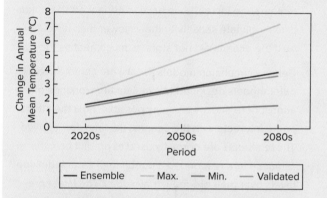

FIGURE 3 | Future projections of annual mean temperature changes at the City of Toronto grid cell in the 2020s, 2050s, and 2080s—extremes (max./min.), ensemble, and validated (change in degrees Celsius from 1961–90 baseline period).

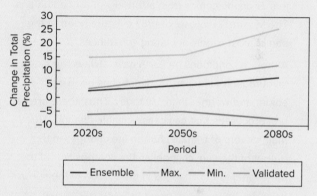

FIGURE 4 | Future projections of annual total precipitation changes at the City of Toronto grid cell in the 2020s, 2050s, and 2080s extremes (max./min.), ensemble, and validated (change in percentage from 1961–90 baseline period).

17.4 | Chapter Summary

1. Air pollutants are substances present in the atmosphere at high enough concentrations to be harmful. They can be in gaseous, liquid, or solid form. Some pollutants are emitted directly into the atmosphere, while others form in the atmosphere. Primary gaseous pollutants include carbon monoxide, nitrogen oxides, and sulphur dioxide. Liquid or solid pollutants are collectively known as particulate matter. Volatile organic compounds can be primary or secondary pollutants; they can also be in gaseous, liquid, or solid form. Ozone, a secondary gaseous pollutant, is the main component of photochemical smog. This smog forms when sunlight reacts with nitric oxide and volatile organic compounds from motor vehicle exhaust. Air pollution concentrations often reach their highest levels under anticyclonic conditions.

2. The ozone layer is being threatened by certain gases associated with human activity. The most serious threat comes from CFCs. These gases do not break down until they reach the stratosphere, where the chlorine they release has been destroying the ozone layer. This is significant because the ozone layer absorbs the sun's ultraviolet radiation, which is harmful to life. Proof that CFCs were destroying the ozone layer came in the mid-1980s, when the ozone hole was discovered and studied. The ozone hole has been shown to be a result of very complex chemistry involving polar stratospheric clouds (PSCs). Nitric oxide and bromine also destroy stratospheric ozone. The source of nitric oxide is mostly nitrous oxide from fertilizer use, but a minor source is the exhaust of supersonic jets. Bromine comes from artificially produced halons. The Montreal Protocol of 1987 eventually led to the ban of most ozone-depleting substances.

3. Although conditions on Earth have always remained favourable for life, Earth's climate has been both colder and warmer than it is today. Thus, there are natural processes that can cause climate change. Three main drivers of climate change are changes in the output of the sun, changes in the composition of the atmosphere, and changes in the albedo of the surface.

4. Since the beginning of the Industrial Revolution, human activities have been causing an increasingly positive radiative forcing on climate. The highest contribution to this positive forcing comes from increases in greenhouse gas concentrations. The burning of fossil fuels and deforestation are responsible for the increases in carbon dioxide, while agriculture is responsible for most of the increases in methane and nitrous oxide. Aerosol concentrations have also been increasing, mostly due to the burning of fossil fuels. The negative forcing contributed by aerosols partially offsets the positive forcing due to greenhouse gases. The increase in surface albedo, caused mostly by deforestation, also contributes a small negative forcing.

5. Our ability to make predictions about anthropogenic climate change is complicated by feedback effects. Two of the most important, and best understood, positive feedbacks are the water vapour feedback and the ice feedback. It is most likely that clouds will create a negative feedback, but the strength of this feedback is very uncertain. These three feedbacks have a direct effect on temperature; other feedbacks impact greenhouse gas concentrations. Our current understanding is that the positive feedbacks will outweigh the negative feedbacks.

6. Climate sensitivity provides the relationship between radiative forcing and temperature change. A large value for climate sensitivity indicates that small forcings can lead to large temperature changes. The temperature increase we have experienced so far indicates a climate sensitivity that is lower than it would be if the oceans did not store as much heat as they do.

7. General circulation models (GCMs) are complex computer models designed to simulate the workings of the atmosphere and, as much as possible, how the atmosphere interacts with the oceans, the land, ice, and life. These models are normally used to predict how the climate might respond to a doubling of carbon dioxide levels over their pre-industrial values. The latest predictions are that temperature will increase somewhere between 2 and 4.5°C before the end of this century.

8. There are signs that Earth's climate is changing. Temperatures have increased about 0.8°C over the last century. Sea level is rising. Glaciers and sea ice are melting. The amount of weather-related natural disasters is increasing. Precipitation patterns are changing.

9. The Kyoto Protocol is an international agreement designed to minimize the dangerous effects of anthropogenic climate change. As part of this agreement, most of the developed countries of the world committed to reducing their greenhouse gas emissions to 5 per cent of their 1990 levels by 2012. Very little action was taken to reduce emissions, and overall global emissions are continuing to increase.

Review Questions

1. How are primary pollutants different from secondary pollutants?

2. What are the major air pollutants, and what are their main sources?

3. What is photochemical smog? Under what conditions will it form? Why does it often reach its highest concentration in the afternoon?

4. How can atmospheric conditions influence pollutant concentrations? Under what conditions is air quality likely to be poor?

5. Why do we want ozone in the stratosphere, but not in the troposphere?

6. Why are CFCs safe in the troposphere but dangerous once they reach the stratosphere?

7. How do CFCs destroy the ozone layer? What other gases can destroy ozone? What are their sources?

8. How did the unique conditions in the atmosphere over Antarctica lead to the formation of the ozone hole?

9. Why will destruction of the ozone layer continue for a long time even though CFCs have been banned?

10. What are the causes of climate change? Which of these causes can be influenced by human activities?

11. What is meant by the term *radiative forcing*? How is it different from global warming potential? What changes create positive forcings? What changes create negative forcings? (Include both natural and anthropogenic changes.) How is the concept of radiative forcing used to help us understand climate change?

12. What are some of the feedback effects associated with global warming? How do these feedback effects work? In the context of anthropogenic climate change, why is it unfortunate that positive feedback effects outweigh negative ones?

13. What is climate sensitivity? How is it used?

14. What evidence do we have that Earth's climate is changing?

15. In what ways has the Kyoto Protocol been successful? In what ways has it not been successful?

Suggestions for Research

1. Investigate ways that we could either cool the planet or remove greenhouse gases from the atmosphere. Provide a critical evaluation of the methods you discover.

2. Find some resources on phenology—the study of the timing of seasonal biological phenomena such as flowering, migration, and length of growing season—and describe how this field of study can offer tools that might be useful for studying the effects of climate change.

3. Research the ways in which climate change is suspected to be affecting agriculture, both in Canada and around the world. Summarize the evidence behind this suspicion, and note how people around the world are adapting their approaches to agriculture in response to climatic changes.

References

Dawson, B., & M. Spannagle. (2009). *The complete guide to climate change*. London, UK: Routledge.

IPCC. (2007a). *Climate change 2007: Synthesis report. Contribution of working groups I, II, and III to the fourth assessment report of the Intergovernmental Panel on Climate Change*. Geneva, Switzerland: IPCC.

IPCC. (2007b). *Fourth assessment report*. Geneva, Switzerland: IPCC.

IPCC. (1995). *IPCC second assessment: Climate Change 1995. A report of the Intergovernmental Panel on Climate Change*. Geneva, Switzerland: IPCC.

Kaufman, T.J., & R.S. Fraser. (1997). The effect of smoke particles on clouds and climate forcing. *Science, 277*, 1636–9.

Kump, L.R., J.F. Kasting, & R.G. Crane. (2004). *The earth system* (2nd edn.). Upper Saddle River, NJ: Pearson Prentice Hall.

Wallace, J.M., & P.V. Hobbs. (2006). *Atmospheric science: An introductory survey* (2nd ed.). Boston, MA: Elsevier Academic Press.

UN. (1992). *United Nations Framework Convention on Climate Change*. Retrieved online at http://unfccc.int/resource/docs/convkp/conveng.pdf

Appendix

Guide to Weather Station Symbols

Key to Surface Weather Station Symbols

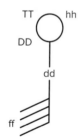

C$_H$	High Cloud Type	PPP	Sea-Level Pressure*
C$_M$	Middle Cloud Type	PP	Pressure Change in the Last 3 Hours**
C$_L$	Low Cloud Type		
N	Cloud Cover	a	Pressure Tendency
TT	Air Temperature (°C)	dd	Wind Direction
T$_d$T$_d$	Dew-Point Temperature (°C)	ff	Wind Speed
ww	Present Weather		

*The initial 9 or 10 has been omitted. For example, a pressure of 996.3 hPa is indicated as 963, and a pressure of 1023.5 hPa is indicated as 235.

**For example, a pressure change of 1.1 hPa is indicated as 11.

Key to Upper Air Weather Station Symbol s

TT	Air Temperature (°C)	ff	Wind Speed
DD	Dew-Point Depression (°C)	hh	Height of Pressure Surfaces
dd	Wind Direction		

*For example, a height of 546 decametres is indicated as 46.

Weather

••	Rain		Squalls
,,	Drizzle		Ice pellets (shower)
	Freezing rain	∞	Haze
✳✳	Snow	=	Mist
	Ice pellets	≡	Fog
	Hail		Lightning
	Rain shower		Thunderstorm
	Heavy rain shower		Thunderstorm with rain and/or snow
	Snow shower		Thunderstorm with hail

Wind Speeds

| < 5 knots
<9 km/h | 5–9 knots
9–17 km/h | 10–14 knots
18–26 km/h | 15–19 knots
27–36 km/h | 20–24 knots
37–45 km/h | 25–29 knots
46–54 km/h |

| 30–34 knots
55–63 km/h | 35–39 knots
64–73 km/h | 40–44 knots
74–82 km/h | 45–49 knots
83–91 km/h | 50–54 knots
92–100 km/h | 55–59 knots
101–110 km/h |

| 60–64 knots
111–119 km/h | 65–69 knots
120–128 km/h | 70–74 knots
129–138 km/h | 75–79 knots
139–147 km/h | 80–84 knots
148–156 km/h | 85–89 knots
157–165 km/h |

| 90–94 knots
166–174 km/h | 95–99 knots
175–184 km/h | 100–104 knots
185–193 km/h | 105–109 knots
194–202 km/h | 110–114 knots
203–212 km/h | 115–119 knots
213–221 km/h |

Cloud Cover

 no clouds

 1/8 coverage or less
1/10 coverage or less

 2/8 coverage
2/10 to 3/10 coverage

 3/8 coverage
4/10 coverage

 4/8 coverage
5/10 coverage

 5/8 coverage
6/10 coverage

 6/8 coverage
7/10 to 8/10 coverage

 7/8 coverage
9/10 coverage or more (but less than 10)

 full coverage (overcast)

 sky obscured or cloud amount cannot
be determined

Cloud Type: High Clouds

Cirrus filaments, strands, or hooks; not increasing

Cirrus, in dense patches or tufts

Cirrus, dense, often in the form of an anvil

Cirrus hooks or filaments; increasing

Cirrus and cirrostratus (or just cirrostratus); increasing and growing denser; the continuous veil does not reach 45° above the horizon

Cirrus and cirrostratus (or just cirrostratus); increasing and growing denser; the continuous veil extends more than 45° above the horizon

A veil of cirrostratus covering the sky

Cirrostratus, not increasing, and not completely covering the sky; cirrus and cirrocumulus may be present

Cirrocumulus, sometimes accompanied by some cirrus and/or cirrostratus

Cloud Type: Middle Clouds

Altostratus, mostly semitransparent

Altostratus, dense enough to hide the sun, or nimbostratus

Altocumulus, mostly semitransparent; clouds all at the same level; appearance changes slowly

Altocumulus in patches, mostly semitransparent; clouds at one or more levels; appearance changes constantly

Altocumulus in bands or in one or more layers, semitransparent; increasing and thickening

Altocumulus resulting from the spreading out of cumulus

Altocumulus in two or more layers, mostly or entirely opaque; not increasing; *also* altocumulus combined with altostratus

Altocumulus with tower-like protrusions or with the appearance of cumuliform tufts

Altocumulus of a chaotic sky; clouds usually at multiple levels

Cloud Type: Low Clouds

Cumulus, little vertical development, appears flattened

Cumulus, moderate or strong vertical development, often with tower- or dome-like protrusions; may be accompanied by other cumulus or stratocumulus; all cloud bases at the same level

Cumulonimbus without clearly defined tops; cumulus, stratocumulus, or stratus may also be present

Stratocumulus resulting from the spreading out of cumulus; cumulus may also be present

Stratocumulus not resulting from the spreading out of cumulus

Stratus in a mostly continuous layer or in ragged shreds

Fractured stratus or cumulus of bad weather, usually below altostratus or nimbostratus

Cumulus and stratocumulus not resulting from the spreading out of cumulus and with the base of the cumulus at a different level from the base of the stratocumulus

Cumulonimbus with clearly defined tops, often in the form of an anvil; may be accompanied by cumulonimbus without clearly defined tops, cumulus, stratocumulus, or stratus

Pressure Tendencies

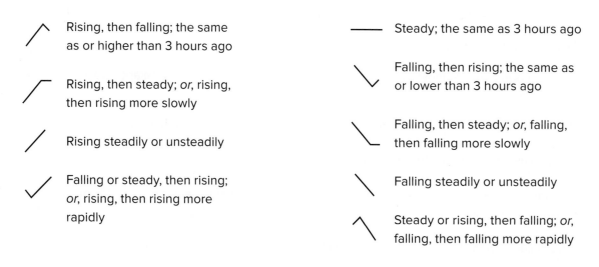

Rising, then falling; the same as or higher than 3 hours ago

Rising, then steady; *or*, rising, then rising more slowly

Rising steadily or unsteadily

Falling or steady, then rising; *or*, rising, then rising more rapidly

Steady; the same as 3 hours ago

Falling, then rising; the same as or lower than 3 hours ago

Falling, then steady; *or*, falling, then falling more slowly

Falling steadily or unsteadily

Steady or rising, then falling; *or*, falling, then falling more rapidly

Glossary

absolute humidity   The mass of water vapour in a unit volume of air. Absolute humidity is usually expressed in units of grams per cubic metre (g/m^3).

absolute vorticity   The sum of planetary vorticity and relative vorticity. If there is no vertical motion, absolute vorticity is conserved.

absolutely stable   The condition of an air layer that is characterized by a temperature profile (ELR) that is less than the saturated adiabatic lapse rate (SALR). An inversion is also absolutely stable. In this environment, an unsaturated or saturated air parcel that is displaced upward will be colder than the surrounding air, so that vertical motion is suppressed.

absolutely unstable   The condition of an air layer that is characterized by a temperature profile (ELR) that is greater than the dry adiabatic lapse rate (DALR). In this environment, an unsaturated or saturated air parcel that is displaced upward will be warmer than the surrounding air, so that vertical motion is favoured.

absorptivity   A measure of a substance's ability to absorb incident radiation.

accretion   A process by which ice crystals grow by colliding with supercooled water droplets that then freeze onto them.

acid rain   Rain with a pH less than about five. Acid rain occurs due to air pollutants, usually sulphur dioxide or nitrogen oxides, that have reacted to form acid droplets.

adiabatic process   A thermodynamic process in which temperature changes without a transfer of heat.

advection   Horizontal transfer across a fluid, caused by movement within the fluid. In the atmosphere, advection is achieved by the winds; in the oceans, it is achieved by the currents.

advection fog   A fog that forms when warm, moist air cools to its dew-point temperature as it is blown over a cold surface.

aerosols   Tiny solid or liquid particles suspended in the atmosphere.

aggregation   A process by which ice crystals grow by colliding with other ice crystals that then stick together.

air mass   A large body of air with similar temperature and moisture levels in the horizontal direction.

air pollutant   Any gaseous, liquid, or solid substance present in the atmosphere in sufficient quantity to negatively impact the health of people and animals, or harm plants or materials.

Air Quality Health Index (AQHI)   An index used to report to the Canadian public about the health risk associated with certain levels of air pollutants. It is calculated based on the amounts of ozone, PM_{10}, $PM_{2.5}$, and nitrogen dioxide present in the atmosphere.

air-mass thunderstorm   See **single-cell thunderstorm**.

albedo   The proportion of the sun's incident radiation that is reflected by a surface.

Aleutian low   The subpolar low over the northern Pacific Ocean. It is particularly strong in winter.

altitude angle   The angle of the sun above the horizon.

anaerobic decomposition   A process of decay that occurs when oxygen is unavailable. The products are methane and carbon dioxide.

analemma graph   A graph giving the latitude at which the sun is directly overhead for any day of the year.

angular momentum   The momentum associated with motion along a circular path. It is expressed as the product of mass, velocity, and the distance from the centre of rotation.

anthropogenic   Related to human activities.

anticyclone   An area of high pressure around which winds blow clockwise in the northern hemisphere and counterclockwise in the southern hemisphere.

anvil top   The top of a cumulonimbus cloud. This top spreads horizontally, as a result of the stable air above, and it is composed of ice crystals.

aphelion   The position in Earth's orbit when Earth is farthest from the sun. Aphelion occurs in early July.

atmosphere   The layer of gases surrounding a planet or other celestial body.

atmospheric pressure The force exerted by the atmosphere on Earth's surface. The average pressure of the atmosphere at sea level is 101.325 kPa.

atmospheric sounding Measurement of the change with height of certain atmospheric variables, such as temperature, pressure, humidity, and wind. Soundings are traditionally measured by radiosondes attached to balloons.

atmospheric stability A measure of the tendency for a parcel of air, once disturbed, to move vertically in the atmosphere due to temperature differences.

atmospheric window The band of wavelengths of radiation, from 8 to 11 μm, that is not absorbed by gases in the atmosphere.

backing Rotating counterclockwise with height; used to describe the rotation of winds.

baroclinic The atmospheric condition in which isotherms cross isobars. Under these conditions, temperature advection is likely to occur.

baroclinic instability A type of instability that occurs when there are strong horizontal temperature gradients that lead to thermal advection.

barotropic The atmospheric condition in which isotherms parallel isobars.

barotropic instability A type of instability that occurs when there are no horizontal temperature gradients. It arises due to the rotation of Earth alone.

beam depletion The increasing depletion of the solar beam by atmospheric constituents—through scattering, reflection, and absorption—as the sun's path length through the atmosphere increases.

beam spreading The spreading of the solar beam over an increasing surface area as the sun's angle decreases.

Bermuda high The subtropical high over the northern Atlantic Ocean. It is particularly strong in the summer.

biogeochemical cycle The model that describes how an element or compound is transferred between the atmosphere, rocks, water, and life of planet Earth.

biosphere Life on Earth.

black body A hypothetical substance that does not reflect or transmit radiation; instead, it absorbs all of the radiation incident on it. It emits the maximum amount of radiation possible at its temperature.

Canadian high A thermal high that forms over Canada in the winter.

carbon cycle The model that describes the processes by which carbon is transferred between the various reservoirs of the Earth-atmosphere system.

carbon dioxide equivalent The amount of carbon dioxide that would have the same global warming potential as a given amount of another greenhouse gas. For example, the carbon dioxide equivalent of 1 tonne of methane is 25 tonnes.

carbonate–silicate cycle The inorganic part of the carbon cycle. In this cycle, carbon dioxide is removed from the atmosphere as silicate rocks weather, and it is returned to the atmosphere hundreds of thousands to millions of years later by volcanic eruptions.

centripetal force The net force, directed toward the centre of rotation, that acts on a rotating object. The centripetal force is not a real force, but a net force or acceleration.

chance of precipitation The chance that measurable precipitation will fall on any random point of the forecast region during the forecast period. *Measurable precipitation* is defined as 0.2 mm of rain or 0.2 cm of snow.

chaos Behaviour, such as that of weather, that is extremely sensitive to initial conditions and very difficult to predict.

chinook The Canadian name for a warm, dry wind that blows down the leeward side of a mountain range.

circle of illumination A circular boundary between Earth's light half and its dark half.

climate The average conditions of the atmosphere.

climate controls Characteristics of a location that combine to create the climate of that location.

climate sensitivity A term that provides the relationship between a radiative forcing and the resulting temperature change. Climate sensitivity is expressed in $°C/W \cdot m^{-2}$.

climatology The study of climate.

cloud A dense mass of water droplets and/or ice crystals suspended in the atmosphere.

cloud condensation nuclei (CCN) Atmospheric aerosols on which water vapour can condense to form water droplets. They tend to be large or giant aerosols that are also hygroscopic.

coalescence efficiency The probability that two colliding cloud droplets will coalesce.

cold cloud A cloud in which the temperature is below 0°C in at least part of the cloud. Such clouds contain a mixture of supercooled water droplets and ice crystals.

cold front A front at which cold air is advancing and replacing warm air.

cold-type occlusion An occlusion in which the air behind the cold front pushes under the air ahead of the warm front.

collision efficiency The probability that two droplets in a cloud will collide.

condensation The process by which a substance, usually water, changes phase from a gas to a liquid.

conditionally unstable The condition of an air layer that is characterized by a temperature profile (ELR) that is less than the dry adiabatic lapse rate (DALR) but greater than the saturated adiabatic lapse rate (SALR). This air will be unstable on the condition that it becomes saturated.

conduction The transfer of heat between molecules in contact with one another.

conductive sensible heat flux The transfer of sensible heat by conduction between the surface and the ground below.

conductivity The property of a substance that describes its ability to conduct heat.

constant gases Gases whose relative proportions do not change up to a height of about 80 km in the atmosphere or over time scales of hundreds of years. These gases have relatively long residence times, so they become well mixed in the atmosphere.

continentality The degree to which a climate is affected by its distance from a body of water. For example, inland locations tend to experience larger annual and daily temperature ranges than do coastal locations.

convection Motions in a fluid that transfer the properties (e.g., heat) of that fluid. In the atmosphere, convection normally refers to vertical motions.

convectional precipitation Precipitation that falls from clouds produced by surface heating. This precipitation is characteristically heavy, localized, and of short duration.

convective available potential energy (CAPE) A measure of the potential intensity of thunderstorms.

convective inhibition (CIN) A measure of the strength of the cap preventing surface air from rising to produce a thunderstorm. The larger is the CIN, the greater is the trigger needed to start a thunderstorm.

convective latent heat flux The transfer of latent heat by convection between Earth's surface and the atmosphere.

convective sensible heat flux The transfer of sensible heat by convection between Earth's surface and the atmosphere.

convectively unstable The condition of an air layer in which the lower air is moist and the upper air is dry. Such an air layer has the potential to become unstable if it is lifted.

convergence The net inflow of air to a region. Speed convergence occurs when winds are slowing down along the direction of flow. Directional convergence occurs when air streams are flowing together.

Coriolis force An apparent force, resulting from the rotation of Earth, that causes objects moving freely above Earth's surface to seem to be deflected from a straight-line path.

cryosphere The ice of Earth.

cumuliform cloud Cloud that is heaped in form and often exhibits strong vertical development.

curvature effect The effect in which increased curvature of a droplet's surface increases the relative humidity required for the droplet to be in equilibrium with its surroundings. Curvatures increase as droplets get smaller.

cutoff high A warm upper-air high that has been cut off from the general westerly flow, so that it lies poleward of this flow.

cutoff low A cold upper-air low that has been cut off from the general westerly flow, so that it lies equatorward of this flow.

cyclogenesis The development or strengthening of a mid-latitude cyclone. This is usually associated with both an increase in cyclonic circulation and a decrease in surface pressure.

cyclone An area of low pressure around which winds blow counterclockwise in the northern hemisphere and clockwise in the southern hemisphere.

December solstice The date on which the north end of the axis points away from the sun so that the sun is directly overhead at the Tropic of Capricorn, 23½° S. This is the first day of winter in the northern hemisphere and the first day of summer in the southern hemisphere.

deepening A decrease in the central pressure of a pressure system. A deepening cyclone is strengthening.

denitrification The process by which bacteria convert nitrogen in the soil to nitrogen gas or nitrous oxide gas (N_2O). This is a source for atmospheric nitrogen because nitrous oxide is eventually converted to nitrogen gas (N_2).

density The amount of mass in a unit volume. The units of density in the MKS system are kilograms per cubic metre (kg/m^3).

deposition   A phase change from gas to solid.

depression of the wet bulb   The difference between the dry-bulb temperature and the wet-bulb temperature.

dew   Atmospheric water vapour that has condensed onto a cool surface during the night.

dew-point depression   The difference between the air temperature and the dew-point temperature.

dew-point temperature   The temperature to which the air must be cooled, at constant pressure, to reach saturation.

diabatic process   A process in which temperature changes as a result of heat transfer.

diffuse radiation   The sun's radiation that reaches Earth's surface after being scattered.

diffusion   The movement of water vapour molecules toward a water droplet or ice crystal upon which they condense or are deposited, respectively.

direct beam radiation   The sun's radiation that reaches Earth's surface without first being scattered.

divergence   The net outflow of air from a region. Speed divergence occurs when winds are speeding up along the direction of flow. Directional divergence occurs when air streams are flowing apart.

downburst   A strong, gusty wind that forms due to downdrafts in thunderstorms.

drizzle   A type of precipitation made up of water droplets smaller than typical rain drops. Drizzle can form in shallow stratiform clouds, or it can result from significant evaporation of rain as it falls through dry air.

dry adiabatic lapse rate (DALR)   The rate of change of temperature of a rising or sinking unsaturated air parcel. This rate is equal to 10°C/km.

dry line   A boundary separating warm, moist air from warm, dry air. Because it is less dense than dry air, the moist air will frequently rise, forming a line of thunderstorms.

dry-bulb thermometer   The thermometer in a psychrometer that is used to measure air temperature.

dynamic pressure systems   High- and low-pressure systems that develop as a result of complex air motions. These pressure systems are deep.

easterly wave   A disturbance, in the shape of a wave, in the general easterly flow of the tropics.

eddies   Irregular whirling motions that characterize turbulent flow.

eddy viscosity   Large-scale resistance to flow, or internal friction, due to the random, irregular motions associated with turbulence.

effective radiating temperature   The temperature at which a system radiates away as much energy as it receives.

electromagnetic spectrum   The continuous spectrum of wavelengths of electromagnetic radiation.

electromagnetic waves   Waves that are formed and propagated by oscillating electric and magnetic fields.

emissivity   The ratio of radiation emitted by a real substance to the amount emitted by a black body at the same temperature.

energy   The property of a substance that gives it the capacity to do work. The units of energy in the MKS system are joules (J).

energy balance   An account of the flows of energy in a system. The energy balance concept can be applied at any spatial or temporal scale. This term is also used to denote the balance between energy entering into and leaving a system.

energy flux density   The rate of the flow of energy per unit area of surface. The units of energy flux density in the MKS system are watts per square metre (W/m^2).

ensemble forecasting   A forecasting technique that involves running forecasting models several times with slightly different initial conditions and/or small changes in the way the model operates. The resulting forecasts can be averaged or used to assess the amount of uncertainty in a forecast.

entrainment   A process by which air surrounding a cloud is drawn into the cloud. This air will usually be colder and drier than the air in the cloud.

environmental lapse rate (ELR)   The change in temperature with height in the atmosphere. The environmental lapse rate is usually measured by a radiosonde.

equation of state   An equation that provides the relationship between the temperature, pressure, and volume of a substance. The ideal gas law is an equation of state.

equatorial low   A region of low pressure that develops at, or near, the equator. It is a thermal low due to strong surface heating at this location.

equilibrium vapour pressure   See **saturation vapour pressure**.

evaporation   The process by which a substance, usually water, changes phase from a liquid to a gas.

extratropical   Relating to regions lying poleward of the tropics. These regions include both the mid-latitudes and the polar latitudes.

extratropical cyclone   See **frontal system**.

fall streaks   Ice crystals falling from a cirrus cloud.

feedback effect   A mechanism that operates within a system to either amplify (in the case of positive feedback) or lessen (in the case of negative feedback) an initial change.

filling   An increase in the central pressure of a pressure system. A filling cyclone is weakening.

first law of thermodynamics   A law stating that a change in the internal energy of a substance is associated with the transfer of energy as heat or by work. It is a statement of the conservation of energy.

fluid   A substance that can flow. Liquids and gases are both fluids.

flux convergence   The result when the input of energy to a volume exceeds the output of energy from that volume. This will produce a temperature increase.

flux divergence   The result when the input of energy to a volume is less than the output of energy from that volume. This will produce a temperature decrease.

fog   Suspensions of water droplets and/or ice crystals in a layer of air at Earth's surface.

force   An action capable of accelerating an object. The units of force in the MKS system are newtons (N).

freezing rain   A type of precipitation that occurs when there is a shallow subfreezing air layer at the surface topped by a layer of air in which temperatures are above freezing. The effect of this temperature stratification is that precipitation will melt then refreeze upon contact with the surface.

friction   The force that resists motion whenever objects move, or try to move, relative to each other.

front   A narrow zone of transition between air of different properties.

frontal fog   See **precipitation fog**.

frontal inversion   An upper-air inversion that forms at a front. Because warm air rises over cold air at fronts, inversions form in and above the frontal zone.

frontal precipitation   Precipitation that falls from clouds produced in frontal systems. This precipitation is characteristically moderate in intensity, and it can be of long duration and fairly widespread.

frontal system   A large area of low pressure that forms due to fronts. These storms usually cover thousands of square kilometres and are unique to the mid-latitudes because it is only there that fronts occur. *Frontal system* is another name for a mid-latitude cyclone or an extratropical cyclone.

frontogenesis   A process that increases a temperature gradient and forms a front.

frontolysis   A process that leads to the dissipation of a front.

frost   Atmospheric water vapour that has collected on a surface as ice crystals.

general circulation   The average patterns of pressure and wind over Earth's surface.

general circulation model (GCM)   A computer program that represents the physics of the atmosphere through a set of equations. These models are used to study past climates and make predictions about climate change.

geostrophic wind   A wind, depicted on a weather map as blowing parallel to straight isobars or height contours, that develops when the pressure gradient force is balanced by the Coriolis force. These winds occur in absence of friction with Earth's surface.

global warming   The increase in Earth's temperature caused by increasing concentrations of greenhouse gases associated with human activities.

Global Warming Potential (GWP)   A measure of the capacity of a greenhouse gas to impact global temperatures, relative to the warming potential of carbon dioxide.

gradient wind   A wind that develops when the pressure gradient force and the Coriolis force are unbalanced. It is depicted on a weather map as flowing parallel to curved isobars, or height contours.

graupel   Lumps of ice that begin as ice crystals but become rounded as supercooled water freezes onto them. A characteristic of graupel is that the original form of the ice crystal is no longer recognizable.

greenhouse effect   An increase in the temperature of a planet due to the presence of greenhouse gases in its atmosphere.

greenhouse gas   A gas that allows the shorter wavelength radiation from the sun to pass through the atmosphere, while it absorbs the longer wavelength radiation leaving Earth's surface.

gust front   The boundary at which the cold air in a thunderstorm downdraft meets warm, moist air at the surface.

Hadley cell A thermally driven convection cell located between latitude 30° N and the equator, or between latitude 30° S and the equator.

hail Frozen precipitation made up of concentric layers of alternating clear and opaque ice.

halo A ring of light appearing around the sun or the moon due to the refraction of light by the ice crystals in a cloud.

halocarbons Substances containing carbon and one or more halogen atoms (i.e., chlorine, bromine, fluorine, iodine). Most halocarbons are artificially produced.

halons A type of halocarbon that contains carbon, bromine, and other halogen atoms. They are used mostly in fire extinguishers.

haze A reduction of visibility caused by the scattering of visible radiation in the atmosphere.

heat Energy in the process of being transferred from one object to another as a result of a temperature difference between them. Heat is transferred from high temperature to low temperature.

heat capacity The amount of heat required to raise a unit volume of a substance by 1 K.

heterogeneous nucleation The formation of water droplets on a nucleus. In the atmosphere, these are aerosols. Droplets that form this way will be composed of water and the nucleus upon which the water condensed.

heterosphere The upper atmosphere, above about 85 km, in which molecular diffusion dominates, so that the molecules settle out with the heaviest on the bottom and the lightest on the top.

homogeneous nucleation The formation of water droplets by the chance collision of water vapour molecules. Such droplets will be composed of water only.

homosphere The layer of the atmosphere, below about 85 km, in which the constant gases are thoroughly mixed.

humidex An index used in Canada to provide a measure of how warm it feels due to a combination of high temperature and high humidity.

humidity The amount of water vapour in a quantity of air.

hurricane A tropical cyclone in which sustained winds are higher than 120 km/h. These storms are called *hurricanes* in the Atlantic and eastern Pacific, *typhoons* in the western Pacific, and *cyclones* in the Indian Ocean. In addition to strong winds, these storms bring extremely heavy rain and can lead to storm surges.

hydrocarbons Substances containing hydrogen and carbon. The most common hydrocarbon is methane.

hydrologic cycle The model that describes the processes by which water is transferred between the various reservoirs of the Earth-atmosphere system.

hydrophilic Readily wettable; used to describes a surface that allows water to form a film on it.

hydrophobic Not wettable; used to describe a surface that is not hydrophilic. Water will form beads rather than a film on a hydrophobic surface.

hydrosphere The water of Earth.

hydrostatic balance The state of a stationary fluid when the vertical forces on it are balanced.

hydrostatics The study of stationary fluids. It is part of a branch of physics known as fluid mechanics.

hygrometer An instrument used to directly measure the amount of water vapour in the air. Hygrometers make use of a variety of different methods.

hygroscopic nuclei Cloud condensation nuclei that both attract water and dissolve in it.

hypothesis A tentative explanation for an observation.

hypsometry The science of measuring heights.

ice nuclei Atmospheric aerosols on which ice crystals can form.

Icelandic low The subpolar low over the northern Atlantic Ocean. It is particularly strong in winter.

ideal gas A gas in which there are *no* attractive forces between molecules.

ideal gas constant The constant, R, in the ideal gas equation. For dry air, $R = 287 \ J \cdot kg^{-1} \cdot K^{-1}$.

ideal gas law A scientific law that provides the relationship between the pressure, temperature, and volume (or density) of a gas.

internal energy The total energy contained within the atoms and molecules of a substance.

Intertropical Convergence Zone (ITCZ) The location in the tropics where the northeasterly trade winds converge with the southeasterly trade winds. This location is closely associated with the latitude of maximum solar input, and it is perpetually cloudy and wet. It migrates north in July and south in January.

inverse square law A general mathematical law used to determine the amount of any physical quantity propagating from a point source at a given distance from that source.

inversion   An increase in temperature with altitude.

isentropes   Lines on a map or graph connecting points of equal potential temperature or entropy.

isobars   Lines on a map or graph connecting points of equal pressure.

isohumes   Lines on a map or graph connecting points of equal atmospheric moisture.

isohyets   Lines on a map connecting points of equal precipitation.

isotachs   Lines on a map connecting points of equal wind speed.

isotherms   Lines on a map or graph connecting points of equal temperature.

jet contrail   A long narrow cloud in the upper troposphere produced by the condensation of the water vapour in aircraft exhaust.

jet streak   A zone within the jet stream in which the winds flow the fastest.

June solstice   The date on which the north end of Earth's axis points toward the sun, so that the sun is directly overhead at the Tropic of Cancer, 23½° N. This is the first day of summer in the northern hemisphere and the first day of winter in the southern hemisphere.

kinetic energy   Energy associated with motion.

kinetic theory of matter   A scientific theory stating that matter is composed of molecules and that these molecules are in constant motion.

Kirchhoff's law   A radiation law stating that the emissivity of a substance at a given wavelength is equal to the absorptivity of that substance at the same wavelength.

Kohler curve   A graph showing the relationship between the size of a solution droplet and the relative humidity required for that drop to be in equilibrium with its surroundings. There are different relationships for different types and amounts of solute.

lake-effect snows   Snowfall that occurs downwind of large, unfrozen lakes. These snowfalls occur as cold, dry air flows over unfrozen lakes, warming from below and picking up moisture.

laminar boundary layer   The layer of air, a few millimetres thick, in contact with Earth's surface, through which heat is transferred by conduction.

laminar flow   A type of flow in which layers of the fluid flow over each other with no disruption between the layers.

land breeze   A breeze that flows from the land to the ocean at night. It develops because the land cools more than the water does at night.

latent heat   Heat associated with a phase change.

latent heat of fusion   The amount of heat associated with the phase change of a substance between solid and liquid. This heat is absorbed during the phase change from solid to liquid and released during the phase change from liquid to solid.

latent heat of vaporization   The amount of heat associated with the phase change of a substance between liquid and gas. This heat is absorbed during the phase change from liquid to gas and released during the phase change from gas to liquid.

law of conservation of energy   A fundamental law of science stating that energy cannot be created or destroyed but can be transformed from one form to another and transferred as heat or work.

lee cyclogenesis   The formation of a mid-latitude cyclone on the leeward side of a mountain range.

Level of Free Convection (LFC)   The height at which air can rise due to its own buoyancy. At this height, rising air becomes warmer than surrounding air.

Lifting Condensation Level (LCL)   The height at which a rising parcel of air will reach its dew-point temperature and the water vapour it contains will begin to condense.

lightning   An electrical discharge within a cloud, between a cloud and the ground, or between a cloud and the surrounding air.

limit of convection   The height at which rising air stops rising. This is often the top of a cloud. At this height, rising air is the same temperature as the surrounding air.

lithosphere   The rocks of Earth.

long waves   See **Rossby waves**.

longwave radiation   The radiation emitted by Earth. It covers the range of wavelengths from roughly 3.0 μm to 100.0 μm; this includes only infrared radiation.

March equinox   A date on which Earth is not tilted with respect to the sun, so that the sun is directly overhead at the equator. This is the first day of spring in the northern hemisphere and the first day of fall in the southern hemisphere.

mechanical convection Convection that is driven by mechanical forces, such as the turbulent eddies associated with wind shear and surface roughness. Cold air can be forced to rise and warm air can be forced to sink.

meridional Relating to or varying along a meridian, or in the north–south direction.

mesocyclone Cyclonically spinning air in a convective system. Mesocyclones are most commonly associated with supercell thunderstorms, and they can develop into tornadoes.

mesopause The top of the mesosphere. The mesopause is located at an altitude of about 85 km.

mesoscale A scale of a few kilometres to a few hundred kilometres.

mesoscale convective complex A group of thunderstorms that operate together as a system. They are identified on infrared satellite imagery as roughly circular areas of very cold cloud tops. Mesoscale convective complexes are an example of a mesoscale convective system.

mesoscale convective system A system driven by vertical circulations (i.e., convection) that is at least 100 km in one direction. Examples are tropical cyclones, squall lines, and mesoscale convective complexes.

mesosphere The layer of the atmosphere that extends from about 50 km to about 85 km above Earth's surface. Temperature decreases with height in this layer.

meteorology The study of the atmospheric processes responsible for weather.

microclimate The climate of a small area at Earth's surface. Microclimates vary with the characteristics of surfaces.

mid-latitude cyclone See **frontal system**.

Mie scattering Scattering of radiation by particles bigger than the wavelengths they scatter. Aerosols and cloud droplets scatter all wavelengths of light.

mixed layer A turbulent air layer that extends from near the surface to an upper inversion. The temperature lapse rate through most of this layer is close to the DALR.

mixing condensation level The height at which water vapour will condense as a result of the mixing of an atmospheric layer.

mixing ratio The ratio of the mass of water vapour to the mass of dry air. Mixing ratio is usually expressed in units of grams per kilogram (g/kg).

model A representation of reality used to help in understanding complex or abstract natural phenomena.

model output statistics (MOS) A statistical method that adjusts the output of a weather forecasting model for local conditions, thus producing more accurate and detailed forecasts for specific locations.

molecular viscosity Small-scale resistance to flow, or internal friction, due to the random motions of molecules in laminar flow. Molecular viscosity is commonly referred to simply as *viscosity*.

momentum The product of mass and velocity.

monsoon A circulation pattern that leads to great seasonal contrasts in precipitation. Summers are very wet and winters are very dry.

mountain wind Air blowing downslope.

multicell thunderstorm A thunderstorm containing more than one cell, each at a different stage of development. These storms form under conditions of vertical wind speed shear. They usually last longer than single-cell thunderstorms and, thus, are likely to become severe.

neutral stability The condition of an air layer that is characterized by a temperature profile (ELR) that is equal to the dry adiabatic lapse rate (DALR) when the air is dry, or the saturated adiabatic lapse rate (SALR) when the air is saturated. In this environment, an air parcel that is displaced upward will be the same temperature as the surrounding air.

nitrogen fixation The process by which nitrogen gas is removed from the atmosphere by soil bacteria and converted to a soluble form of nitrogen that can be taken up by plants. Nitrogen can also be fixed by lightning. Nitrogen fixation is the sink for atmospheric nitrogen.

occluded front A front that separates a cold air mass from a cool one, and the low pressure centre from the warm-sector air in a mid-latitude cyclone. An occluded front is noticeable as a small temperature change at the surface, but it is associated with a trough of warm air aloft, or an upper front, that produces clouds and precipitation.

occlusion A process that separates the centre of low pressure from the warm-sector air in a mid-latitude cyclone. As a result of this process, an occluded front is produced at the surface and a trough of warm air is produced aloft.

omega high A ridge of high pressure that forms in the shape of the Greek letter *omega* (Ω) in the upper airflow.

An omega high blocks the normal eastward movement of weather systems and can stay in place for a number of days.

orographic lift   The process by which air is forced to rise up a slope.

orographic precipitation   Precipitation that falls from clouds produced by air rising along a topographic barrier. As a result of orographic precipitation, the windward sides of mountains can be very wet.

outgassing   The release of gases dissolved in rock.

oxidation   The addition of oxygen to a compound. This process is accompanied by a loss of electrons.

ozone hole   A roughly circular area in the ozone layer where the amount of stratospheric ozone is about 50 per cent less than it is in the surroundings. The ozone hole is centred over Antarctica and occurs during the southern hemisphere's spring.

Pacific high   The subtropical high over the northern Pacific Ocean. It is particularly strong in the summer.

parameterization   An approximation used in a model either when we don't understand a process well enough to represent it with equations or when the process is so complex that it would require too much computer time to adequately represent it.

partial pressure   The pressure contributed by a single gas in a mixture of gases. The sum of the partial pressures of the gases in the atmosphere should equal the total pressure exerted by the atmosphere.

particulate matter   Atmospheric pollutants in solid or liquid form. PM_{10} refers to particulate matter smaller than 10 μm in diameter, while $PM_{2.5}$ refers to particulate matter smaller than 2.5 μm in diameter.

path length   The distance that the sun's rays must travel through the atmosphere to reach Earth's surface.

perihelion   The position in Earth's orbit when Earth is closest to the sun. Perihelion occurs in early January.

permafrost   A subsurface layer of soil and rock material that remains frozen throughout the year.

photochemical smog   A mixture of toxic gases that forms when gases emitted by motor vehicles react in sunlight. Ozone is a major component of photochemical smog.

photodissociation   A process in which a molecule is split apart by the absorption of radiation.

photosynthesis   The life process in which energy from the sun is used to convert carbon dioxide and water to oxygen and carbohydrates.

Planck's curve   The graphical representation of Planck's law. The graph has a characteristic shape. It shows that the rate of emission of radiation per wavelength rises rapidly with increasing wavelength, reaches a peak at the wavelength of maximum emission, and then decreases gradually with further increases in wavelength.

Planck's law   A radiation law stating that objects will emit radiation over a range of wavelengths. The rate of emission will not be the same at all wavelengths. The rate of emission per wavelength increases with temperature, and the lengths of the waves emitted decrease with temperature.

planetary boundary layer   The layer of air closest to the surface that is influenced by friction from the surface. During the day, this layer can reach a depth of 1 to 2 km, but at night it can shrink to less than 100 m.

planetary vorticity   Vorticity that arises due to Earth's rotation. It is positive in the northern hemisphere and negative in the southern hemisphere.

Poisson's equation   An equation relating temperature and pressure for an adiabatic process.

polar front   An idealized front that represents the meeting of cold polar air with warm tropical air.

polar front jet stream   A narrow band of very fast westerly wind that occurs in the mid-latitudes, in the upper portion of the troposphere. It occurs above a polar front and, like a polar front, separates cold polar air from warm tropical air.

polar front theory   A theory developed to explain the formation and development of mid-latitude cyclones. It states that mid-latitude cyclones form along the polar front and have a recognized life cycle.

polar high   A region of high pressure occurring at the poles. These are thermal highs due to the low temperatures of this region.

potential energy   Energy associated with position.

potential evaporation   The amount of evaporation that would occur if moisture were unlimited.

potential temperature   The temperature an unsaturated air parcel would have if brought adiabatically to a pressure of 100 kPa.

power   The rate at which energy is transferred or work is done. The units of power in the MKS system are watts (W).

precipitation   Any liquid or solid water particles that fall from the atmosphere and reach the ground. Precipitation can take several forms including rain, snow, sleet, hail, and freezing rain.

precipitation fog   A fog that forms while it is raining. Water vapour resulting from the evaporation of raindrops causes saturation as it mixes into cold air. These fogs are also called frontal fogs because they often occur at warm fronts, where precipitation commonly falls through cold air.

pressure   The force per unit area. The units of pressure in the MKS system are pascals (Pa).

pressure gradient force   A force that occurs due to differences in pressure. The magnitude of this force is proportional to the pressure gradient, and its direction is from high pressure to low pressure.

pressure surface   An imaginary surface in the atmosphere upon which the pressure is the same everywhere.

primary aerosols   Aerosols that are emitted directly into the atmosphere.

primary pollutants   Pollutants emitted directly into the atmosphere.

prognosis   A weather forecast.

psychrometer   An instrument used to measure the amount of water vapour in the air, based on the cooling produced by evaporation.

radar   A system that transmits radio waves and receives their reflection.

radiant energy   See **radiation**.

radiation   The emission of energy as electromagnetic waves. This term is also used to denote the energy that travels in this way, in which case it is also called *radiant energy*. These waves can transfer energy through a vacuum.

radiation fog   A fog that forms when a layer of moist air at the surface cools radiatively to its dew-point temperature.

radiation inversion   A surface-based inversion that forms as air near the surface cools by emitting more radiation than it is absorbing. These inversions form overnight, or over very cold surfaces, and are best developed under clear skies.

radiative forcing   A change in a climate system's energy balance that will ultimately lead to climate change. Such changes can have natural or anthropogenic origins, and they usually involve changes in the output of the sun, the composition of the atmosphere, or the surface albedo.

radiosonde   A package of instruments that is carried up through the atmosphere attached to a balloon. The instruments in the package measure pressure, temperature, and moisture, and they transmit this information back to the surface through radio transmissions.

rain-shadow deserts   Deserts that form on the leeward sides of mountains.

Rayleigh scattering   Scattering of radiation by particles smaller than the wavelengths they scatter. Gas molecules in the atmosphere selectively scatter blue light.

reflectivity   A measure of a substance's ability to reflect incident radiation.

relative humidity   The ratio of the actual amount of water vapour in the air to the saturation value at the air's temperature.

relative vorticity   Vorticity occurring due to a fluid's own motion. Relative vorticity can occur due to curvature or shear in the flow.

reservoir   In the context of biogeochemical cycles, a storage place. Reservoirs include the atmosphere, biosphere, hydrosphere, and lithosphere.

residence time   The average amount of time that a substance might be expected to remain in a reservoir of the Earth-atmosphere system.

respiration   The life process in which oxygen is removed from the atmosphere and carbon dioxide is returned.

ridge   An elongated area of high pressure.

Rossby waves   Waves in the upper-air westerly flow of the mid-latitudes with wavelengths of several thousand kilometres. These waves are sometimes called long waves.

saturated adiabatic lapse rate (SALR)   The rate of change of temperature of a rising or sinking saturated air parcel. This rate decreases as the moisture content of the air increases.

saturation   The maximum amount of water vapour that can exist at a given temperature.

saturation vapour pressure   The maximum water vapour pressure that can exist at a given temperature.

scale height   The height the atmosphere would have if the density throughout was the same as it is at the surface.

scattering   The process by which atmospheric gases and aerosols reflect radiation in multiple directions.

scientific law A precise statement that describes the behaviour of nature and is believed to always hold true.

scientific method A series of steps followed in scientific investigation: making an observation, asking a question about the observation, formulating a hypothesis to explain the observation, making further observations to test the hypothesis, and reaching a conclusion.

scientific theory A body of knowledge that provides a detailed explanation for a set of observations.

sclerophyll Woody plants with small, waxy leaves. They are found in regions of Mediterranean climates.

sea breeze A breeze that flows from the ocean to the land during the day. It develops because the land warms more than does the water during the day.

secondary aerosols Aerosols that form in the atmosphere.

secondary pollutants Pollutants that form in the atmosphere.

sensible heat Heat associated with a temperature change. It can be felt, and it can be measured by a thermometer.

September equinox A date on which Earth is not tilted with respect to the sun, so that the sun is directly overhead at the equator. This is the first day of fall in the northern hemisphere and the first day of spring in the southern hemisphere.

short waves Waves in the upper airflow with wavelengths of about 1000 km.

shortwave radiation The radiation emitted by the sun. It covers the range of wavelengths from roughly 0.15 to 3.0 μm. This range includes ultraviolet, visible, and infrared radiation.

Siberian high A thermal high that forms over Siberia in the winter.

sine law of illumination An equation used to calculate the amount of radiation incident on a surface. This relationship shows that the amount of radiation received increases with the altitude angle of the incident radiation.

single-cell thunderstorm A thunderstorm that contains one cell. These storms form due to strong surface heating when there is little to no wind shear. They last half an hour to an hour and are unlikely to become severe. They are also known as air-mass thunderstorms.

sink A process by which a substance leaves a reservoir.

sky-view factor A measure of the amount of sky that can be "seen" from a point on the ground.

sleet A type of frozen precipitation that does not have a crystal structure. Sleet occurs when there is a subfreezing layer at the surface topped by a layer of air in which temperatures are above freezing. The effect of this temperature stratification is that precipitation will melt and then refreeze before reaching the surface.

snow A type of precipitation in the form of ice crystals.

solar constant The amount of energy that strikes the top of the atmosphere, on a surface perpendicular to the solar beam, when Earth is at an average distance from the sun.

solute effect The effect in which a dissolved substance reduces the relative humidity required for a droplet to be in equilibrium with its surroundings.

source A process by which a substance enters a reservoir.

source region A very large area of uniform surface type over which air can remain stagnant long enough to form an air mass.

specific heat The amount of heat, in joules, required to raise the temperature of 1 kg of a substance by 1 K.

specific humidity The ratio of the mass of water vapour to the total mass of air. Specific humidity is usually expressed in units of grams per kilogram (g/kg).

squall line A line of thunderstorms that forms at or just ahead of a cold front.

standard atmosphere A set of values that represents the average vertical distribution of pressure, temperature, and density in the atmosphere.

stationary front A front along which there is no significant movement for at least a few hours.

steady state In the context of biogeochemical cycles, a condition that exists when the inflows to a reservoir are equal to the outflows from the reservoir.

steam fog A fog that forms when water vapour evaporating from a warm, moist surface is mixed into colder air above that surface.

Stefan–Boltzmann law A radiation law stating that the rate of emission of radiation by a substance will increase with the temperature of the substance.

storm surge A rise in water level that can cause flooding along coasts. These surges are usually caused by a combination of strong winds and very low atmospheric pressure.

stratiform cloud Cloud that is layered in form.

stratopause The top of the stratosphere, located at an altitude of about 50 km.

stratosphere The layer of the atmosphere extending from, on average, 11 km to 50 km above Earth's surface. Temperature increases with height in this layer.

sublimation A phase change from solid to gas.

subpolar low A region of average low pressure occurring in the mid-latitudes. In the northern hemisphere, subpolar lows usually form two distinct cells in winter known as the Aleutian low and the Icelandic low. In the southern hemisphere, this region of low pressure tends to take the form of a ring of low pressure surrounding the polar high. These areas of low pressure develop as a result of convergence and lifting of air at the polar fronts.

subsidence inversion An upper-air inversion that commonly forms in the subsiding air of a high-pressure area, normally about 1 or 2 km above the surface. These inversions form under anticyclonic conditions as air sinks from above and warms.

subsolar point The latitude at which the sun is directly overhead at noon.

subtropical high A region of high pressure occurring in the subtropics. In the northern hemisphere, subtropical highs usually form two distinct cells over the oceans in summer known as the Pacific high and the Bermuda high. These areas of high pressure develop as a result of convergence and sinking of air aloft.

subtropical jet stream A narrow band of very fast westerly wind that occurs near the top of the troposphere at about 30° N or S. The jet streams are associated with the conservation of angular momentum.

supercell thunderstorm A thunderstorm with one cell consisting of a strong rotating updraft. This type of thunderstorm is capable of creating severe weather and even tornadoes.

supercooled Cooled below the temperature at which a substance would normally freeze, while remaining in a liquid state.

supersaturated The condition that occurs when the amount of water vapour in the air is higher than the saturation value.

surface roughness The degree of irregularity of a surface. Surfaces with a large degree of irregularity are rough.

synoptic scale A scale of a few hundred kilometres to a few thousand kilometres.

synoptic weather map A weather map that gives a visual synopsis of the weather conditions that are occurring at a given time.

system An interrelated set of parts.

teleconnections Linkages that result because atmosphere–ocean interactions in one part of the world affect the large-scale atmospheric circulation and, thus, weather in another part of the world.

temperature A measure of the average kinetic energy of the molecules in a substance.

tephigram The thermodynamic diagram used in Canada.

terminal velocity Constant fall speed in still air reached when the force of gravity equals the opposing force created by the resistance of the air.

thermal advection The flow of either warmer or colder air into a region. Thermal advection is one way in which temperature can change at a location, and it leads to vertical motions.

thermal convection Convection that is driven by the density differences that result from temperature differences. Warm, less dense air will rise and cold, denser air will sink.

thermal energy That part of the internal energy of a substance that is associated with the kinetic energy of the molecules.

thermal pressure systems Areas of high or low pressure that are created by cooling or warming, respectively. These pressure systems are shallow.

thermal wind The change in geostrophic wind with height that is due to the horizontal variation of temperature.

thermodynamic diagrams Diagrams, or graphs, that show thermodynamic processes. Several different types are used in meteorology to show the variation in the state of the atmosphere with height. They are useful in weather forecasting.

thermosphere The top layer of the atmosphere. Its base is located at an altitude of about 85 km; it has no well-defined top. Temperature increases with height in this layer.

thickness The difference in height between two pressure surfaces in the atmosphere. The thickness of an atmospheric layer increases with the average temperature of the layer.

thunder The sound produced as the heat from lightning causes a sudden and rapid expansion of the surrounding air.

thunderstorm A convective storm that produces thunder and lightning. Such storms also usually produce heavy rain, hail, and strong winds, and they sometimes produce tornadoes.

tornado A violently rotating column of air that can develop in association with severe thunderstorms. They hang from the bottom of cumulonimbus clouds and touch the ground. The very strong winds of tornadoes are extremely damaging.

trade winds   The steady winds of the tropical region. They are northeasterly in the northern hemisphere and southeasterly in the southern hemisphere.

transmissivity   A measure of a substance's ability to transmit incident radiation.

transpiration   The process by which water vapour is released through a plant's stomata.

tropical cyclone   A low-pressure area that forms in the tropics. These storms range in intensity from tropical depressions through to hurricanes or typhoons.

tropical depression   A small cluster of thunderstorms that forms in association with easterly waves over tropical oceans. Tropical depressions differ from tropical disturbances in that they have begun to develop cyclonic rotation and appear on a map with at least one closed isobar.

tropical disturbance   A small cluster of thunderstorms that forms in association with easterly waves over tropical oceans. Wind speeds do not exceed 60 km/h.

tropical storm   A tropical depression in which sustained winds are greater than 60 km/h but less than 120 km/h. Once storms reach this intensity, they are named.

tropopause   The top of the troposphere. The average height of the tropopause is 11 km.

troposphere   The layer of the atmosphere extending from Earth's surface to an average height of 11 km. Temperature decreases with height in this layer.

trough   An elongated area of low pressure.

TROWAL   An acronym for "TROugh of Warm air ALoft." The TROWAL symbol marks the location of an upper trough of warm air, or front, projected onto the surface. TROWALS are labelled instead of occluded fronts on Canadian weather maps.

turbulence   Random, irregular motions in a fluid.

typhoon   See **hurricane**.

ultraviolet radiation   Radiation with wavelengths ranging from 0.1 to 0.4 μm.

unsaturated   The condition that occurs when the amount of water vapour in the air is lower than the saturation value.

upslope fog   A fog that forms as air rising up a slope cools adiabatically and condenses.

urban heat island   A microclimate created by a city, in which temperatures are significantly higher than they are in the surrounding region.

valley wind   Air blowing upslope.

vapour pressure   Partial pressure exerted by water vapour.

vapour pressure deficit   The difference between the saturation vapour pressure and the actual vapour pressure.

variable gases   Gases whose concentrations change in space and/or time.

vector   A concept used to represent phenomena that have both speed and direction.

veering   Rotating clockwise with height; used to describe the rotation of winds.

virga   Water particles that fall from a cloud but do not reach the ground.

virtual temperature   The temperature dry air would need to have to be the same density as moist air. This temperature is used in the ideal gas law to account for the fact that moist air is less dense than dry air.

viscosity   Resistance to flow. See **molecular viscosity** and **eddy viscosity**.

volatile organic compounds (VOCs)   Carbon-containing compounds that easily vaporize. Many VOCs are hydrocarbons. The sources of VOCs include emissions from motor vehicles and vegetation, and the evaporation of paints and solvents.

vorticity   A measure of spin in a fluid. Air spinning counterclockwise is defined as having positive vorticity, while air spinning clockwise is defined as having negative vorticity.

warm cloud   A cloud in which the temperature is above 0°C throughout. Such clouds contain only water droplets.

warm front   A front at which warm air is advancing and replacing cold air.

warm-sector air   The warm air lying between the cold front and the warm front in a mid-latitude cyclone.

warm-type occlusion   An occlusion in which the air behind the cold front rides over the air ahead of the warm front.

wavelength   The distance between any two like points on a wave—for example, the distance from crest to crest.

wavelength of maximum emission   The wavelength at which the rate of emission of radiation is highest. This wavelength will decrease with increasing temperature.

weather   The state of the atmosphere at a given place and time.

weather station symbols   Symbols plotted on a weather map to provide information about observed weather elements.

wet-bulb potential temperature   The wet-bulb temperature that an air parcel would have if brought adiabatically back to 100 kPa.

wet-bulb temperature   The temperature to which air will cool by evaporating water into it.

wet-bulb thermometer   The thermometer in a psychrometer that is used to measure the temperature to which air will cool by evaporating water into it.

wettable   In relation to atmospheric aerosols, able to allow water to form a film. See also **hydrophilic** and **hydrophobic**.

Wien's law   A radiation law stating that the wavelength at which a substance emits the most energy, as well as the wavelengths emitted, will decrease as the temperature of the substance increases.

wind   The mostly horizontal movement of air relative to Earth's surface.

wind chill   A measure of how cold it feels due to a combination of low temperature and high wind.

wind shear   The change in wind speed and/or direction in any direction across the flow.

work   The transfer of energy by mechanical means. When work is performed, a force is moved through a distance. The units of work in the MKS system are joules (J).

zenith angle   The angle of the sun from the zenith. When the sun is at the zenith, it is directly overhead.

zonal   Relating to or varying in the east-west direction. For example, easterly or westerly winds are zonal winds.

Index

Photo Credits

Chapter 1 opener: NASA/Science Photo Library
Fig. 1.1: SPL/Photo Researchers, Inc.
Fig. 1.2: © Fred Greenslade/Reuters/Corbis
Fig. 1.4: (*top*) © Stefano Politi Markovina/Alamy; (*bottom*) © Bruce Obee/All Canada Photos/Corbis
Fig. 1.8: © Getty/Bobby Model;
Fig. 1.9: Stephen & Donna O'Meara/Photo Researchers, Inc.

Chapter 2 opener: © Martin Rietze/Westend61/Corbis
Fig. 2.4: © Picture Press / Alamy
Fig. 2.9: © Barrett & MacKay/All Canada Photos/Corbis
Fig. 2.10: © Ron Erwin/All Canada Photos/Corbis
Fig. 2.11: © Andrew McConnell / Alamy
Fig. 2.12: NASA/Science Photo Library

Chapter 3 opener: © Robert Bartow/Design Pics/Corbis
Fig. 3.2: George Bernard/Science Photo Library
Fig. 3.4: Sheila Terry/Science Photo Library
Fig. 3.7: © Christopher J. Morris/Corbis
Fig. 3.19: Stephen J. Krasemann/Photo Researchers, Inc.
Fig. 3.20: NASA, Earth Science Office, http://www.ghcc.msfc.nasa.gov/GOES, 7 May 2012
Fig. 3.21: British Antarctic Survey/Science Photo Library

Chapter 4 opener: Louise Murray/Science Photo Library
Fig. 4.1: Detlev Van Ravenswaay/Science Photo Library
Fig. 4.3: Andrew Lambert Photography/Science Photo Library
Fig. 4.9: © Richard du Toit/Corbis
Fig. 4.13: © kzenon/iStockphoto
Fig. 4.14: © Aurora Photos/Alamy
Fig. 4.17: John Chumack/Photo Researchers, Inc.

Chapter 5 opener: George D. Lepp/Photo Researchers, Inc.
Fig. 5.3: © Mike Grandmaison/Corbis
Fig. 5.5: T-Service/Science Photo Library
Fig. 5.8: © Radius Images/Corbis
Fig. 5.9: NASA, http://ceres.larc.nasa.gov/science_information.php?page=EBAFclrsky, 5 June 2012
Fig. 5.12: Babak Tafreshi/Photo Researchers, Inc.
Fig. 5.25, b): Source: http://apod.nasa.gov/apod/ap020709.html; © Vasilij Rumyantsev, Crimean Astrophysical Obsevatory
Fig. 5.31: © Getty/Alan Marsh

Chapter 6 opener: © Peter Muller/cultura/Corbis
Fig. 6.4: Science Photo Library
Fig. 6.12: © Marc Dozier/Hemis/Corbis
Fig. 6.13: (*top left*) © Megapress / Alamy; (*top right*) © Gunter Marx Photography/Corbis; (*bottom left*) © RalphWilliam/Alamy; (*bottom right*) © Don Hammond/Design Pics/Corbis
Fig. 6.14: (*left*) © Dustin Goodspeed; (*right*) cover by Bastien Liutkus from Binary Mind
Fig. 6.20: (*top*) Dr Carleton Ray/Photo Researchers, Inc; (*bottom*) Hoa-Qui/Photo Researchers, Inc.

Chapter 7 opener: © Mike Grandmaison/Corbis
Figure 7.1: NASA/Goddard Space Flight Center/Science Photo Library
Fig. 7.6: © Roman Sigaev/iStockphoto
Fig. 7.8: Garry Black/All Canada Photos/Corbis
Fig. 7.11: AP / Back Page Images / Rex Features
Fig. 7.12: © John E Marriott/All Canada Photos/Corbis
Fig. 7.14: © Pgiam/iStockphoto
Fig. 7.16: Damien Lovegrove/Science Photo Library

Chapter 8 opener: © Carol Hughes; Gallo Images/ Corbis
Fig. 8.4: John Mead/ Science Photo Library
Fig. 8.17: Thomas & Pat Leeson/Photo Researchers, Inc.
Fig. 8.18: Nathan Denette/The Canadian Press
Fig. 8.22: © Bill Gozansky/Alamy
Fig. 8.26: © Dirk Baltrusch/iStockphoto

Chapter 9 opener: © Mark Karrass/Corbis
Fig. 9.8: © Frank Krahmer/Corbis
Fig. 9.9: (*top*) Colin Cuthbert/Photo Researchers, Inc.; (*bottom*) Karl G. Vock/Photo Researchers, Inc.
Fig. 9.10: Adam Jones/Photo Researchers, Inc.
Fig. 9.11, a): John R. Foster / Photo Researchers, Inc.
Fig. 9.12, a): John Mead/Photo Researchers, Inc.
Fig. 9.14: Sinclair Stammers/Science Photo Library
Fig. 9.15: (*top*) Pekka Parviainen/Science Photo Library; (*bottom*) Kajr. Svensson/Science Photo Library
Fig. 9.16: Bildarchiv Okapia/Photo Researchers, Inc.
Fig. 9.17: © Igor Trepeshchenok/Alamy
Fig. 9.18: © Martin Lladó/Gaia Moments/Alamy
Fig. 9.20: © Layne Kennedy/Corbis
Fig. 9.22: © Gunter Marx Photography/Corbis

Fig. 9.23: © Getty/Guido Barberis Photographer
Fig. 9.25: © Rick Doyle/Corbis
Fig. 9.27: © Pavel Cheiko/iStockphoto
Fig. 9.29: © Getty/Bill Miles Photography

Chapter 10 opener: Georg Gerster/Photo Researchers, Inc.
Fig. 10.1: © Lowell Georgia/ Corbis;
Fig. 10.3: University Corporation for Atmospheric Research/Science Photo Library
Fig. 10.7: Ted Kinsman / Photo Researchers, Inc.
Fig. 10.9: © Kathryn Hatashita-Lee/iStockphoto
Fig. 10.10: James Steinberg/Photo Researchers, Inc.
Fig. 10.12: *The Canadian Press–Charlottetown Guardian–Heather Taweel*
Fig. 10.13: Ted Kinsman/Photo Researchers, Inc.

Chapter 11 opener: © Thomas Kitchin & Victoria Hurst/All Canada Photos/Corbis
Fig. 11.16: (*left*) © Chloe Dulude/Corbis; (*right*) © Radius Images/Alamy
Fig. 11.26: © Donald Enright/Alamy
Fig. 11.27: Mark Boulton/Photo Researchers, Inc.
Fig. 11.28: © Radius Images/Corbis

Chapter 12 opener: © 2009 Jeff Lynch Photography
Fig. 12.6: NASA/Science Photo Library
Fig. 12.9: © Frank Krahmer/Corbis
Fig. 12.17: (*left*) Pekka Parviainen/Science Photo Library; (*right*) NASA/Science Photo Library

Chapter 13 opener: © Kazuyoshi Nomachi/Corbis
Fig. 13.2: (*top left*) Bryan and Cherry Alexander/Photo Researchers, Inc.; (*top right*) © Getty/Blend Images/PBNJ Productions; (*bottom left*) © Getty/Raul Touzon; (*bottom right*) © Hamilton, Barry/SuperStock/Corbis
Fig. 13.6: NASA (Unspecified Center)
Fig. 13.7: NASA GOES Project, Dennis Chesters

Chapter 14 opener: © Robert Postma/First Light/Corbis
Fig. 14.2: Science Source/Photo Researchers, Inc.
Fig. 14.3: NASA Goddard MODIS Rapid Response Team
Fig. 14.6, a): A.J. Henry. (August 1922). J. Bjerknes and H. Solberg on meteorological conditions for the formation of rain. Monthly Weather Review, 50(8), 404.
Fig. 14.6, b): A.J. Henry. (September 1922). J. Bjerknes and H. Solberg on the life cycle of cyclones and the

polar front theory of atmospheric circulation. Monthly Weather Review, 50(9), 469.
Fig. 14.23: © U.S. Coast Guard–digital ve/Science Faction/Corbis
Fig. 14.27: Science Source/Photo Researchers, Inc.
Fig. 14.29: © Mike Hollingshead/Science Faction/Corbis
Fig. 14.33: Cooperative Institute for Meteorological Satellite Studies, University of Wisconscin-Madison; http://cimss.ssec.wisc.edu/goes/blog/archives/6337
Fig. 14.36: David He

Chapter 15 opener: © Ryan McGinnis/Alamy
Fig. 15.3: (*left*) © British Crown Copyright, The Met Office/Science Photo Library; (*right*) Environment Canada
Fig. 15.4: http://www.esrl.noaa.gov/psd/spotlight/2007/past-weather.html
Fig. 15.5: Michael Donne/Photo Researchers, Inc.
Fig. 15.7: Pierre cb/Wikimedia Commons
Fig. 15.8: NOAA/Photo Researchers, Inc.
Fig. 15.10: NASA
Fig. 15.11: NASA/Science Photo Library
Fig. 5.12: Science Source/Photo Researchers, Inc.

Chapter 16 opener: James Warwick/Getty Images
Fig. 16.1: (*top*) James Steinberg / Photo Researchers, Inc.; (*bottom*) © Mike Grandmaison/Corbis
Fig. 16.11: © Getty/Gary Braasch
Fig. 16.14: © brytta/iStockphoto
Fig. 16.16: © Ludovic Maisant/Corbis
Fig. 16.18: © Bill Bachman/Alamy
Fig. 16.20: © Getty/Tim Graham
Fig. 16.23: (*top*) © GetStock/Karinclaus/Dreamstime.com; (*bottom*) Michael P. Gadomski/Photo Researchers, Inc.
Fig. 16.24: © Matthias Breiter/Minden Pictures/Corbis
Fig. 16.26: Stephen J. Krasemann/Photo Researchers, Inc.
Fig. 16.29: (*top*) © Dave Reede/All Canada Photos/Corbis; (*bottom*) © Ron Erwin/All Canada Photos/Corbis
Fig. 16.31: © Getty/Alan Majchrowicz
Fig. 16.32: © Wayne Lynch/All Canada Photos/Corbis
Fig 16.34: © Image Source/Corbis

Chapter 17 opener: © John Short/Design Pics/Corbis
Fig. 17.3: © Getty/Noel Hendrickson
Fig 17.8: © Bill Brooks/Alamy
Fig. 17.10: © Jon Bower/Loop Images/Corbis
Fig 17.13: NASA/Science Photo Library